QUANTIFICATIONAL INFERENCE

UNIVERSAL INSTANTIATION (UI)
From a universal quantification, one may infer any instantiation, provided that the instantiation was produced by uniformly replacing each occurrence of the variable that was bound by the quantifier with a constant or John Doe name.

EXISTENTIAL GENERALIZATION (EG)
From a sentence containing a constant or a John Doe name, you may infer any corresponding existential generalization, provided that: (a) the variable used in the generalization does not already occur in the sentence generalized upon; (b) the generalization results by replacing at least one occurrence of the constant or John Doe name with the variable, and no other changes are made.

EXISTENTIAL INSTANTIATION (EI)
From an existential quantification, you may infer an instantiation, provided that (a) each occurrence of the variable bound by the quantifier in the existential quantification is uniformly replaced with a John Doe name and no other changes are made; (b) the John Doe name does not appear in any earlier line of the deduction.

UNIVERSAL GENERALIZATION (UG)
From a sentence containing a John Doe name, one may infer the corresponding universal generalization, provided that (a) the John Doe name that is replaced by a variable does not occur in any preceding line derived by EI; (b) the generalization results by replacing each occurrence of the John Doe name with the variable (and no other changes are made); (c) the variable you use in the generalization does not already appear in the sentence you are generalizing from; (d) the John Doe name does not appear in any assumed premise that has not already been discharged.

QUANTIFICATIONAL REPLACEMENT RULES

QUANTIFIER EXCHANGE (QE)
If **P** contains either a universal or an existential quantifier, **P** may be replaced by or may replace a sentence that is exactly like **P** except that one quantifier has been switched for the other in accord with (a) - (c) below:
>   (a) Switch one quantifier for the other.
>   (b) Negate each side of the quantifier.
>   (c) Cancel out any double negatives that result.

IDENTITY

IDENTITY A (ID A)
At any step in a proof, you may assert $(x)(x = x)$.

IDENTITY B (ID B)
If **c** and **d** are two constants or John Doe names and a line of a proof asserts that the individual designated by **c** is identical with the individual designated by **d**, then you may carry down and rewrite any available line of the proof replacing any or all occurrences of **c** with **d** or any or all occurrences of **d** with **c**. A line of a proof is available unless it is within the scope of a discharged assumption.

P9-DUN-453

THE
MANY WORLDS
— OF —
LOGIC

SECOND EDITION

PAUL HERRICK

Shoreline Community College

New York Oxford
OXFORD UNIVERSITY PRESS

Oxford University Press

Oxford   New York
Auckland   Bangkok   Buenos Aires   Cape Town
Chennai   Dar es Salaam   Delhi   Hong Kong   Istanbul   Karachi
Kolkata   Kuala Lumpur   Madrid   Melbourne   Mexico City   Mumbai   Nairobi
São Paulo   Shanghai   Singapore   Taipei   Tokyo   Toronto

and an associated company in Berlin

Copyright © 2000 by Oxford University Press, Inc.

Published by Oxford University Press, Inc.
198 Madison Avenue, New York, New York 10016
http://www.oup-usa.org

Oxford is a registered trademark of Oxford University Press

All rights reserved. No part of this publication may be reproduced,
stored in a retrieval system or transmitted, in any form or by any
means, electronic, mechanical, photocopying, recording, or otherwise,
without the prior permission of Oxford University Press.

ISBN: 0-19-515503-3

Printing number: 9 8 7 6 5 4 3 2 1

Printed in the United States of America
on acid-free paper

To the Instructor

This second edition of *The Many Worlds of Logic* embodies a number of changes. Many of these changes were suggested to me by instructors who used the first edition in their classrooms. Numerous other changes were suggested to me by students in my classes. During the five years that I used the first edition in my own classroom, ideas for even more changes came to me. First, I added chapters on the main topics in informal logic and critical thinking. The text may now be used to teach a logic course that blends critical thinking, informal logic, and formal logic. The new chapters on critical thinking and informal logic also allow for a course organized around the theme of blending traditional and contemporary logical theory.

Second, the book now contains shorter chapters so that instructors can more easily pick and choose material among the topics. The shorter, tighter chapters also will be easier for the student to digest.

Third, I meticulously edited every page, with the goal of making the text easier to learn from and easier to teach with. As a result, many explanations have been clarified; the truth-functional logic unit has been broken up into smaller modules; the quantificational deduction system has been revised and improved; the quantificational unit has also been broken up into smaller modules; and the modal deduction system has been revised and streamlined.

The following are suggestions for organizing various compact units within a logic course.

- For a short unit on basic truth-functional logic, I recommend Chapters 2, 3, 7, 8.

- For a short unit on basic truth-functional natural deduction (inference rules and no replacement rules), I suggest Chapters 7 and 8. (For a slightly longer unit, add Chapter 9 to this.)

- For a short unit on basic quantificational logic (monadic predication and no overlapping quantifiers), I recommend Chapters 16 and 18.

- For a short unit on basic modal logic: Chapter 23 up to Exercise 23.3 and Chapter 24 up to Exercise 24.1.

The following are suggestions for teaching the several different logic courses that can be effectively taught with *The Many Worlds of Logic*:

1. A standard symbolic logic course that covers truth-functional and quantificational logic:

- Chapters 1–11 and 16–22.

Chapters 4, 10, 11, 20, 21, 22 can be cut from this list if time is short. Also, Aristotelian logic can be taught as an entry to modern quantificational logic by adding Chapters 14 and 15 to the list.

2. A comprehensive symbolic logic course that covers truth-functional logic, basic quantificational logic, and modal logic:

• Chapters 1–10, 16, 18, 23, and selected parts of 24.

3. A critical thinking course:

• Chapters 1, 12, 13, 14, 15, 25, 26.

I believe these seven chapters will suffice for a very good critical thinking course if they are supplemented by examples from newspapers, magazines, advertisements, contemporary political discourse, and also by student presentations, classroom debates, group projects, and class discussion.

4. A logic course that blends formal and informal theory or traditional and contemporary theory:

• Chapters 1–3, 7, 8, 12–16, 18, 25, 26.

Most chapters conclude with a helpful glossary. In the back of the book are appendix materials on truth-trees and truth. The Appendices are followed by Answers to Selected Exercises for students who wish to self-test and monitor how well they are grasping various concepts. In addition to this main text, available ancillary materials include an instructor's manual, a test bank, and study guide. Finally, if you have any comments or suggestions regarding this text, I would appreciate hearing from you. You may write me at:

Shoreline Community College
16101 Greenwood Avenue, North
Shoreline, WA 98133

To the Student

I have three pieces of advice for students of logic. First, in most cases, each section of a logic course builds on previous sections. Consequently, if you fail to understand one section, you may be unable to understand subsequent sections. My first piece of advice is: try not to fall behind.

Second, logic courses typically emphasize problem solving. With most people, problem solving is a skill that only develops with practice. Your instructor will show you how to solve logic problems in class, but if you do not regularly practice solving the problems on your own, you may do poorly on the tests and you may come away with little understanding when the course is over.

Third, regular attendance is especially important in a class such as logic, because many of the ideas and techniques are difficult to grasp without a teacher's personal explanations.

WHY STUDY LOGICAL THEORY?

Logic is a field of study that has engaged the minds of many of the world's greatest thinkers since ancient times, and for many centuries, it has been considered an important ingredient of a university education. A basic understanding of logical theory will therefore acquaint you with an important part of humankind's intellectual history. Furthermore, the study of logic should contribute to your development of the important skill of thinking critically. This skill includes:

- Your ability to identify the parts and the logical structure of the arguments others present to you
- Your ability to evaluate those arguments and decide for yourself which arguments are good and which are not
- Your ability to construct and present good arguments of your own

In addition, the study of logic should contribute to the development of:

- Your problem-solving skills
- Your ability to think abstractly
- Your ability to think systematically and in a disciplined, precise way
- Your ability to learn a technical field of thought
- Your ability to work with a symbolic language
- Your understanding of the logical aspects of language
- Your understanding of an important branch of modern philosophy

Acknowledgments

Professor Sandra Johanson was kind enough to meet with me to discuss her experiences using the first edition of the text, and gave me many excellent suggestions for improving the book. Professor Karl Hillstrom also shared his experiences using the text. Karl gave me a number of very valuable suggestions that improved the book in several places. Professor Andrew Jeffrey was also kind enough to meet with me to discuss his classroom experiences using the text, and he gave me very good suggestions for improving the text. In addition to discussing his experiences teaching with the text, Dr. Steven Duncan, my colleague, read the new chapters and offered many helpful suggestions. And I have benefited from discussions with Professors S. Marc Cohen, Robert Coburn, Dick Burton, Lawrence Stern, Richard Purtill, Norman Swartz, Robert Kirk, Richard McClelland, C. Stephen Layman, Daniel Howard-Snyder, Rob Brady, and Mark Story. I have also benefited from discussions with Professor Michael Matriotti concerning the relative merits of traditional-versus-modern logical theory. I feel grateful to these individuals who took their time to offer me needed advice. I also feel indebted to the many students over the years who took the time to offer me feedback and suggestions that helped me improve the text. I also feel a deep gratitude to the six reviewers. Each wrote excellent and extremely helpful analyses of the text, and each provided valuable suggestions: Beth Huffer, University of Colorado; Sandra Johanson, Green River Community College; Bredo C. Johnson, University of Houston; Robert Pielke, El Camino College; David A. Spieler, University of South Florida; and Dona Warren, University of Wisconsin-Stevens Point.

I also wish to personally thank the developmental editor for this project, Scott Spoolman, who was helpful at many crucial points. I also wish to thank the editorial and production teams at Harcourt Brace College Publishers, including David Tatom, executive editor; Travis Tyre and Joyce Fink, project editors; Linda McMillan, production manager; Laura Brennan, market strategist; and Chris Morrow, art director. All did an excellent job of helping make this edition possible. Joyce Fink, the project editor who served as my main contact at Harcourt Brace, deserves special mention. Joyce was so helpful, and so patient. She did her job well, and it was nice to work with such a kind and easygoing person.

Paul Herrick

Contents

To the Instructor   v
To the Student   vii
Acknowledgments   ix

CHAPTER 1   FUNDAMENTALS   1
   Recognizing Arguments   2
   Some Typical Conclusion Indicators   3
   Some Typical Premise Indicators   3
   Distinguishing Sentences and Statements   5
   Two General Categories of Argument: Deductive and
      Inductive Arguments   6
   Deductive and Inductive Indicator Words   8
   Evaluating Inductive Arguments: The Strong,
      the Weak, and the Inductively Sound   9
   Evaluating Deductive Arguments: The Valid, the Invalid,
      and the Deductively Sound   11
   Deciding Whether an Argument is Valid or Invalid   13
   Consistency and Inconsistency   20
   Implication   21
   Logical Equivalence   21
   Necessity   22
   Ideals   23
   Appendix: Some Logical Puzzles   24
   Glossary   25

PART I   TRUTH-FUNCTIONAL LOGIC   29

CHAPTER 2   INTRODUCTORY TRUTH-FUNCTIONAL LOGIC   31
   Simple and Compound Sentences, Sentence Operators,
      and the Conjunction   32
   Negation   35
   Disjunction   37
   Truth-Functions and Truth-Functional Compound Sentences   39
   Conditional Sentences   40
   Biconditional Sentences   42
   Glossary   46

CHAPTER 3   TRANSLATING ENGLISH INTO LOGICAL SYMBOLS   47
   Symbolizing Sentences Containing More than One Operator   48
   Throwing the Tilde into the Mix   51
   From And to Or and Back Again-With a Few Nots Thrown In   52
   Some General Hints on Symbolizing   55
   Translating Conditionals and Biconditionals   57
   Symbolizing Necessary and Sufficient Conditions   59
   Glossary   65

CHAPTER 4   OUR NEW LANGUAGE GETS A NAME AND
            A FORMAL SYNTAX   66
            The Language TL   66
            How to Calculate the Truth-Value of the Whole
               from the Values of the Parts   69
            Glossary   74

CHAPTER 5   TRUTH-TABLE ANALYSIS   77
            Constructing a Truth-Table for a Formula   78
            Eight-Row Tables   79
            How to Make Your Own Tautology Detector
               Using Just Paper and Pencil   81
            How to Make an Inexpensive Contradiction Detector
               for Home or Office   82
            The Contingency Detector: Don't Leave Home without It   83
            Testing an Argument for Validity   85
            Showing an Argument Invalid with a Partial Truth-Table   92
            Testing a Pair of Sentences for Equivalences   97
            Glossary   99

CHAPTER 6   THE CONCEPT OF LOGICAL FORM   101
            Sentence Forms   101
            Argument Forms   105
            The Disjunctive Syllogism Form   105
            The Modus Ponens Form   106
            The Modus Tollens Form   107
            The Hypothetical Syllogism Form   108
            Valid Argument Forms   108
            Invalid Argument Forms   110
            Concluding Comments   112
            Appendix: Contradictory and Tautological Sentence Forms   114
            Glossary   116

CHAPTER 7   TRUTH-FUNCTIONAL NATURAL DEDUCTION   118
            The Disjunctive Syllogism Rule   119
            The Modus Ponens Rule   121
            The Modus Tollens Rule   122
            The Hypothetical Syllogism Rule   124
            Proving that a Conclusion Validly Follows   125
            The System TD   128
            Proofs   128
            Glossary   139

CHAPTER 8   FOUR MORE INFERENCE RULES   140
            The Simplification Rule   140
            The Conjunction Rule   141
            The Addition Rule   143
            The Constructive Dilemma Rule   144
            Some Unsolicited Advice on Learning to Construct Proofs   146
            Proof Strategies   147
            Some Additional Suggestions Concerning Strategy   151
            Appendix: Some Common Deduction Errors   157

CHAPTER 9   INDIRECT PROOFS AND CONDITIONAL PROOFS   160
The Indirect Proof Rule   162
The Conditional Proof Rule   168
Nested Proofs   175
Proving Sentences Tautological   178
The Law of Noncontradiction   181
Glossary   182

CHAPTER 10   REPLACEMENT RULES   183
The Commutative Rule   184
The Associative Rule   185
The Double Negation Rule   186
DeMorgan's Rule   187
The Distribution Rule   190
Five More Replacement Rules   197
The Transposition Rule   197
The Implication Rule   197
The Exportation Rule   198
The Tautology Rule   199
The Equivalence Rule   200
Are Replacement Rules Worth the Bother?   201
Glossary   209

CHAPTER 11   INDIRECT AND CONDITIONAL PROOFS
WITH REPLACEMENT RULES   210
Indirect Proofs with Replacement Rules   210
Conditional Proof with Replacement Rules   214
Proving Tautologies   216
Glossary   219

PART II   TWO INFORMAL TOPICS   221

CHAPTER 12   DEFINITION   223
The Purposes of Definition   224
Five Types of Definition   226
Two Types of Meaning   230
Constructing a Definition: Techniques   233
Rules for Intensional Definitions   237
Glossary   239

CHAPTER 13   INFORMAL FALLACIES   242
Fallacies of No Evidence   243
Fallacies of Little Evidence   255
Fallacies of Language   263
Glossary   270
A Summary of the Fallacies   271

PART III   ARISTOTELIAN CATEGORICAL LOGIC   275

CHAPTER 14   THE LOGIC OF CATEGORICAL STATEMENTS   277
Categorical Sentences   278
Quality and Quantity   280

The Traditional Square of Opposition   281
Translating English Sentences into Standard Categorical Forms   287
Equivalence Rules for Aristotelian Logic   294
Dropping the Assumption of Existential Import   297
The Modern Square of Opposition   299
Glossary   300

CHAPTER 15   CATEGORICAL SYLLOGISMS   303
Logical Form   303
Venn Diagrams   306
Testing a Categorical Syllogism for Validity with Venn Diagrams   308
Diagramming Aristotelian Categorical Syllogisms   310
Diagramming from the Boolean Standpoint   315
The Sorites   320
Testing a Sorites with Venn Diagrams   322
Enthememes   325
Refutation by Logical Analogy   327
Appendix: Rules for Evaluating Categorical Syllogisms   329

PART IV   MODERN QUANTIFICATIONAL LOGIC   335

CHAPTER 16   QUANTIFICATIONAL LOGIC I: THE LANGUAGE QL   337
Two Types of Sentences   338
General Sentences   342
A Syntax for our New Language   348
The Vocabulary of QL   348
Symbolizing General Sentences   349
Categorical Sentences   349
The Old "Quantifier Switch" Trick   354
Switching Quantifiers on Categoricals   356
Symbolizing Complicated General Sentences   358
Denying Existence   360
The Only Way to Go   360
What is a Cat-Dog?   362
Glossary   364

CHAPTER 17   THE LANGUAGE OF QUANTIFICATIONAL LOGIC II:
RELATIONS   365
Sentences with a Quantifier-Dyadic Predicate Combo   366
Any and Every   367
Reflexive Sentences   368
Sentences with Overlapping Quantifiers   369
"What Are You Talking About?" The Universe of Discourse   372
Dean Martin, Universal Love, and a Summary of Logic Relations   374
To Be or Not To Be: The Logic of Identity   376
The Identity Sign   378
Appendix: Properties of Relations   382
Glossary   385

CHAPTER 18   PROOFS WITH MONADIC PREDICATES   386
The Universal Instantiation Rule   386
Existential Generalization   389
Existential Instantiation   390
Memories of Geometry Class: Universal Generalization   393
One New Replacement Rule: Quantifier Exchange   400

Naming Our System   404
Glossary   405

CHAPTER 19   INTERPRETATIONS, INVALIDITY, AND SEMANTICS   **406**
Interpretations of Multiply Quantified Sentences   408
Using Interpretations to Show Invalidity   409
Semantics   414
The Monadic Predicate Test   414
Glossary   417

CHAPTER 20   CONDITIONAL AND INDIRECT QUANTIFIER PROOFS   **418**
Adding Truth-Functional Replacement Rules to the Mix   420
Putting QD on a Diet: A Reduced Set of Quantifier Rules   423
Proving Logical Truths   424

CHAPTER 21   PROOFS WITH OVERLAPPING QUANTIFIERS   **427**
Properties of Relations   430

CHAPTER 22   PROOFS WITH IDENTITY   **433**
Properties of the Identity Relation   436
Glossary   439

PART V   MODAL LOGIC   441

CHAPTER 23   INTRODUCTORY MODAL LOGIC   **443**
To Shave or Not to Shave: That Is the Question   443
Five Modal Properties   444
Possible Truths, Possible Falsehoods, Contingencies   445
Necessary Truths   446
Necessary Falsehoods   448
Putting Statements into Symbols   448
Translating English Sentences into Modal Symbols   455
A Name and Syntax for our Modal Language   457
The Vocabulary for ML   457
The Grammar for ML   457
Linking Modal Operators   459
"It Ain't Necessarily So," Or, Trading a Diamond for a
   Box and a Box for a Diamond   460
Modal Operators Need Scope, Too   462
Modal Relations   464
Scopes of the Dyadic Modal Operators   469
Symbolizing with Dyadic Operators   470
Modal Operators Are Not Truth-Functional   472
Appendix: There's Nothing New under the Sun   473
Glossary   476

CHAPTER 24   MODAL LOGIC: METHODS OF PROOF   **478**
Five Modal Principles   478
Six Inference Rules   480
The Possibility to Necessity Rule   484
The Necessitation Rule   487
Four Modal Replacement Rules   491
Validity in S5   494
Proving Theorems of S5   499
Another Inference Rule: The Tautology Necessitation Rule   500

Appendix 1: Putting an S5 Formula on a Diet: S5 Reduction   502
Appendix 2: The Modal Fallacy   505
Glossary   508

PART VI   INDUCTION   509

CHAPTER 25   INDUCTIVE REASONING   511
Analogical Reasoning   512
Evaluating Analogical Arguments   513
Analogies as Models   518
Enumerative Induction   522
Statistical Inductive Generalization   525
Inference to the Best Explanation   531
What Makes One Explanation Better than Another?   532
Glossary   534

CHAPTER 26   SCIENTIFIC REASONING   535
Scientific Reasoning   536
Comments on the Steps   536
Confirming and Disconfirming Scientific Hypotheses   538
The Confirmation of a Scientific Hypothesis   541
The Disconfirmation of a Scientific Hypothesis   544
The Fact of the Cross   547
What Makes One Hypothesis Better than Another?   548
Case Studies   549
Cause and Effect and Mill's Methods   553
Cause and Effect   554
Mill's Method of Agreement   555
Mill's Method of Difference   558
The Joint Method of Agreement and Difference   560
Mill's Method of Residues   561
Mill's Method of Concomitant Variation   561
Glossary   566

APPENDIX 1   TRUTH-TREES   569

APPENDIX 2   TRUTH   592

Answers to Selected Exercises   599
Index   669

Chapter

1   FUNDAMENTALS

S ometimes when we state a belief or make a claim, others either disagree or ask for evidence. In such cases, we usually give *reasons* in support of our belief. When we give someone a reason for a claim, we are reasoning with them. Logic is the systematic study of reasoning. In logic, one of the goals is to identify and understand the principles involved in reasoning, principles that link reasons with the conclusions they would support or prove.

However, people sometimes have inadequate reasons for a conclusion. And sometimes people draw the wrong conclusion. Logic also seeks to identify the principles that distinguish good reasoning from bad and to understand what makes some reasoning stand out as good or correct reasoning.

When we give reasons for a belief or claim, we are giving an argument. In everyday life, we sometimes use the word *argument* to refer to an angry verbal dispute. However, within logic, an **argument** is nothing more than reasons offered in support of a claim or belief. The reasons offered in support are the **premises** of the argument, and the belief they are intended to support or justify is the **conclusion** of the argument. Thus, an argument has three parts:

1. One or more premises offered as evidence in support of a conclusion.

2. One conclusion.

3. An explicit or implicit *claim* that the premises provide logical support for the conclusion, that is, they provide reasons to believe the conclusion.

Premises are sometimes called the "grounds" for the conclusion, and the conclusion is sometimes said to "rest" on the premises. When the premises provide adequate logical support for the conclusion, we say that the conclusion "follows from" the premises.

2 Chapter 1 *Fundamentals*

However, not all arguments are equal. Some arguments have inadequate premises, some draw the wrong conclusion, and so on. Consequently, one important goal of logic is to understand the principles that distinguish good or correct argumentation from bad or incorrect argumentation.

RECOGNIZING ARGUMENTS

Of course, not all groups of sentences are arguments. Sometimes a group of sentences constitutes a story, sometimes it is an explanation, other times questions are being asked. In such cases, because no evidence is offered for a conclusion, no argument exists. Other times, facts are presented with no attempt to prove anything, or a warning is given, or an opinion is offered. These also are nonarguments, for no one is offering reasons in support of a conclusion. A group of sentences constitutes an argument only when it offers reasons for a claim. The first of the following groups of sentences is an argument, the rest are not.

1. All aardvarks are mammals. All mammals are warm-blooded. Therefore, all aardvarks are warm-blooded. (This is an argument.)

2. One day, Elaine went to buy some soup at the "soup Nazi's" shop. However, because she didn't follow the proper procedure when she placed her order, the soup Nazi told her, "No soup for you." (This is a story, not an argument. No reasons are offered in support of a conclusion.)

3. I believe that everything that happens to you happens for a purpose. I also believe you were smart to take that job offer. (This is an opinion, not an argument. Reasons are not offered for a conclusion, although these statements could form a conclusion if they were supported by reasons.)

4. Watch out for the wild illy-pies! (This is a warning, not an argument. No reasons are offered for a conclusion.) ("Illy-pies" are mythical beasts said to inhabit the wilds of the Pacific Northwest.)

5. It has been raining for seven days and seven nights. (This is a report—a Seattle weather report—not an argument, although it could become a premise if a conclusion were to be drawn.)

6. Your computer shut itself off because it detected that a program performed an illegal operation. (This is an explanation, not an argument. No evidence is offered in support of a conclusion.)

7. Go directly to jail, do not pass go, do not collect two hundred dollars. (This is a command, not an argument.)

8. If were you, I would buy Starbucks stock. (This is advice, not an argument.)

Again, a group of statements constitutes an argument only when it offers reasons in support of a claim.

Let us say that an argument is in **basic form** when the premises are listed one after the other and the conclusion is listed last. For example:

1. All Vulcans are logical and control their emotions.

2. Spock is a Vulcan.

3. Therefore, Spock is logical and controls his emotions.

However, when we encounter arguments in ordinary circumstances—in speeches, newspapers, conversations, and so on—they are seldom organized so simply and clearly. Often, people will put the conclusion first with the premises afterward, or they will give premises, draw a conclusion, and then add more premises. Arguments in everyday contexts can appear all jumbled up. This can raise a problem, for in order to understand an argument, you must first distinguish the premises from the conclusion (otherwise you won't know what is being offered in support of what).

Fortunately, the English language has many expressions that may be used to indicate which statements are offered as premises and which is offered as the conclusion. **Conclusion indicators** are words or expressions used in the context of an argument to signal the conclusion. If we see one of these words or expressions in an argumentative context—a context in which reasons are being offered for a claim—then we can expect that a conclusion will follow.

SOME TYPICAL CONCLUSION INDICATORS

therefore	this shows that	this implies that	thus
hence	consequently	in conclusion	It necessarily follows that
so	it follows that	accordingly	This entails that

For example, if someone is giving an argument, and after presenting a number of statements says "consequently," or "therefore," you naturally expect the conclusion to follow.

Premise indicators are words or expressions used in the context of an argument to indicate that a premise is about to be given.

SOME TYPICAL PREMISE INDICATORS

because	since	given that
for	may be inferred from	due to the fact that
for the reason that	as shown by	on the assumption that

For example, if someone makes a questionable claim and then says "because" or "since," you naturally expect the next statement to serve as a premise for the claim.

Some arguments contain no indicator words. Nevertheless, as long as reasons are being offered in support of a claim, an argument is being given—even if no indicator words are present. In the absence of indicator words, how do you decide which of the statements are premises and which is the conclusion? You must interpret the argument as best you can. It is helpful to begin by asking a question: What is the author trying to prove? If you answer this question correctly, you will have identified the conclusion of the argument. Next, you should ask what evidence the author offers in support of the conclusion. On what does she base her conclusion? In answering this, you will be identifying the premises.

It is important to remember that the premise and conclusion indicator words listed above, words such as *because, since,* and so on, do not always signal the premises and conclusions of arguments; in nonargumentative contexts they can be used to assert other kinds of connections besides argumentative connections. Here are some examples:

- The engine quit *because* it ran out of gas.

Here *because* indicates a cause–effect connection rather than an argumentative (a reason–conclusion) connection.

- We haven't eaten *since* noon.

Here *since* indicates a *temporal* connection rather than an argumentative one.

- I ate, *so,* I'm not hungry.

Here the word *so* indicates an *explanatory* connection—an explanation is being offered.

These are not arguments, for none of these three cases offer reasons in support of a conclusion. In these examples, the words *because* and *since* do not signal an argument's premise and *so* does not signal an argument's conclusion. As noted previously, indicator words only point to premises and conclusions within the context of an argument.

►   **EXERCISE 1.1**   ◄

For each passage, decide whether or not an argument is being presented. If a passage contains an argument, state the conclusion of the argument. (Problems marked with an asterisk are answered in the Selected Answers section at the back of the book.)

1.*I have loved classical music ever since I was in junior high. My seventh-grade teacher said she was related to Beethoven.

2. Susan is hungry because she didn't eat before she started working.

3. So what if I don't like lima beans! You eat them if you think they are so yummy.

4. I heard that the reason Skip left early was because he got sick.

5.*The aardvark is warm-blooded, because aardvarks are mammals and all mammals are warm-blooded.

6. The Blues Brothers are on a mission from God.

7.*That must be a DOS-based computer, for it is running WordPerfect 4.2.

8. For reasons we don't quite understand, Steve sold his car and now takes the bus.

9. Maguire is going to lose the election, for the latest poll showed him trailing the incumbent by twenty points.

10.*Since Susan started college, her grades have improved dramatically.

11. The meeting lasted four hours and concluded with a prayer.

12. I'm casting my vote for Jones for mayor.

13. Spiders are not insects, for insects only have six legs.

DISTINGUISHING SENTENCES AND STATEMENTS

The sentence is a useful invention. Without the sentence we could not adequately put our reasoning into spoken or written form (and thus communicate our reasoning easily to others). However, some kinds of sentence are commonly used for the expression of reasoning, while others are not. Consequently, logicians distinguish between several different types of sentences. A **declarative sentence** makes a claim about the world and thus expresses something that is true or false. For example, "The Moon has mountains" is declarative because it makes a claim that is either true or false. (Sentences that express something true or false are also called "cognitive" sentences.) **Imperative sentences**, such as "Close the door!" express commands, which are neither true nor false. If someone said to you, "Close the door," it would be weird to reply, "That's true." **Interrogatory sentences** such as "What time is it?" express questions, not truth or falsehood. If someone asked you, "What time is it?" it would be odd to say, "How true!" **Exclamatory sentences** such as "Ouch" are used to express emotions or feelings. (Thus, imperative, interrogatory, and exclamatory sentences are "noncognitive.")

Now, the ultimate goal of reasoning is to discover the truth. An argument's conclusion is either true or it is false. And a given premise is either true or false. When we put our reasoning into spoken or written form—that is, when we present an argument—we typically use declarative sentences. In short, reasoning is normally expressed with declarative sentences.

It is important that we draw another distinction. Logicians also distinguish between a sentence and a statement or proposition. Here is one way to introduce the distinction. Two different sentences can express the same thing, as when a sheet of paper has written on it:

Pat likes Jan.

Jan is liked by Pat.

These are two different sentences, for one contains three words and the other contains five words. Suppose Pat does like Jan. The above two sentences therefore express one truth. And so there seems to be a distinction here between the sentence and the truth expressed by the sentence. Moreover, two different sentences from two different languages can express one and the same truth:

The moon has craters.

La luna tiene cráteres.

These two sentences also express the same truth.

It has become customary in logic to call that which is expressed by a declarative sentence, that is, that which is true or false, a "statement" or "proposition." The sentence "The Earth is round" expresses a true statement (or a true proposition), whereas the sentence "The Earth is square" expresses a false statement (or a false proposition). We shall use the terms "statement" and "proposition" to refer to that which can be designated true or false, namely, that which is typically expressed with a declarative sentence. In everyday terms, the proposition expressed by a sentence is sometimes called the "meaning" or "thought" conveyed by the sentence, sometimes it is called the "content" of the sentence.

Propositions or statements generally are expressed by declarative sentences. The grammarian is usually interested in the sentences themselves, while the logician is interested in the logical relationships between the propositions expressed by the sentences. Logical confusions can result if we fail to note the distinction between the sentence and the statement or proposition expressed by the sentence. Consequently, in the pages that follow, we will often distinguish between sentences and the statements or propositions expressed by those sentences.

TWO GENERAL CATEGORIES OF ARGUMENT: DEDUCTIVE AND INDUCTIVE ARGUMENTS

Logicians distinguish two general categories of argument: deductive arguments and inductive arguments. Every argument makes a *claim* concerning the relation between the premises and the conclusion. A **deductive argument** claims, explicitly or implicitly, that *if* the premises are true, then the conclusion *must* be true. Put differently, a deductive argument claims that it is *impossible* for the conclusion to be false *if* the premises are all true. For example, consider the following simple argument in basic form:

1. All aardvarks are mammals.

2. All mammals have lungs.

3. Therefore, it necessarily follows that all aardvarks have lungs.

The words, "it *necessarily* follows that" constitute an explicit claim that the conclusion—"All aardvarks have lungs"—*must* be true if the premises are true. This is therefore a deductive argument. A deductive argument has an air of certainty, an air of definiteness. If the premises are true, it is claimed, we can be *certain* that the conclusion is true. In a deductive argument, the claim is that the premises, *if* accepted, provide conclusive evidence that the conclusion is true.

An inductive argument, on the other hand, claims no such certainty or conclusiveness. An **inductive argument** claims, explicitly or implicitly, that *if* the premises are true, it is *probable* that the conclusion is true, although the conclusion might nevertheless be false. In other words, inductive arguments claim that it is *improbable* that the conclusion is false if the premises are true. Here is an example of an inductive argument:

1. Kramer has ordered a caffé latte every morning for the past two months.

2. So, he will probably order a caffé latte tomorrow, too.

Compared to the deductive argument above, an inductive argument makes a more modest claim, namely, that if the premises are true, it is *probable* that the conclusion is true. It does not claim that the conclusion must be true or is certain to be true.

Here are two more examples of an inductive argument:

- We surveyed one hundred people at random, and seventy-two plan to vote for Jones for mayor. Therefore, Jones will probably be elected mayor.

- Your car is making exactly the same noise mine made just before my car blew its transmission. Therefore, your car is probably going to blow its transmission.

Here are two more examples of a deductive argument:

1. They will serve us either hamburgers or hot dogs.

2. They will not serve hot dogs.

3. So, it *must* be that they will serve hamburgers.

(Continued)

1. If it rains the concert will be canceled.

2. It will rain.

3. So, it necessarily follows that the concert will be canceled.

In sum, if an argument claims—explicitly or implicitly—that no possibility exists of true premises and a false conclusion, the argument is a deductive argument. If an argument claims—explicitly or implicitly—that the premises make the conclusion probable, the argument is an inductive argument.

DEDUCTIVE AND INDUCTIVE INDICATOR WORDS

If you want to claim explicitly that your argument is deductive, you should introduce your conclusion with a phrase indicating that you think your conclusion *must* be true if the premises are true. Phrases such as "it necessarily follows that," "it must follow that," or "it is therefore certain that" usually serve this purpose. Placing single words such as *necessarily, certainly,* and *must* in the conclusion at an appropriate spot also indicates a deductive claim.

On the other hand, if you want to claim in an explicit way that your argument is inductive, you should introduce your conclusion with a phrase indicating that you believe your conclusion is probably true if the premises are true. Phrases such as *it's therefore probably true that, it's therefore likely that,* or *it's therefore plausible to suppose that* usually suffice. Also, placing single words such as *probably* or *likely* in the conclusion at an appropriate spot will indicate an inductive claim.

▶ **EXERCISE 1.2** ◀

Identify each of the following arguments as deductive or inductive.

1.*All whales are mammals. All mammals have lungs. So, necessarily, all whales have lungs.

2. The K-man hasn't worked in more than two years. So, he probably won't be working tomorrow.

3.*It rarely rains in Seattle in August, so, Charlie will likely have sunny weather for his picnic this August in Seattle.

4. The sign says "Closed." So, they most likely are not open.

5.*All certified financial consultants must pass a test. She is certified, so, she must have passed the test.

6. The figure has four equal sides and four equal angles, so, it necessarily follows that it is a square.

7. The substance is an acid, so, it certainly must turn blue litmus paper red.

8. "Each snow crystal contains about a quintillion water molecules, and these molecules can be arranged in countless numbers of ways. Of course, no one can prove it, but it is quite likely that every snowflake is unique." –Scott Camazine, *The Naturalist's Year.*

9.*Dr. Frasier Crane, Seattle's most popular radio psychologist, will be broadcasting from the event, so, I'll bet a lot of people will be there.

10. Fred is frowning, so he is most likely unhappy.

11. Angle A is larger than angle B, and angle B is larger than angle C, so angle C must be smaller than angle A.

12. The good guys always win. Austin Powers is the good guy and Dr. Evil is the bad guy. So, Austin Powers will certainly defeat Dr. Evil.

EVALUATING INDUCTIVE ARGUMENTS: THE STRONG, THE WEAK, AND THE INDUCTIVELY SOUND

An inductive argument *claims* that if the premises are true, the conclusion is *probably* true. But not all claims are correct. Some claims are, some claims are not. Now, if an inductive argument's claim is correct, we call the argument an *inductively strong argument.* If the claim is incorrect, we call the argument *inductively weak.* An inductively strong argument has the following feature:

If the premises were to be true, the conclusion would be probably true.

Another way to put this is:

The truth of the premises would make the conclusion probable.

For example, the following argument is a strong inductive argument:

1. Jones has ordered a cheeseburger for lunch every day for the past year.

2. He is about to order his lunch.

3. Therefore, probably he will order a cheeseburger.

It is possible the conclusion of this argument is false even if the premises are all true; however, it is unlikely. The premises support the conclusion in the following sense: *if* the premises are true, the conclusion is probable. If the premises are true, conclusion is not certain or conclusively established, but it is made likely or probable. The support provided by the premises is "strong" but not conclusive.

In order to see this, suppose—just for the sake of argument—that the premises are true. Now, given that the premises are true, what follows? It follows that the conclusion is probably true. That is, the truth of the premises would indeed make the conclusion highly probable. The argument's inductive claim is therefore correct, and this is an inductively strong argument.

On the other hand, if an inductive argument's claim is incorrect, we say that the argument is a weak inductive argument.

> *A weak inductive argument has the following feature: The truth of the premises would not make the conclusion probably true.*

Here is an example of a weak inductive argument:

1. I've known three people from New York, and all three were Episcopalians.

2. The Stern brothers are from New York.

3. Therefore, the Stern brothers are probably Episcopalian.

The premises do not make the conclusion probable. That is, if all we know is that the premises are true, that alone would not make the conclusion probable.

If an argument is a strong inductive argument, and in addition it has *true premises*, we say that it is an **inductively sound argument**. The following strong inductive argument has an undeniably true premise:

1. The Sun has risen every morning for the past two hundred years.

2. Therefore, it will probably rise tomorrow morning.

Because it is strong and also has true premises, this is an inductively sound argument.

On the other hand, here is an inductive argument that is strong but not inductively sound:

1. The Beatles are going to hold a reunion concert at the Hollywood Bowl next week.

2. Therefore, it is probable that at least one hundred people will attend the concert.

Now, this argument is incredibly *strong*, that is, if the premises are true, the conclusion is extremely probable. However, unfortunately, the premise is false. An argument must be strong *and* have true premises in order to qualify as an inductively sound argument; this argument is therefore not inductively sound.

EVALUATING DEDUCTIVE ARGUMENTS: THE VALID, THE INVALID, AND THE DEDUCTIVELY SOUND

In general, a deductive argument *claims* that if its premises are true, its conclusion must be true. However, not all claims are true. Some claims are true, some claims are false. There are two types of deductive arguments. A *valid* deductive argument is a deductive argument whose claim is correct. Thus, an argument is a valid deductive argument (or "valid" argument) if the following is true of it: it would be impossible for the conclusion to be false *if* the premises were all true. Another way to put this is:

> *A valid deductive argument has the following feature: It would not be possible that the premises are all true and yet the conclusion is false.*

For example:

1. All shrews are nocturnal.

2. All shrews are insectivores.

3. Therefore at least some insectivores are nocturnal.

Notice that *if* the premises are true, the conclusion *must* be true. This is a deductively valid argument.

A valid deductive argument is such that

- If all of the premises were true, the conclusion would have to be true as well; or

- It would be impossible for its premises all to be true while its conclusion is false; or

- It is not possible for the conclusion to be false when all the premises are true; or

- It is not possible that the premises are true and yet the conclusion is false.

An invalid deductive argument, on the other hand, is a deductive argument whose claim is incorrect. Thus, an invalid deductive argument has the following feature: even if its premises are all true, it is still possible that its conclusion is false. Another way to put this is: It is possible that its premises are true and yet its conclusion is nevertheless false. For example:

1. Susan is Nancy's cousin.

2. Janice is Nancy's cousin.

3. Therefore, Susan and Janice *must* also be cousins.

Notice that even if the premises are true, the conclusion might be false. (For instance, Susan and Janice might be sisters.)

> A deductively invalid argument is such that
>
> • Even if all of the premises are true, the conclusion might nevertheless be false; or
>
> • Its premises might all be true and yet its conclusion false; or
>
> • The conclusion might be false when all the premises are true; or
>
> • The premises might be true and yet the conclusion false.

The following are examples of valid arguments.

1. All dogs are mammals.

2. All mammals are warm-blooded.

3. Therefore, all dogs are warm-blooded.

1. The computer is either a PC or a Mac.

2. It is not a PC.

3. So, it must be a Mac.

1. If Elaine places her order incorrectly, the "Soup Nazi" will refuse to serve her.

2. If the Soup Nazi refuses to serve her, she won't bring home any "Muligatoni" soup.

3. Therefore, necessarily if Elaine places her order incorrectly, she won't bring home any Muligatoni soup.

In each of these arguments, the claim—that the conclusion must be true if the premises are true—is correct. In each case, the conclusion simply *must* be true if the premises are true.

The following are examples of invalid arguments.

1. Fred is a senior in high school.

2. Jan is a junior in high school.

3. So, Fred must be older than Jan.

1. Whenever Ann goes swimming, Bob goes swimming.

2. Ann won't swim today.

3. So, it necessarily follows that Bob won't swim today.

1. Banya is a friend of Jerry's.

2. George is a friend of Jerry's.

3. Therefore, it necessarily follows that Banya is a friend of George's.

The claim in each of these invalid arguments—that if the premises are true the conclusion must be true—is incorrect in each case, because in each case it is *possible* that the premises are true while the conclusion is false.

DECIDING WHETHER AN ARGUMENT IS VALID OR INVALID

To decide whether or not an argument is valid, ask yourself the following question: Are there any possible circumstances in which the premises are true and the conclusion is false? If the answer is "no," the argument is valid. If the answer is "yes," the argument is invalid. For example, consider the following argument.

1. Every member of the Jerry Lewis fan club is a senior citizen.

2. Some Starbucks customers are senior citizens.

3. So, some Starbucks customers are members of the Jerry Lewis fan club.

Now, would it at least be *possible* that the premises are true but the conclusion is false? Surely. It could be that although all Jerry Lewis fan club members are senior citizens, and although many Starbucks customers are senior citizens, no Starbucks customers are members of the Jerry Lewis fan club. We have found a possible circumstance in which the premises of the argument are true while the conclusion is false. This shows that the argument is invalid.

It is extremely important to observe the following: we figured out that this argument is invalid without knowing whether the premises are actually true or not. We proceeded by considering, abstractly, the possibilities. Thus, two questions were kept separate:

A. Are the premises true?

B. Is the argument valid or invalid?

We answered question B without having answered question A. The information one might uncover by answering A would not contribute in any way to the answer to B. And, of course, the reverse is also true.

Now consider the following argument:

1.   Frasier is taller than Niles.

2.   Niles is taller than Mr. Crane.

3.   So, Frasier must be taller than Mr. Crane.

Is this argument valid or invalid? Could the premises possibly be true while the conclusion is false? If the premises were true, might the conclusion nevertheless be false? Clearly, no such possibility exists. The argument is therefore deductively valid. Incidentally, are the premises true? That's a completely separate issue. To resolve that question, we'd have to measure the cast of the TV show *Frasier*. In most cases, logical theory alone is insufficient to determine whether the premises are true or not.

► COUNTEREXAMPLES

If an argument is deductively invalid, then it is possible that the premises are true and the conclusion is false. One way to show that an argument is deductively invalid is to describe a possible situation in which the premises are true and the conclusion is false. Such a situation is said to be a **counterexample** to the argument. For instance, suppose someone proposes the following deductive argument:

1.   Craig is Karen's son.

2.   Karen is married to Tom.

3.   Therefore, Craig *must* be Tom's son.

This is a deductive argument: this person is claiming if the premises are true the conclusion can't be false. How could you convince this person that the argument is invalid? You could describe a counterexample to the argument—a possible circumstance in which the premises of the argument are true and the conclusion is false. Thus, you might ask the author of the argument to consider the following possibility:

Possibly, Craig is Karen's son from her first marriage.

Karen married Tom years later, when Craig was 12.

Craig is Tom's stepson.

In this situation, Craig would not be Tom's son, and the conclusion would be false. This mere *possibility* constitutes a counterexample to the argument, for it is a possible circumstance in which the premises of the argument are true and its conclusion is false. When we describe for someone a counterexample to an argument, we help them see a way in which the premises of the argument could be true and the conclusion false. Thus, the presentation of a counterexample establishes that an argument is invalid.

Now, look back at the argument above involving Frasier, Niles, and Mr. Crane. Can you describe a counterexample? Try to describe a situation in which the premises of that argument are true while its conclusion is false. Of course, this can't be done, for the argument is valid. There is no such thing as a counterexample to a valid argument.

The Relation Between Validity, Invalidity, and Truth

In everyday life, we sometimes use the word *valid* to mean "true," as when we say, "You have a valid point" or "Your idea is not valid." Consequently, upon being introduced to the terminology of logic, some people begin using *true argument* in place of *valid argument*. However, within the field of logic, *valid* does not mean "true," and, as the various terms of argument appraisal have been defined above, truth is not a property of an argument. There are true statements, true premises, true conclusions, but—as defined in the field of logic—there is no such thing as a "true argument." There are, of course, valid arguments.

This raises an important yet somewhat difficult point. If you reflect on the definition of deductive validity, you will notice that an argument's validity has nothing to do with whether or not the premises are actually true. When we say that a particular argument is deductively valid, we are not saying that its premises are actually true, nor are we saying that its conclusion is actually true. Rather, we are merely saying that it is not possible for the premises to be true and the conclusion false. So, an argument can have false premises and nevertheless still be valid. It is not required that an argument have true premises in order to be valid. For example:

1. All people who drink Frapuccinos also drink beer.

2. All people who drink beer also drink bottled water.

3. So, it necessarily follows that all people who drink Frapuccinos also drink bottled water.

Notice that this argument is indeed valid, even though its premises are false and its conclusion is false. If the premises were to be true, the conclusion would have to be true. If this puzzles you, remember that validity is concerned with a purely *hypothetical* matter—the relationship between the truth of the premises and the truth of conclusion—rather than with the *actual* truth of the premises.

(Continued)

Here is a related point: The mere fact that the premises and the conclusion of a given argument are all true does not make the argument deductively valid. Some arguments have true premises and a true conclusion but are, nonetheless, deductively invalid. For example, consider the following argument:

1. Bob Dylan played a concert at The Gorge.

2. Van Morrison played a concert at The Gorge.

3. Therefore, surely Bob Dylan and Van Morrison performed together at The Gorge.

This argument does have true premises and a true conclusion, yet it is nevertheless deductively invalid. True premises and a true conclusion do not, by themselves, make an argument valid.

Consider next a curious feature of deductive invalidity. When we say that an argument is deductively invalid, we are not saying that its premises are false, nor are we saying that its conclusion is false. Remember that some invalid arguments have true premises and a true conclusion (for example, the argument about Dylan immediately above). Furthermore, the mere fact that an argument has false premises and a false conclusion does *not* mean that the argument is deductively invalid. Remember, some have false premises and a false conclusion, and yet are deductively valid (for example, the argument about the Frapuccinos above).

► **EXERCISE 1.3** ◄

Part 1. Which of the following deductive arguments are valid and which are invalid?

1.*Because some Fords are purple and some Fords are trucks, it necessarily follows that some Fords are purple trucks.

2. No car is a fruit. All lemons are fruits. So, no car is a lemon.

3.*The United States House of Representatives has more members than there are days in a leap year. Therefore, at least two members of the House of Representatives must have the same birthday.

4.*Boys Town is located in Nebraska. Nebraska is located in the United States. So, Boys Town must be located in the United States.

5. Every member of the Progressive Labor Party must study the writings of Chairman Mao. Bob studied the writings of Chairman Mao. So, Bob must be a member of the Progressive Labor Party.

6. Listening to loud music ruins your hearing. Joe is hard of hearing. Therefore, Joe has been listening to loud music.

7. The Monterey Pop Festival (July 1967) featured some of the greatest bands of the 1960s. The Beatles did not perform at the Monterey Pop Festival. So, the Beatles were not among the greatest bands of the 1960s.

8. The Monterey Pop Festival featured some of the greatest bands of the 1960s. Big Brother and the Holding Company performed at the Monterey Pop Festival. So, Big Brother and the Holding Company was one of the greatest bands of the 1960s.

9.*Only Hell's Angels live in the Brentwood apartments. Hank belongs to the Hell's Angels. Therefore, Hank must live in the Brentwood apartments.

10. All bikers are independent spirits. Chris is a biker. So, Chris must be an independent spirit.

11.*All Hell's Angels live in California. Pat lives in California. So, Pat certainly belongs to the Hell's Angels Motorcycle Club.

12. Figure A has four equal sides. It therefore must be a square.

13. All snowboarders are skateboarders. Joe is a skateboarder. So, Joe must also be a snowboarder.

Part 2. Go back to Exercise 1.2 and determine which of the arguments are deductively valid and which are inductively strong.

▶ **DEDUCTIVELY SOUND ARGUMENTS**

Suppose we know that a particular argument is deductively valid. Does it follow that the conclusion of the argument is true? As we have seen, the answer is no. An argument need not have true premises in order to qualify as valid. To say that an argument is deductively valid is only to say that it is not possible for the premises to be true and the conclusion false. However, when we evaluate an argument, we usually want to know more than whether it is deductively valid or not. We also want to know whether the premises and conclusion are *true*. Truth is the ultimate goal of arguing, of using our capacity to reason. So, we are looking for more than deductive validity when we evaluate an argument—we are looking for truth. A **deductively sound** argument has these two features:

1. It is deductively valid.

2. Its premises are all true.

That is, a deductively sound argument is a valid argument whose premises are all true.

In general, logic is not concerned with determining whether an argument is deductively sound or not, for, generally, logic alone cannot determine whether the premises are true or not. (An exception is when a premise is a logical truth, a concept we will investigate in later chapters.)

Will the conclusion of a deductively sound argument be true or false? Because any sound deductive argument is deductively valid, it follows that if the premises are true, the conclusion must be true. Furthermore, the premises of a deductively sound argument are true. The conclusion of a deductively sound argument must therefore be true as well. Every deductively sound argument is valid, has true premises, and consequently has a true conclusion. Because truth is, or should be, the ultimate goal of arguing, it is deductive soundness, rather than mere deductive validity, that we are ultimately interested in when we evaluate arguments such as the ones above.

►   **EXERCISE 1.4**   ◄

Identify which of the following statements are true and which are false.

1. *An argument may have true premises and yet be invalid.

2. An argument may be invalid and yet have a true conclusion.

3. An argument may have false premises and a false conclusion and yet be valid.

4. *Some deductively sound arguments have false premises.

5. Some invalid arguments have true premises.

6. All deductively sound arguments have true premises and are also valid.

7. *An argument may have false premises and yet be valid nevertheless.

8. Some deductively sound arguments are invalid.

9. *All valid arguments have true premises.

10. All invalid arguments have false premises.

11. All inductively strong arguments have true premises.

12. Some inductively sound arguments have false premises.

13. If the premises are all true and the conclusion is true, then the argument must be valid.

14. *If the premises are all false and the conclusion is false, then the argument must be invalid.

Enthememes

Sometimes, when people give an argument, they leave out a premise. Perhaps the missing premise is an item of common knowledge that is assumed. Perhaps it is an overlooked step in the reasoning process. Also, sometimes the conclusion of an argument is left unstated because the author wants the reader to draw the conclusion for himself or herself. An **enthememe** is an argument that is missing one or more premises or a conclusion. (*Enthememe* is derived from a Greek word meaning "have in mind," which is appropriate, because an enthememe leaves unstated something we have in mind.)

Here is an enthememe: "All chemists are scientists, so, Miss Hathaway is a scientist." The unstated premise in this case, assuming the arguer intends the argument to be valid, is obviously "Miss Hathaway is a chemist." Notice that without the unstated premise, the argument is invalid; with the missing premise added, the argument is valid.

▶	**EXERCISE 1.5**	◀

This exercise should help deepen your understanding of the concept of deductive validity. Each of the following enthememes is missing a premise. In each case, add a premise so as to produce a valid argument.

1. *All reptiles are cold-blooded, so, your pet is cold-blooded.
2. If Ann swims today, Bob will swim today. So, Bob will swim today.
3. *We will either take our bikes or we will walk. So, we'll be walking.
4. The universe exists. Therefore, the universe has a cause.
5. Human thoughts are not spatial objects. Therefore, human thoughts are not physical objects.
6. Each human thought is an existent entity. Therefore, each human thought has a location in space.
7. *All human actions have causes. So, no human action is an act of free will.
8. Spiders are not insects, for insects have six legs.
9. Ed must be unhappy, for he is scowling.
10. That is a particle of matter. Therefore it has mass.
11. Ann lives in Las Vegas. So, Ann lives in Nevada.
12. *Liberace was a great pianist. Therefore, Liberace was a great musician.

CONSISTENCY AND INCONSISTENCY

Logical reasoning has another important purpose. Besides its utility in constructing an argument or justifying a belief, logical reasoning provides a means of evaluating **consistency** and **inconsistency** and can lead us to reject a belief when we realize some of our beliefs stand in logical conflict. For example, suppose Pat, a drummer in a combination heavy-metal/polka band, has long believed that

Ringo was the only drummer the Beatles ever had.

One day he learns:

Pete Best was the drummer for the Beatles before Ringo.

Pat's two beliefs now stand in logical conflict. They cannot both be true. He now must give up at least one of his beliefs.

Or suppose Pat believes all three of the following statements:

All politicians are honest.

Fred is a politician.

Taking a bribe is dishonest.

Then one day Pat finds out that:

Fred took a bribe.

These four statements are logically interconnected, and they cannot all be true. At least one is false. Again, Pat must give up one or more beliefs.

The same principles that are used in argumentation are used to decide if a set of statements stands in logical conflict. A set of statements is **logically inconsistent** if and only if it is not possible that all of the statements in the set are true together. The two sets of statements above, the one about the Beatles and the one about politicians, are each logically inconsistent.

On the other hand, a set of statements is said to be **logically consistent** if and only if there is any possibility—no matter how unlikely—that all the statements in the set are true together. For example, the following set of statements is logically consistent, even though it is unlikely all the statements are true.

Muhammad Ali will reenter the ring next year and regain his old heavyweight title.

A write-in candidate will be elected the next president of the United States.

The remaining Beatles will hold a reunion concert next summer.

It is important that you understand the following point: When we say that a particular set of statements is consistent, we are *not* saying that all of the statements in the set are true. We are merely saying it is possible that all of the statements in the set are true. When we say that a set of statements is inconsistent, we are not saying that all the statements in the set must be false. We are merely saying no possibility exists that all of the statements in the set are true, which implies that at least one statement in the set must be false.

IMPLICATION

If an argument is valid, we say that the premises *imply* the conclusion. There are two cases to consider:

- A group of statements **implies** a further statement if and only if there is no possible situation in which all the statements in the group are true and in which the further statement is false.

- One statement **implies** a second statement if and only if there is no possibility that the first statement is true and the second is false.

In the following example, statements 1 and 2 together imply statement 3:

1. All human beings are mortal.
2. Pat is a human being.
3. Pat is mortal.

In each of the following examples, the first statement implies the second:

1. Julio is sixteen.
2. Julio is under twenty-one.

1. Margarita is thirty.
2. Margarita is legally an adult.

In each of the following examples, the first statement does *not* imply the second.

1. Julio is rich.
2. Julio is materialistic.

1. Julio is religious.
2. Julio is Jewish.

LOGICAL EQUIVALENCE

If two statements imply each other, we say they are **logically equivalent.** For example, consider the following two statements:

No birds are insects.

No insects are birds.

Because they imply each other, these two statements are logically equivalent. If two statements imply each other, then if one is true then the other must be true and if one is false then the other must be false. So, two statements are logically equivalent just in case it is not possible for the two to differ in terms of truth or falsity. In other words, if two statements are equivalent, then in any possible circumstance, if one is true, the other will also be true and if one is false, the other will also be false. Two equivalent statements will always match in terms of truth and falsity. Each of the following pairs of statements is a pair of equivalent statements:

1. Sue is married to Ron.

2. Ron is married to Sue.

1. Sue is taller than Ron.

2. Ron is shorter than Sue.

1. Sue is exactly as old as Ron.

2. Ron is exactly as old as Sue.

1. The charge on a returned check is twice the value of the check, or one hundred dollars, whichever is larger.

2. The charge on a returned check is twice the value of the check, with a one-hundred-dollar minimum charge.

NECESSITY

In everyday reasoning, when we say that something is "necessarily" a particular way, we mean that it could not possibly have been otherwise. In logic, an individual statement is said to be a **necessarily true statement** if and only if it could not possibly have been false. In other words, a statement is necessarily true just in case there's no possible circumstance in which it is false. Consider the following statements:

Nothing is red all over and green all over at the same time.

The Space Needle either is one hundred years old or is not one hundred years old.

If today is Tuesday, today is not Monday.

Notice that in each of these three cases, the statement—given its standard meaning—cannot possibly be false. Therefore, each statement—if given its standard meaning—expresses a necessary truth.

A statement is said to be a **necessarily false statement** if and only if there's no possible circumstance in which it is true. Consider the following examples:

Jan and Pat are each older than the other.

Pat is thirty, and it's not the case that Pat is thirty.

Some circles are square.

Notice that in each of these three cases, there's no possibility that the statement—given its standard meaning—is true. That is, each statement—if given its standard meaning—expresses a necessary falsehood.

In everyday discourse, we generally use the term *contingent* to mean "dependent." For example, we may say the sale of a house is contingent upon the sale of the buyer's house. A statement that is neither necessarily true nor necessarily false is going to be true in some possible circumstances and false in others. That is, its truth or falsity will vary from circumstance to circumstance. Such a statement is a **contingent statement,** because its truth or falsity depends on the circumstance. Here are some examples of contingent statements:

Red Delicious apples are fifty-nine cents a pound.

Abe Lincoln wore a beard.

Rita has a job as a meter maid.

In future chapters, we will investigate the concepts of necessity and contingency and their relationship to deductive validity.

IDEALS

We associate ideals with each type of human activity. Let's reflect for a moment upon ideals relative to the activity of rational discussion and thought. When you have intellectual discussions with others, what ideals should you try to live up to? You might begin thinking about this by asking yourself, "When engaging in discussions, how do I want others to treat me?" The following two ideals of intellectual interaction are presented merely as suggestions to start you thinking about the question.

> *Intellectual Fairness: Ideally, when you consider an issue, you ought to try to understand the main reasons for and against each of the different points of view on the matter. Someone who tries to understand and evaluate all sides of an issue before drawing conclusions may be described as an "intellectually fair" individual.*

Intellectual Kindness: Some people, when engaged in reasoned discussion, can become angry, mean-spirited, and even abusive. In the heat of argument, they sometimes treat opposing points of view with contempt. In short, some people are just no fun to argue with. In contrast, an "intellectually kind" individual treats each point of view with equal respect. This individual is listening for valuable points put forward by others.

▶ TWO BRANCHES OF LOGIC

Deductive logic is the branch of logic concerned with the study of deductive validity and related concepts. **Inductive logic** is concerned with the study of inductive strength and related concepts. Within inductive logic, theorists develop methods to measure the degree of probability or rational support conferred by the premises upon the conclusion. Within deductive logic, techniques are developed to help us determine whether or not an argument is deductively valid.

CHAPTER 1 APPENDIX: Some Logical Puzzles

The following problems are adapted from puzzles presented by Raymond Smullyan in his delightful book titled *What is the Name of this Book?*

Imagine that you have been shipwrecked on a magical island, the Isle of Knights and Knaves. Every inhabitant of this island is either a knight or a knave. The island is governed by two absolutes: Knights always speak the truth, and knaves always lie. The problem is that knights and knaves look alike, that is, you can't tell them apart in any visual way. Now, solve each of the following problems independently from the others.

1.* One morning, you are out riding your rented unicorn and you come across two individuals, one tall and one short. You can tell that they want to test your logical abilities. The tall one says, "Either I am a knave or the short one is a knight." Is the tall one a knight or a knave? Is the short one a knight or a knave?

2. Later that day, you come across three strangers, one dressed in black, one in white, and one in red. The one in red remains silent, but the other two speak:

Black: All of us are knaves.

White: Exactly one of us is a knight.

What are these individuals? Knights? Knaves? Which is which?

3.* The next day, you meet two more strangers coming out of an enormous castle. One is old and the other is very young. The old one says: "I am a knave, but the young one isn't." What are they?

4. All of a sudden, three strangers ride up on enormous horses. The strangers eye you suspiciously and challenge you to a logical contest. The first one says, "The second one of us is a knave." The second one says, "The first and the third of us are of the same type." (On the Isle of Knights and Knaves, two individuals are of the same type only if they are both knights or both knaves.) The question you must answer, or you will lose your life, is: What is the third individual? Knight or knave?

5.*The next day, you meet three new strangers at the local vampire cave. You nickname them "Happy," "Goofy," and "Grumpy" because of their distinct personalities. Happy says, "Goofy and Grumpy are of the same type." Someone asks Grumpy, "Are Happy and Goofy of the same type?" What does Grumpy answer?

6. That evening, you come across an individual who slowly circles around you and quietly says, "Either I am a knave or 2 + 3 = 5." What is he? Knight or knave?

7. The next day, you go to return your rented suit of armor and the cashier says, "Either I am a knight or I am not a knight." What is he?

8. At the local castle, the guard says to you, "I am a knight and I am not a knight." What is he?

9. You go to the local zoo and the zookeeper at the aardvark cage says to you, "Most aardvarks eat ants and most aardvarks do not eat ants." Is the zookeeper a knight or a knave?

10.*A stranger comes up to you and says, "If I am a knave then I am a knave." What is he? Knight or knave?

11. Is the following event possible? An inhabitant of the island speaks the words "I am a knave."

12. Two strange looking individuals cautiously approach you. The first says, "Neither of us is a knight." The second is silent. What are they?

CHAPTER 1 GLOSSARY

Argument   One or more statements, called "premises," offered as reasons or evidence in support of a further statement, called the "conclusion."

Basic form   (for an argument) An argument is in basic form when the premises are listed one after the other and the conclusion is listed last.

Conclusion   In an argument, the statement which the premises are said to support or justify.

Conclusion indicators   Words or expressions used in the context of an argument to signal the conclusion.

Consistency   A set of statements is consistent if and only if it is possible that all of the statements in the set are true together.

Contingent statement   A statement is contingent if there are circumstances in which it would be true and there are circumstances in which it would be false.

Counterexample to an argument   A description of a possible situation in which the premises are true and the conclusion is false (which shows that the argument is not deductively valid).

Declarative sentence   (also called a "cognitive" sentence) A sentence typically used to make a claim about the world and thus to express something true or false.

Deductive argument   An argument that claims, explicitly or implicitly, that *if* the premises are true, the conclusion *must* be true, which is to claim that it would be *impossible* for the conclusion to be false if the premises were all to be true.

- **Invalid deductive argument**   (or "deductively invalid" argument) A deductive argument whose claim is incorrect. Thus, a deductive argument with the following feature: even if its premises were all true, its conclusion might be false, that is, its premises might be true and yet its conclusion false.

- **Valid deductive argument**   (or "deductively valid" argument) A deductive argument whose claim is correct. Thus, a deductive argument that has the following feature: the conclusion cannot be false *if* the premises are all true, that is, the premises cannot all be true and yet the conclusion false.

- **Deductively sound argument**   A deductive argument that is valid and that also has true premises.

Deductive logic   The branch of logic concerned with the study of deductive validity and related concepts.

Enthememe   An argument that is missing one or more premises or a conclusion.

Equivalence   Two statements are equivalent if and only if it would be impossible for the two to differ in terms of truth and falsity, so that if one is true, the other is true, and if one is false, the other is false. In short, two statements are equivalent if and only if they imply each other.

Exclamatory sentence   A sentence that expresses an emotion or feeling.

Imperative sentence   A sentence that expresses a command.

Implication   A *group* of statements implies a further statement if and only if no possible situation exists in which all the statements in the group are true and in which the further statement is false. *One* statement implies a second statement if and only if there is no possibility that the first statement is true and the second is false.

Inconsistency   A set of statements is inconsistent if and only if it is impossible that all of the statements in the set are true together.

Inductive argument   An argument that claims, explicitly or implicitly, that *if* the premises are true, it is *probable* that the conclusion is true, although the conclusion might nevertheless be false.

- **Strong inductive argument**   (or "inductively strong" argument) An inductive argument whose claim is correct. Thus, an inductive argument that has the following feature: If the premises were true, the conclusion would be probably true, that is, the truth of the premises would make the conclusion probable.

- **Weak inductive argument** (or "inductively weak" argument) An inductive argument whose claim is incorrect. Thus, an inductive argument that has the following feature: If the premises were true, the conclusion would not be probably true, that is, the truth of the premises would not make the conclusion probable.

- **Inductively sound argument** An inductive argument that is strong and that also has true premises.

Inductive logic The branch of logic concerned with the study of inductive strength and related concepts.

Interrogatory sentence A sentence that expresses a question.

Necessarily false statement An individual statement is necessarily false if and only if it is false and could not possibly have been true, that is, no possible circumstances exist in which it is or would be true.

Necessarily true statement An individual statement is necessarily true if and only if it is true and could not possibly have been false, that is, no possible circumstances exist in which it is or would be false.

Premise In an argument, a statement offered as evidence in support of a conclusion.

Premise indicators Words or expressions that are used in the context of an argument to indicate that a premise is about to be given.

P a r t

I

TRUTH-FUNCTIONAL LOGIC

INTRODUCTORY TRUTH-FUNCTIONAL LOGIC

F ields such as music and mathematics use special symbolic languages to simplify the expression of complex ideas. Those symbolic languages, called *artificial languages,* differ in important ways from natural languages such as English, French, and Swahili. A natural language is typically learned during the first few years of life. Artificial languages, on the other hand, are learned after we first learn a natural language, and indeed are learned only with the aid of a natural language. A natural language serves as a general tool of communication, whereas an artificial language is designed to express a restricted body of ideas, not as a tool for general communication. Artificial languages are designed on purpose for a special subject matter; natural languages are not designed by a person or a committee, rather they evolve (without central direction) over long periods of time.

Artificial symbolic languages serve an important purpose: they allow us to express complex ideas more simply than if we used only a natural language. Consider, for example, the symbolic language of musical notation. Beethoven could have written the musical scores for his compositions entirely in words, without using any musical notation. However, if he had spelled out in words every note to be played, one composition would take up hundreds of pages of text, and the members of an orchestra would find it difficult to read the "sheet" music. By using the symbols of musical notation, Beethoven was able to express his ideas in a more compact and easier to read form. Similarly, a calculus book could be written entirely in English, without the use of a single mathematical symbol. However, when compared to an ordinary calculus text, a "natural language" calculus book would be lengthy and difficult to understand. In short, artificial languages facilitate communication.

Artificial symbolic languages serve another purpose. The words of a natural language typically have multiple meanings and often push emotional "buttons" as well. This may be good for certain types of communication such as politics, poetry, or drama, but it generally leads to confusion and error within a technical field of thought such as mathematics, computer science, or logic. In a technical field of thought, ideas must be expressed in precise and unambiguous terms or else communication breaks down. The symbols of an artificial language can be assigned the precise meanings required for the communication of technical ideas.

This chapter introduces the basic elements of a symbolic logical language that is designed to represent the logical properties of an important type of argumentation—truth-functional argumentation. Later chapters will introduce additional logical languages representing the logical properties of additional types of argumentation. In the next chapter, after our first symbolic language has been introduced, we will translate arguments out of English and into this artificial language. From there we will employ procedures that allow us to solve a variety of logical problems in a precise and systematic way, much as a mathematician solves an algebra problem.

From here on, we will frequently attribute truth and falsity to statements. Two terminological conventions will be used. First, if a sentence in a given context expresses a truth, we will sometimes simplify matters by saying that the sentence itself is true. However, it should be understood that when we call a sentence "true," this is just a shorthand way of saying that the sentence expresses a truth. Similarly, if a sentence in a given context expresses a falsehood, we'll sometimes simplify things by saying that the sentence itself is false. Second, instead of saying that a sentence is true, it will sometimes be convenient to say that the sentence has the "truth-value" *true;* if a sentence is false we shall sometimes say that it has the "truth-value" *false.*

SIMPLE AND COMPOUND SENTENCES, SENTENCE OPERATORS, AND THE CONJUNCTION

Some sentences contain within themselves shorter sentences, and other sentences do not. A **compound sentence** is a sentence that contains within itself one or more shorter sentences. A sentence that is not compound is called a **simple sentence**. A compound sentence is formed by joining a **sentence connective** (also called a "sentence operator") to one or more shorter sentences. The sentence connective links the component or components together into a compound sentence.

For instance, if we join together the following two simple sentences:

Ann is home

Bob is home

by placing the sentence connective *and* between them, the result is a compound sentence, namely:

Ann is home and Bob is home.

A sentence connective such as *and* that joins together *two* sentences is called a **dyadic connective.** Notice that in this context *and* has, in a sense, two "hooks"—one on each side—by which it links together two sentences to form a compound sentence. The two sentences within the compound are called **component** or **embedded** sentences because they are part of (or are embedded within) the compound sentence.

A compound sentence formed with the connective *and* is called a **conjunction**, and the two component sentences joined together by the connective *and* are called the **conjuncts** of the conjunction. In the sentence above, "Ann is home and Bob is home," the left conjunct is "Ann is home," and the right conjunct is "Bob is home." When it is used to form a conjunction, the connective *and* is called a **conjunction operator**.

A *compound sentence* is any sentence that contains within itself a shorter sentence. A *simple sentence* is any sentence that does not contain within itself a shorter sentence. *Sentence connectives* are words or phrases that can be connected to one or more sentences to produce a compound sentence. Sentence connectives are also called *sentence operators* and are said to "operate" on sentences to form compound sentences.

Now, in the case of our sentence, "Ann is home and Bob is home," let us think of all the possible truth-values that the two conjuncts and the compound could have. If Ann and Bob are both home, both conjuncts are true. That is, if they are both home, "Ann is home" is true and "Bob is home" is true. If both conjuncts are true, the compound as a whole is true. That is, if Ann really is home and Bob really is home, the conjunction is true. However, if instead the first conjunct is true but the second is false, the compound as a whole is false. That is, if Ann really is home but Bob actually is not home, the conjunction is false. Similarly, if the first conjunct is false but the second is true, the conjunction is false. And if both conjuncts are false, the conjunction is false.

Let us summarize these truth-value possibilities:

• If both conjuncts are true, the conjunction as a whole is true.

• If the first conjunct is true and the second is false, the conjunction is false.

• If the first conjunct is false and the second is true, the conjunction is false.

• If both conjuncts are false, the conjunction is false.

In short:

• The conjunction is true if both conjuncts are true, otherwise it is false.

Notice that once you know the truth-values of the conjuncts, you can figure out the exact truth-value of the compound. In other words, the truth-value of the compound is determined by—it is a *function* of—the truth-values of the conjuncts.

This information can be put into a table of truth-values called a **truth-table**. If we let the letter "A" abbreviate or stand for "Ann is home," and if we let "B" stand for "Bob is home," and if we abbreviate the conjunction's connective *and* with the symbol "&" (called the ampersand), the table looks like this:

A	B	A & B
T	T	T
T	F	F
F	T	F
F	F	F

The first row, which contains three **T**'s, indicates that in any circumstance in which A is true and B is true, the conjunction as a whole is true. The second row indicates that when A is true and B is false, the conjunction is false, and so on.

Generalizing This The truth-table above expresses the relationships between the truth-values of the specified sentences, "Ann is home" and "Bob is home," and the conjunction of the two. However, the above table expresses something that is true not only of those particular sentences (represented by the letters A and B) but of *any* sentences, namely:

> *In general, a conjunction asserting just that both conjuncts are true will itself be true if and only if both conjuncts are indeed true. So, if one or both conjuncts are false, the conjunction as a whole is false as well.*

This general principle may be expressed on the following truth-table, with the understanding that **P** is any sentence and **Q** is any sentence:

P	Q	P & Q
T	T	T
T	F	F
F	T	F
F	F	F

Truth-Table for Conjunction

The first row indicates that in any circumstance in which **P** is true and **Q** is true, the conjunction of the two is true; the second row indicates that in any circumstance in which **P** is true and **Q** is false, the conjunction of the two is false; and so on. Notice that the four rows of the table represent every possible combination of truth-values that the components could have, and the table specifies the truth-value of the compound for each combination of truth-values the components might have.

The symbols **P** and **Q** in the above table are *variables*. Whereas a **constant** is a symbol that stands for one specified thing, a **variable** is a symbol that stands for anything from a group of things. Thus, in the previous table, when we let A abbreviate or stand for the specific sentence "Ann is home," we were using A as a constant. In the current table, we are using **P** to stand for any sentence, and **P** is thus serving as a variable. To distinguish our variables from our constants, we will use boldfaced capital letters from the middle of the alphabet, **P**, **Q**, **R**, **S**, as sentence-variables and we will use normal capital letters as sentence constants.

NEGATION

The phrase "It is not the case that" is a sentence operator (or connective); it can be joined to a sentence to form a compound sentence. When we connect the operator "It is not the case that" to a sentence, we produce a compound that "negates" or contradicts the original sentence. For example, consider the following:

Ann owns a pet aardvark.

Let us add to this the prefix "It is not the case that":

It is not the case that Ann owns a pet aardvark.

This new sentence, called the **negation** of the original sentence, is a compound sentence, since it has at least one shorter sentence within it. Notice that the new sentence contains within itself just one component sentence, namely, "Ann owns a pet aardvark." The expression "It is not the case that," which also lies within the new sentence, does not qualify as a component sentence, because it is not a sentence in its own right. (As noted, it serves only as a sentence connective because it is connected to a sentence to form a compound sentence out of a simpler sentence.) When we use it to form a negation, the expression "It is not the case that" is called a **negation operator**.

A sentence connective that connects to just one sentence to form a compound is called a **monadic connective**. Notice that the negation operator has, in a sense, a "hook" on its right side by which it attaches to one sentence to form a compound. This operator is therefore monadic.

Let us consider the possible truth-values of this negation. If the original sentence "Ann owns a pet aardvark" is true, the negation, "It is not the case that Ann owns a pet aardvark" is false; and if the original sentence is false, the negation is true.

Let us summarize the possible truth-values of this negation:

• The negation is true if the component sentence is false.

• The negation is false if the component sentence is true.

This information can also be put into a truth-table. Let us abbreviate the negation operator, "It is not the case that" with the symbol ~, called the **tilde** (pronounced "till-da"). And we shall let A stand for the sentence "Ann owns a pet aardvark." Thus, "It is not the case that Ann owns a pet aardvark" will be abbreviated or symbolized as: ~A. The truth-table is:

A	~A
T	F
F	T

The first row of the table (running horizontally across the table) indicates that when A is true, ~A is false. The second row indicates that when A is false, ~A is true.

Generalizing This   The truth-table above expresses the relationship between the truth-value of the specified sentence, "Ann owns a pet aardvark" and its negation. However, the table expresses something that is true not only of that sentence (represented by the constant A) but of *any* sentence, namely: If a sentence is true, its negation is false and if a sentence is false, its negation is true. We need to represent this general claim in a symbolic way. With the understanding that **P** is any sentence, the general principle may be put this way:

P	~P
T	F
F	T

Truth-Table for Negation

The first row of the table specifies that in any circumstance in which **P** is true, the compound ~**P** is false. The second row specifies that in any circumstance in which **P** is false, the compound ~**P** is true.

DISJUNCTION

When two sentences are linked together by *or*, the resulting compound sentence is called a **disjunction**. Thus, if we take the following two simple sentences:

The starter motor is kaput.

The battery is dead.

and join them together with the word *or*, we form the disjunction

The starter motor is kaput or the battery is dead.

Notice that this is a compound sentence. The two component sentences within a disjunction are called **disjuncts**. Thus, "The starter motor is kaput" is the left disjunct, and "The battery is dead" is the right disjunct. Notice that *or* functions here as a dyadic operator, for it joins two sentences into a compound. When *or* is used to form a disjunction, it is called a **disjunction operator.**

The word *or* is used in two different ways in English. When we use *or* in its **exclusive** sense, we mean to assert that one or the other of the two disjuncts is true *but not both*. For example, restaurant meals typically come with a choice of soup *or* salad. This is always an exclusive *or*: you can have soup or salad but not both. However, when we use *or* in its **inclusive** sense, we mean to assert that one, or the other, *or both*, of the disjuncts is true. For instance, suppose Joe can see that someone is home at Ann and Bob's place, but he can't tell whether it is Ann alone, Bob alone, or whether both are home. Because he is sure that at least one of the two is home, Joe says, "Ann is home or Bob is home." If he means that one or the other *or both* are home, Joe is using the *or* in an inclusive sense.

Now, let us stay with the inclusive disjunction "Ann is home or Bob is home" for a moment, and let us think of all the possible truth-values that the two disjuncts and the compound could possibly have. If both disjuncts are true, the compound as a whole is true. That is, if Ann really is home and Bob really is home, the compound as a whole is true. If the first disjunct is true and the second is false, the compound is still true. Likewise, if the first is false but the second is true, the compound is still true. However, if both disjuncts are false, the compound is false.

Here is a summary of those truth-value possibilities:

- If both disjuncts are true, the disjunction as a whole is true.

- If the first disjunct is true and the second is false, the disjunction is true.

- If the first disjunct is false and the second is true, the disjunction is true.

- If both disjuncts are false, the disjunction is false.

In short:

- The disjunction is false if both disjuncts are false, otherwise it is true.

This information can be put into a truth-table. If we let A abbreviate or stand for "Ann is home," and if we let B stand for "Bob is home," and if we abbreviate the disjunction's connective *or* with the symbol "v" (called the wedge), we get the following:

A	B	A v B
T	T	T
T	F	T
F	T	T
F	F	F

The first row indicates that in any circumstance in which A is true and B is true, the disjunction as a whole is true. The second row indicates that when A is true and B is false, the disjunction is also true, and so on.

Generalizing This   In general, when we use *or* in its inclusive sense, the disjunction is true just in case one or both of the disjuncts is true, which is to say that it is false only when both disjuncts are false. This general principle may be expressed with variables on a truth-table. If **P** is any sentence and **Q** is any sentence, the truth-table for the inclusive disjunction of the two is:

P	Q	P v Q
T	T	T
T	F	T
F	T	T
F	F	F

Truth-Table for Inclusive Disjunction

You may be wondering, "What about the exclusive sense of *or*? As you will see, we will not need to build into our logical language symbols for both the inclusive and exclusive disjunction operators, for as long as one type of *or* is symbolized, the other type of *or* can be expressed using a combination of other symbols. This will be demonstrated later. It is customary, in introductory logic, to provide a symbol for inclusive disjunction, and to symbolize exclusive disjunctions using a combination of symbols. Consequently, from here on, let the term "disjunction" refer to inclusive disjunction, and unless an *or* is specifically identified as an exclusive disjunction operator, let us treat it as an inclusive disjunction operator and symbolize it with the wedge.

George Burns and his wife Gracie Allen were one of the great comedy teams in the golden age of radio and during the early years of television. In one of their radio skits from the 1930's, Gracie seems to have the truth-table for disjunction in mind:

Gracie: My cousin had a baby.

George: Was it a boy or a girl?

Gracie: Why, yes, what else would it have been?

TRUTH-FUNCTIONS AND TRUTH-FUNCTIONAL COMPOUND SENTENCES

A **function** is essentially a rule that relates one set of values to another set of values. A **truth-function** is a rule that relates one set of truth-values to another set of truth-values. If you will examine the truth-tables for negation, conjunction, and disjunction, you will notice that in each case the truth-value of the compound sentence as a whole is determined precisely by the truth-values of the component sentences. That is, if you were to be given only the truth-values of the components, you could calculate the truth-value of the compound by following the directions given by the tables. Thus, a truth-table specifies a truth-function. The negation table specifies the negation function; the conjunction table specifies the conjunction function, and so on.

A **truth-functional compound sentence** is a compound sentence whose truth-value is a function of the truth-value (or values) of its component (or components). Compound sentences such as the negations, conjunctions, and disjunctions above are truth-functional compound sentences because the truth-value of the compound as a whole is a function of the truth-values of the components. Any sentence operator that forms a truth-functional compound when attached to one or more component sentences is a **truth-functional sentence operator.**

CONDITIONAL SENTENCES

Consider the following sentence:

If Ann quits her job, then Bob will quit his job.

Notice that this sentence, called a **conditional sentence**, is a compound containing two sentences within itself. The component sentence following the word *if* is the **antecedent** of the conditional, and the component following the word *then* is the **consequent** of the conditional. The expression *if, then* functions as a dyadic sentential operator, for it links two sentences into a compound. Specifically, the *if, then* operator links together the antecedent, "Ann quits her job" and the consequent, "Bob will quit his job." When it forms a conditional sentence, *if, then* is a **conditional operator**.

Because conditionals play an important role in argumentation, we need to represent the *if, then* operator within truth-functional logic. First, let the symbol $\supset$, called the "horseshoe," represent the conditional or *if, then* operator. If we abbreviate "Ann is home" with A and "Bob is home" with B, then our conditional may be abbreviated or symbolized as

$$A \supset B$$

This formula is pronounced "A horseshoe B" or "If A then B." When we symbolize an English conditional sentence, the symbolized antecedent will always be placed in front of the horseshoe, and the symbolized consequent will always be placed after the horseshoe. For one more example, consider the sentence "If Kramer tries to defraud the post office, then Newman will bust him." Using K for the antecedent and N for the consequent, this goes into symbols as $K \supset N$.

▶ A TRUTH-FUNCTION FOR THE HORSESHOE

The next step in the process of representing *if, then* within truth-functional logic is not as simple. We need to assign a truth-function to the horseshoe operator. Unfortunately for our purposes, *if, then* is rarely used in everyday discourse in a purely truth-functional way. That is, the conditional sentences we construct using *if, then* almost never display a purely functional relationship between the truth-value of the compound and the truth-values of the components. How then can a truth-function be associated with the *if, then* operator?

First, we shall isolate an element of meaning common to *all* natural language uses of *if, then*. This core element constitutes an essential *part* of what we normally mean when we write or utter a conditional sentence. Next, we will specify a truth-function that precisely represents this core element of meaning, and we will assign this truth-function to the horseshoe symbol. Finally, we'll go on to develop truth-functional logic in such a way that the representation of *if, then* is based only on this core element of meaning.

To begin with, whenever anyone asserts an English language conditional, and we find that the antecedent is true and the consequent is false, we automatically say that the conditional as a whole is false. For instance, suppose at a party the teenage host says, "If my parents come home, then my dad will ask everyone to leave." The antecedent of this conditional is "My parents come home" and the consequent is "My dad will ask everyone to leave." Now, suppose at midnight the parents come home (and so the antecedent is true), but the father does *not* ask everyone to leave (and so the consequent is false). We would naturally say that the host was wrong when he said, "If my parents come home, then my dad will ask everyone to leave." That is, we would spontaneously say that the host's statement was false.

For another example, suppose someone says:

If Wimpy eats one more hamburger, then he will throw up.

Suppose Wimpy *does* eat one more hamburger but does *not* get sick. The conditional is false, isn't it? So, someone who asserts this conditional as true is asserting at least:

It is not the case that: Wimpy will eat one more hamburger and yet not throw up.

These examples illustrate (but do not prove) the following general point: if someone asserts a conditional sentence **P ⊃ Q**, where **P** is the antecedent and **Q** is the consequent, at least *part* of what the speaker means to assert is:

It is not the case that: **P** is true while **Q** is false.

This is part of what is asserted when any conditional is asserted, and this core element happens to be truth-functional in nature, as we shall see momentarily. Truth-functional logic abstracts this core partial meaning of *if, then* and associates it with a corresponding truth-table.

This core meaning, a common partial meaning, is this: To say "If **P** then **Q**" is at least to say "It's not the case that: **P** but not **Q**." And this core meaning is represented precisely by the following truth-table, which is assigned to the horseshoe:[1]

P	Q	P ⊃ Q
T	T	T
T	F	F
F	T	T
F	F	T

Truth-Table for the Horseshoe

[1] In Chapter 3, you will learn how to translate an expression such as "It's not the case that: **P** but not-**Q**" into symbols and in Chapter 5 you will learn how to construct the corresponding truth-table. You will then be able to prove to yourself that this table is indeed the correct table for the above formula.

This truth-table represents the truth-functional part of what we mean when we assert a conditional sentence. Notice that the table assigns **F** to **P ⊃ Q** only when **P** is **T** and **Q** is **F**. Thus:

- If the antecedent is true and the consequent is true, the conditional as a whole is true.

- If the antecedent is true and the consequent is false, the conditional as a whole is false.

- If the antecedent is false and the consequent is true, the conditional as a whole is true.

- If the antecedent is false and the consequent is false, the conditional as a whole is true.

In short:

- If the antecedent is true and the consequent is false, the conditional as a whole is false, otherwise, the conditional is true.

The horseshoe is called the **material conditional operator**. A sentence whose main operator is the horseshoe and whose truth-value is a function of its components according to the truth-table for ⊃ is called a **material conditional sentence**. So, when we take an English conditional and translate it into a material conditional sentence, we have two different sentences: the English conditional with its (complex) meaning, and the associated material conditional with its truth-table-specified meaning. However, as noted, the two meanings are not entirely different, for the material conditional represents a core of meaning found in every English conditional.

Upon first learning the truth-table for the horseshoe operator, it is natural to feel a little puzzled. The table doesn't seem to represent exactly what we mean when we utter a conditional sentence. That is, the truth-functional relationships specified by the horseshoe's table don't seem to fit ordinary English usage. This is because when we write or utter an ordinary English conditional, we almost never mean to assert only what the corresponding material conditional asserts; we normally assert more than this. Remember, the horseshoe table only captures the truth-functional *part* of the meaning of *if, then*.

BICONDITIONAL SENTENCES

The last truth-functional operator we shall introduce is the **material biconditional**. Suppose that a club has the following rule:

The vote may be taken *if and only if* at least ten official members are present.

This is a compound sentence, for it contains two simpler sentences within itself, namely, "The vote may be taken" and "at least ten official members are present." The phrase in the middle, *if and only if,* is serving as a dyadic connective (or operator) joining the two simpler sentences into a compound. In order to represent this sentence symbolically, the operator *if and only if,* the **material biconditional operator,** will be assigned the symbol ≡, called a **triple bar**. Thus, if we let V stand for the component "The vote may be taken" and if we let O stand for the component "at least ten official members are present," we may symbolize the sentence as follows:

$$V \equiv O$$

In ordinary discourse, the operator *if and only if* is rarely used; however, it is used in several technical fields, including mathematics, law, and philosophy. In these fields, *if and only if* is typically used in a way closely related to the ordinary conditional operator. Typically, to say something like, "Ann swims if and only if Bob swims" is just to say, "If Ann swims then Bob swims and if Bob swims then Ann swims." In general, a sentence of the form "**P** if and only if **Q**" is equivalent to a sentence of the form "If **P** then **Q** and if **Q** then **P**." This type of sentence is called a "biconditional" because it contains *two* conditionals within itself.

The truth-table for "If **P** then **Q** and if **Q** then **P**" assigns **T** to the compound when the components are both true, it also assigns **T** to the compound when the components are both false, and it assigns **F** to the compound in the other two cases. (In Chapter 5, you will learn precisely how these values are calculated.) Thus, the following table is assigned to the triple bar:

P	Q	P ≡ Q
T	T	T
T	F	F
F	T	F
F	F	T

Truth-Table for the Triple Bar

Notice that a material biconditional compound is true when the components *match* in terms of truth-value, and it is false otherwise.
Here is a summary of the truth-value possibilities for the triple bar.

- If both components are true, the biconditional is true.

- If both components are false, the biconditional is true.

- If one component is true while the other is false, the biconditional is false.

In short:

• A biconditional is true if the components have matching truth-values, it is false otherwise.

> Why does logic require so many definitions? And why do those definitions have to be so technical? In the seventeenth century, as modern science was first taking shape, the philosopher Francis Bacon (1561–1626) wrote:
>
> > *Although we think we govern our words . . . certain it is that words, as a tartar's bow, do shoot back upon the understanding of the wisest, and mightily entangle and pervert the judgment. So that it is almost necessary, in all controversies and disputations, to imitate the wisdom of the mathematicians, in setting down in the very beginning the definitions of our words and terms, that others may know how we accept and understand them, and whether they concur with us or no. For it cometh to pass, for want of this, that we are sure to end there where we ought to have begun, which is in questions and differences about words.*
>
> If we try to communicate using undefined terms with imprecise meanings, the result will be confusion. Definitions and terminology are crucial to every field of thought, which is why the early part of just about any course of study involves the introduction of terminology.

▶ **EXERCISE 2.1** ◀

Identify the sentence operators and the embedded sentences in each of the following compound sentences.

1. *Juan is a carpenter and Rita is a teacher.

2. If Bubba had the "all you can eat" special at Skipper's Fish and Chips Restaurant again, then his wife Ann will lecture him on his heart condition.

3. Either Herman Munster will feed the ravens or Lilian Munster will feed the spiders, or Eddie will take the snakes out for a walk.

4. It is not the case that Gilligan likes caviar and it is not the case that the Skipper likes coconut.

5. *If First Officer Spock is presented with a problem then it is certain that he will solve it logically, and if Captain Kirk is attacked, then it is certain he will win the fight in the end.

6. Schultz knows nothing if and only if Schultz gets a bribe.

7. It is not the case that Colonel Klink is vain.

8. If the Blues Brothers play the theme from Rawhide, then those cowboys will stop throwing bottles and the band might get out of the joint alive.

9. Either it is not the case that Kramer is hungry or it is not the case that Jerry's refrigerator has any food in it.

10. *If Archie talks politics with Michael tonight, then Archie will be sure to become angry and Michael will become exasperated, and Archie will call Michael "Meathead."

► **EXERCISE 2.2** ◄

Refer to the truth-tables and complete the following sentences.

1. *If the left conjunct is true and the right conjunct is false, the conjunction as a whole is . . .

2. If the left conjunct is false and the right conjunct is false, the conjunction as a whole is . . .

3. If the left conjunct is true and the right conjunct is true, the conjunction as a whole is . . .

4. *If the left disjunct is true and the right disjunct is false, the disjunction as a whole is . . .

5. If the left disjunct is false and the right disjunct is false, the disjunction as a whole is . . .

6. If the left disjunct is true and the right disjunct is true, the disjunction as a whole is . . .

7. *If the antecedent is true and the consequent is false, the conditional as a whole is . . .

8. If the antecedent is false and the consequent is false, the conditional as a whole is . . .

9. If the antecedent is true and the consequent is true, the conditional as a whole is . . .

10. If the left component of a biconditional sentence is true and the right component is false, the biconditional as a whole is . . .

11. *If the left component of a biconditional sentence is false and the right component is false, the biconditional as a whole is . . .

12. If the left component of a biconditional sentence is true and the right component is true, the biconditional as a whole is . . .

CHAPTER 2 GLOSSARY

Ampersand   The symbol ("&") for the conjunction operator.

Component (or embedded) sentence   A sentence that is a part of a compound sentence.

Compound sentence   A sentence that contains within itself one or more shorter sentences.

Constant   A symbol that stands for one specified thing.

Dyadic connective   (or "dyadic operator") A connective that joins together two sentences to form a compound sentence.

Exclusive *or*   The *or* that asserts that one or the other but not both disjuncts are true.

Function   A rule that relates one set of values to another set of values.

Horseshoe   The symbol (" ⊃ ") for the conditional ("if-then") operator.

Inclusive *or*   The *or* that asserts that one or the other or both disjuncts are true.

Monadic connective   (or "monadic operator") A connective that is joined to just one sentence to form a compound sentence.

Sentence connective   (or "sentence operator") A word or phrase that can be connected to one or more sentences to produce a compound sentence.

Simple sentence   A sentence that does not contain within itself one or more shorter sentences.

Tilde   The symbol (" ~ ") for the negation operator.

Triple bar   The symbol (" ≡ ") for the biconditional ("if and only if") operator.

Truth-function   A rule, usually defined in terms of a table of values called a "truth-table," that relates one set of truth-values to another set of truth-values.

Truth-functional compound sentence   A compound sentence with the following feature: the truth-value of the compound as a whole is a function of the truth-values of the components.

Truth-functional operator   An operator that forms a truth-functional compound sentence when attached to one or more component sentences.

Truth-table   A table that relates one set of truth-values to another set of truth-values. More specifically, a truth-table specifies the truth-value for a formula (or group of formulas) for each possible assignment of truth-values to the components of the formula or formulas.

Variable   A symbol that stands for anything from a specified group of things.

Wedge   The symbol (" v ") for the inclusive disjunction operator.

TRANSLATING ENGLISH INTO LOGICAL SYMBOLS

3

B efore we put the truth-tables from the previous chapter to work solving various logical problems, we must first learn to translate English sentences into logical symbolism. As noted earlier, we shall use sentence constants to abbreviate specified English sentences. For example, instead of writing

Hogan is a colonel.

we may write simply "H" with the understanding that H abbreviates or stands for the entire sentence "Hogan is a colonel." Likewise, instead of writing

It is not the case that Hogan is a colonel.

we may simply write ~H with the understanding that the H still represents the sentence "Hogan is a colonel" and the tilde represents the negation operator "It is not the case that." When we use a constant to abbreviate a sentence, the constant serves as a name of the sentence.

Now let us consider the following compound sentence:

Schultz knows nothing and Klink has a rather large ego.

This is easily symbolized as:

S & K

with the understanding that S abbreviates the left conjunct, "Schultz knows nothing," and K abbreviates the right conjunct, "Klink has a rather large ego." The ampersand, of course, abbreviates the conjunction operator *and*. It is important that you understand this: the letter S does not abbreviate just

"Schultz." Rather, the letter S abbreviates the entire component sentence "Schultz knows nothing."

SYMBOLIZING SENTENCES CONTAINING MORE THAN ONE OPERATOR

It is relatively simple to symbolize a sentence when it contains only one sentential operator. However, sentences that contain two or more operators present a new type of problem. In order to see this, consider the following. Suppose you have been invited to a party, and the invitation includes the following:

Michael Jordan will be there and Dennis Rodman will be there or Kelsey Grammer will be there.

What does this sentence tell you? If you examine it closely, you will see that this is an **ambiguous sentence**—there are at least two ways to interpret it. If you read it to yourself and emphasize the "or" with your voice, you will hear one interpretation, and if you read it and emphasize the "and" you will hear a different interpretation.

Interpretation 1   Michael Jordan will be there and Dennis Rodman will be there **or** Kelsey Grammer will be there.

Interpretation 2   Michael Jordan will be there **and** Dennis Rodman will be there or Kelsey Grammer will be there.
    The second interpretation is more specific than the first interpretation. The second interpretation conveys that Michael Jordan will definitely be there (and then, in addition, Dennis or Kelsey will be there). However, in the first interpretation, no one specific individual is guaranteed to be there. Unfortunately, the original sentence, as written, does not indicate which of the two interpretations is intended.
    Now, suppose we try to place the sentence into symbols. If we let M abbreviate "Michael Jordan will be there," D abbreviate "Dennis Rodman will be there," and K abbreviate "Kelsey Grammer will be there," we get:

$$M \ \& \ D \lor K$$

However, this *formula* is also ambiguous. Does the ampersand join the M with the D, or does it join the M with the disjunction D v K? Compare this problem with a similar problem in arithmetic. Consider the unpunctuated expression

$$2 + 3 \times 4$$

In the absence of any prior rules regarding punctuation, does 3 go with 2, or does 3 go with 4? Two different interpretations are possible:

1. First, add 2 and 3, and then multiply this by 4.

2. First, multiply 3 times 4, and then add this to 2.

Mathematicians resolve this type of ambiguity by using parentheses to establish an order of operations in such formulas. If we intend the first interpretation, we write

$$(2 + 3) \times 4$$

and if we intend the second interpretation, we write

$$2 + (3 \times 4)$$

In logic, we will use parentheses in much the same way. The two interpretations of our sentence about Michael Jordan may be expressed symbolically as

Interpretation 1

$$(M \& D) \lor K$$

Interpretation 2

$$M \& (D \lor K)$$

In the symbolized interpretation 1, the parentheses tell us two things: (1) the ampersand joins together M with D; and (2) this conjunction—as a whole—is joined by the wedge to K. In the symbolized interpretation 2, the parentheses tell us that: (1) the wedge joins together D with K, and (2) this compound is joined by the ampersand to the M.

Our original Michael Jordan sentence must be rewritten and clarified if we are to translate it into our logical symbols. Recall interpretation 1:

Michael Jordan will be there and Dennis Rodman will be there **or** Kelsey Grammer will be there.

We can rewrite the original sentence in several ways to indicate this interpretation. Each will be symbolized by the formula (M & D) ∨ K. Here are a few possibilities:

- Michael Jordan will be there and Dennis Rodman will be there, or Kelsey Grammer will be there.

Notice that a comma, an important grouping device, has been added to divide the sentence into two parts and also to group the Michael Jordan and Dennis Rodman components together.

- Michael Jordan and Dennis Rodman will be there or Kelsey Grammer will be there.

This sentence has no comma, but notice that the Michael Jordan and Dennis Rodman components now share the same verb phrase ("will be there") and this fact alone groups them into a unit.

- Both Michael Jordan and Dennis Rodman will be there or Kelsey Grammer will be there.

In this sentence, the coordinate phrase "both . . . and" has been added to group the first two components of the sentence together. Thus, the word *both* helps group the Michael Jordan component and the Dennis Rodman component together.

- Either Michael Jordan and Dennis Rodman will be there or Kelsey Grammer will be there.

In this sentence, the coordinate phrase "either . . . or" was added to group the first two components of the sentence together and thus to divide the sentence in half at the *or*.

The second interpretation of our original sentence was

Michael Jordan will be there **and** Dennis Rodman will be there or Kelsey Grammer will be there.

We can rearrange the grammatical grouping indicators to indicate this second interpretation. The resulting sentence in each case will be symbolized by the formula M & (D v K). Here are several possibilities:

- Michael Jordan will be there, and Dennis Rodman will be there or Kelsey Grammer will be there.

Notice again that the comma divides the sentence into two parts; however, this time it groups the Dennis Rodman and Kelsey Grammer components together.

- Michael Jordan will be there and Dennis Rodman or Kelsey Grammer will be there.

The Rodman and Grammer components share the same verb phrase ("will be there"), and this groups them into a unit.

- Michael Jordan will be there and either Dennis Rodman or Kelsey Grammer will be there.

The word *either* has been added. Notice that *either* groups the Dennis Rodman and Kelsey Grammer components together into a unit.

When you symbolize a sentence that contains several operators, you must begin by figuring out which parts of the sentence are linked by which operators. This requires that you pay attention to grouping devices such as the ones we have been examining—commas, shared verbs, and coordinating words such as *both* and *either*. When you translate sentences from English into logical symbols, pay attention to the ways that these grammatical devices function.

Here are several further examples:

Either Ann swims and Bob swims, or Pete jogs and Ed jogs.

In this sentence, everything from the *either* to the *or* is grouped by the comma into a logical unit, and the rest of the sentence forms a unit of its own. To symbolize this, each of these two large units will have to be wrapped with parentheses:

Either (Ann swims and Bob swims), or (Pete jogs and Ed jogs).

Next, each *and* will have to be traded for an ampersand, and the *or* will have to be traded for a wedge. The resulting symbolization is thus:

Either (Ann swims & Bob swims), v (Pete jogs & Ed jogs).

Finally, if we abbreviate "Ann swims" with A, "Bob swims" with B, and so on, we get:

(A & B) v (P & E)

This English sentence, "Either Ann swims and Bob swims, or Pete jogs and Ed jogs," could also have been written more concisely as

Either Ann and Bob swim or Pete and Ed jog.

Notice that in this shorter sentence, *Ann* and *Bob* share the same verb *(swim)* and *Pete* and *Ed* share a common verb *(jog)*. This shorter sentence requires the same parentheses and symbols as the previous example, namely: (A & B) v (P & E).

THROWING THE TILDE INTO THE MIX

When a tilde is added to the mix, a new set of issues emerge. Let us proceed in stages. First, consider the following sentence:

It is not the case that Ann is home and Bob is home.

As this sentence is written, it is not entirely clear whether the negation operator applies just to "Ann is home" or to the whole conjunction "Ann is home and Bob is home." Two interpretations are possible.

Interpretation 1   It is not the case that Ann is home, but Bob is home.

Interpretation 2   It is not the case that both Ann and Bob are home, in other words, it is not the case that Ann and Bob are both home together.

Let us symbolize each interpretation separately.

▶ INTERPRETATION 1

In English, the negation operator typically applies to the component to its immediate right. Likewise, in a formula, the tilde applies to whatever is to its immediate right. Now, notice that the comma in the English sentence (for interpretation 1) confines the influence of the English negation operator to the first conjunct "Ann is home." Consequently, when we write the formula, the formula's tilde should only apply to the A, just as the English negation operator only applied to "Ann is home." Notice also that within the English sentence, the word *but* is serving as a conjunction operator, for it asserts that both components are true. (We sometimes use the word *but* in place of *and* when there is a contrast between the conjuncts or when the conjunction is surprising in some way.) Consequently, in the formula for interpretation 1, *but* will be symbolized with &. Interpretation 1 thus goes into symbols as: ~A & B. In this formula, notice that the tilde applies only to A.

▶ INTERPRETATION 2

Notice that here the English negation operator applies to the conjunction as a whole, "Both Ann and Bob are home." Consequently, the tilde should apply to the conjunction of A and B, not just to A alone, just as the English negation operator applied to the conjunction and not to Ann alone. When we symbolize the sentence, we first let A stand for "Ann is home" and we assign B to "Bob is home." Next, we join the A to the B with an ampersand and wrap this conjunction in parentheses. Finally, we apply the tilde to all of this. The sentence is thus symbolized with the formula ~ (A & B). Notice that the tilde applies to the entire conjunction (A & B), because the parenthesis forming the conjunction sits to the tilde's immediate right.

FROM *AND* TO *OR* AND
BACK AGAIN—WITH A FEW *NOTS* THROWN IN

Now, consider the second interpretation again:

It is not the case that both Ann and Bob are home.

In other words:

It's not the case that Ann and Bob are both home.

As we saw, the negation operator applies to a conjunction as a whole, and the sentence is symbolized

$$\sim (A \ \& \ B)$$

Suppose this is true. Ann and Bob are *not both* home. Then, who is home? Does the sentence tell us that Ann is not home *and* that Bob is not home? No. It only tells us that we won't find them both home together. Given this, Ann might be home (and Bob gone). Or Bob might be home (and Ann gone). Perhaps neither is home. The sentence just tells us that we won't find them both home together.

Most languages possess at least two ways to convey an idea. Notice that instead of saying "Ann and Bob are not both home," someone could say instead:

Either Ann is not home or Bob is not home

This is just another way of saying that Ann and Bob are not both home.

If one of these two sentences is true, the other will be true; if one is false, the other will be false. In other words, the two sentences are *logically equivalent.*

The sentence "Either Ann is not home or Bob is not home" is symbolized as follows. With A abbreviating "Ann is home" and B abbreviating "Bob is home," we place the A and B on opposite sides of a wedge, and give each letter its own tilde (because each component in the English sentence has its own negation operator). The result is: ~ A v ~ B. So, these two symbolizations, namely ~ (A & B) and ~ A v ~ B, represent two different ways of saying the same thing.

Consider next the sentence, "Both Ann and Bob are not home." This sentence tells us something entirely different. Indeed, this sentence indicates two definite things: Ann is not home, and Bob is not home. That is, each is not home. So, nobody is home in this case, assuming, of course, that nobody else lives at their house. Consequently, when we symbolize this sentence, the A and the B each get their own tildes, and the two are then conjoined. The resulting symbolization is: ~ A & ~ B. This says that each is not home.

Again, there is usually more than one way to say a given thing. Instead of saying "Both Ann and Bob are not home," the sentence could have been put this way:

It's not the case that either Ann or Bob is home.

which is abbreviated in English as

Neither Ann nor Bob is at home.

This is just another way of saying that both Ann and Bob are not home. For consider: if both Ann and Bob are not home, then neither Ann nor Bob is home, which is to say that they are both gone. And on the flip side, if neither Ann nor Bob is home, then both Ann and Bob are not home. In short, "both are not home" is the same as saying, "neither is home." If one is true, the other will be true, if one is false, the other will be false. In other words, the two sentences are logically equivalent.

When we symbolize "It is not the case that either Ann or Bob is home," we first must notice that the sentence has two major parts: a disjunction "Either Ann or Bob is home, " and a negation operator, which is applied to the disjunction. Consequently, we need to symbolize the disjunction *inside* the sentence, that is, "Ann or Bob is home," with A v B. In order to keep this unit intact, we next wrap it in parentheses, which produces (A v B). We then apply a tilde to this disjunction as a whole, because the English negation operator is applied to the disjunction as a whole. This produces ~ (A v B). The parentheses indicate that the tilde applies to the disjunction as a whole.

Because "Both Ann and Bob are not home" and "It is not the case that either Ann or Bob is home" are equivalent, these two symbolizations, namely ~ A & ~ B and ~ (A v B), represent two different ways of saying the same thing.

Here is a point that needs emphasis. Notice, in the examples above, that when *both* operates on *not*, as in "Both Ann and Bob are not home," *both* distributes *not* onto each of the subjects equally. If both are not home, then Ann is not home and Bob is not home. Each letter gets its own *not*. However, matters differ when *not* operates on *both,* as in "Ann and Bob are not both home." In this case, *not* does not distribute onto each subject. If Ann and Bob are not both home, it does not follow that Ann is not home and Bob is not home. It does not follow that each letter gets its own *not*.

Finally, consider the sentence "Either Ann or Bob will not go home." Notice that in this sentence, we have a disjunction operator between *Ann* and *Bob* and each equally shares the sentence's only verb "will *not* go home." Because each subject term equally shares the one verb, the sentence is equivalent to "Ann will not go home or Bob will not go home." In symbols, we place a wedge between the A and the B and each gets its own tilde: ~ A v ~ B.

English:	Symbols:
It is not the case that both Ann and Bob will go.	~ (A & B)
Ann and Bob will not both go.	~ (A & B)
Both Ann and Bob will not go.	~ A & ~ B
Ann and Bob both will not go.	~ A & ~ B
It is not the case that either Ann or Bob will go.	~ (A v B)
Neither Ann nor Bob will go.	~ (A v B)
Either Ann or Bob will not go.	~ A v ~ B

The Tilde Functions Differently from a Mathematical Negative Sign

Upon first being exposed to these symbols, some people suppose that ~ (A & B) is equivalent to ~ A & ~ B. That is, they suppose the tilde "distributes" as a negative sign might in arithmetic. However, in order to see the error in this thinking, compare the following two sentences:

1. Ann and Bob are not both home.

2. Both Ann and Bob are not home.

These two sentences look similar, but they actually say two different things. In sentence 1 the "not" comes before the "both." In sentence 2 the "both" comes *before* the "not." This makes a big difference. Sentence 2 tells us two definite things: Ann is not home and Bob is not home. *Each is not home.* In contrast, recall that sentence 1 tells us only that the two are not both home together. One might be home, or the other might be home (or perhaps nobody is home). Sentence 2 was symbolized with ~A & ~ B, and sentence 1 was symbolized earlier as ~ (A & B), a symbolization with different logical consequences. Thus, ~ A & ~ B and ~ (A & B) do *not* represent equivalent sentences.

SOME GENERAL HINTS ON SYMBOLIZING

The words *both* and *either* are important English punctuational devices. When symbolizing a sentence, it is important that you use your understanding of elementary English grammar to spot to which parts of the sentence these words apply. For example, suppose you are symbolizing the following:

Either both Ann and Bob will swim or both Darla and Ed will swim.

In this example, the first *both* joins the two components following it into a conjunction, the second *both* joins the last two components into a conjunction, and the word *either* joins these two conjunctions into an overall disjunction. So, "both Ann and Bob will swim" is symbolized with (A & B). We symbolize "both Darla and Ed will swim" with (D & E). Finally, we place a wedge between these two sub-units to get the correct symbolization:

$$(A \& B) \text{ v } (D \& E)$$

Just as punctuation marks such as commas are of crucial importance in written languages, tone and voice inflection serve as punctuation devices in spoken languages. In the symbolic logical language we are building, the only punctuation markers will be parenthetical devices of various kinds, including parentheses, brackets ([ ]) and braces ({ }).

In general, when translating an English sentence into logical symbols, we recommend that you follow this three-step procedure:

Step 1   First, identify the simplest sentential components and replace these with constants.

Step 2   Next, identify the sentence operators and replace these with their symbolic counterparts.

Step 3   Finally, use the logical structure of the English sentence to determine which parts are to be grouped together, and then put parenthetical devices in the appropriate places.

For example, consider the following sentence:

Either Jim swims and Pat swims or else Ruth will not swim and Andy will not swim.

Applying the three steps in stages, we get:

Step 1   Either J and P or else not-R and not-A.

Step 2   J & P v ~R & ~A

Step 3   (J & P) v (~R & ~A)

► SYMBOLIZING EXCLUSIVE DISJUNCTION

Suppose someone asserts an exclusive disjunction, such as

I'll be in Chicago at noon or I'll be in New York at noon.

Assuming that the speaker does not have the power to bilocate, her claim is an exclusive disjunction and may be put more explicitly as

I'll be in Chicago at noon or I'll be in New York at noon, but I won't be in both Chicago and New York at noon.

This exclusive disjunction is easily symbolized as

$$(C \lor N) \ \& \sim(C \ \& \ N)$$

►	**EXERCISE 3.1**	◄

Imagine a mythical, magical, ethereal reunion of 1960s rock groups and symbolize the following sentences.

1. *The Animals will do three songs and the Beatles will do six songs, or the Doors will do four songs.

2. The Animals will do three songs, and the Beatles will do six songs or the Doors will do four songs.

3. Either the Yardbirds will perform two songs and the Beach Boys will do five songs or Santana will perform three numbers.

4. *The Jefferson Airplane will do three songs or the Grateful Dead and the Byrds will play.

5. Donovan will do a song and either Cream will perform or Otis Redding will perform.

6. Jimi Hendrix will perform and the Beach Boys and the Grateful Dead will perform.

7. *Ike and Tina Turner won't both sing.

8. Grace Slick and Janis Joplin both will not do encores.

9. Grace Slick and Janis Joplin will not both do encores.

10. The Lovin' Spoonful will not perform an extra song but the Moody Blues will perform an extra song.

11. *Both the Stones and Procol Harem will perform but Creedence Clearwater Revival will not play.

12. Either the Monkees or the Stones will not perform.

13. Either Canned Heat performs or Moby Grape performs, or The Young Rascals and Buffalo Springfield perform.

14. Neither the Wailers nor the Sonics will perform, but the Bards and the Natural Gas Company will perform.

15. *Either the Association or The Dave Clark Five have not performed or Tiny Tim and Janis Ian both have not been up.

16. Although the Hollies and the Supremes both won't do encores, either Van Morrison or Jesse Colin Young will perform encores.

17. The group The Mamas and The Papas will perform and the group Don and the Goodtimes will perform, and the group Gerry and the Pacemakers will perform.

18. Slim Whitman, the yodeling cowboy, and Boxcar Willie, the famous hobo singer, will crash the concert but Zamfir, the master of the pan flute, will not attend.

TRANSLATING CONDITIONALS AND BICONDITIONALS

The sentence, "If it's sunny, then we will jog," translates easily into our logical symbolism as

$$S \supset J$$

with S abbreviating "It's sunny" and J abbreviating "We will jog." However, consider the sentence "We will jog if it's sunny." It won't work to symbolize this as

$$J \supset S$$

Here is why. In this context, $J \supset S$ symbolizes

If we jog, then it will be sunny.

This states that our jogging will in some way make the sun come out. Perhaps there is a sun-god and, if we jog, he will see us and flood us with his light. Which is absurd. So, "We will jog *if* it is sunny" should not be translated $J \supset S$.

Consider again the above two conditionals, "If it's sunny, then we will jog" and "We will jog if it's sunny." If you will carefully compare them, you will see that both say exactly the same thing. In English, we usually place the antecedent ("it's sunny") first, as in the first of the two. However, we sometimes place the antecedent second, as in the second of the two. In each case, the antecedent is always introduced with the word *if*. Because the two sentences, "If it's sunny, then we will jog" and "We will jog if it's sunny," amount to the same claim, both should be symbolized with $S \supset J$.

When translating conditionals, keep two points in mind:

- In English, the antecedent is typically introduced by the word *if*.
- In our logical symbolism, the abbreviated antecedent is always placed to the left of the horseshoe.

Suppose we wish to represent the following sentence in our logical symbolism:

If Elaine orders soup and Jerry orders soup, then either George orders soup or Newman orders soup.

The first thing to notice is that everything in this sentence from *if* to *then* forms a logical unit, namely the antecedent of a conditional. If we let E stand for "Elaine orders soup" and J for "Jerry orders soup," this part is symbolized with (E & J). Second, everything from *then* to the end forms another unit, the consequent of a conditional. Using obvious abbreviations, this part is symbolized with (G v N). *Then* sits between these two units, and must be traded in for a horseshoe. So, we place a ⊃ between (E & J) and (G v N). The result is: (E & J) ⊃ (G v N). In sum, the sentence may be symbolized in the following three steps:

1. If E and J then either G or N.
2. If (E & J) then (G v N)
3. (E & J) ⊃ (G v N)

Here are several interesting examples of conditional sentences. Suppose an insurance company has the following policy: If you accumulate an accident and a ticket on your record, your insurance will be canceled. In other words:

If you have an accident and you have a ticket, then your policy will be canceled.

Notice that the comma breaks the sentence into two parts. Using obvious abbreviations, the symbolization is

$$(A \ \& \ T) \supset C$$

Now, this implies that *if* you have an accident, you had better watch out, for *then* it will follow that if you get a ticket, your insurance will be canceled. In other words:

If you have an accident, then if you get a ticket then your policy will be canceled.

Notice that in this sentence, the comma divides the sentence after *accident*. Also, the second part of the sentence, the part after *accident*, is itself compound, and therefore must be enclosed in parentheses. Furthermore, each *then* will need a horseshoe. The symbolization of this sentence is thus:

$$A \supset (T \supset C)$$

If **P** is any sentence, **Q** is any sentence, and **R** is any sentence, the *general* logical structure of the insurance policy's restriction is as follows: If it is the case that **P**, then if **Q** then **R**.

Suppose you invite a friend to drive across the country with you. Your friend says, "If I go with you, I have to take Ralphie with me." Ralphie happens to be your friend's two-hundred-pound Saint Bernard, a dog that could easily win the world championship in the dog slobbering category of next year's Dog Olympics. Thinking of how hot the trip will be and thinking also of your upholstery, you reply, "If that's the case, then I don't think so." Your reply may be put this way:

If it's the case that if you go then Ralph goes, then you're not going.

The comma divides the sentence after *goes*. So, the first two components within the sentence will need to be wrapped in parentheses. Furthermore, each *then* will need to be traded in for a horseshoe. The proper symbolization is thus:

$$(Y \supset R) \supset \sim Y$$

where Y abbreviates "You go," and R abbreviates "Ralph goes." The overall logical structure of this sentence is: If it is the case that if **P** then **Q**, then **R**.

We need to symbolize a sentence containing the biconditional operator. Suppose a country has the following law:

A person has the right to vote if and only if the person is an adult and the person is also a citizen.

It follows that

Bart has the right to vote if and only if Bart is an adult and Bart is also a citizen.

The sentence divides in half at the *if and only if* operator, and the half to the right of the *if and only if* is clearly a conjunction of "Bart is an adult" and "Bart is a citizen." To symbolize this, we must give the *if and only if* a triple bar, and we must enclose the conjunction on the right side in parentheses. In symbols, using obvious abbreviations, we get: B ≡ (A & C).

SYMBOLIZING NECESSARY AND SUFFICIENT CONDITIONS

When a jet crashes, scientists from the FAA are always called in to track down the likely *cause*. When scientists track down causes, the procedures they use are often based on a set of principles first formulated by the British philosopher John Stuart Mill (1806–73) in his *System of Logic* published in 1843. We will not examine these principles, now known as Mill's Methods, until Chapter 26. However, here we will take a brief look into the way cause-and-effect statements can be symbolized within truth-functional logic.

When scientists search for causes, they seek to discover the conditions under which a specific effect will occur and the conditions under which it will not. A key question of causal reasoning is, under which conditions will the effect occur and under which conditions will the effect be absent? Thus, philosophers have found it illuminating to analyze causes in terms of underlying or antecedent *conditions,* and specifically in terms of two types of antecedent or underlying conditions: necessary conditions and sufficient conditions.

A condition N is called a **necessary condition** for an event E just in case event E is not possible without N. In the absence of condition N, E cannot occur. In other words, condition N must be present if E is to occur.

> N is a necessary condition for effect E if and only if the following is true:
>
> E will not be present unless N is present; or
>
> In the absence of N, E will not be present; or
>
> Without N, E won't occur.

For example, the presence of oxygen is a necessary condition for the operation of a gasoline engine, reaching the age of eighteen is a necessary condition for voting, a bachelor's degree is usually a necessary condition for entering graduate school.

A condition S is called a **sufficient condition** for an event E just in case S is all that is required for E to occur. For example, jumping into Seattle's most famous lake, Green Lake, is sufficient for getting wet, for jumping into a lake is all that is required for getting wet. A sufficient condition for an effect E has the following feature: when the sufficient condition is present, E must occur.

> S is a sufficient condition for effect E if and only if the following is true:
>
> If S is present, E will occur.
>
> If S is present, E is certain to occur.
>
> In presence of S, E is certain to occur.

Notice that although gasoline is necessary for a gasoline engine to run, it is not sufficient, for some engines will not run even though they have a full tank of gasoline. Notice also that although jumping in Green Lake is sufficient for getting wet, it is not necessary, for there are other ways to get wet besides jumping in a lake.

The concept of a "cause" is not a precise, clearly defined concept. Some philosophers consider the cause of an effect E to be the sufficient condition of

E. In this sense of the word *cause*, the cause of (for instance) the crash of KAL Flight 107 was a Soviet missile striking the jetliner and exploding. However, other philosophers hold that a cause is a necessary condition for an effect. The complete set of all the necessary conditions for an effect E, on this interpretation, is the "complete cause" of E, while one necessary condition is a "partial" cause of E. Thus, oxygen is a partial cause of a fire, the fuel is another partial cause, and so on.

In logic, statements about necessary and sufficient conditions are treated as conditional sentences and are consequently symbolized with horseshoes. Suppose we want to symbolize

Gasoline is necessary for the engine's operation.

We shall need to paraphrase this sentence slightly. If we let G abbreviate "The engine's fuel tank has gasoline" and if E abbreviates "The engine will operate," we could try to represent this sentence by placing the statement about the necessary condition in the antecedent of a conditional:

$$G \supset R$$

This says, "If the engine's fuel tank has gasoline, then the engine will operate." It is obvious that this way of representing a necessary condition is incorrect. The mere presence of gasoline does not guarantee that an engine will run. However, if we represent the necessary condition in the *consequent*, the symbolism seems correct:

$$R \supset G$$

This indicates that if the engine is running, it has gasoline in it. This seems right. If a gasoline engine is running, it does have gasoline in it, but it is not the case that if a gasoline engine has gas in it, it will run. In general, a statement of a necessary condition typically goes in the consequent and that for which the necessary condition is necessary is typically represented in the antecedent.

Next, suppose we want to symbolize

Jumping in Green Lake is sufficient for getting wet.

First, in order to allow this to fit into our symbolism, we shall have to slightly rewrite the sentence. We could try representing the sufficient condition in the consequent as follows:

$$W \supset J$$

where W abbreviates "You are wet" and J abbreviates "You jumped into Green Lake." The formula indicates: "If you are wet, then you jumped into Green Lake." This is obviously incorrect. The mere fact that someone is wet doesn't imply that she or he jumped into Seattle's Green Lake. Clearly, the way to symbolize a claim about a sufficient condition is to represent the sufficient condition in the antecedent:

$$J \supset W$$

This indicates that if you jumped into Green Lake you are wet. This seems true.

> In general, a statement of a necessary condition typically goes in the consequent of a conditional, and that for which the necessary condition is necessary is typically represented in the antecedent. In general, a statement of a sufficient condition is typically represented in the antecedent of a conditional, and that for which the condition is sufficient is represented in the consequent.
>
> A useful mnemonic device for remembering this is the word SUN. Now, picture the U pointing to the right so that it looks like this: ⊃. Next, let S represent "sufficient condition" and let N represent "necessary condition." SUN now reminds you that a sufficient condition is represented in the antecedent of a conditional, and a necessary condition is represented in the consequent.

Statements concerning necessary conditions are sometimes expressed using the connective *only if.* For example, suppose a group of coal miners is trapped deep inside a mine shaft and the rescue crew is about to search for survivors. Here is one statement that is surely true:

Human life is present only if oxygen is present.

The standard interpretation of such a sentence is that the statement following *only if* records a necessary condition for that which is referred to by the statement to the left of the *only if.* The sentence, "Human life is present only if oxygen is present" could thus also be put this way: "Oxygen is a necessary condition for human life."

In general, then, a statement of the form "**P** only if **Q**" typically reports that **Q** is a necessary condition for **P**. Thus, the earlier statement of a necessary condition, "Gasoline is necessary for the engine to run" could also be put as, "The engine will run only if it has gasoline." If the words *only if* serve to introduce a necessary condition, and if, as we have seen, necessary conditions are represented in the consequent of a conditional, we must symbolize a statement of the form "**P** only if **Q**" as: **P** ⊃ **Q**.

Because "Human life is present only if oxygen is present" reports that oxygen is necessary for human life, the sentence goes into symbols as

$$H \supset O$$

where H abbreviates "Human life is present" and O abbreviates "Oxygen is present."

It would certainly be incorrect to symbolize this sentence the other way around, that is, as O ⊃ H. This symbolization would indicate, "If oxygen is present, then human life is present." But the mere presence of oxygen does not guarantee the presence of human life.

So, in general, a statement fitting the general form "**P** only if **Q**" must be symbolized as: **P ⊃ Q.** Earlier, it was noted that the word *if* typically introduces the antecedent, which is always symbolized to the left of the horseshoe. A statement of the form "**P** only if **Q**" is an exception to this general rule.

Two more commonly used sentence connectives are also translated with horseshoes, and it will be worth your while to understand how this is accomplished. Suppose the medics have arrived at the scene of a car accident and one of the medics says:

The victim will die *unless* he gets a blood transfusion.

This is equivalent to

The victim will die *if he does not* get a blood transfusion.

Now, this is equivalent to

If the victim does not get a blood transfusion, then the victim will die.

If V abbreviates "The victim will die" and if B abbreviates "The victim gets a blood transfusion," this sentence is symbolized ~ B ⊃ V. In general, a statement of the form "So and so unless such and such" is equivalent to one of the form "If not such and such, then so and so."

Notice that when the medic says, "The victim will die *unless* he gets a blood transfusion," he does *not* mean

If the victim gets a transfusion, then he will live.

Perhaps the victim will get a transfusion but die from other complications. The medic is indicating that the blood transfusion is necessary but not sufficient for the victim's survival.

Next, consider a sentence such as

Provided that the rental car is returned, the VISA surcharge will be refunded.

If R abbreviates "The rental car is returned" and V abbreviates "The VISA surcharge will be refunded," this sentence is obviously abbreviated R ⊃ V. In general, a sentence of the form "Provided that so and so, such and such" is equivalent to one of the form "If so and so, then such and such."

A sentence such as "We will sign the papers *provided that* the lien is removed" is equivalent to

Provided that the lien is removed, we will sign the papers.

If L abbreviates "The lien is removed" and S abbreviates "We will sign the papers," the proper symbolization of "We will sign the papers *provided that* the lien is removed" is L ⊃ S. In general, a sentence of the form "So and so provided that such and such," is equivalent to one of the form "Provided that such and such, so and so."

▶   **EXERCISE 3.2**   ◀

Imagine a mythical, ethereal, magical reunion of artists from 1930's Harlem, from the Big Band era, and from the 1950's. Now symbolize the following sentences.

1. *If Fats Domino performs one song then either Jerry Lee Lewis or the Big Bopper will perform one song.

2. If both Buddy Holly and Little Richard perform, then Chuck Berry will perform.

3. Either Cab Calloway will perform or if Duke Ellington performs then Ella Fitzgerald will perform.

4. *If Ethel Waters performs and either Louis Armstrong or Lena Horne performs, then Mel Torme will sing.

5. If it's not the case that Bill "Bojangles" Robinson will be there, then either Cab Calloway or Duke Ellington will be there.

6. Either Benny Goodman's band or Glenn Miller's orchestra will perform if Lena Horne will not be available.

7. *It is not the case that if Nat King Cole performs then ticket prices will be raised.

8. If Johnny Mathis sings, then it's not the case that Frank Sinatra or Johnny Ray will miss the show.

9. Elvis and Patsy Cline will sing if and only if admission is free and the performance is not televised.

10. If Elvis and Ricky Nelson perform, then the Brothers Four and the Drifters will perform.

11. *Provided that Joe buys a ticket, Joe will be admitted to the concert.

12. The concert is in San Francisco only if the concert is in California.

13. Bill Haley and the Comets will perform provided that Richie Valens performs.

At the concession stand...

14. *This hot dog will taste good only if ketchup and mustard are both present.

15. Ann will order a Dick's Deluxe burger if and only if Bob and Sue both order Deluxes. (Dick's Drive-In is a Seattle institution founded in 1954.)

16. Relish is necessary for this to be a good hamburger.

17.*Four hamburgers and a fry will be sufficient to fill me up, or three giant hot dogs will also do the trick.

18. It is not the case that mayonnaise is necessary for this to be a good hamburger.

19. Joe and Pete will each order pizza provided that Sue and Bob each order pizza.

20. Joe will go hungry unless Bob has money.

CHAPTER 3 GLOSSARY

Ambiguous sentence   A sentence that can be interpreted in more than one way.

Necessary condition   Condition N is a necessary condition for an event E just in case event E is not possible without N, that is, in the absence of condition N, E cannot occur.

Sufficient condition   A condition S is a sufficient condition for an event E just in case S is all that is required for E to occur, that is, once S occurs, E will occur.

OUR NEW LANGUAGE GETS A NAME AND A FORMAL SYNTAX

A language can be specified in terms of its syntax and its semantics. The **syntax** of a language consists of its vocabulary and its rules of grammar. The **semantics** is the theory of meaning for the language. Essentially, the vocabulary gives us the elements we can use to construct expressions of the language, the grammar tells us how to construct properly formed expressions, and the semantics supplies the meanings for the various expressions.

THE LANGUAGE TL

The beginning of Chapter 2 discussed the distinction between artificial and natural languages. Our artificial logical language will be named **TL** (for "truth-functional language"). While the syntax of a natural language such as English has hundreds of rules and takes years to learn, the syntax of TL can be specified precisely by a set of three simple rules and can be learned in a few minutes.

▶ THE VOCABULARY FOR TL

Sentence constants: A, B, . . . Z

Sentence operators: ~ v & ⊃ ≡

Parenthetical devices: ( ) [ ] { }

▶ THE GRAMMAR FOR TL

Rule 1. Any sentence constant is a sentence of TL

Rule 2. If **P** is a sentence of TL, then ~**P** is a sentence of TL.

Rule 3. If **P** and **Q** are sentences of TL, then (**P** v **Q**), (**P** & **Q**), (**P** ⊃ **Q**), and (**P** ≡ **Q**) are sentences of TL.

Any expression that contains only items drawn from the vocabulary of TL and that can be constructed by a finite number of applications of the rules of TL's grammar is a sentence of TL. Nothing else counts as a sentence of TL. Sentences of TL will also be called **well-formed formulas of TL** ("wffs" for short) or "formulas."

The sentences specified by TL's first rule of grammar are called **atomic sentences of TL**. An atomic sentence contains no operators. Sentences constructed out of atomic sentences according to the second and third rules of grammar are **molecular sentences of TL**. The atomic sentences out of which a molecular sentence is built are the atomic components of the molecular sentence.

We can establish that a given sentence is indeed a sentence of TL by beginning with the atomic components of the sentence and applying the rules of grammar until the molecular sentence is constructed. For example, consider the expression

$$\sim(\sim A \ v \sim B)$$

We can show that this is a well-formed formula of TL (or "wff" for short) with the following line of reasoning. First, A by itself is a wff according to rule 1. Next, B is a wff according to rule 1. Consequently, ~A is a wff according to rule 2 and ~B is also a wff according to rule 2. Therefore, if we put these two wffs together into (~A v ~B) , the result is a wff according to rule 3, because we have merely joined two wffs together with a wedge and enclosed the whole in parentheses. And finally, if we apply a tilde to this to produce ~(~A v ~B) the result is a wff according to rule 2, because we have merely applied a tilde to a wff.

Consider the wff [(A v B) & (E v J)]. As you will see shortly, if no tilde has been applied to the outer parentheses of a formula, the outer parentheses are actually redundant—they add no new information to the formula and can be dropped. Consequently, when writing such a formula, the bare outer parentheses may be dropped without any loss of information. When outer parentheses are dropped, formulas become easier to read. This will all become more obvious shortly. Consequently, we will add the following helpful rule:

> *You may drop the outer parentheses when they are not encumbered by a tilde.*

When we drop the outer parentheses, we are in effect abbreviating a formula with no loss of meaning. So, using this "abbreviatory" rule, we can do the following:

(A v B)	may be simplified to	A v B
[(A v B) & C]	may be simplified to	(A v B) & C
[(G & J) v ~ (E & ~ S)]	may be simplified to	(G & J) v ~ (E & ~ S)

Notice that while you may drop the outer parentheses in [(A v B) & C], you are not allowed to drop the outer parentheses in the formula ~ [(A v B) & C], for they are encumbered by a tilde on the left.

It is important that you be able to tell the difference between an expression that is a wff and one that is not a wff. If an expression could not be constructed out of TL's vocabulary using only TL's rules of grammar, the expression is not a wff. Here are some examples:

Expression: Reason why this is not a wff:

AB v G   TL's grammar does not allow two constants side by side with nothing between them.

A ~ H   TL's grammar does not allow you to place just a tilde between two constants.

H &v E   TL's grammar does not allow two dyadic operators side by side with nothing in between.

G & S~   TL's grammar does not allow a tilde to dangle on the right.

P v **Q**   TL's vocabulary does not include variables such as the bold P or the bold Q.

A v B & C   This expression needs a pair of parentheses (according to rule 3 plus the abbreviatory rule).

A (B v C)   This expression needs a dyadic operator between the A and the first parenthesis.

~ &B   The rules do not allow you to apply a tilde directly to a dyadic operator.

▶   **EXERCISE 4.1**   ◀

Which of the following are wffs and which are not wffs?

1. * (A & ~B) v (~E & ~H)

2. ~(A v S) & [(H v B) v S]

3. ~[~(~A v ~S)~(H v G)]

4. A v (B v C) v E

5. * ~(~H & R)

6. R v S v (B & G)

7. H & ~(P & S)

8. ~(~S v ~G)

9. *~(A & B) v (v S v G)

10. A & (~& R v G)

11. P v (QA)

12. R & ~ v H

13. *(~A v B) & ~(H v S)

14. A v & G

15. AB

16. *~[~(A v B) v ~E] v ~S

▶ OBJECT LANGUAGE AND METALANGUAGE

When we use one language to talk about a second language, the first language is called the **metalanguage** and the language being talked about is called the

object language. Thus, in Spanish 101, if the teacher speaks in English about a Spanish sentence, in that context English is the metalanguage and Spanish is the object language. We have used the English language to define TL. TL, our artificial logical language, is in this context the object language. English, the language we use to talk about the object language, is the metalanguage.

We have been using boldfaced capital letters **P, Q, R,** and **S** as variables ranging over sentences. These variables belong to the metalanguage, for the language TL contains no variables. Because our variables belong to the metalanguage (English) and because they range over linguistic entities—sentences—they are called "metalinguistic variables."

HOW TO CALCULATE THE TRUTH-VALUE OF THE WHOLE FROM THE VALUES OF THE PARTS

First, some necessary terminology. The **scope of an operator** occurring in a sentence of TL is the operator itself along with the part of the sentence that the operator applies to or links together. For example, consider the following formula:

$$(H \text{ v } S) \supset (A \text{ \& } B)$$

In this formula, the scope of the & is the entire (A & B) part of the formula. (The parentheses around the A & B tell us this.) The scope of the v is the entire (H v S) part. (The parentheses indicate this.) The scope of the $\supset$ is the whole formula.

Next, the **main operator** of a sentence of TL is the operator of largest scope. The main operator will be the operator whose scope spans the entire sentence. In the above formula, the $\supset$ is therefore the main connective or main operator. The truth-value determined by the main operator will always be the truth-value for the compound as a whole. This will become clearer as we work through several examples.

We will employ the **replacement method** to calculate the truth-values of compound sentences. This will involve replacing parts of formulas with truth-values as we calculate the values of whole formulas. Let us begin with the sentence ~ A & B. Suppose the truth-value of A is false (**F**) and the truth-value of B is true (**T**). What, then, is the truth-value of ~A & B? At step 1, in order to keep track of our calculations, we will copy the formula but we will insert an **F** in place of A and we will insert a **T** in place of B. This will remind us that A is assigned false and that B is assigned true:

Start: ~A & B
Given: A is false (**F**), B is true (**T**)
Step 1:   ~**F** & **T**

It should be noted that when we put an **F** in place of A and a **T** in place of B, this is not meant to suggest that A is **F** and that B is **T**; rather, it is merely a "bookkeeping" device that records the fact that A has the truth-value **F** and that B has the truth-value **T**.

In ~A & B, the negation sign applies only to that part of the compound to its immediate right. So, in the formula at step 1, when we calculate the value, we must apply the tilde only to the **F** to its immediate right. Now, the negation truth-table tells us that when a tilde applies to a component that has a truth-value of **F**, the compound has a truth-value of **T**. Therefore, at step 2 we "cancel out" the ~**F** and replace ~**F** with **T**:

Start: ~A & B
Given: A is false (**F**), B is true (**T**)
Step 1:   ~**F** & **T**
Step 2:   **T** & **T**

Now we can see that the ampersand actually joins a unit that has the truth-value **T** with a unit that also has the truth-value **T**. According to the truth-table for the &, when & joins that which is true to that which is also true, the compound is true. So, at step 3 we replace the **T** & **T** with **T**:

Start: ~A & B
Given: A is false (**F**), B is true (**T**)
Step 1:   ~**F** & **T**
Step 2:   **T** & **T**
Step 3:   **T**

Because the ampersand is the main operator, it determines the truth-value of the compound as a whole. So, the truth-value of the compound as a whole is true.

Let's look at a slightly more complex example. Suppose someone symbolizes an English sentence with the formula ~ (A & B). Furthermore, assume A has the truth-value false and B has the value true. Before we begin the calculation, notice that the tilde applies to the parenthesis to its immediate right and not just to A alone. That is, the tilde applies to the conjunction (A & B) as a whole. The first step, then, is to calculate the value of (A & B). After that step is completed, we apply the tilde. Here are the steps:

Formula: ~ (A & B)
Given: A is **F**, B is **T**
Step 1:   ~ (**F** & **T**)

At step 1, the (**F** & **T**) part has the overall truth-value **F**, for a false component conjoined to a true component is false. So, at the next step, we replace just the (**F** & **T**) part with **F**:

Formula: ~ (A & B)
Given: A is **F**, B is **T**
Step 1:   ~ (**F** & **T**)
Step 2:   ~ **F**

Because a tilde applied to a false component produces a compound with the truth-value of true, we replace ~**F** with **T**:

Formula: ~ (A & B)
Given: A is **F**, B is **T**
Step 1:   ~ (**F** & **T**)
Step 2:   ~ **F**
Step 3:   **T**

The value of the compound is thus true.

    Try this: If A is assigned false and B is assigned true, what is the truth-value of ~[~(A & ~B) v ~(A v B)] ?

Formula: ~ [~ (A & ~ B) v ~ (A v B)]
Given: A is **F**, B is **T**
Step 1:   ~[~ (**F** & ~ **T**) v ~ (**F** v **T**)]

We begin our calculations with the connectives of smallest scope. The smallest scope possible is a tilde applied to a single atomic component with no intervening bracket. This is the ~**T** inside the first parenthesis at step 1 above, and we begin by calculating that. A tilde applied to a **T** produces an **F**, and so the ~**T** is replaced by an **F**:

Formula: ~ [~ (A & ~ B) v ~ (A v B)]
Given: A is **F**, B is **T**
Step 1:   ~[~ (**F** & ~ **T**) v ~ (**F** v **T**)]
Step 2:   ~[~ (**F** & **F**) v ~ (**F** v **T**)]

Next, we evaluate the function of next largest scope (and if there is a tie, the order doesn't matter). Here, we evaluate the (**F** v **T**) component. The truth-table for the wedge tells us that a false component disjoined to a true component produces a true compound, that is, **F** v **T** is true. We replace the **F** v **T** with a **T**:

Formula: ~ [~ (A & ~ B) v ~ (A v B)]
Given: A is **F**, B is **T**
Step 1:   ~[~ (**F** & ~ **T**) v ~ (**F** v **T**)]
Step 2:   ~[~ (**F** & **F**) v ~ (**F** v **T**)]
Step 3:   ~[~ (**F** & **F**) v ~ **T**]

The function of next largest scope is the (**F** & **F**) component. The truth table for & tells us that a false component conjoined to a false component yields a false compound. Thus, we replace the (**F** & **F**) with an **F**:

Formula: ~ [~ (A & ~ B) v ~ (A v B)]
Given: A is **F**, B is **T**
Step 1:   ~[~ (**F** & ~ **T**) v ~ (**F** v **T**)]
Step 2:   ~[~ (**F** & **F**) v ~ (**F** v **T**)]
Step 3:   ~[~ (**F** & **F**) v ~ **T**]
Step 4:   ~[~ **F** v ~ **T**]

Next, we follow the negation table and replace ~ **F** with **T** and we replace ~ **T** with **F**:

Formula: ~ [~ (A & ~ B) v ~ (A v B)]
Given: A is **F**, B is **T**
Step 1:   ~[~ (**F** & ~ **T**) v ~ (**F** v **T**)]
Step 2:   ~[~ (**F** & **F**) v ~ (**F** v **T**)]
Step 3:   ~[~ (**F** & **F**) v ~ **T**]
Step 4:   ~[~ **F** v ~ **T**]
Step 5:   ~[**T** v **F**]

Because a disjunction with a true left disjunct and a false right disjunct is true, we replace [**T** v **F**] with **T**:

Step 6:   ~ **T**

Finally, the remaining tilde—the sentence's main operator—is evaluated. A tilde applied to a truth produces falsity:

Step 7:   **F**

This is the final truth-value. Remember that the final value, the value for the compound as a whole, is always the value determined by the main operator, and the main operator is always the last operator to be evaluated.

Notice the way in which the truth-values "flow" from the atomic components down through the structure and emerge at the end from the main operator. Notice also that as we calculate the truth-values, we move from the inside to the outside, starting with the functions of smallest scope and graduating in stages to larger functions, until we reach the function of greatest scope, the main operator.

Here are two more calculations, in compact form:

Formula: ~ (~ A ⊃ ~B) v ~ C
Given: A is **T**, B is **F**, C is **F**
Step 1:   ~ (~ **T** ⊃ ~ **F**) v ~ **F**
Step 2:   ~ (**F** ⊃ ~ **F**) v ~ **F**
Step 3:   ~ (**F** ⊃ **T**) v ~ **F**
Step 4:   ~ (**F** ⊃ **T**) v **T**
Step 5:   ~**T** v **T**
Step 6:   **F** v **T**
Step 7:   **T**

Formula: ~ [(~ H v ~ R) & (~ R ⊃ G)]
Given: H is **T**, R is **F**, G is **T**
Step 1:   ~ [(~ **T** v ~ **F**) & (~ **F** ⊃ **T**)]
Step 2:   ~ [(**F** v ~ **F**) & (~ **F** ⊃ **T**)]

Step 3: ~ [(**F** v **T**) & (~ **F** ⊃ **T**)]
Step 4: ~ [(**F** v **T**) & (**T** ⊃ **T**)]
Step 5: ~ [**T** & (**T** ⊃ **T**)]
Step 6: ~ [**T** & **T**]
Step 7: ~ **T**
Step 8: **F**

▶ CATEGORIZING COMPOUND SENTENCES

Truth-functional compound sentences may be categorized in terms of their main operators as follows.

- If the main operator of a sentence is a conjunction operator, the sentence is a **conjunction**. For example, the following sentence is a conjunction: (A v B) & (E v H)

- If the main operator is a disjunction operator, the sentence is a **disjunction**. For example: (A & B) v ~ (E & S)

- If the main operator is a negation operator, the sentence is a **negation**. For example: ~ (A & H)

- A **conditional** is any sentence whose main operator is a horseshoe. For example: (A & H) ⊃ F

- If the main operator is the triple bar, the sentence will count as a **biconditional**. For example: (A v D) ≡ (E & S)

| ▶ | **EXERCISE 4.2** | ◀ |

Identify the main connective in each of the following.

1. *~ (A & S) & ~ (G v B)

2. (A v B) v (H & ~ S)

3. ~ [~ (A v B) v (E & F)]

4. *~ A v B

5. ~ (~ A v ~ B) & E

6. [(A v B) v C] v (~ S & G)

7. *A v ~ (H v B)

8. D ⊃ (~H v ~ J)

9. B ≡ ~ (J ⊃ E)

10. [(B & ~K) ⊃ (G ≡ U)] ⊃ D

► | **EXERCISE 4.3**
FUN WITH THE WEDGE, TILDE, AND AMPERSAND | ◄

Let A abbreviate "Ann is home," let B abbreviate "Bob is home," and let C abbreviate "Connie is home." Assume that A is true, B is false, C is true, and determine whether each of the following is true or false.

1. *A & B

2. ~(A v B)

3. ~ A & ~B

4. *~A & B

5. ~(A & B)

6. (A & B) v C

7. *~(A v B) & C

8. A & (B v C)

9. ~A v ~B

10. (A v B) v C

11. *A v (B v C)

12. (A & B) & C

13. A & (B & C)

14. (B v A) v C

15. (C v A) v B

16. *A v (A & B)

17. (~A & ~B) & ~C

18. ~[(A v B) v C]

19. ~[(A & B) & C]

20. *(~A v ~B) v ~C

► | **EXERCISE 4.4** | ◄

Remember in algebra class when you had to solve equations with one unknown? The following problems are similar. Suppose you know that A and B are true and that C and D are false. However, suppose you do not know the truth-values of P and Q. Nevertheless, determine the truth-values of the following.

1. * P v A

2. P & D

3. P & ~P

4. * P v ~P

5. ~(Q v B)

6. ~(C & P)

7. * (A v P) v (P & Q)

8. (D & P) & (P & Q)

9. ~(P & Q) v P

10. P v (A v ~A)

11. *P & (A & ~A)

12. ~P & ~(A v ~A)

13. ~P v (~Q v P)

14. ~B & P

15. ~P v (Q v P)

16. *~(C & P) v ~(~P & Q)

▶ **EXERCISE 4.5**
FUN WITH HORSESHOES AND TRIPLE BARS ◀

Let A abbreviate "Ann swims," let B abbreviate "Bob swims," and let E abbreviate "Ernest swims." Assume A, B, and E are each true, and determine truth-values of the following.

1. *A ⊃ B

2. B ⊃ A

3. ~B ⊃ ~A

4. *(A ⊃ B) ⊃ E

5. A ⊃ (B ⊃ E)

6. A ⊃ (B & E)

7. *A ⊃ (B ≡ E)

8. (A ⊃ B) ⊃ (B ⊃ A)

9. ~A ⊃ (~B & ~E)

10. ~E ⊃ ~(A ≡ B)

11. *(A & B) ⊃ E

12. ~(A & B) ⊃ ~ E

13. (A & B) ≡ (A & E)

14. A ≡ E

15. E ≡ A

16. *(~A & ~B) ⊃ ~E

17. B ⊃ (A ⊃ E)

18. A ⊃ [(B v E) v (B & E)]

19. *(A ≡ B) v (A ≡ E)

▶ **EXERCISE 4.6**
FORMULAS WITH ONE OR MORE UNKNOWNS ◀

Suppose A and B are true, C and D are false, but the truth-values of P and Q are unknown. Determine the truth-values of the following.

1. * C ⊃ P

2. P ⊃ A

3. ~A ⊃ Q

4. * Q ⊃ ~C

5. P ⊃ (Q ⊃ P)

6. D ⊃ (P ⊃ Q)

7. * (P v Q) ⊃ A

8. Q ⊃ (P ⊃ A)

9. P ⊃ (P ⊃ P)

10. A ≡ (C & P)

11. *A ⊃ (P v ~D)

12. ~(P ⊃ P)

13. (Q ⊃ A) ⊃ C

14. (C ⊃ P) ⊃ (D ⊃ Q)

15. (P & C) ⊃ Q

16. *[P ⊃ (Q ⊃ P)] ⊃ D

17. (P & A) ⊃ B

18. *~[P ⊃(~B ⊃ Q)]

CHAPTER 4 GLOSSARY

Atomic sentence of TL   A TL sentence containing no operators.

Biconditional   A sentence whose main operator is a biconditional operator.

Conditional   A sentence whose main operator is a conditional operator.

Conjunction   A sentence whose main operator is a conjunction operator.

Disjunction   A sentence whose main operator is a disjunction operator.

Main operator   The operator with the largest scope.

Metalanguage/Object Language   When one language is used to talk about a second language, the first language is called the metalanguage and the language being talked about is called the object language.

Molecular sentence of TL   A TL sentence containing one or more operators.

Negation   A sentence whose main operator is a negation operator.

Scope of an operator   The operator itself along with the part of the sentence that the operator applies to or links together.

Semantics   The theory of meaning for a language. The semantics supplies the meanings for the various expressions of the language.

Syntax   The vocabulary and rules of grammar of a language. Essentially, the vocabulary gives us the elements to construct expressions of the language, and the rules of grammar tell us how to construct properly formed expressions of the language.

TL   The name of the formal language used in this text to represent truth-functional logical properties and logical relationships.

Well-formed formula of TL   (or "wff of TL" or "sentence of TL") Any expression that contains only items drawn from the vocabulary of TL and that can be constructed by a finite number of applications of the rules of TL's grammar.

Chapter

5

TRUTH-TABLE ANALYSIS

A truth-table can provide a visual display of all the possible truth-values a particular formula might have in all possible circumstances. Such a display of *possibilities* has numerous uses within logical theory, as we shall see shortly. Let us begin with the English sentence "Ann is home or Ann is not home." This is symbolized as: A v ~A. What is the truth-value of the compound as a whole? "Well," someone might say, "it depends on the values of the components." Let's consider the possibilities. There are only two: (1) A is true, and (2) A is false. Beginning with the first, if A has the truth-value **T**, then:

A v ~ A

T v ~T

T v F

T

The compound is true according to this truth-value assignment. Turning to the second possibility, if A has the truth-value **F**, then:

A v ~ A

F v ~F

F v T

T

The compound is still true. Notice that in both cases, the final truth-value—the truth-value of the whole—is true.

CONSTRUCTING A TRUTH-TABLE FOR A FORMULA

However, the above calculations take up a lot of space. Both of the above possibilities may be compactly displayed on a truth-table as follows. First, on top of a table we write the formula we are evaluating. To the side of this we fill in all possible truth-values the letters (the atomic components) might be assigned. In this case, the formula only has one atomic component, namely A:

A	A v ~A
T	
F	

Next, we calculate the truth-value of the compound for each assignment of truth-values:

A	A v ~ A
T	T T FT
F	F T TF
	*

On each row of this table, the truth-values were calculated just as they were in the diagrams above, except that truth-values were entered horizontally along a row of the table. The table indicates that if A is assigned **T**, then A v ~A is true and if A is assigned **F** then A v ~A is true. The column of truth-values directly under the main operator is called the **final column**. In the table above, a star has been placed under the final column.

Next, consider a slightly more complex formula:

$$\sim [(A \,\&\, B) \,\&\, \sim (A \,\&\, B)]$$

The table for the formula must list all the different combinations of truth-values that could possibly be assigned to the atomic components of the formula. In this case we have only four possible combinations, and so we have only four **truth-value assignments** to consider:

1. Perhaps A is true and B is true.

2. Perhaps A is true and B is false.

3. Perhaps A is false and B is true.

4. Perhaps A is false and B is false.

These four possibilities may be displayed on a truth-table as follows:

A B	~ [(A & B) & ~ (A & B)]
T T	
T F	
F T	
F F	

This truth-table displays all combinations of truth-values the atomic components could possibly be assigned. Next, we calculate the value of the compound for each possible assignment of truth-values:

A B	~	[(A	&	B)	&	~	(A	&	B)]
T T	T	T	T	T	F	F	T	T	T
T F	T	T	F	F	F	T	T	F	F
F T	T	F	F	T	F	T	F	F	T
F F	T	F	F	F	F	T	F	F	F

The completed table shows how each combination of truth-values determines the truth-value of the compound as a whole.

Each row of a truth-table represents a different truth-value assignment. That is, each row of a table represents a different possible combination of truth-values that could be assigned to the atomic components of the formula or formulas on top of the table. Between them, the rows of a table collectively represent all possible combinations of truth-values that could be assigned to the atomic components of the formula or formulas on top of the table.

EIGHT-ROW TABLES

Suppose we have a formula containing *three* different atomic components:

$$(A \lor B) \supset C$$

How many different truth-value assignments are there for this formula? The formula contains three atomic components, namely A, B, and C. A fairly simple process of trial and error reveals only eight possible combinations of truth-values for three components considered three at a time:

Possibility 1   A is **T**, B is **T**, C is **T**

Possibility 2   A is **T**, B is **T**, C is **F**

Possibility 3   A is **T**, B is **F**, C is **T**

Possibility 4   A is **T**, B is **F**, C is **F**

Possibility 5   A is **F**, B is **T**, C is **T**

Possibility 6   A is **F**, B is **T**, C is **F**

Possibility 7   A is **F**, B is **F**, C is **T**

Possibility 8   A is **F**, B is **F**, C is **F**

The truth-table for the formula above begins with the following truth-value assignments:

	A	B	C
1.	T	T	T
2.	T	T	F
3.	T	F	T
4.	T	F	F
5.	F	T	T
6.	F	T	F
7.	F	F	T
8.	F	F	F

The table now displays all the different combinations of truth-values that could be assigned to the atomic components. When the truth-value assignments have been filled in, the rows of the table may be calculated as follows:

A	B	C	(A	v	B)	⊃	C
T	T	T	T	T	T	T	T
T	T	F	T	T	T	F	F
T	F	T	T	T	F	T	T
T	F	F	T	T	F	F	F
F	T	T	F	T	T	T	T
F	T	F	F	T	T	F	F
F	F	T	F	F	F	T	T
F	F	F	F	F	F	T	F

The Number of Rows on a Truth-Table

In general, if you place a sentence containing n (any number of) atomic components on a truth-table, the resulting table will have 2^n rows. Thus:

If the sentence has:	For example:	The table will have:
1 atomic component	A v ~ A	2 rows
2 atomic components	A & B	4 rows
3 atomic components	(A v B) ⊃ (E v A)	8 rows
4 atomic components	(G v N) & (S ≡ J)	16 rows
5 atomic components	(D & W) ⊃ [(I v R) ≡ H]	32 rows

and so on...

HOW TO MAKE YOUR OWN TAUTOLOGY DETECTOR USING JUST PAPER AND PENCIL

Consider the sentence

It's not the case that Annette is both forty years old and not forty years old.

In TL, this is easily symbolized as: ~ (A & ~ A). What's the truth-value of the compound? That depends on the truth-values assigned to A. The truth-table allows us to look at all the possible combinations of truth-values.

A	~ (A & ~ A)
T	T T F F T
F	T F F T F
	*

Notice that only two possibilities exist: A is true (row 1), A is false (row 2). The table shows us two things: if A is true, the compound is true; if A is false, the compound is true. In both cases, the final truth-value—the truth-value of the whole—is **T**.

The sentence ~ (A & ~ A) thus has an interesting feature. On every row of its truth-table, the formula computes true at the main operator. This indicates that the sentence has the truth-value true in every possible circumstance. In no circumstance will it ever be false. A sentence whose main connective shows true on every row of its truth-table is called a **tautology**. Recall that the final column on a truth-table is the column of truth-values determined by and placed directly underneath the main operator of the formula. Thus, we can say that a formula of TL symbolizes a tautology if and only if the formula's *final column* shows all T's.

Here is another example:

Either today is Saturday or today is not Saturday.

When symbolized, this becomes

$$S \text{ v} \sim S$$

where S abbreviates "Today is Saturday." The truth-table for this formula is

S	S	v	~	S
T	T	T	F	T
F	F	T	T	F
		*		

Again the formula computes true on every row of its table. S v ~ S is thus a tautology. It cannot possibly be false.

The Tautology Test

A formula of TL is a tautology if and only if the final column of the formula's truth-table shows all T's.

HOW TO MAKE AN INEXPENSIVE CONTRADICTION DETECTOR FOR HOME OR OFFICE

Imagine a chemist-turned-politician who is trying to please all sides and who says, "Aluminum is an element and it's not the case that aluminum is an element." This politician is trying to get the pro-aluminum vote *and* the anti-aluminum vote. In symbols, using obvious abbreviations, this political profundity is: A & ~ A. The truth-table is as follows:

A	A & ~A
T	T F F T
F	F F T F
	*

The final column on this table contains all **F**'s, which indicates that the sentence has the truth-value false in every possible circumstance. In no possible circumstance is the sentence true. In truth-functional logic, a sentence whose main connective shows false on every row of the truth-table is a **contradiction**. Thus:

The Contradiction Test

A formula of TL is a contradiction if and only if the final column on the formula's truth-table shows all **F**'s.

Another example: Consider the symbolized sentence ~ (A v ~ A). The table shows it to be a contradiction:

A	~ (A v ~ A).
T	F T T F T
F	F F T T F
	*

The table shows that the formula cannot possibly be true.

THE CONTINGENCY DETECTOR: DON'T LEAVE HOME WITHOUT IT

In truth-functional logic, a sentence is said to be **contingent** if the final column of its truth-table contains at least one **T** and at least one **F**. For example, take the sentence "Gilligan is homesick or the Skipper is hungry." This goes into symbols as G v S, and the corresponding table shows a mix of **T**'s and **F**'s in its final column:

G S	G v S
T T	T T T
T F	T T F
F T	F T T
F F	F F F
	*

Because each row of a truth-table represents a possible circumstance, a contingent sentence is thus one that is true in some circumstances and false in others. In the case of any contingent sentence, whether the sentence is true or false depends on (is "contingent upon") the circumstance or row of the table under consideration.

> ### *The Contingency Test*
>
> A formula of TL is contingent if and only if the final column on the formula's truth-table shows at least one **T** and at least one **F**.

These first three truth-table tests may be summarized as follows:

- A sentence of TL is a tautology if the final column on its truth-table contains all **T**'s.

- A sentence of TL is a contradiction if the final column on its truth-table contains all **F**'s.

- A sentence of TL is contingent if the final column on its truth-table contains at least one **T** and at least one **F**.

▶ LOGICAL STATUS AND SOME ALTERNATIVE TERMINOLOGY

Tautologies are also called **logical truths** and are said to be *logically true* because they can be proven true using only the procedures of logical theory—without investigating the physical world. Similarly, contradictions are called **logical falsehoods** and are said to be *logically false* because they can be proven false using just the procedures of logic alone—without investigating the physical world.

Tautologies are *necessarily true* because they cannot possibly be false, and contradictions are *necessarily false* because they cannot possibly be true. If we specify whether a sentence is necessarily true, necessarily false, or contingent, we are specifying the sentence's **logical status**.

▶   **EXERCISE 5.1**   ◀

Use truth-tables to determine which of the following are tautological, which are contradictory, and which are contingent.

1. *(A & B) ⊃ (A ∨ B)

2. ~(A ⊃ A)

3. (A ⊃ B) ⊃ (~A ∨ B)

17. [(A ⊃ B) & A] ⊃ B

18. [(A ∨ B) & ~A] ⊃ B

19. (A & ~A) ⊃ B

4. *(A ⊃ B) ⊃ (~ B ⊃ ~ A)

5. (A ∨ B) ⊃ (A & E)

6. A ⊃ (A ∨ B)

7. *~(A ≡ A)

8. ~A ⊃ A

9. A ⊃ ~A

10. A ≡ (A ∨ B)

11. * A ∨ (B ⊃ B)

12. A ≡ ~A

13. A ≡ (A ∨ A)

14. (A & B) ⊃ B

15. (A & B) ⊃ (~A ⊃ ~B)

16. *[(A ⊃ B) & ~B] ⊃ ~A

20. *[(A & B) & E] ⊃ [A & (B & E)]

21. (A ⊃ B) ≡ (~A ∨ B)

22. (A ⊃ B) ≡ (~B ⊃ ~A)

23. (A & B) ⊃ ~(~A ∨ ~B)

24. *~(A & B) ⊃ (~A ∨ ~B)

25. ~(A ∨ B) ⊃ (~A & ~B)

26. A ⊃ (~B ⊃ A)

27. ~A ⊃ (A ⊃ B)

28. *~B ⊃ (A ⊃ B)

29. ~(A & ~A)

30. A ∨ (A & ~A)

31. (A & ~A) ∨ (B & ~B)

32. *(A & ~A) ∨ B

TESTING AN ARGUMENT FOR VALIDITY

Recall from Chapter 1 that an argument is valid if and only if there is no possibility that its premises are true and its conclusion is false. Consider the following argument:

1. If Ann is swimming then Bob is swimming.

2. Bob is not swimming.

3. Therefore, Ann is not swimming.

In symbols, this is

1. A ⊃ B

2. ~ B / ~ A

In this symbolized argument, the premises are numbered and the symbolized conclusion follows the slanted slash mark. Now, is this argument valid? To many people, the argument looks invalid. Perhaps you can see, intuitively, that it is indeed valid. However, we can use a truth-table to decide the issue in a precise way. The procedure involves the following three steps:

Step 1 Create a truth-table for the argument and place the symbolized version of the argument, in order, across the top of the table. This argument only has two letters, A and B. We place these letters on the left, assign the four possible combinations of truth-values in the standard order, and then write the argument across the top:

	First premise:	Second premise:	Conclusion:
A B	A ⊃ B	~B	~A
T T			
T F			
F T			
F F			

Step 2   Compute the truth-values of the premises and conclusion on each row. The final values for our current argument, that is, the values of the main connectives, are filled in below:

	First premise:	Second premise:	Conclusion:
A B	A ⊃ B	~B	~A
T T	T	F	F
T F	F	T	F
F T	T	F	T
F F	T	T	T

Step 3   This step has two instructions:

- If at least one *row* of the table shows a **T** under the main operator of each premise but shows an **F** under the main operator of the conclusion, the argument on top of the table is an invalid argument.

- If the table contains *no row* showing all true premises and a false conclusion, the argument is valid.

This argument is valid; no row shows the true premises–false conclusion combination.

Next, consider an argument that looks similar:

1. If Ann is swimming then Bob is swimming.

2. Ann is not swimming.

3. Therefore, Bob is not swimming

In symbols, this is:

1. A ⊃ B

2. ~ A / ~ B

Is this argument valid? Let us construct a truth-table. This argument only has two letters, A and B, and so we place these letters on the left and list the four possible combinations of truth-values in the standard order:

	First premise:	Second premise:	Conclusion:
A B	A ⊃ B	~A	~B
T T			
T F			
F T			
F F			

Next, we compute the truth-values of the premises and conclusion on each row. The final values for our current argument are filled in below:

	First premise:	Second premise:	Conclusion:	
A B	A ⊃ B	~A	~B	
T T	T	F	F	
T F	F	F	T	
F T	T	T	F	←
F F	T	T	T	

Remember, if at least one *row* of the table shows a **T** under the main operator of each premise but shows an **F** under the main operator of the conclusion, the argument on top of the table is an invalid argument. This argument shows the "all true premises-false conclusion" combination on the third row down, so the argument is shown to be invalid.

Here is the rationale behind this truth-table test for validity. An argument is valid if there are no possible circumstances in which the premises are true and the conclusion is false. Each row of the table represents a collection of possible circumstances. Between them, the rows of a table represent all possible combinations of truth-values the argument's components might have. The truth-table allows us, then, to see how the truth-values of the premises and conclusion vary across all possible circumstances. A row showing true premises and a false conclusion represents a *possibility* of true premises and false conclusion. If a table contains such a row, the argument represented on top of the table can have true premises and a false conclusion. If there is any possibility that the premises are true and the conclusion is false, an argument is invalid. If no row of a table shows all true premises with a false conclusion, we know that in no possible circumstance are the premises true and the conclusion false. If it is not possible for the premises to be true and the conclusion false, an argument is valid.

The next argument may seem valid to some people at first glance, but the truth-table shows it to be invalid:

1. If Ann swims today, then Bob swims today.

2. Bob swims today.

3. So, Ann swims today.

In TL, this is

1. A ⊃ B

2. B / A

Notice row 3 on the argument's truth-table:

A	B	A ⊃ B	B	A	
T	T	T	T	T	
T	F	F	F	T	
F	T	T	T	F	←
F	F	T	F	F	

Row 3 shows us the possibility of true premises and a false conclusion, which shows that the argument is invalid.

Here is a more complex example. Consider the following English argument:

1. Ann and Bob are not both swimmers.

2. Ann is a swimmer.

3. Therefore, Bob is not a swimmer.

In symbols, this is

1. ~ (A & B)

2. A / ~ B

The table is set up as follows:

	First premise:	Second premise:	Conclusion:
A B	~ (A & B)	A	~ B
T T			
T F			
F T			
F F			

Next, we compute the truth-values of the premises and conclusion on each row. The values for our current argument are filled in below:

A B	~ (A & B)	A	~ B
T T	F T T T	T	F T
T F	T T F F	T	T F
F T	T F F T	F	F T
F F	T F F F	F	T F
	*	*	*

Because the table contains *no* row that shows all true premises with a false conclusion, the argument is valid.

The truth-table for the following argument shows that the argument is invalid:

1. If Hank does not go swimming then John will not go swimming.

2. Hank will go swimming.

3. So, John will go swimming.

In TL, this argument is

1. ~H ⊃ ~ J

2. H / J

Here is the argument's table:

H J	~H ⊃ ~ J	H	J
T T	FT T FT	T	T
T F	FT T TF	T	F
F T	TF F FT	F	T
F F	TF T TF	F	F

Notice that row 2 shows true premises and a false conclusion, which shows that the argument is invalid.

▶ TRUTH-FUNCTIONAL VALIDITY

In this chapter, we are examining arguments whose validity or invalidity depends on the various truth-functional relationships that connect the premises to the conclusion. These truth-functional relationships are revealed in the configuration of **T**'s and **F**'s in the argument's truth-table. In later chapters, we will study arguments whose validity relies not on truth-functional relationships but on logical relationships of other types. To distinguish the valid arguments we have been examining from valid arguments of other types, we can say that an argument whose validity is established with the methods of truth-functional logic alone is a *truth-functionally valid* argument. An argument that is shown to be valid with a truth-table is sometimes said to be *truth-table valid*.

> *The great German philosopher and mathematician*
> *Gottfried Wilhelm Leibniz (1646–1716) once wrote:*
>
> *I feel that controversies can never be finished, nor silence imposed on the Sects, unless we give up complicated reasoning in favor of simple calculation, words of vague and uncertain meaning in favor of fixed symbols . . . When controversies arise, there will be no more . . . disputation between two philosophers than between two accountants. Nothing will be needed but that they should take pen in hand, sit down with their counting tables and . . . say to one another: Let us calculate.*

▶ **EXERCISE 5.2** ◀

Use truth-tables to determine which of the arguments below are valid and which are invalid. In each problem, the premises are listed first and the conclusion follows the slash.

1. *A v B

B v C / A v C

2. A v B

~ A / ~ B

3. *A ⊃ B / B ⊃ A

4. A ⊃ B / ~A ⊃ ~B

5. A ⊃ B

B ⊃ C / A ⊃ C

6. A ⊃ (B ⊃ C) / C ⊃ (B ⊃ A)

7. *(A & B) ⊃ C / A ⊃ (B ⊃ C)

8. A ⊃ B / ~ B ⊃ ~A

9. A ⊃ B

~B / ~ A

10. A ⊃ B

A / B

11. *(A & B) ⊃ C

~C / ~A v ~B

12. A ⊃ B

B / A

13. ~ [A & (B & C)] / ~ A & (~ B & ~ C)

14. ~ A v ~ B / ~ (A & B)

15. ~ (A v B) / ~A & ~ B

16. *~ (A & B) / ~ A & ~ B

▸ Truth-Functional Logic Is Analytic

The truth-table test for validity is an example of an **analytic method.** One way to understand something is to *analyze* it (from the Greek word *analytkos* meaning "to resolve into elements"). This involves breaking the subject down into its constituent parts and seeing how those parts fit together to constitute the whole. Imagine, for example, taking a watch apart and putting it back together to see how it works. The analytic method has been successful in numerous fields of thought, especially the physical sciences, where its application led to the discovery of photons, protons, neutrons, electrons, and various other subatomic particles.

Logic is concerned with arguments, and we use the analytic method in logic to break arguments down into their component parts and study how those parts fit together to produce a chain of reasoning. In truth-functional logic, we can analyze an argument by symbolizing it and running a truth-table test on it. The truth-table tests are thus analytic in nature.

SHOWING AN ARGUMENT INVALID WITH A PARTIAL TRUTH-TABLE

Upon inspection, the following argument looks invalid:

1. A ⊃ B

2. B ⊃ C

3. C ⊃ D

4. D ⊃ E / E ⊃ A

However, to prove this argument invalid using the truth-table method introduced above would require a huge table. The argument has five components, so the table would have thirty-two rows. A far simpler way to prove this argument invalid is to employ a **partial truth-table**.

First, across a horizontal line, write down the premises and conclusion of the argument. Next, for the purposes of the partial table, suppose hypothetically that the argument is invalid. In line with that supposition, assign **T** to the main operator of each premise and assign **F** to the main operator of the conclusion. This assignment of truth-values represents a *hypothesis* that the premises are true and the conclusion is false. Next, try to fill in the rest of the values based on the hypothesis that the argument is invalid. If this cannot be done without producing a contradictory assignment of truth-values to one or more components, the hypothesis—that the premises are true and the conclusion is false—is contradictory or impossible. If so, it is not possible for the premises to be true and the conclusion false, which proves the argument valid. (A contradictory assignment of truth-values is an assignment that assigns **T** and **F** to the same letter, that is, to the same atomic component.)

However, if it is possible to assign the truth-values in such a way that the premises are true and the conclusion is false, with no contradictory assignment of truth-values, the hypothesis—that the premises are true and the conclusion is false—is a possibility. In this case, the argument must be invalid, for an argument is invalid if it is even possible that it has true premises and a false conclusion.

So, the first step in constructing a partial table for the argument above is to assign **T** to the main connective of each premise and **F** to the main connective of the conclusion:

A ⊃ B	B ⊃ C	C ⊃ D	D ⊃ E	/ E ⊃ A
T	T	T	T	F

Let us focus next on the conclusion, E ⊃ A. If the conditional E ⊃ A is false (as assumed), then E must be assigned **T** and A must be assigned **F**, for a conditional is false only when its antecedent is true and its consequent is false. Thus, we place a **T** under the E and an **F** under the A:

A ⊃ B	B ⊃ C	C ⊃ D	D ⊃ E	/ E ⊃ A
T	T	T	T	T F F

Now, if we are going to be consistent, we must assign **T** to each occurrence of E and we must assign **F** to each occurrence of A throughout the rest of the argument:

A ⊃ B	B ⊃ C	C ⊃ D	D ⊃ E	/ E ⊃ A
F T	T	T	T T	T F F

From here, it is easy to assign values in such a way that the premises are true and the conclusion is false. If we add that B is true, C is true, and D is true, that is, if we assign **T** to B, **T** to C, and **T** to D, the premises will all be true while the conclusion is false:

A ⊃ B	B ⊃ C	C ⊃ D	D ⊃ E	/ E ⊃ A
F T T	T T T	T T T	T T T	T F F

This partial table presents us with a *possibility*—the possibility of the argument's premises being true while its conclusion is false. Thus, the argument is proven *invalid*, for an argument is invalid if it is *possible* for it to have true premises and a false conclusion.

In effect, we have simply reproduced a single row from the argument's full truth-table—a row on which the premises are all true and the conclusion is false. In this case, that row is the row on which A is assigned **F**, and B, C, D, and E are each assigned **T**. Remember that an argument is invalid if its table has at least one row showing all true premises and a false conclusion.

Here's another way to understand a partial table. Recall that a **counter-example** to an argument is a possible circumstance in which the argument's premises would be true and its conclusion false. If there is a counterexample to an argument, the argument is invalid. To show someone that his argument is invalid, we can present or describe a counterexample to the argument. The partial table for an argument presents a counterexample to the argument.

> To show an argument invalid using a partial table, write down the premises and conclusion and assign truth-values in such way that the premises are made true and the conclusion is made false.

> ▶   **EXERCISE 5.3**   ◀

Each of the arguments below is invalid. In each case, symbolize the argument and use a partial table to show that it is invalid.

1. *If Sue has a hamburger, then if Joe will not have one, then Rita will have one. So, if Rita will have a hamburger, then Joe will not have one.

2. If Arnold quits his job, then Betty will quit her job. Arnold won't quit his job. So, neither Arnold nor Betty will quit.

3. If Pat goes swimming, then neither Quinn nor Rita will swim. If Rita doesn't go swimming, then Sue will swim, but Dolores won't swim. So, if Sue swims, then Pat swims.

4. If Janet swims, then Bill will swim. Janet won't swim. So, Bill won't swim.

5. *If Angie sings, then Randy will sing. If Randy sings, then either Chris or Pat will sing. If Pat sings, then Roberta will sing. Therefore, if Angie sings, then Roberta will sing.

6. Craig and Faith won't both order burritos. If Craig doesn't order a burrito, then Darla will order one. So, Darla and Faith will both order burritos.

7. Neither Ann nor Bob will order tacos, but Chris and Darla will. If either Ann or Chris orders a taco, then Roberta will order a burrito. So, if Roberta orders a burrito, then Ann will order a taco.

8. Both Ann and Bob won't order french fries. If Bob doesn't order french fries, then either Chris or Darla will order fries. If Robert doesn't want to eat, then Chris won't order fries and neither will Darla. Either Robert will not want to eat or Chris will order fries.

9. If neither Ann nor Bubba orders the Bob's Brontosaurus Burger Bar Special Combo Plate, then Cherie and Zack will both be disappointed. If Cherie is disappointed, then she will order four hamburgers. So, if Ann orders the Bob's Brontosaurus Burger Bar Special Combo Plate, then Cherie won't order four hamburgers.

> ▶   **EXERCISE 5.4**   ◀

Use partial truth-tables to prove that each of the following symbolized arguments is invalid. In each case, the argument's symbolized premises are written first and the symbolized conclusion appears after the slash mark.

1. *A ⊃ (B ∨ W)

   B ⊃ S / A ⊃ S

2. A ⊃ B

W ⊃ S

~A v ~W / ~B v ~S

3. A v B

A / ~B

4.*~(A & B)

~A / B

5. A ⊃ B

B ⊃ W

W ⊃ S

S ⊃ H / H ⊃ A

6. A ⊃ B / B ⊃ A

7. ~(A & B) / ~A & ~B

8. ~(A v B) / ~ (~ A v ~B)

9.*~(A & B) / ~ A

10. A v B

A / B

11. A ⊃ B

W ⊃ B / A ⊃ W

12.*A ⊃ B

W ⊃ S

B v S / A v W

13. A ⊃ B

B / A

14. A ⊃ B

~A / ~B

►   **EXERCISE 5.5**   ◄

Symbolize the following arguments and test each for validity with a truth-table of your choice.

1. *If Betsy takes Philosophy 110, then John will take Philosophy 110. John won't take Philosophy 110. So Betsy won't take Philosophy 110.

2. If Plektus IV invades Ruritania, then the Ruritania Senate will flee. Plektus IV will not invade Ruritania. Therefore, the Ruritania Senate will not flee.

3. Either Herman will be home or Lillian will be home. Herman won't be home, so therefore Lillian won't be home.

4. Gomez and Morticia won't both be home. But Morticia will be home. So, Gomez won't be home.

5. *If either Granny or Jethro is home, then Mr. Drysdale will be happy. Mr. Drysdale won't be happy. So Granny won't be home and Jethro won't be home.

6. It is not the case that both Kirk and Spock will beam down. If Kirk doesn't beam down, then McCoy will. If Spock doesn't beam down, then McCoy will. So, McCoy will beam down.

7. If Ralph wins the game, then Alice will be happy. If Alice will be happy, then Norton will be happy. So, if Norton will be happy, then Alice will be happy.

8. If either Ralph or Alice is home, then Norton will come in and make himself comfortable. Therefore, either Ralph is not home or Norton won't come in and make himself comfortable.

9. If Moe slaps Larry up the side of the head, then Larry slaps Curly up the side of the head, and Curly slaps Moe up the side of the head. Since Moe will not slap Larry up the side of the head, either Larry won't slap Curly up the side of the head or Curly won't slap Moe up the side of the head.

10. Either the sun will shine and there will be no wind, or the sun will shine and the race will be canceled. The sun will shine and the race will not be canceled. So there will be no wind.

11. If Moe slips, then if Curly slips, then Larry will slip. If Curly slips, then if Moe slips, then Larry will slip. However, neither Curly nor Moe will slip. So, Larry won't slip.

12. Dobie will be pleased if Maynard gets a job and saves some money. But if Maynard gets a job and doesn't like the job, then Dobie won't be pleased. Maynard will get a job only if he wants to. So, Maynard won't both get a job and save some money.

TESTING A PAIR OF SENTENCES FOR EQUIVALENCE

Recall from Chapter 1 that two sentences are **equivalent** just in case they never differ as far as truth and falsity are concerned. If one is true, the other will be true, if one is false, the other will be false. Consider the following two sentences:

1. It is not the case that either Ann or Bob is home.

2. Ann is not home and Bob is not home.

In symbols, these are:

1. ~ (A v B)

2. ~ A & ~ B

We saw in Chapter 3 that these two sentences are equivalent. We can now show this on a truth-table. Let us place both formulas side by side on a single table and let us fill in the final column for each formula:

A B	~ (A v B)	~ A & ~ B
T T	F	F
T F	F	F
F T	F	F
F F	T	T

Notice that these two formulas have identical or matching final columns. That is, on each row, the final values match each other. Whenever, on a row, one formula computes true, the other—on that row—also computes true; and whenever on a row one formula computes false, the other—on that row—computes false as well. When two formulas match in this way, we say they are equivalent to each other.

The Equivalence Test

To see if two formulas are equivalent, place the two formulas side by side on a single table. If the two formulas have matching final columns, they are equivalent.

▶   **EXERCISE 5.6**   ◀

Use truth-tables to determine, for each pair of sentences below, whether the sentences are equivalent or not equivalent. Note that a comma separates the members of each pair.

1. *~(P v Q) , ~P & ~Q

2. ~ ~P , P

3. ~(P & Q) , ~ P v ~Q

4. ~(P v Q) , ~ P v ~ Q

5. *~(P & Q) , ~ P & ~ Q

6. P v Q , ~ (~ P & ~ Q)

7. *~(P & ~P) , (Q & ~ Q)

8. P & (Q v R) , (P & Q) v (P & R)

9. P ⊃ Q , Q ⊃ P

10. P ⊃ Q , (~Q ⊃ ~P)

11. * P ⊃ Q , ~P v Q

12. ~(~ P v Q) , ~ (P & ~ Q)

13. ~(P ⊃ Q) , (~P ⊃ ~Q)

14. *P & P , P

15. P v P , P

16. P & Q , Q & P

17. *P v Q , Q v P

18. P ≡ Q , [(P ⊃ Q) & (Q ⊃ P)]

19. P ≡ Q , [(P & Q) v (~P & ~Q)]

20. *P ⊃ (Q ⊃ R) , (P ⊃ Q) ⊃ R

21. (P v Q) v R , P v (Q v R)

22. (P & Q) & R , P & (Q & R)

23. ~(P & Q) & R , ~P & (Q & R)

24. P v (Q & R) , (P v Q) & (P v R)

25. (P & Q) v R , P & (Q v R)

Decision Procedures

A **decision procedure**, also called an *algorithm,* is a problem-solving method with precisely specified steps. Following the steps is a matter of following precise rules, requiring no creativity or ingenuity, and the method is guaranteed to give a definite solution in a finite number of steps. The rules for "long division," for example, constitute a decision procedure that most people learn in grade school.

The truth-table technique provides us with decision procedures for a number of logical problems, including the determination of logical status, validity, equivalence, and so on. We test sentences and arguments for the various logical properties by filling in the T's and F's on a truth-table according to definite rules. In this sense, truth-functional logic is mechanical. However, the use of the truth-table, as a decision procedure, has its limits. It cannot be applied to certain types of arguments, which we will examine in future chapters.

CHAPTER 5 GLOSSARY

Analytic method   A method that involves breaking a subject down into its constituent parts and seeing how those parts fit together to constitute the whole.

Contingent sentence   (in truth-functional logic) A sentence with the following feature: the final column of its truth-table shows at least one **T** and at least one **F**.

Contradiction (in truth-functional logic) A sentence with the following feature: the final column of its truth-table shows all **F**'s.

Counterexample (to an argument) A possible circumstance in which the argument's premises would be true and its conclusion false.

Decision procedure (or "algorithm") A problem-solving method that has the following features: (a) the method has precisely specified steps; (b) following the steps is a matter of following exact rules requiring no creativity or ingenuity; and (c) the method is guaranteed to give a definite solution in a finite number of those steps.

Equivalence Two sentences are equivalent just in case they never differ as far as truth and falsity are concerned. If one is true, the other will be true, if one is false, the other will be false.

Final column In a truth-table, the column of truth-values under the main operator.

Logical falsehood A sentence that can be known to be false using only the methods of logical theory—without investigating the physical world. (Such a sentence is said to be "logically false.")

Logical truth A sentence that can be known to be true using only the methods of logical theory—without investigating the physical world. (Such a sentence is said to be "logically true.")

Tautology (in truth-functional logic) A sentence with the following feature: the final column of its truth-table shows all **T**'s.

Truth-value assignment An assignment of truth-values to the letters of a formula or group of formulas.

THE CONCEPT OF LOGICAL FORM

O ne of the goals of this chapter is to define and explain the concept of logical form in precise terms. In this chapter, it is extremely important that you keep in mind the distinction between two different kinds of symbols—variables and constants. Recall that a variable stands for anything from a specified group of things whereas a constant is a symbol that stands for one specified thing. For example, in the commutative law of addition,

$x + y = y + x$

x and y are variables that stand for, in each case, any number whatsoever. Along with a variable goes the understanding that anything from a specified group of things may be named and the name inserted in place of the variable. In truth-functional logic, a constant is used to abbreviate or stand for one specifically identified sentence and the metalinguistic variables we are using—**P, Q, R,** and **S**—stand for any sentences whatsoever.

SENTENCE FORMS

With the distinction between sentence variables and constants in mind, we may now turn to the concept of logical form. Consider the following three sentences:

a. It will rain or it will snow.

b. Jan is from Michigan or Jan is from Arkansas.

c. It is Wednesday or this is Belgium.

Symbolized within TL, these become

a. R v S

b. M v A

c. W v B

What do these three sentences of TL have in common? Each consists of a left disjunct joined by a disjunction operator to a right disjunct. This general pattern is expressed by

$$P \text{ v } Q$$

This expression, **P** v **Q**, represents a general pattern exhibited by the three TL sentences above.

The expression **P** v **Q** is called the **form** of the three TL sentences, and those three sentences are said to be **substitution instances** or *instantiations* of the form **P** v **Q**. A TL sentence will qualify as a substitution instance of a form if and only if the TL sentence can be generated from the form by replacing *only the variables* in the form with TL sentences and, in addition, making any necessary parenthetical adjustments. This will become clearer after we work through several examples.

Consider the form **P** v **Q**. Suppose we abbreviate the English sentence "Jean attended the Monterey Pop Festival in 1967" with the letter J and "Chris attended the Monterey Pop Festival in 1967" with the letter C. If we begin with the form **P** v **Q** and replace **P** with the constant J and **Q** with the constant C, we generate the substitution instance

$$J \text{ v } C$$

To record the replacements, let us write

$$\mathbf{P} / J$$
$$\mathbf{Q} / C$$

Here, **P** / J simply indicates that each occurrence of the variable **P** was replaced by the constant J and **Q** / C indicates that every occurrence of the variable **Q** was replaced by the constant C.

If we return to the form **P** v **Q** and replace **P** with the TL formula A, and if we replace **Q** with the TL formula B, we generate a new substitution instance:

$$A \text{ v } B$$

To record the replacements, we write

$$\mathbf{P} / A$$
$$\mathbf{Q} / B$$

Substitution instances can grow quite large. For instance, remaining with the same form, if we replace **P** with (A & B) and if we replace **Q** with (C v D), we generate the substitution instance

$$(A \& B) \text{ v } (C \text{ v } D)$$

To record the replacements, we write

P / A & B

Q / C v D

Next, let us replace **P** with ~ (A & ~ B) and **Q** with ~ (~ A & E):

$$\sim (A \& \sim B) \text{ v } \sim (\sim A \& E)$$

Again, we record the replacements by writing

P / ~ (A & ~ B)

Q / ~ (~ A & E)

Consider now a different form, the form ~ (**P** v **Q**). If we replace **P** with (A & B) and if we replace **Q** with (C v D), after we make the appropriate parenthetical adjustments, we derive the substitution instance

$$\sim [(A \& B) \text{ v } (C \text{ v } D)]$$

If instead we replace **P** with ~ (A & ~ B) and **Q** with ~ (~ A & E) we have a different substitution instance:

$$\sim [\sim (A \& \sim B) \text{ v } \sim (\sim A \& E)]$$

Other TL substitution instances of this form include the following:

INSTANCE:	REPLACEMENTS:	
i. ~ [(A & E) v ~(E v F)]	**P** / (A & E)	**Q** / ~ (E v F)
ii. ~ (S v ~ G)	**P** / S	**Q** / ~ G
iii. ~ (H v ~ R)	**P** / H	**Q** / ~ R

Consider the form below and the list of three substitution instances under it:

Form:	~**P** v ~ **Q**
Some instances:	~A v ~ B
	~E v ~ G
	~(A & B) v ~ (H ⊃ R)

Notice that when you go from a form to one of its substitution instances, the arrangement of the form's operators—their relative positions and scopes—remains in place. The only change occurs when the form's *variables* are replaced. The arrangement of the form's operators and their associated scope-indicating devices gives the form its character, its identity. It is this character, this arrangement of operators, that the form shares with or passes on to all of its instantiations or substitution instances. Thus, the arrangement of the form's operators constitutes the abstract pattern or structure that its instances have in common. Within the constraints of that abstract structure, the instances of a form may vary. A form therefore constitutes a sort of symbolic template, and substitution instances of a form are sentences that "fit" that template.

Notice also how the arrangement of the form's operators carries over into the instance. The relationship between a form and its instances is similar in some ways to the relationship between a cookie cutter and the cookies it produces. The cookie cutter imposes a common form on each cookie in the batch. However, within the constraints of that form, the individual cookies may vary. For instance, within the common form some might have chocolate chips, others might have cherries, and so on. Similarly, all the instances of a form share its arrangement of operators, but *within* that structure each instance of the form may have different TL sentences.

▶   **EXERCISE 6.1**   ◀

For each sentence form (a–o) list which of the numbered sentences (1–18) is an instance *of the form.*

a. **P v Q**

b.* **P & Q**

c. **~P**

d.* **~(P & Q)**

e. **~(P v Q)**

f. **~P & ~Q**

g. **P v P**

h. **~P & Q**

i.* **~P v ~Q**

j. **P ⊃ Q**

k. **~P ⊃ ~Q**

l. **(P v Q) & R**

m. **~(P & Q) v R**

n.* **P ⊃ (Q ⊃ R)**

o. **P ≡ Q**

1. ~(A & B)

2. ~A v ~B

3. (A & B) ⊃ ~C

4. ~(H ≡ S) v ~(G & B)

5. (H v B) ≡ [S ⊃ (R & H)]

6. ~A & ~B

10. ~A ≡ (S ⊃ H)

11. ~(A & B) v (H ⊃ Q)

12. [(H & B) v S] & S

13. ~(A & B) ⊃ ~(E v H)

14. ~(H v S) & (F ≡ G)

15. ~[(A v B) v C]

7. [(R v H) v G] v [(R v H) v G]   **16.** ~(A ≡ B)

8. ~[(E & S) v (F ≡ G)]   **17.** (A & B) ≡ (C v E)

9. ~[~(A v B) & ~(H v S)]   **18.** (H ⊃ S) ⊃ (F ⊃ R)

ARGUMENT FORMS

An **argument form** may be defined as a set of two or more sentence forms with one of these designated as the conclusion and the other (or others) designated as a premise (or premises). An argument form can be transformed into a symbolized argument by replacing the variables with TL sentences, and an argument formed in this way from an argument form is a **substitution instance** (or instantiation) of the argument form.

The relationship between an argument form and a symbolized argument is similar to the relationship between a sentence form and a symbolized sentence. The distinction should become clear as we identify and examine four important argument forms.

THE DISJUNCTIVE SYLLOGISM FORM

Consider the argument:

1. We'll either swim or jog.

2. We won't swim.

3. So, we'll jog.

In TL this is

1. S v J

2. ~S / J

(The premises are numbered and the symbolized conclusion is placed after the slash.)

Now consider the next argument:

1. We'll either eat or drink.

2. We won't eat.

3. So, we'll drink.

In TL this is

1. E v D

2. ~E / D

These two arguments look similar, don't they? As you can see by simple inspection, these arguments display the same abstract logical structure or logical form. We can see that form when we replace the constants with variables. In each case, if we replace constants with variables, the result is the same:

1. **P v Q**

2. **~P / Q**

This argument form appears so often in everyday reasoning that it has been given a name: the **Disjunctive Syllogism form**. (The term *syllogism* is an old-fashioned term meaning "deductive argument." This form's name comes from the fact that its first step is disjunctive in nature and the argument form is syllogistic in nature.).

Another substitution instance of the Disjunctive Syllogism logical form is

1. (A & B) v (E & F)

2. ~(A & B) / E & F

where the replacements are: **P** / (A & B), **Q** / (E & F). Each of the following is also a substitution instance of the Disjunctive Syllogism form:

1. E v G	1. (A v B) v (E ⊃ S)
2. ~E / G	2. ~(A v B) / E ⊃ S

In the first, the replacements were: (**P** / E, **Q** / G). In the second, the replacements were: (**P** / (A v B), **Q** / E ⊃ S)

Two arguments having an argument:
Argument 1 to argument 2: "That argument over there has poor form."
Argument 2 to argument 1: "I wouldn't be too critical if I were you. Your form isn't all that valid either."
Argument 1 to argument 2: "You know, it's really poor form to call my form invalid. I'm leaving you for good. May The Form be with you."

THE MODUS PONENS FORM

Consider next this argument:

1. If it is raining over the roof, then the roof is wet.

2. It is raining over the roof.

3. The roof is wet.

In TL this is symbolized as follows:

1. R ⊃ W

2. R / W

This argument is a substitution instance of the argument form

1. **P ⊃ Q**

2. **P / Q**

This form, known as the **Modus Ponens** form, appears frequently in everyday reasoning. (*Modus ponens* is Latin for "method of affirmation." This form's name reflects the fact that the second premise affirms the antecedent of the first premise.) Additional instances of the Modus Ponens form follow:

1. (A v B) ⊃ (E v F)

2. (A v B) / (E v F)

Here, the variable **P** was replaced with (A v B), and **Q** was replaced with (E v F).

1. ~A ⊃ ~(H & G)

2. ~A / ~(H & G)

Here, **P** was replaced with ~A, and **Q** was replaced with ~ (H & G).

Notice that an argument form constitutes a general form or pattern of reasoning that many individual arguments may instantiate.

THE MODUS TOLLENS FORM

Consider the following argument:

1. If it's raining, then the roof is wet.

2. The roof is not wet.

3. So, it is not raining.

In TL, with obvious abbreviations, this is:

1. R ⊃ W

2. ~W / ~R

This argument is an instance of the argument form known as **Modus Tollens**:

1. **P ⊃ Q**

2. **~Q / ~P**

(*Modus Tollens* is Latin for "method of denial." The form's name derives from the fact that the second premise is a denial of the consequent of the first premise.)

THE HYPOTHETICAL SYLLOGISM FORM

In everyday reasoning we often link several things together in a chain of if-then links. For instance:

1. If it rains, then the roof gets wet.

2. If the roof gets wet, then the ceiling leaks.

3. So, if it rains, then the ceiling leaks.

In TL, with obvious abbreviations, this becomes

1. R ⊃ W

2. W ⊃ C / R ⊃ C

This argument instantiates the **Hypothetical Syllogism** argument form:

1. **P ⊃ Q**

2. **Q ⊃ R / P ⊃ R**

Here is another instance of this form:

1. If Moe slaps Curley, then Curley will jab Larry.

2. If Curley jabs Larry, then Larry will hit Moe.

3. So, if Moe slaps Curley, then Larry will hit Moe.

In TL, with the usual abbreviations, this is:

1. M ⊃ C

2. C ⊃ L / M ⊃ L

VALID ARGUMENT FORMS

Some argument forms are valid forms, others are invalid forms. In truth-functional logic, an argument form is a **valid argument form** if its truth-table has no row showing all true premises and a false conclusion. To test an argument form for validity, construct a truth-table for the form, write the form across the top of the table, and fill in the truth-values for each cell of the table. Look for a row showing the possibility of true premises and false conclusion. If you find *no* such row, the form is a valid argument form. If you find one or more such rows, the form is an **invalid argument form**.

The Modus Ponens and Disjunctive Syllogism argument forms are both valid argument forms, as the two tables below show:

Disjunctive Syllogism				Modus Ponens			
P Q	P v Q	~P	Q	P Q	P ⊃ Q	P	Q
T T	T	F	T	T T	T	T	T
T F	T	F	F	T F	F	T	F
F T	T	T	T	F T	T	F	T
F F	F	T	F	F F	T	F	F

Notice that neither table contains a row showing true premises and a false conclusion, which proves that each is a valid form.

The Modus Tollens form is proven valid by the following table:

P Q	P ⊃ Q	~Q	~P
T T	T	F	F
T F	F	T	F
F T	T	F	T
F F	T	T	T

The table contains no row with the true premises–false conclusion combination; the form is a valid form.

The Hypothetical Syllogism form is shown valid by the following table:

P Q R	P ⊃ Q	Q ⊃ R	P ⊃ R
T T T	T	T	T
T T F	T	F	F
T F T	F	T	T
T F F	F	T	F
F T T	T	T	T
F T F	T	F	T
F F T	T	T	T
F F F	T	T	T

The table contains no row with the true premises–false conclusion combination. This form, too, is a valid form.

Testing an Argument Form for Validity

Write the form across the top of a truth-table and fill in the rows. If no row shows the all true-premises-false conclusion combo, then the form is a valid form. If at least one row shows the all true premises and a false conclusion combo, the form is an invalid form.

▶ **IF THE FORM FITS, YOU MUST ADMIT, IT'S VALID**

If a particular argument instantiates a valid argument form, the argument itself must be valid, for *every substitution instance of a valid argument form is a valid argument.* Here is why: The arrangement of operators in the form carries over into every substitution instance of the form. This arrangement of operators determines the final columns on the truth-table. So, the arrangement of operators determines the final columns of the form's table as well as the final columns for the tables of each instance of the form. Thus, the final columns on the truth-table of any instance of the form will match the final columns on the form's truth-table. One consequence of this is that every instantiation of a valid form will itself be a valid argument.

In logic, we study valid argument forms because these serve as guides to help us reason properly. We can be sure our reasoning is valid if it instantiates a valid form of argument.

INVALID ARGUMENT FORMS

Invalid argument forms also exist, and because some of these appear deceptively similar to certain valid forms, invalid forms are easy to confuse with valid forms. For instance, the following invalid argument form looks deceptively like the Modus Ponens form:

1. $P \supset Q$

2. $Q \; / \; P$

Here is an instance of this form of reasoning:

1. If Freddie is a member of the Mafia, then Freddie is Italian.

2. Freddie is Italian.

Therefore, Freddie is a member of the Mafia.
Symbolized in TL, this argument is

1. $F \supset I$

2. $I \; / \; F$

You can easily verify the invalidity of this argument form by performing a simple truth-table test. Notice that the third row of the form's truth-table shows true premises and a false conclusion:

P Q	P ⊃ Q	Q	P
T T	T	T	T
T F	F	F	T
F T	T	T	F
F F	T	F	F

In the case of an invalid argument form, some (but not all) substitution instances will be invalid arguments.

A **fallacy** is an error in reasoning that may nevertheless appear correct to some even though it is incorrect. The above invalid form is named the **fallacy of affirming the consequent,** because the second premise "affirms" the consequent of the first premise. Be careful not to confuse Modus Ponens with the fallacy of affirming the consequent.

For another example, the following invalid argument form, the **fallacy of denying the antecedent,** looks deceptively similar to Modus Tollens:

1. **P ⊃ Q**

2. **~P / ~Q**

For instance, consider the argument:

1. If Chris swims, then Dave will swim.

2. But Chris won't swim.

3. So, Dave won't swim.

Symbolized in TL, this is an instance of the fallacy of denying the antecedent:

1. C ⊃ D

2. ~C / ~D

Once again, you may easily prove the invalidity of this form with a simple truth-table. Be careful not to confuse this form with the Modus Tollens form.

These two fallacies, the fallacy of affirming the consequent, and the fallacy of denying the antecedent, are **formal fallacies** because the logical error has to do with the general *form* of the argument rather than the content of the argument. (*Informal* fallacies are the topic of Chapter 13.) When arguing within a natural language, one must constantly be on the lookout for invalid or fallacious patterns of reasoning, because what appears at first glance an instance of a valid argument form sometimes turns out to be an invalid pattern of reasoning.

CONCLUDING COMMENTS

▶ THE VALUE OF ABSTRACTION

An argument form is an abstraction. It is a pattern abstracted from an argument. When we study an argument form after it has been abstracted from a specific natural language argument, we can see characteristics we might not see if it were left embedded within that particular stretch of language. By studying abstract logical forms, modern logic—sometimes called "formal" logic—can teach us much about the inner structure of an argument that we might not learn otherwise.

When we abstract the *form* from a specific natural language argument, we move from the particular to the general. That is, we shift our analysis from the particular features of an individual argument, expressed within a particular language and within a particular context, to a general form that many arguments in many languages and in many contexts might share. The particular features of a natural language argument are those that distinguish one specific argument, expressed within a particular context, from other arguments. The process of going from the particular to the general is an important part of modern thought; it can be found in just about every subject. It is also an important part of modern formal logic.

▶ TRUTH-FUNCTIONAL LOGIC IS FORMAL

Most everyday arguments are expressed within a natural language. That is, we normally put our reasoning into the words and sentences of the particular linguistic community we belong to. In **informal logic,** the student learns how to recognize and evaluate arguments as they appear within a natural language. Consequently, in an informal logic course few logical symbols are used. However, in **formal logic**, arguments are studied from a more abstract or general perspective, one that involves concentrating on abstract argument forms rather than on specific natural language arguments. By studying argument forms, we can analyze patterns of reasoning common to many arguments across many natural languages. However, argument forms are highly abstract. In order to study them efficiently, we must use an artificial symbolic language. Thus, modern formal logic employs special symbols and is sometimes also called "symbolic" logic.

Before the nineteenth century, logicians used few symbols in their theorizing. Because logic is concerned with arguments, and because arguments are typically expressed within natural languages, logicians mainly studied arguments that were expressed within a natural language. However, in the late nineteenth century and during the first part of the twentieth century, Gottlob Frege and Bertrand Russell pioneered the development of symbolic artificial languages for logical theory. These languages allowed logicians to simplify expressions of extremely complex ideas, and this led to enormous advances in logical theory. Incidentally, the study of symbolic logic also greatly advanced our understanding of language and it also laid the conceptual foundations of computer science. Thus, linguistics makes heavy use of symbolic logical languages, and computer science simply could not exist as a subject of study were it not for the development of symbolic logical languages.

► **EXERCISE 6.2** ◄

Use truth-tables to determine which of the following are valid argument forms and which are not.

1. * P ⊃ Q

    Q / P

2. P ⊃ Q / Q ⊃ P

3. P v Q

    Q / P

4. P ⊃ Q

    ~P / ~ Q

5. * P ⊃ Q / ~Q ⊃ ~P

6. P ⊃ Q / ~ P v Q

7. * P v Q / Q v P

8. (P & Q) ⊃ R / P ⊃ (Q ⊃ R)

9. ~(P v Q) / ~P

10. P ⊃ Q

    P ⊃ R / Q ⊃ R

11. * P ⊃ Q

    Q ⊃ P / P ≡ Q

12. P ⊃ Q

    P ⊃ R / P ⊃ (Q & R)

13. ~ P v ~Q / ~ (P v Q)

14. ~P & ~Q / ~ (P & Q)

15. ~ (P & Q) / ~ P & ~ Q

16. * P v (Q & R) / (P v Q) & (P v R)

17. P ⊃ R

   Q ⊃ R / (P v Q) ⊃ R

18. ~(P & Q) / ~ P v ~ Q

19. ~(P v Q) / ~ P & ~ Q

20. P & (Q v R) / (P v Q) & (P v R)

21. * P ⊃ Q

   R ⊃ S

   P v R / Q v S

22. P ≡ Q / (P ⊃ Q) & (Q ⊃ P)

23. P & ~P / Q

24. * P v Q

   Q v R / P v R

CHAPTER 6 APPENDIX: *Contradictory and Tautological Sentence Forms*

Consider the truth-table for the sentence form **P v ~P**:

P	P v ~ P
T	T T FT
F	F T TF
	*

An individual sentence form is a **tautological form** if and only if the final column of its truth-table displays all **T**'s. In this case, the table proves that **P v ~P** is a tautological form.

   Another tautological form is the form ~ (**P & ~P**):

P	~(P & ~P)
T	T T F F T
F	T F F T F
	*

Consider next the truth-table for the sentence form **P** & ~**P**:

P	P & ~P
T	T F F T
F	F F T F
	*

An individual sentence form is a **contradictory form** if and only if the final column of its truth-table displays all **F**'s. The table above thus proves that **P** & ~**P** is a contradictory form.

IF THE FORM FITS

If you reflect upon the relationship between a sentence form and its instances, you will see that every substitution instance of a tautological form will be tautological and every instance of a contradictory form will be contradictory. This is because the final column of the form's truth-table is determined by the configuration of the form's operators. This configuration of operators carries over into each of the form's instances in such a way that the final column of each instance of a form will always match the form's final column. So, if a sentence form's final column is all **T**'s (or all **F**'s), each of its instances will have a final column that is all **T**'s (or all **F**'s).

Let us now put these ideas to work. What is the logical status of the following complicated TL sentence?

{[(A & ~B) v ~(C & D)] v (E & U)} v ~{[(A & ~B) v ~(C & D)] v (E & U)}

That is, is this sentence tautological, contradictory, or contingent? The truth-table method requires that we construct a table for the sentence and see whether the final column contains all **T**'s, all **F**'s, or a mixture of **T**'s and **F**'s. If we were to construct a table for this sentence, the table would have sixty-four rows. That's a lot of **T**'s and **F**'s. However, our understanding of forms makes possible a simpler solution. A careful inspection of the sentence reveals that it is an instance of the form **P** v ~**P**. As we saw above, this form is tautological. Since we know that every instance of a tautological form is a tautology, it follows that the very complicated TL sentence above is tautological. It thus has the truth-value true.

What is the logical status of the following sentence?

{[(~A v ~B) & (~C v ~S)] & (~E & ~U)} &
~{[(~A v ~B) & (~C v ~S)] & (~E & ~U)}

If we were to construct a table to test this sentence, the table would again have sixty-four rows. However, inspection of this sentence reveals that it is an instance of the form **P** & ~**P**. A little two-row table shows that this form is a contradictory form. Because this long sentence is an instance of a contradictory form, it follows that it is itself contradictory.

Sentence forms that are neither tautologous nor contradictory are said to be **indeterminate forms**. The final column on the truth-table of an indeterminate form will have a mixture of **T**'s and **F**'s. We do not call such forms "contingent" because not every substitution instance of such a form will be a contingent sentence.

| ▶ | **EXERCISE 6.3** | ◀ |

Without using truth-tables, determine in each case whether each sentence is a tautology or a contradiction.

1.   {[(A v B) & ~D] v N} v ~{[(A v B) & ~D] v N}

2.* {[(A v B) & ~D] v N} & ~{[(A v B) & ~D] v N}

3.   ~{[(A v B) & ~D] v ~[(A v B) & ~D]} v {[(A v B) & ~D] v ~[(A v B) & ~D]}

CHAPTER 6 GLOSSARY

Argument form   A general logical pattern or structure that many arguments may have in common. More specifically, a set of two or more sentence forms, one of which is designated as the conclusion and the other (or others) designated as a premise (or premises). An argument form may be expressed by writing the formulas for its component sentence forms and designating the conclusion and the premise or premises.

- **Invalid argument form**   An argument form for which the following is the case: not every substitution instance of the form is valid. In truth-functional logic, an invalid argument form has a truth-table that contains at least one row with true premises and a false conclusion.

- **Valid argument form**   An argument form for which the following is the case: every substitution instance of the form is valid. In truth-functional logic, a valid argument form has a truth-table that contains no row with true premises and a false conclusion.

Fallacy   An error in reasoning that may appear correct to some even though it is incorrect.

Sentence form   A general logical pattern or structure that many sentences may have in common. A sentence form is expressed using a combination of variables and operators.

- **Contradictory sentence form**   A sentence form all of whose instances are contradictions.

- **Indeterminant sentence form**   A sentence form that is neither contradictory nor tautologous.

- **Tautological sentence form**   A sentence form all of whose instances are tautologies.

- **Substitution instance**   (of an argument form) An argument is a substitution instance of an argument form if and only if the argument can be generated from the form by replacing only the variables in the form with TL sentences and, in addition, making any needed parenthetical adjustments.

- **Substitution instance**   (of a sentence form in TL) A TL sentence is a substitution instance of a sentence form if and only if the TL sentence can be generated from the form by replacing *only the variables* in the form with TL sentences and, in addition, making any necessary parenthetical adjustments.

Chapter 7

TRUTH-FUNCTIONAL NATURAL DEDUCTION

The truth-table method from Chapter 5 provides us with decision procedures for truth-functional logic, procedures so mechanical that they are easily written into computer programs. However, truth-tables have one major drawback. As we evaluate bigger and bigger formulas, the size of the truth-table increases exponentially. A formula with two atomic components may require only a four-row table, but a formula containing four atomic components requires a sixteen-row table. Six components call for a sixty-four-row table, and so on. Thus, when we try to evaluate complex formulas and complicated arguments, the truth-tables become so large that the method becomes impractical.

Several alternative methods have been developed that accomplish some of the same purposes but that are easier to use when it comes to complicated arguments. Two alternatives to the truth-tables are: (a) the method of natural deduction and (b) the truth-tree method. Logic instructors typically introduce truth-tables first, and then, in order to deal with the more complicated arguments, employ one of the two alternatives—either truth-trees or natural deduction.

The **natural deduction method,** which is the subject of this chapter, gets its name from the fact that with this method we *deduce* a conclusion from a set of premises through a series of valid inferences corresponding to *natural* patterns of reasoning. These patterns of reasoning are "natural" in the sense that they resemble patterns we follow in everyday thought. The patterns of reasoning we will follow are spelled out in the form of inference rules.

We say a person "draws an inference" when the person asserts a conclusion on the basis of one or more premises. An inference is a **valid inference** just in case the argument formed out of the premises and conclusion of the inference constitutes a valid argument. In other words, in the case of a valid inference, if the inference's premises are true, the inference's conclusion must be true. An

inference rule is a rule specifying that a particular conclusion may be inferred when certain premises are given. A **valid inference rule** is an inference rule that has the following feature: When the rule is followed, the result is always a valid inference. So, if an inference is made according to a valid inference rule, the inference's conclusion must be true if the inference's premises are true.

In Chapter 6, we studied the concept of a valid argument form and we looked at four valid forms: Disjunctive Syllogism, Modus Ponens, Modus Tollens, and Hypothetical Syllogism. Each of these argument forms was shown valid on a truth-table. The following four valid inference rules correspond to these four valid argument forms.

THE DISJUNCTIVE SYLLOGISM RULE

Consider the following valid argument:

1. Either we will watch the Austin Powers movie or we'll watch the Blues Brothers movie.

2. We won't watch Austin Powers.

3. Therefore, we will watch the Blues Brothers movie.

In TL, using obvious abbreviations, this becomes

1. A v B

2. ~A / B

The symbolized premises are numbered and the symbolized conclusion has been placed after the slanted slash mark. This argument is an *instance* of the valid argument form Disjunctive Syllogism:

1. **P v Q**

2. **~P / Q**

Recall from Chapter 6 that if a particular argument is a substitution instance (an "instance") of a valid form, the argument must be valid. Every instance of a valid form is valid. The inference rule in the following box reflects the valid argument form Disjunctive Syllogism.

> ### *The Disjunctive Syllogism Rule (DS)*
>
> From a sentence of the form **P v Q** and a corresponding sentence **~P**, you may infer the corresponding sentence **Q**.

In other words, from a disjunction and the negation of the left disjunct, you may infer the other disjunct. The Disjunctive Syllogism rule will be abbreviated as follows:

$$P \vee Q$$

$$\frac{\sim P}{Q}$$

This symbolic abbreviation can simply be read, "From a sentence of the form **P** v **Q** and a corresponding sentence ~**P**, you may infer the corresponding sentence **Q**." The virtue of the symbolic abbreviation is that it is compact and it closely mirrors the actual inferences you will be making—inferences that move in a downward direction.

In each of the following examples, the Disjunctive Syllogism rule was applied to the first two sentences (the two above the line), and the third sentence (below the line) was then deduced in accord with the rule.

1. A v B	**1.** (A & B) v (E & F)	**1.** ~ (A ⊃ B) v ~ (H ≡ S)
2. ~ A	2 ~ (A & B)	**2.** ~ ~ (A ⊃ B)
3. B	**3.** E & F	**3.** ~(H ≡ S)

We can represent this rule in other ways, too. For instance, if we let a square represent the left disjunct and a circle the right disjunct, we could put the rule this way:

From: □ v ○

          ~ □

Infer: ○

With this diagram in mind, begin with the following two sentences:

1. E v (A & B)

2. ~ E

E is the square, (A & B) is the circle, and the rule tells you to derive the circle, that is, (A & B):

1. E v (A & B)

2. ~ E

3. (A & B)

It's as simple as 1, 2, 3!

We naturally reason in accord with this inference rule—whether we are aware of it or not—from an early age. For instance, suppose a little child is told that today the family will either stay home *or* the family will go to the circus. If Mom or Dad announces, "We are not going to stay home," nobody needs to tell the child what inference to draw. The child immediately gets ready to go to the circus.

THE MODUS PONENS RULE

This argument is clearly valid:

1. If it's raining over the roof, then the roof is wet.

2. It's raining over the roof.

3. Therefore, the roof is wet.

Using obvious abbreviations, this argument may be expressed in TL as

1. $R \supset W$

2. R / W

This instantiates the valid argument form known as Modus Ponens:

1. $\mathbf{P} \supset \mathbf{Q}$

2. $\mathbf{P} / \mathbf{Q}$

The inference rule in the following box corresponds to the Modus Ponens argument form.

The Modus Ponens Rule (MP)

From a sentence of the form $\mathbf{P} \supset \mathbf{Q}$ and a corresponding sentence $\mathbf{P}$, you may infer the corresponding sentence $\mathbf{Q}$.

In other words, from a conditional and the antecedent of that conditional, we may infer the consequent of the conditional. The Modus Ponens rule will be abbreviated as:

$$\mathbf{P} \supset \mathbf{Q}$$
$$\frac{\mathbf{P}}{\mathbf{Q}}$$

In each of the following examples, we applied the Modus Ponens rule to the two sentences above the line and deduced the sentence below the line in accord with the rule.

1. $A \supset B$	**1.** $(A \& B) \supset (E \vee F)$	**1.** $(A \supset E) \supset (S \vee H)$
2. A	**2.** $(A \& B)$	**2.** $(A \supset E)$
3. B	**3.** $E \vee F$	**3.** $S \vee H$

We can represent this rule in other ways, too. For instance, if we let a square represent the antecedent and a circle the consequent, we could put the rule this way:

From: □ ⊃ ○
           □
Infer: ○

With this diagram in mind, begin with the following two sentences:

1. (A & B) ⊃ C

2. (A & B)

The formula (A & B) is the square, C is the circle, and the rule tells you to derive the circle, that is, C:

1. (A & B) ⊃ C

2. (A & B)

3. C

We also naturally think in accord with this inference rule from an early age. For instance, suppose a little child is told that *if* the family stays home and works in the yard today, everyone gets to go on a picnic tomorrow. So, the family works in the yard, and after the yard is finished, the child naturally expects to go on a picnic.

THE MODUS TOLLENS RULE

The following argument is valid, although it may not appear to be valid at first glance:

1. If it's raining over the roof, then the roof is wet.

2. The roof is *not* wet.

3. So, it is *not* raining over the roof.

In TL, with obvious abbreviations, this is:

1. R ⊃ W

2. ~ W / ~ R

If the two premises are both true, the conclusion must be true. This argument is an instance of the argument form Modus Tollens:

1. P ⊃ Q

2. ~Q / ~P

This valid argument form gives rise to the valid inference rule shown in the following box:

The Modus Tollens Rule (MT)

From a sentence of the form **P ⊃ Q** and a corresponding sentence ~ **Q,** you may infer the corresponding sentence ~ **P.**

In other words, from a conditional along with the negation of the consequent of the conditional, we may infer the negation of the antecedent. The Modus Tollens rule will be abbreviated as

$$\mathbf{P \supset Q}$$

$$\frac{\mathbf{\sim Q}}{\mathbf{\sim P}}$$

In each of the following examples, we applied the Modus Tollens rule to the first two sentences and deduced the third sentence in accord with the rule.

1. A ⊃ B	**1.** (A & B) ⊃ (E & F)	**1.** ~ (A & B) ⊃ ~ (S v H)
2. ~ B	**2.** ~ (E & F)	**2.** ~ ~ (S v H)
3. ~ A	**3.** ~ (A & B)	**3.** ~ ~ (A & B)

We can represent this rule in other ways, too. For instance, if we let a square represent the antecedent and a circle the consequent, we could put the rule this way:

From: □ ⊃ ○

            ~ ○

Infer: ~ □

Keeping this diagram in mind, begin with the following two sentences:

1. A ⊃ E

2. ~E

The formula A is the square, E is the circle, and the rule tells you to derive the negation of the square, that is, ~A:

1. A ⊃ E

2. ~E

3. ~A

We do often reason this way. Imagine reports that the King of Ruritania has stepped down and has fled the country. Furthermore, the government and state-run media are rumored to be in the hands of his opponents. Now, assume it is true that, if the King remains in power, he will appear on the six o'clock state television program tonight. Assume next that you turn on your television and the king is not on the six p.m. state television program. It definitely follows that the king no longer remains in power. This natural inference is an example of Modus Tollens.

THE HYPOTHETICAL SYLLOGISM RULE

In everyday reasoning, we often weave several things together in a chain of if–then links. For instance:

1. If it rains, then the roof gets wet.

2. If the roof gets wet, then the ceiling leaks.

3. So, if it rains, then the ceiling leaks.

In TL this is

1. $R \supset W$

2. $W \supset C / R \supset C$

This argument instantiates the natural and valid argument form Hypothetical Syllogism:

1. $P \supset Q$

2. $Q \supset R / P \supset R$

The inference rule corresponding to this argument form is presented in the following box.

The Hypothetical Syllogism Rule (HS)

From a premise of the form $P \supset Q$ and a corresponding premise $Q \supset R$, you may infer $P \supset R$.

In other words, if we begin with two conditionals that are such that the consequent of the first one matches the antecedent of the second, we may infer a third conditional whose antecedent is the antecedent of the first conditional and whose consequent is the consequent of the second conditional. The Hypothetical Syllogism rule will be abbreviated as follows:

$$P \supset Q$$
$$\underline{Q \supset R}$$
$$P \supset R$$

In each of the following examples, we applied the Hypothetical Syllogism rule to the first two sentences and deduced the third sentence in accord with the rule.

1. A ⊃ B	1. (A & B) ⊃ (E v F)	1. ~ (A v B) ⊃ ~ (S & H)
2. B ⊃ C	2. (E v F) ⊃ (G & H)	2. ~ (S & H) ⊃ R
3. A ⊃ C	3. (A & B) ⊃ (G & H)	3. ~ (A v B) ⊃ R

Another way to put this rule is:

From: □ ⊃ ○
           ○ ⊃ △
Infer: □ ⊃ △

Keeping this diagram in mind, begin with the following two sentences:

1. A ⊃ E

2. E ⊃ S

The formula A is the square, E is the circle, S is the triangle, and the rule tells you to derive square ⊃ triangle, that is, A ⊃ S:

1. A ⊃ E

2. E ⊃ S

3. A ⊃ S

We reason in accord with this form in everyday life. For instance, suppose that *if* George orders a double cheeseburger, *then* Joe will order one. Suppose also that if Joe orders a double cheeseburger, Fred will order one, too. If so, what can you expect if George does order a double cheeseburger? Let us now put these rules to work.

PROVING THAT A CONCLUSION VALIDLY FOLLOWS

In the following symbolized argument, the premises are numbered and the conclusion is the formula placed after the slanted slash mark at the end.

1. A ⊃ B

2. B ⊃ (C v D)

3. A

4. ~C / D

This happens to be a valid argument—the conclusion must be true *if* the premises are true. To prove the conclusion must be true if the premises are true, we might naturally reason as follows: First, assuming premises 1 and 3, it follows that B must be true. Indeed, this inference is in perfect accord with the MP rule. So we write

5. B (Reasoning from 1 and 3 in accord with the MP rule.)

Next, assuming lines 2 and 5, it follows that C v D is true. This inference also follows the MP rule. So we write

6. C v D (Reasoning from 2 and 5 in accord with the MP rule.)

Assuming lines 4 and 6, it follows that D is true. This inference follows the DS rule. So we write

7. D (From 4 and 6 in accord with the DS rule.)

Now, each of these steps—5 through 7—follows in a valid way from one or more of the preceding steps. Therefore, if the premises are true, each succeeding step must be true. It follows that D must be true if the premises are true, which is to say that *if* the premises are true the conclusion must be true. The argument is therefore valid. We deduced the conclusion from the premises in accord with valid rules of inference and this proved that the argument is valid.

Here is another example. Consider the following short argument: Either Annette will swim or Bobby will swim. But Annette won't swim. If Bobby swims, then Darlene will swim. And if Darlene will swim then Ernest won't swim. If Henrietta will be the lifeguard, then Ernest will swim. Therefore, Henrietta will not be the lifeguard.

Using obvious abbreviations, this argument may be expressed within TL as

1. A v B

2. ~A

3. B ⊃ D

4. D ⊃ ~E

5. H ⊃ E / ~ H

The premises are numbered, and the symbolized conclusion follows the slash mark. We can reason our way through the argument as follows:

Step 1 Look at the first two premises of the argument. What logically follows from just these two sentences? From the sentence A v B and the sentence ~ A, the sentence B validly follows according to the Disjunctive Syllogism rule. That is, if we apply the Disjunctive Syllogism rule to premises 1 and 2, the rule allows us to derive the sentence B. The inference from A v B and ~ A, to B is therefore a valid inference.

Step 2 We just deduced sentence B from the first two premises. Now, consider premises 3 and 4. What validly follows from these two? From the sentence

B ⊃ D and the sentence D ⊃ ~ E the sentence B ⊃ ~ E validly follows according to the Hypothetical Syllogism rule. That is, if we apply the Hypothetical Syllogism rule to premises 3 and 4, we are allowed to derive B ⊃ ~ E. We've now deduced B ⊃ ~ E from the premises. So far, the line of reasoning we've developed is the following:

1. A v B

2. ~ A

3. B ⊃ D

4. D ⊃ ~E

5. H ⊃ E / ~ H

6. B   (from lines 1, 2 according to DS)

7. B ⊃ ~ E   (from lines 3, 4 according to HS)

When we write down the rule by which an inference is made and the lines the rule was applied to when we made the inference, we are stating the **justification** for the inference. The justifications for the inferences made at lines 6 and 7 are written in parentheses to the right of those two lines.

Step 3   Next, consider lines 6 and 7. From the sentence B and the sentence B ⊃ ~ E, what follows? Sentence ~E follows according to the Modus Ponens rule. That is, when we apply the Modus Ponens rule to lines 6 and 7 we deduce the sentence ~ E. Therefore, the inference from sentence B and sentence B ⊃ ~ E, to sentence ~ E, is a valid inference. We've now deduced ~E from the premises.

1. A v B

2. ~ A

3. B ⊃ D

4. D ⊃ ~ E

5. H ⊃ E / ~ H

6. B   (from 1, 2 by DS)

7. B ⊃ ~ E   (from 3, 4 by HS)

8. ~ E   (from 6, 7 by MP)

Step 4   Finally, notice that lines 5 and 8 match the premise section of the Modus Tollens rule. If we apply Modus Tollens to lines 5 and 8, we may derive ~ H. Thus, from lines 5 and 8, ~ H validly follows according to Modus Tollens:

9. ~ H (from 5, 8 by MT)

We began with the premises of the argument and, following only valid rules of inference, reasoned to the argument's conclusion, ~H. This proves that the argument is valid. Here is why: If the premises of the argument—lines 1 through 5—are true, certainly lines 1 and 2 are true, because 1 and 2 are premises of the argument. If lines 1 and 2 are true, line 6 must be true, because 6 follows from 1 and 2 according to a valid rule of inference. So, if 1 through 5 are true, 6 must be true. If 1 through 6 are true, 7 must be true, because 7 follows from just lines 3 and 4 by a valid rule of inference. If 1 through 7 are true, 8 must be true, because 8 follows from 6 and 7 by a valid rule of inference. Finally, if 1 through 8 are true, 9, which is the argument's conclusion, must be true, because 9 follows from just 5 and 8 by a valid rule of inference. Therefore, if the premises are true, the conclusion must be true, which is to say that the argument is a valid argument. In general:

> *If the conclusion of an argument is deduced from the premises, using only valid rules of inference, then the argument must be a valid argument.*

THE SYSTEM TD

In this unit, you will learn how to use a natural deduction system to prove truth-functional arguments valid and to prove sentences tautological. A **natural deduction system** consists of two elements: a formal language and a set of deduction rules. The natural deduction system **TD** (for "truth-functional deduction"), with which we will work here, consists of two parts:

1. The language TL

2. A set of deduction rules

The set of deduction rules for TD consists of the inference rules and replacement rules introduced in Chapters 7–10.

A **proof in TD** is a sequence of sentences of TL, each of which is either a premise or an assumption or follows from one or more previous sentences according to a TD inference or replacement rule, and in which (1) every line (other than a premise) has a justification and (2) any assumptions have been discharged. (We will explain the use of assumptions in Chapter 9.) The conclusion of a proof is the last line of the proof. An argument is **valid in TD** if and only if there exists a proof in TD whose premises are the premises of the argument and whose conclusion is the conclusion of the argument. An argument is **invalid in TD** if and only if it is not valid in TD.

PROOFS

We are finally ready to begin constructing proofs. The following are examples of proofs that employ the first four inference rules. In each of the following proofs, the numbered sentences above the line are the premises and the sentence after the slash is the argument's conclusion. Each line derived from the premises is written under the line and the justifications are written to the right of each derived line.

(1) 1. A ⊃ B

  2. B ⊃ C

  3. ~C / ~A

  4. A ⊃ C HS 1, 2

  5. ~A   MT 3, 4

Notice that this proof began (at line 4) with the application of HS to lines 1 and 2. Look at HS, then look at just lines 1 and 2 of this proof, and see that the lines fit the HS rule's premise section. Next, look at line 4, the line being justified, and notice that it has been derived simply by following the HS rule and deriving the specified conditional. At the next line, (5), MT has been applied to lines 3 and 4 and ~A has been derived. Since this is the argument's conclusion, the proof is complete.

(2) 1. A v B

  2. ~H ⊃ ~ A

  3. H ⊃ J

  4. ~J / B

  5. ~ H   MT 3, 4

  6. ~ A   MP 2, 5

  7. B   DS 1, 6

In this proof, the opening move was made on line 5, where ~ H was deduced from lines 3 and 4. Lines 3 and 4 fit the MT pattern: line 3 is an instance of **P ⊃ Q** and line four instantiates the ~ **Q**. This allows us to infer ~H, which instantiates the ~**P**. Line 6 was derived by applying MP to lines 2 and 5. Notice that 2, 5, and 6 fit the MP pattern. The proof is finished at line 7, because the conclusion has been successfully derived.

(3) 1. A ⊃ (B ⊃ E)

  2. S ⊃ (E ⊃ J)

  3. F v A

  4. ~ F

  5. S / B ⊃ J

  6. A   DS 3, 4

  7. B ⊃E   MP 1, 6

  8. E ⊃ J   MP 2, 5

  9. B ⊃ J   HS 7, 8

It is important that you understand the following: When these inference rules are applied, they must be applied to *whole* lines only. For example, if MP is to be applied to two lines, one of those lines must—as a whole—be an instance of the form **P ⊃ Q** and the other line—as a whole—must be the antecedent of the first line. Notice that in proof (3) above, B ⊃ E and E ⊃ J on lines 1 and 2 "fit" the Hypothetical Syllogism pattern. However, when HS is applied, it must be applied to whole lines. Therefore, if we are to apply HS to B ⊃ E and E ⊃ J, those formulas cannot be parts of bigger formulas, as they are on lines 1 and 2. Rather, each must constitute an entire line by itself. Consequently, in proof (3) we first derived A by DS. Then, at line 7, we applied Modus Ponens to lines 1 and 6, detached B ⊃ E from line 1, and brought it down to line 7. We then used Modus Ponens to bring E ⊃ J down onto its own line, line 8. When B ⊃ E and E ⊃ J were situated on lines of their own (lines 7, 8), we then applied Hypothetical Syllogism to those two lines to derive the conclusion, B ⊃ J.

Also, when we apply an inference rule, the order in which the premises make their appearances in the argument does not matter. That is, HS requires that we have a sentence of the form **P ⊃ Q** and a corresponding sentence **Q ⊃ R**. But the **P ⊃ Q** might appear above the **Q ⊃ R**, or the **Q ⊃ R** might appear above the **P ⊃ Q**. In other words, HS is applied correctly in the following two arguments:

1. A ⊃ B	1. B ⊃ C
2. B ⊃ C / A ⊃ C	2. A ⊃ B / A ⊃ C
3. A ⊃ C HS 1,2	3. A ⊃ C HS 1,2

▶   **EXERCISE 7.1**   ◀

In each proof below, the conclusion is the formula that follows the slant-ed slash mark and the premises are the numbered formulas that precede the conclusion. These proofs are complete except that the justifications for the derived lines have not been filled in. Using the first four rules, supply a justification for each derived line by filling in the appropriate rules and line numbers.

(1)*1. J ⊃ (I & R)

  2. (I & R) ⊃ S

  3. (J ⊃ S) ⊃ ~G

  4. G v H / H

  5. J ⊃ S

  6. ~G

  7. H

(2) **1.** J v ~ (I v R)

    2. S ⊃ ~ J

    3. H v S

    4. ~H

    5. ~(I v R) ⊃ A / A

    6. S

    7. ~ J

    8. ~(I v R)

    9. A

(3) **1.** J ⊃ (S & I)

    2. ~ J ⊃ (B v R)

    3. ~ (S & I)

    4. (B v R) ⊃ (I v H) / I v H

    5. ~ J

    6. B v R

    7. I v H

(4) **1.** S v ~ (H & S)

    2. S ⊃ G

    3. J ⊃ ~ G

    4. M ⊃ (H & S)

    5. J / ~ M

    6. ~ G

    7. ~ S

    8. ~ (H & S)

    9. ~ M

(5)*1. (H & S) ⊃ ~ (F v G)

    2. M v (H & S)

    3. M ⊃ R

    4. ~ R / ~ (F v G)

    5. ~ M

    6. H & S

      7. ~ (F v G)

(6) **1.** (H & G) ⊃ (F ⊃ R)

     2. J & I

     3. A v ~ (F ⊃ R)

     4. ~A / ~ (H & G)

     5. ~ (F ⊃ R)

     6. ~ (H & G)

(7)* **1.** ~ H v ~ Z

     2. ~ (S ≡ G) ⊃ (F v S)

     3. ~ ~ H

     4. ~ (F v S) ⊃ Z / ~ ~ (F v S)

     5. ~ Z

     6. ~ ~ (F v S)

(8) **1.** A ⊃ (J v S)

     2. B ⊃ ~ J

     3. S ⊃ (G ⊃ H)

     4. H ⊃ R

     5. A

     6. B / G ⊃ R

     7. J v S

     8. ~ J

     9. S

    10. G ⊃ H

    11. G ⊃ R

(9) **1.** (A & B) ⊃ ~ S

     2. A & B

     3. S v [H ⊃ (F ⊃ M)]

     4. ~ M

     5. H / ~ F

     6. ~ S

     7. H ⊃ (F ⊃ M)

    8. F ⊃ M

    9. ~ F

(10)*** 1.** ~ A

    2. ~ A ⊃ [~ A ⊃ (E ⊃ A)] / ~ E

    3. ~ A ⊃ (E ⊃ A)

    4. E ⊃ A

    5. ~ E

(11) 1. A ⊃ B

    2. B ⊃ R

    3. (A ⊃ R) ⊃ G / G

    4. A ⊃ R

    5. G

(12) 1. J

    2. ~ G

    3. J ⊃ (A ⊃ ~ B)

    4. G ∨ (~ B ⊃ E) / A ⊃ E

    5. A ⊃ ~ B

    6. ~ B ⊃ E

    7. A ⊃ E

(13)*** 1.** (A & B) ∨ (E ⊃ S)

    2. G ⊃ E

    3. ~ (A & B) / G ⊃ S

    4. E ⊃ S

    5. G ⊃ S

(14) 1. A ∨ E

    2. ~ A

    3. E ⊃ [~ A ⊃ (E ⊃ S)] / S

    4. E

    5. ~ A ⊃ (E ⊃ S)

    6. E ⊃ S

    7. S

(15)*1. (S & E) ⊃ ~ G

   2. H ⊃ (S & E)

   3. (H ⊃ ~ G) ⊃ ~ R

   4. B ⊃ R / ~ B

   5. H ⊃ ~ G

   6. ~ R

   7. ~ B

(16) 1. H ⊃ S

   2. S ⊃ ~ R

   3. ~ B ⊃ F

   4. (H ⊃ ~ R) ⊃ (L ⊃ ~ B) / L ⊃ F

   5. H ⊃ ~ R

   6. L ⊃ ~ B

   7. L ⊃ F

(17)*1. H

   2. H ⊃ (F v S)

   3. (F v S) ⊃ (H ⊃ B) / B

   4. H ⊃ (H ⊃ B)

   5. H ⊃ B

   6. B

(18)*1. A v (B ⊃ G)

   2. G v ~ A

   3. ~ G

   4. G ⊃ R / B ⊃ R

   5. ~ A

   6. B ⊃ G

   7. B ⊃ R

(19) 1. (A v B) ⊃ (E & S)

   2. (E & S) ⊃ G

   3. H ⊃ (A v B)

   4. M ⊃ ~ (H ⊃ G) / H ⊃ G

    5. (A ∨ B) ⊃ G

    6. H ⊃ G

(20) **1.** (A ⊃ B) ⊃ (E ⊃ A)

    2. (E ⊃ B) ⊃ [(E ⊃ ~ H) ⊃ J]

    3. ~ H ⊃ B

    4. A ⊃ ~ H / J

    5. A ⊃ B

    6. E ⊃ A

    7. E ⊃ ~ H

    8. E ⊃ B

    9. (E ⊃ ~ H) ⊃ J

    10. J.

►   **EXERCISE 7.2**   ◄

Each of the following symbolized arguments is valid. Using the first four rules, provide a proof for each argument.

(1) * **1.** (A & B) ⊃ S

    2. H ⊃ R

    3. (A & B) / S

(2) **1.** G ∨ R

    2. H & (J ⊃ I)

    3. ~ G / R

(3) **1.** ~ (F ≡ S)

    2. R ∨ M

    3. ~ (F ≡ S) ⊃ D / D

(4) **1.** ~ (A & B)

    2. R ⊃ (A & B)

    3. H / ~ R

(5) * **1.** F ⊃ (H & B)

    2. A ⊃ ~ (H & B)

     3. A / ~ F

(6) **1.** (B & C) v A

     2. A ⊃ F

     3. ~ (B & C) / F

(7)*1. J v ~ S

     2. ~ J

     3. S v G / G

(8) **1.** (E ≡ F) v ~ (A & B)

     2. H ⊃ (A & B)

     3. ~ (E ≡ F) / ~ H

(9) **1.** F ⊃ S

     2. S ⊃ G

     3. (F ⊃ G) ⊃ M / M

(10)*1. R ⊃ H

     2. H ⊃ S

     3. S ⊃ G

     4. (R ⊃ G) ⊃ F / F

(11) **1.** ~ F ⊃ ~ S

     2. ~ S ⊃ ~ G

     3. (~ F ⊃ ~ G) ⊃ H / H

(12) **1.** B ⊃ (A & G)

     2. R ⊃ B

     3. [R ⊃ (A & G)] ⊃ S / S

(13)*1. (H v B) ⊃ ~ (S ≡ F)

     2. R ⊃ (H v B)

     3. ~(S ≡ F) ⊃ I / R ⊃ I

(14) **1.** J ⊃ I

     2. I ⊃ ~ R

     3. J

     4. A ⊃ ~ B

     5. A

      6. R v (B v S)

      7. S ⊃ L / L

(15)* 1. A ⊃ (B ⊃ E)

      2. A

      3. ~ E / ~ B

(16) 1. A ⊃ B

      2. B ⊃ E

      3. B ⊃ R

      4. ~ R

      5. A v B / E

(17)* 1. I v A

      2. (I v A) ⊃ ~ S

      3. ~ J ⊃ [J v (I ⊃ S)]

      4. ~S ⊃ [S v (J ⊃ S)] / A

(18)* 1. A ⊃ B

      2. B ⊃ W

      3. ~ W

      4. J ⊃ I

      5. S ⊃ ~ I

      6. ~ A ⊃ (J v Z)

      7. S / Z

▶   **EXERCISE 7.3**   ◀

Each of the following English arguments is valid. Symbolize each argument in TL and use natural deduction to prove each valid. Suggested abbreviations are provided.

  1.*If Ann goes swimming, then Bob will go swimming. Either Ann or Chris will go swimming. Bob won't go swimming. So, Chris will go swimming. (A, B, C)

  2. If Ann goes swimming, then Bob will go swimming. If Bob goes swimming, then Chris will go swimming. But Chris won't swim. So, Ann won't swim. (A, B, C)

3.*If either Ann or Bob is home, then it's the case that either Bob or Chris is home. Either it's the case that Ann is home, or if Bob is home then Chris is home. If it's the case that if Bob is home then Chris is home, then either Ann or Bob is home. Ann is not home. So, either Bob or Chris is home. (A, B, C)

4. If either Pat or Quinn works late, then Rita and Winona will work late. If it's the case that Herman is on vacation, then either Pat or Quinn will work late. So, if Herman is on vacation, then Rita and Winona will work late. (P, Q, R, W, H)

5.*If Ralph tells Alice he's sorry, then Alice will be forgiving. If Norton has a talk with Ralph, then Ralph will tell Alice he's sorry. Alice will not be forgiving. So, Norton won't have a talk with Ralph. (R, A, N)

6. If Pat gets a ticket, then if Pat has an accident then Pat's insurance will be canceled. Pat's insurance won't be canceled. If Pat doesn't slow down, then Pat will get a ticket. Either Pat's insurance will be canceled or Pat won't slow down. So, Pat won't have an accident. (T, A, I, S)

7.*Elroy won't go jogging. If Lulu jogs, then Chris will jog. Either Elroy will jog or Lulu will jog. Therefore, Chris will jog. (E, L, C)

8. If Donovan doesn't perform tonight, then either Fever Tree or Jefferson Airplane will perform. Fever Tree will not perform. If the Jefferson Airplane performs then Suzie will get to hear her favorite song. Donovan won't perform tonight. So, Suzie will get to hear her favorite song tonight. (D, J, F, S)

9. If it's the case that if Ann swims then Bubba will swim, then Chris will swim. Either Ann will swim or Dave won't swim. If it's not the case that Dave swims, then if Ann swims then Bubba will swim. Ann won't swim. So, Chris will swim. (A, B, C, D)

10. Ann and Bob won't both sing. Either Chris or Rob will sing or Mary will sing. If Lucy doesn't sing, then it's not the case that either Chris or Rob will. Either Ann and Bob will both sing or Lucy won't sing. Consequently, Mary will sing. (A, B, C, R, M, L)

11. If Wimpy eats a hamburger, then Popeye will eat another can of spinach. Either Olive Oyl will eat a peach or Popeye won't eat another can of spinach. Olive Oyl won't eat a peach, so Wimpy won't eat a hamburger. (W, P, O)

12. Either Gilligan won't show up for dinner, or either Mr. Howell will be upset, or Mrs. Howell won't show up for dinner. If Gilligan doesn't show up for dinner, then Mr. Howell will be upset. Mr. Howell won't be upset. So, Mrs. Howell won't show up for dinner. (G, H, M)

13.*If Moe makes a face, then Curly will make a face. If it's the case that if Moe makes a face then Larry won't make a face, then it follows that if Curly makes a face then the director will not be happy. If Curly makes a face, then Larry won't make a face. So, if Moe makes a face, then the director will not be happy. (M, C, L, D)

14. Either Hogan bribes Schultz or if Schultz tells Klink then General Burkhart will find out. If it's the case that if Schultz tells Klink then General Burkhart will find out, then if General Burkhart finds out, then Hogan will get in big trouble. Hogan won't bribe Schultz. And Hogan won't get in big trouble. So, Schultz won't tell Klink. (H, S, B, T)

15. If Gilligan is late, then the skipper will be worried about his little buddy. If the skipper is worried about his little buddy, then the skipper won't eat. If the skipper won't eat, then Mrs. Howell will become concerned. If it's the case that if Gilligan is late, then Mrs. Howell will become concerned, then the professor will argue that Mrs. Howell's peace of mind is dependent on Gilligan's punctuality. So, the professor will argue that Mrs. Howell's peace of mind is dependent on Gilligan's punctuality. (G, S, E, M, P)

CHAPTER 7 GLOSSARY

Inference   A person "draws an inference" when he or she asserts a conclusion on the basis of one or more premises.

- **Valid Inference**   An inference is a valid inference just in case the argument formed out of the premises and conclusion of the inference constitutes a valid argument.

- **Inference rule**   A rule specifying that a conclusion of a certain form may be inferred when certain premises are given.

- **Valid inference rule**   An inference rule that has the following feature: When the rule is followed, the result is always a valid inference.

Justification   (of an inference) In a proof, a citation of the rule used in drawing the inference and the lines to which the rule was applied.

Natural deduction method   A method in which we deduce a conclusion from a set of premises through a series of valid inferences corresponding to more or less natural patterns of reasoning.

Natural deduction system   A system consisting of (a) a formal language that can be used to symbolize information; and (b) a set of natural deduction rules that can be used to deduce conclusions from premises.

Proof in TD   A sequence of sentences of TL, each of which is either a premise or an assumption or follows from one or more previous sentences according to a TD inference or replacement rule, and in which (1) every line (other than a premise) has a justification and (2) any assumptions have been discharged. The conclusion of a proof is the last line of the proof.

TD   The name of the truth-functional natural deduction system used in Chapters 7–11.

Valid in TD   An argument is valid in TD if and only if there exists a proof in TD whose premises are the premises of the argument and whose conclusion is the conclusion of the argument. (An argument is *invalid* in TD if and only if it is not valid in TD.)

Chapter

8

FOUR MORE INFERENCE RULES

T here is no limit to the number of valid inference rules that can be formulated. However, as you will later see, once we have formulated certain valid rules, any additional rules are redundant, for the additional rules merely allow inferences that the original group of rules already allows. For practical purposes, and as far as proofs of validity are concerned, twenty rules are all we will need. Here are four more valid rules of inference:

THE SIMPLIFICATION RULE

The Simplification rule (abbreviated Simp), summarized in the following box, is obviously a valid inference rule:

The Simplification Rule

From a sentence of the form **P & Q,** you may infer the corresponding sentence **P.**

Also, from a sentence of the form **P & Q,** you may infer the corresponding sentence **Q.**

In other words, given a conjunction, you may infer the left conjunct. And, given a conjunction, you may infer the right conjunct. We shall abbreviate this rule as

P & Q	P & Q
P	Q

In each of the following examples, we applied the Simplification rule to the first sentence and deduced the second sentence in accord with the rule.

1. A & B	**1.** (A v B) & (E v F)	**1.** ~ A & ~ B
2. A	**2.** (A v B)	**2.** ~ B

Here is an English argument that fits the Simplification pattern:

1. Ann is home and Bob is home.

2. Therefore, Ann is home.

In TL, this is:

1. A & B / A

The Simplification rule can also be represented with geometric shapes:

The Simplification Rule:

□ & ○   □ & ○
Infer: □   Infer: ○

Here is a short proof that employs Simplification:

1.	A & B	
2.	A ⊃ E	/ E
3.	A	Simp 1
4.	E	MP 2, 3

In this proof, at line 3, we applied Simp to 1 and pulled down (inferred) the A. At line 4, we applied MP to lines 2 and 3 to prove E. Because E is the conclusion, the proof is complete at line 4, and the argument is proven valid.

THE CONJUNCTION RULE

The next rule, the Conjunction rule (Conj), is summarized in the box below, and is also obviously valid:

The Conjunction Rule

From a sentence **P** and a sentence **Q**, you may infer the corresponding sentence **P & Q**.

We shall abbreviate the Conjunction rule as

$$P$$

$$\frac{Q}{P \ \& \ Q}$$

In each of the following examples, we applied the Conjunction rule to the first two sentences and deduced the third sentence in accord with the rule.

1. A	1. (A v B)	1. ~ A
2. B	2. (E v F)	2. ~ B
3. A & B	3. (A v B) & (E v F)	3. ~ A & ~ B

Here is an English argument that fits the Conjunction pattern:

1. Ann is home.

2. Bob is home.

3. Therefore, Ann is home and Bob is home.

That is,

1. A

2. B / A & B

The Conjunction rule can also be represented with geometric shapes:

The Conjunction Rule:

1. □

2. ○

Infer: □ & ○

Here is a short proof that employs Conjunction:

1. A

2. B

3. (A & B) ⊃ E / E

4. A & B   Conj 1, 2

5. E   MP 3, 4

On line 4, we applied Conj to 1 and 2 and inferred A & B. We simply stuck the A and the B together with an ampersand to generate line 4. Next, we applied

MP to lines 3 and 4 and pulled down (inferred) E. This gave us line 5. Because E is the conclusion, the proof is complete at line 5 and the argument is proven valid.

THE ADDITION RULE

The next rule, the Addition rule (Add), strikes many as invalid at first:

> ### *The Addition Rule*
>
> From a sentence **P**, you may infer **P v Q**.

That is, from a sentence **P**, one may infer the disjunction of **P** with *any* sentence **Q**. For example, from B one may infer B v G, from B one may infer B v (G & D), from B one may infer B v J, and so on. The Addition rule will be abbreviated

$$\frac{P}{P \text{ v } Q}$$

In each of the following examples, we applied the Addition rule to the first sentence and deduced the second sentence in accord with the rule.

1. A	1. A	1. (A & B)	1. J ⊃ E	1. B
2. A v H	2. A v S	2. (A & B) v (H & S)	2. (J ⊃ E) v W	2. B v ~ R

Here is an English argument that fits the Addition pattern:

1. Ann is home.

2. Therefore, Ann is home *or* Bob is home.

That is,

1. A / A v B

And here is a short proof employing Addition:

1. A

2. (A v B) ⊃ H / H

3. A v B   Add 1

4. H   MP 2, 3

At line 3, we simply inferred A v B from line 1 by applying the Addition rule to line 1. We took line 1, copied it onto line 3, added a wedge to the right, and added a formula of our choice, namely B. This gave us A v B on line 3. We then applied MP to 2 and 3 to prove line 4, namely H. Because this is the conclusion, the proof is finished, and the argument is proven valid.

If the Addition rule doesn't seem valid to you, think of it this way: Imagine you are on a game show and you will win a forty-eight-volume series on the history of symbolic logic if you correctly guess the truth-value of the disjunction behind the curtain. You are given only one piece of information: the left disjunct of the hidden disjunction is a true sentence. Now, the question is: what's the truth-value of the disjunction as a whole? Clearly, the disjunction must be true, for if one disjunct is true, then no matter what the truth-value of the other disjunct, the whole disjunction must be true. Thus, if a sentence **P** is true, **P** disjoined to any sentence **Q** produces a true disjunction. In other words, if a sentence **P** is true, **P** v **Q** must also be true *no matter what* **Q**'s *truth-value is*. So, if **P** is true, **P** disjoined to any sentence will produce a true disjunction.

THE CONSTRUCTIVE DILEMMA RULE

The next rule, Constructive Dilemma (CD), is the most difficult to apply of our first eight rules. First, consider the following argument:

1. If it rains, then the roof will get wet.

2. If it snows, then the lawn will get white.

3. It will either rain or snow.

4. So, either the roof will get wet or the lawn will get white.

If we symbolize this English argument in TL, we have

1. R ⊃ W

2. S ⊃ L

3. R v S / W v L

This TL argument is an instance of the following argument form:

1. **P ⊃ Q**

2. **R ⊃ S**

3. **P v R / Q v S**

A truth-table test would reveal this to be a valid argument form. This valid form is reflected in the Constructive Dilemma rule, as summarized in the following box.

> *The Constructive Dilemma Rule*
>
> From a sentence of the form **P ⊃ Q**, and a sentence **R ⊃ S**, and a corresponding sentence **P v R**, you may infer the corresponding sentence **Q v S**.

The Constructive Dilemma rule is abbreviated as

$$P \supset Q$$
$$R \supset S$$
$$\underline{P \vee R}$$
$$Q \vee S$$

In each of the following examples, we applied the Constructive Dilemma rule to the first three sentences and deduced the fourth sentence in accord with the rule.

1. A ⊃ B	**1.** H ⊃ (G & B)	**1.** R ⊃ ~ S
2. E ⊃ F	**2.** S ⊃ (A & M)	**2.** H ⊃ ~ G
3. A v E	**3.** H v S	**3.** R v H
4. B v F	**4.** (G & B) v (A & M)	**4.** ~ S v ~ G

Notice that the Constructive Dilemma rule requires the presence of three elements:

1. a conditional

2. another conditional

3. a disjunction whose left disjunct is the antecedent of the one conditional and whose right disjunct is the antecedent of the other conditional.

From this, you are permitted to infer a disjunction whose left disjunct is the consequent of the one conditional and whose right disjunct is the consequent of the other conditional.

We now have eight *valid* inference rules to work with. Here are some examples of proofs using various combinations of these rules.

(1) 1.	P ⊃ H	
2.	(H v C) ⊃ (D & E)	
3.	P / D	
4.	H	MP 1, 3
5.	H v C	Add 4
6.	D & E	MP 2, 5
7.	D	Simp 6

At line 5, we used Add to derive H v C. In effect, we "added" C to H in order to produce the formula H v C. This matched the antecedent of line 2, which allowed us to apply MP to lines 2 and 5. We could have added any other TL sentence to H at line 5, but we chose C because we needed H v C.

(2) 1. F ⊃ L

    2. (M ⊃ N) & S

    3. N ⊃ T

    4. F v M / L v T

    5. M ⊃ N   Simp 2

    6. M ⊃ T   HS 3, 5

    7. L v T   CD 4, 1, 6

We applied Simp to line 2 because the main connective of 2 is the ampersand. This allowed us to bring down the M ⊃ N, which made possible the application of HS. Notice the application of the Constructive Dilemma rule to lines 1, 4, and 6.

(3) 1. ~ M & N

    2. P ⊃ M

    3. Q & R

    4. (~ P & Q) ⊃ A / A v T

    5. ~ M   Simp 1

    6. ~ P   MT 2, 5

    7. Q   Simp 3

    8. ~ P & Q   Conj 6, 7

    9. A   MP 4, 8

    10. A v T   Add 9

Notice that the last line of the proof is an application of Add to line 9. The T that appears on line 10 does not appear anywhere in the proof prior to line 10, and so at line 9 it was apparent that the T would only enter the proof through an application of the Add rule.

SOME UNSOLICITED ADVICE ON LEARNING TO CONSTRUCT PROOFS

When first learning to work proofs, it is natural to wonder if there are mechanical rules we can follow that will tell us how to begin and just what steps to take. After all, mechanical rules—decision procedures—tell us precisely how to construct a truth-table. We could formulate decision procedures for the construction of proofs, but they are complex and extremely difficult to work with. Consequently, no practical set of step-by-step instructions tells you precisely and mechanically how to construct any and every valid proof. Constructing a proof is therefore something of an art, and building a proof thus requires a certain amount of creativity. Complicated proofs also require a certain amount of skill. Because skills typically improve with practice, you can expect your abilities to improve as you successfully complete more and more proofs.

The first step in learning to build proofs is this: Develop an ability to spot the inference rule *patterns* when you look at the premises of an argument. Once you recognize a pattern, you will be able to apply the appropriate rule and deduce the next line in the proof. And when you have deduced or brought down a sentence, a new pattern will form that will enable you to apply another rule and bring down another sentence, which will form a new pattern and make possible another step. Eventually, this should take you to the conclusion. This process involves the mental skill of pattern recognition, a skill that develops only with practice.

The second step in learning to build proofs is to learn to apply the rules accurately. You must apply the rules precisely or else incorrect sequences of formulas will result. If you apply a rule incorrectly at, say, step 3 and bring down the wrong formula, this can throw off every subsequent step in your proof.

The third step in the process of learning to build proofs involves learning to strategize. Someone who is good at strategizing can look at an argument for a few minutes and plan out *in advance* the steps that will likely lead to the conclusion. This is a skill that develops only after extended practice and only after one fully understands the natural deduction process. We discuss strategy in the next section.

Ultimately, the only way to become skilled at constructing proofs is to practice them on your own. Most people find logic proofs a bit overwhelming at first. However, if you refuse to get discouraged, and if you put some work into it, you will find that your ability to complete a proof increases with each one that you solve. You will also learn a lot by simply watching as proofs are worked on the board in class.

PROOF STRATEGIES

When constructing proofs for the first time, it is natural to ask: Where do I start? How do I know which steps to take? Without practical decision procedures that tell you how to construct a proof, you will have to proceed by a combination of trial and error, intuition, pattern recognition, general problem-solving skills, and a strategy we will call "working backward from the conclusion." Although they are not decision procedures, these strategies will help you construct proofs.

▶ THE TRIAL AND ERROR STRATEGY

Suppose you begin a proof and are unsure where to start or where to go. Scan the premises, looking for *any* of the inference rule patterns. When you spot a pattern, apply the appropriate rule and make the required move, whether or not it seems to take you to the conclusion. Often, if you make enough moves like this, even though you have no overall plan or direction in mind, you will eventually arrive at the conclusion. Let us call this the "trial and error strategy." For example, consider the following argument:

1. (A v E) ⊃ ~ G
2. G v (R & H)
3. (A v E) / H

When you have become familiar with the inference rule patterns, you should be able to spot the MP pattern on lines 1 and 3. This gives you

1. (A v E) ⊃ ~ G
2. G v (R & H)
3. A v E / H
4. ~ G   MP 1, 3

Next, look for another pattern. Notice that lines 2 and 4 instantiate the premise section of DS. Thus:

1. (A v E) ⊃ ~ G
2. G v (R & H)
3. A v E / H
4. ~ G   MP 1, 3
5. R & H   DS 2, 4

Finally, notice that line 5 fits the premise section of Simp, which allows you to infer the conclusion:

6. H Simp 5

If you are proceeding by the trial and error strategy, the most important piece of advice is this: If you spot a pattern and see a "move" you could make, but you don't see where it would take you, don't hesitate . . . just do it! When you infer another line, another pattern will probably open up, which will probably lead to another inference, and another, until you eventually reach the conclusion.

▶ THE DECOMPOSITION STRATEGY

Another proof strategy may be called the decomposition strategy. Break larger sentences down into smaller sentences by deriving parts of the larger sentences. Use these smaller parts to derive additional lines, until you finally derive the conclusion. For example, consider the following symbolized argument:

1. (H v C) ⊃ [(E v F) v (A & B)]

2. ~ (E v F)

3. H / B

Let us begin by deriving something that we can use to break up ("decompose") line 1:

4. H v C   Add 3

Using this, plus MP, we can now bring down the consequent of 1:

5. (E v F) v (A & B)   MP 1, 4

Notice that line 2 is the negation of the left disjunct of line 5. Consequently, we can use DS to break up 5 as follows:

6. A & B   DS 2, 5

The last move is obvious:

7. B   Simp 6

Notice that in each step after line 4 we broke a larger sentence down by deriving one of its parts.

► WORKING BACKWARD FROM THE CONCLUSION

Let's suppose we are trying to prove the following symbolized argument valid:

1. ~H ⊃ (B ⊃ A)

2. S v B

3. H ⊃ E

4. ~S

5. ~ E / A

Ultimately, we want to reach—at the bottom of the proof—the conclusion A. Let's begin with the conclusion and trace a series of steps backward to the first step in the proof. First, where in the premises is the conclusion formula? Notice that A is at the end of line 1. How might we derive it? On line 1 the sentence A is the consequent of a conditional. What rule would let us derive the consequent of a conditional? Modus Ponens. What would MP require? First, the conditional must be all by itself on a line. Second, another line must consist of just the antecedent of the conditional. So, first we will need to bring the conditional B ⊃ A down onto a line by itself, and then we will need a line consisting of B by itself. From those two lines, A will follow. The inference would look like this:

$$B \supset A$$

$$\dfrac{B}{A}$$

How might we "detach" or bring down B ⊃ A? Notice that on the first line, B ⊃ A is itself the consequent of a conditional. If we could get the antecedent of that conditional on a line by itself, we could use MP to bring down B ⊃ A. The inference would look like this:

$$\sim H \supset (B \supset A)$$

$$\dfrac{\sim H}{B \supset A}$$

But how might we get ~H on a line by itself? First, find H in the premises. In line 3, H is the antecedent. Modus Tollens would let us infer ~H from line 3, if we also had on another line the negation of the consequent of line 3. The inference would look like this:

$$H \supset E$$

$$\dfrac{\sim E}{\sim H}$$

So we need ~ E on a line by itself. How will we get that? We locate ~E on line 5. Next, we need to derive B in order to apply MP to B ⊃ A. How will we get B? Notice that B is the right disjunct of the disjunction on line 2. If we can derive or find, on a line by itself, the negation of the left disjunct of that disjunction, we can infer B, which we may then use to bring down our conclusion, A. The negation of the left disjunct of line 2 is ~S. So, we will need ~S in order to break B out of line 2. Looking around, we find ~S on line 4. Putting this together in the proper order, we get:

1. ~H ⊃ (B ⊃ A)

2. S v B

3. H ⊃ E

4. ~S

5. ~ E / A

6. ~H   MT 3, 5

7. B ⊃ A   MP 1, 6

8. B   DS 2, 4

9. A   MP 7, 8

Approach a proof the way you would a game of strategy such as chess or Monopoly. Constantly think about the various possibilities open to you at each step. Look ahead to where you want to be and think of ways to get there from where you are. In a game of strategy, a skilled player develops the ability to look several steps ahead and plans moves in terms of that goal.

SOME ADDITIONAL SUGGESTIONS CONCERNING STRATEGY

1. If one of the lines in the proof is a conditional, try to find or derive the antecedent. When you have the antecedent, apply MP and derive the consequent of the conditional. Or, try to find or derive the negation of the consequent. When you have that, derive the negation of the antecedent through an application of MT.

2. If one of the lines in the proof is a disjunction, try to find or derive the negation of the left disjunct. By applying DS to the disjunction and the negation of the left disjunct, you may derive the right disjunct.

3. If your conclusion is a conjunction, try using the Conjunction rule.

4. If the *main* connective is &, it may be wise to apply Simp and derive one of the conjuncts.

5. If you have a disjunction and can't apply DS, try to spot a CD pattern. This would require two conditionals that have as antecedents the two disjuncts of the disjunction. CD would then allow you to infer the disjunction of the two consequents of the two conditionals.

6. If you have more than one conditional, watch for HS.

7. If you have two conditionals along with a disjunction, watch for CD.

8. If a letter or formula appears just once in the premises and doesn't link with anything else in the proof, it might be a "useless" component. Not every element of the premises must be used in the derivation of the conclusion.

9. If the conclusion contains an element not found in the premises, you will have to use Add to derive the conclusion.

10. A premise may be used more than once. Also, it is not always necessary to use every premise.

11. Try to isolate an atomic component on a line by itself. In many cases, if you can derive a single letter, or a single negated letter, you can use it to break down other lines in the proof.

Using any of the first eight rules, supply justifications for the following derived lines.

(1)* 1. (A v B) ⊃ (G v S)

   2. (G v S) ⊃ H

   3. (H v S) ⊃ M

   4. (A v E) ⊃ I

   5. A & R / M & I

   6. A

   7. A v B

   8. G v S

   9. H

   10. H v S

   11. M

   12. A v E

   13. I

   14. M & I

(2) 1. H ⊃ S

   2. B

   3. R

   4. (B & R) ⊃ ~ S

   5. ~ H ⊃ Z / Z v I

   6. B & R

   7. ~ S

   8. ~ H

   9. Z

   10. Z v I

(3) 1. (H v J) ⊃ (S v W)

   2. S ⊃ O

    3. W ⊃ K

    4. H / O v K

    5. H v J

    6. S v W

    7. O v K

(4)   **1.** A v B

    2. C

    3. [(A v B) & C] ⊃ ~ I

    4. S ⊃ I

    5. (~ S v H) ⊃ X / X v E

    6. (A v B) & C

    7. ~ I

    8. ~ S

    9. ~ S v H

    10. X

    11. X v E

(5)* **1.** (S v J) ⊃ ~ Z

    2. Z v H

    3. H ⊃ A

    4. S & G

    5. A ⊃ M

    6. O ⊃ N / M v N

    7. S

    8. S v J

    9. ~ Z

    10. H

    11. A

    12. A v O

    13. M v N

(6)* **1.** J & R

    2. (J v ~ S) ⊃ I

   3. (R & I) ⊃ H / H & J

   4. J

   5. J v ~ S

   6. I

   7. R

   8. R & I

   9. H

   10. H & J

(7) **1.** J ⊃ I

    2. J v R

    3. R ⊃ (R ⊃ S)

    4. ~ I / S v I

    5. ~ J

    6. R

    7. R ⊃ S

    8. S

    9. S v I

► **EXERCISE 8.2** ◄

   Each of the following arguments is valid. Using any of the eight inference rules, supply proofs.

(1)* **1.** A ⊃ B

   2. B ⊃ R

   3. ~ R / ~ A & ~ B

(2) **1.** (A v B) ⊃ G

   2. A

   3. G ⊃ S / S

(3) **1.** (H & S) ⊃ ~ (F ≡ S)

   2. B ⊃ (F ≡ S)

   3. H

    4. S / ~ B

(4) 1. A & (R ⊃ S)

    2. S ⊃ I

    3. (R ⊃ I) ⊃ O / O

(5)*1. B ⊃ (S v R)

    2. S ⊃ J

    3. R ⊃ G

    4. H & B / J v G

(6) 1. H ⊃ ~ (S v G)

    2. ~(R v W) ⊃ (S v G)

    3. F & H

    4. ~ W ⊃ (J & A)

    5. W ⊃ M

    6. ~ M / A & ~ ~ (R v W)

(7)*1. H v ~ S

    2. [(H v ~ S) v G] ⊃ ~ M

    3. M v R / R v X

(8) 1. ~ F ⊃ ~ S

    2. H & F

    3. F ⊃ B

    4. B ⊃ G / G

(9) 1. F v B

    2. [H & (F v B)] ⊃ ~ S

    3. I & H / ~ S

(10)*1. A ⊃ (J & S)

    2. B ⊃ F

    3. A

    4. [(J & S) v F] ⊃ G / G

(11) 1. (H v S) ⊃ D

    2. (R v M) ⊃ I

    3. (D & I) ⊃ ~ A

    4. H & R / ~ A

(12) **1.** A v (B ⊃ C)

    2. ~ A ⊃ (~ H ⊃ J)

    3. ~ H v B

    4. ~ A & I / J v C

(13)*1. (A ⊃ B) & (A ⊃ C)

    2. A

    3. (B v C) ⊃ Z / Z

(14) **1.** (S ⊃ I) & (S ⊃ J)

    2. (I v J) ⊃ G

    3. H & S / G

(15)*1. (A ⊃ B) & (S ⊃ I)

    2. J ⊃ R

    3. (A v J) & (S v E) / B v R

(16) **1.** (A ⊃ J) & X

    2. F v (J ⊃ F)

    3. ~F / ~A v Z

(17)*1. S ⊃ [J & (I v S)]

    2. [J v (A & B)] ⊃ (~ N & ~O)

    3. S & Z / ~ N & J

(18) **1.** (J ⊃ I) & (R ⊃ S)

    2. J v R

    3. (I ⊃ R) & (S ⊃ A)

    4. (R v A) ⊃ E / E

(19) **1.** (J ⊃ I) & (~ J ⊃ S)

    2. I ⊃ B

    3. [(J ⊃ I) & (I ⊃ B)] ⊃

       [(J & B) v (~ J & ~ B)]

    4. (J & B) ⊃ A

    5. (~ J & ~ B) ⊃ H / A v H

(20) * **1.** (A & S) ⊃ Z

   2. ~ J ⊃ A

   3. ~ J ⊃ S

   4. ~ J & B / Z v I

(21) **1.** A ⊃ B

   2. ~ B & I

   3. ~ S v ~ G

   4. (~ A & ~ B) ⊃ [(~ S ⊃ A) &

   (~ G ⊃ X)] / X

(22) **1.** I v J

   2. ~ I

   3. (~ I v F) ⊃ (J ⊃ M)

   4. (M v H) ⊃ (S ⊃ I) / ~ S

(23) **1.** (A v B) ⊃ [(E ⊃ G) &

   (J ⊃ I)]

   2. (A v E) ⊃ (E v J)

   3. A & B / G v I

(24) **1.** ~ A & B

   2. ~ A & (J v ~ I)

   3. (J ⊃ S) & (~ I ⊃ ~ H)

   4. (S ⊃ A) & (~ H ⊃ ~ Z) / ~ Z

CHAPTER 8 APPENDIX: *Some Common Deduction Errors*

Students are prone to certain common mistakes when they first learn to construct proofs. The following is a list of common errors.

Error: Reason Why this is an Error

~(A & B)

   A   Simp (error)   You cannot apply the Simplification rule to a
                           negated conjunction.

Error:		Reason Why this is an Error
A ⊃ B B <hr> A	MP (error)	If you are to apply Modus Ponens to A ⊃ B, the second line in this case would have to be A, and B would follow by MP.
A ⊃ B ~A <hr> ~B	MT (error)	If you are to apply Modus Tollens to A ⊃ B, the second line in this case would have to be ~ B, and ~A would follow by MT.
A v B A <hr> B	MP (error)	You cannot apply MP when a wedge is the main connective.
A <hr> A & E	Add (error)	You cannot apply Add and stick on an ampersand.
~ (A & B) <hr> ~ A	Simp (error)	You cannot simplify from a negated conjunction.
~ (~ A & ~ S) <hr> ~ ~A	Simp (error)	You cannot simplify from a negated conjunction.
G v (A ⊃ B) S ≡ (B ⊃ C) <hr> A ⊃ C	HS (error)	The inference rules must be applied to *whole* lines, not to formulas that are *part* of a whole line. In this case, you cannot apply HS to the A ⊃B and the B ⊃ C unless they are each on a line alone, as they are here:

1. A ⊃ B

2. B ⊃ C

G v (A ⊃ B)

A
_____

B   MP (error)   The inference rules must be applied to *whole* lines, not to formulas that are *part* of a whole line. In this case, you cannot apply MP to the A ⊃ B and the A unless they are each on a line alone, as they are here:

1. A ⊃ B

2. A

A v B
_____

A   Simp (error)   You may only apply Simp when the main connective is &.

Chapter 9

INDIRECT PROOFS AND CONDITIONAL PROOFS

A reductio ad absurdum proof ("reduction to absurdity" proof) employs the following logical strategy: Suppose you want to prove some claim **P**. You begin your proof by assuming the negation of the claim you wish to prove. Thus, you assume ~ **P**. Next, demonstrate that this assumption combined with your premises implies a **contradiction.** Now, if a set of statements implies a contradiction, the statements in the set cannot all be true, because only that which is contradictory can imply a contradiction. So, if an assumption is added to the premises and a contradiction is derived, then the set consisting of the premises plus the assumption cannot all be true. Assuming the premises are true, the assumption must be false. This shows that the assumption, ~ **P**, is itself false if the argument's premises are true. This is said to "reduce" the assumption to an "absurdity." It then follows that the opposite of the assumption, namely, **P**—the claim you originally sought to prove—must itself be true. Reductio ad absurdum proofs are also called *indirect proofs* because, in such a proof, the conclusion is derived indirectly, by first deriving a contradiction from the negation or denial of the conclusion and then inferring the conclusion from that.

In ancient times, Greek mathematicians proved that if we assume that the square root of two is rational, this assumption implies a contradiction, which shows that the opposite of the assumption must be true, namely, that the square root of two must be *ir*rational. This was one of the first recorded reductio ad absurdum proofs in the history of human thought. The following is a modern version of that ancient reductio that is both interesting and instructive.

1. Assume, for the sake of argument, that √2 is rational.

2. Because every rational number may be expressed as a ratio of two mutually prime numbers, it follows that

$$a \,/\, b = \sqrt{2}$$

where a and b are mutually prime. (Two numbers are mutually prime just in case they have no common factor other than 1.)

3. If we perform the same arithmetical operation on both sides of an equality, the result remains an equality. Thus, multiplying each side by b, we get:

$$a = \sqrt{2}b$$

4. Squaring each side:

$$a^2 = 2b^2$$

5. Any number that is equal to twice some number must itself be an even number. Consequently, because a^2 has been shown to be twice some number, it follows that

$$a^2 \text{ is even}$$

6. If 2 is a factor of a^2, then 2 is a factor of a as well. That is, if a^2 is twice some number, then a is also twice some number. Therefore, if a^2 is even, then a is also even. Thus:

$$a \text{ is even}$$

7. Because a and b are mutually prime, they have no common factor other than one. So, if one is even then the other is odd. We have already established that a is even. Therefore:

$$b \text{ is odd}$$

8. Because a is an even number, we know a is twice some number. Call this number k. Therefore:

$$a = 2k$$

9. Substituting 2k for a in step 4:

$$(2k)^2 = 2b^2$$

10. If we square 2k:

$$4k^2 = 2b^2$$

11. Dividing each side by 2:

$$2k^2 = b^2$$

12. It follows that

b^2 is even, because b^2 is twice some number.

13. Because any factor of b^2 is a factor of b, if 2 is a factor of b^2, then 2 is a factor of b. That is, if b^2 is even, b is also even. Thus:

b is even

14. Combining 7 and 13, we have proven that on the basis of the assumption, it follows that b is odd and b is even—that is,

b is odd and it is not the case that b is odd.

15. The consequence reached at step 14 is, of course, contradictory. This proves that our initial assumption, namely, that $\sqrt{2}$ is rational, implies a contradiction. But only that which is itself contradictory can imply a contradiction. So, our initial assumption must itself be contradictory. Therefore, the opposite of our assumption must be true. Consequently: $\sqrt{2}$ must be irrational.

THE INDIRECT PROOF RULE

The inference rule summarized in the box below embodies the reductio ad absurdum (indirect proof) method:

The Rule of Indirect Proof

Anywhere in a proof, you may do the following:

1. Indent and assume the negation of the conclusion you seek to prove, writing AP (for "assumed premise") as justification.

2. Using the assumption plus any previous lines that are available, derive a contradiction, that is, a whole line of the form **P & ~P.**

3. End the indentation and draw a vertical line in front of the indented steps (to mark them off from the other lines). On the next line assert the original conclusion that you sought to prove, that is, the assumption at step 1 above except with the negation removed. (This counts as "discharging" the assumption.) As a justification, write IP (for "indirect proof") and cite the indented lines.

Proviso: The indented lines produced are called an "indirect proof sequence." Once an assumption has been discharged and the conclusion has been derived, the lines in the indirect proof sequence may no longer be used to derive future lines in the proof. In other words, once the assumption has been discharged, the indented lines are no longer "available."

Let us apply the Indirect Proof rule (IP) to the following argument:

1. ~ A ⊃ ~ H
2. H v G
3. G ⊃ H   / A

Our first step is to indent and assume the negation of our conclusion. We indent and assume ~A:

4.   ~ A   AP

Next, we can apply MP to lines 1 and 4 to derive ~H:

5.   ~ H   MP 1, 4

Now, we apply DS to lines 2 and 5 to bring down G:

6.   G   DS 2, 5

Applying MP to lines 3 and 6, we derive H:

7.   H   MP 3 ,6

Conjoining 5 and 7 produces a contradiction, that is, a sentence of the form **P & ~P**:

8.   H & ~ H   Conj 5, 7

Because we have derived a contradiction, a line of the form **P & ~ P**, we may "dis-indent" and assert our original conclusion:

9. A   IP 4–8

We began at line 4 by assuming the negation of our conclusion. From that, we used the rules to derive H & ~ H, which is an explicit contradiction. Because our assumption implied a contradiction, we discharged that assumption and inferred its opposite, which was the conclusion we originally sought to prove.

Understand that in the above proof, the indented lines 4 through 8 do *not* follow from the premises; they merely follow from the assumption, which itself does *not* follow from the premises. However, for reasons given earlier, line 9 *does* follow from the premises—that's what the indirect proof sequence estab-lishes.

In the following indirect proof, the conclusion we seek to prove is ~ A v ~ B. Because the method requires that we assume the *negation* of our conclusion, our assumed premise will be ~ (~ A v ~ B):

1. (~ A v ~ B) v ~ G

2. G v R

3. R ⊃ G /	~ A v ~ B	
4.	~ (~ A v ~ B)	AP
5.	~ G	DS 1, 4
6.	R	DS 2, 5
7.	G	MP 3, 6
8.	G & ~ G	Conj 5, 7
9. ~ A v ~ B	IP 4–8	

At line 8, we derived a contradiction, for G & ~ G fits the form **P** & ~ **P**. So, we dis-indented, moved to line 9, and wrote down our original conclusion, ~ A v ~ B.

In the following example, we begin, as usual, by negating our conclusion:

1. J ⊃ (B & ~ I)		
2. B v (J & ~ I)	/ B	
3.	~B	AP
4.	J & ~ I	DS 2, 3
5.	J	Simp 4
6.	B & ~ I	MP 1, 5
7.	B	Simp 6
8.	B & ~ B	Conj 3, 7
9. B	IP 3–8	

The Indirect Proof rule may be abbreviated:

> *To prove* **P**: *Indent, assume* ~ **P**, *derive a contradiction, end the indentation, assert* **P**.

The assumption for an Indirect Proof sequence need not be the first line following the premises, and the conclusion established at the end of an Indirect Proof sequence need not be the last step in a proof. For example, in the next proof, HS is applied before the IP sequence begins:

1. A ⊃ B

2. B ⊃ E

3. (A ⊃ E) ⊃ R

4. G v ~ R / G

5. A ⊃ E	H S 1,2	
6.	~ G	AP
7.	~R	DS 4, 6
8.	R	MP 3, 5
9.	R & ~R	Conj 7, 8
10. G	IP 6-9	

In the following proof, the conclusion established by IP is not the conclusion of the proof. IP is used to derive E, and E is used to derive the argument's conclusion:

1. H v E

2. H ⊃ E

3. E ⊃ S / S

4.	~ E	AP
5.	~ H	MT 2, 4
6.	E	DS 1, 5
7.	E & ~ E	Conj 4, 6
8. E	IP 4–7	
9. S	MP 3, 8	

Also, the Indirect Proof rule may be used more than once in a single proof:

1. (~W v ~ E) ⊃ ~ E

2. (~ E v W) ⊃ B

3. B ⊃ ~ B / W

4.	~ E	AP
5.	~ E v W	ADD 4
6.	B	MP 2, 5
7.	~ B	MP 3, 6
8.	B & ~ B	Conj 6, 7
9. E	IP 4–8	
10.	~W	AP
11.	~W v ~ E	Add 10

12.		~ E	MP 1, 11
13.		E & ~ E	Conj 9, 12
14. W		IP 10–13	

Notice that at line 10 of this proof, the assumption made at line 4 has already been discharged. (It was discharged at line 9.) This means that lines 4–8 are no longer available for use; they are now out of the loop, so to speak, and cannot be referred to in justification of any future lines. Thus, after line 9, no justification should refer to lines 4–8. Also, at line 10, notice that a whole new indirect proof is started.

> Remember, when an assumed premise is discharged and the indirect proof sequence has been concluded, the indented lines within that sequence may not be appealed to if further steps in the proof are deduced.

► EXERCISE 9.1 ◄

Part I. Use the Indirect Proof method to prove the following:

(1)* 1. J v (I & E)

    2. J ⊃ E / E

(2) 1. (H v B) ⊃ (A & E)

    2. (E v S) ⊃ (G & ~ H)

    3. ~ ~ H ⊃ H / ~ H

(3) 1. (~J v I) ⊃ (A & B)

    2. (A v Z) ⊃ (B ⊃ J) / J

(4) 1. A v (L & P)

    2. A ⊃ L / L

(5)* 1. ~B ⊃ C

    2. C ⊃ B / B

(6) 1. G ⊃ N

    2. ~ A

    3. ~ G ⊃ A / N

(7) **1.** ~ ~M ⊃ (N ⊃ O)

    2. ~O ⊃ (M ⊃ N)

    3. ~ M ⊃ O / O

(8) **1.** A ⊃ B

    2. ~B / ~ A

(9)* **1.** A

    2. A ⊃ B

    3. (A & B) ⊃ K

    4. K ⊃ G / G

(10) **1.** (A v B) ⊃ (G & W)

    2. ~G

    3. ~ ~ A ⊃ A / ~A

(11) **1.** A v ~ (B & C)

    2. (B & C) v W

    3. ~ (A v W) ⊃ (~ A & ~ W) / A v W

(12) **1.** A v ~ ~B

    2. ~B v A / A

(13) **1.** (A v E) ⊃ (B & H)

    2. ~B

    3. ~ ~A ⊃ A / ~A

(14) **1.** ~ A ⊃ ~ B

    2. ~ A

    3. (~A & ~B) ⊃ ~ (A v B)

    4. ~ (A v B) ⊃ E

    5. E ⊃ S / S

Part II. Go back to Exercise 7.2 and work proofs 1, 3, 10, and 11 using the Indirect Proof rule.

Part III. Examine the following argument and state in English the strategy you would use to prove it valid.

    1. A ⊃ (B ⊃ C)

2. G ⊃ (C ⊃ E)

3. ~ H & A

4. A ⊃ G / (B ⊃ E)

THE CONDITIONAL PROOF RULE

Suppose you are given the following argument:

1. A ⊃ B

2. A ⊃ C / A ⊃ (B & C)

You could show this valid by reasoning as follows. If we assume A is true, then given the premise A ⊃ B, it follows that B must be true, and given the premise A ⊃ C it follows that C must be true. So, given the premises, *if* we assume A is true, then B and C must be true too. But that is the same as saying that if the premises are true, then A ⊃ (B & C) must be true. And A ⊃ (B & C) is the argument's conclusion. So *if* the premises are assumed true, then the conclusion must be true. This reasoning shows that the argument is valid. The **Conditional Proof** rule (CP) incorporates this general pattern of reasoning.

Rule of Conditional Proof

If, at any point in a proof, you wish to derive a conditional **P ⊃ Q**, you may do the following:

1. Indent and assume the antecedent of the conditional you seek to prove. (Write AP as justification.)

2. Using this assumption plus, if needed, any available lines occurring earlier, derive the consequent of the conditional.

3. Draw a vertical line in front of the indented steps (to mark them off from the other lines). The indented lines constitute a conditional proof sequence. The completion of this step is called discharging the assumption.

4. End the indentation and infer the conditional whose antecedent is the assumption with which you began the conditional proof sequence and whose consequent is the latest line in that sequence. As a justification, write CP (for "Conditional Proof") and cite the line numbers of the indented lines within the sequence.

Proviso: Once an assumption has been discharged and the conditional has been derived, the indented lines in the assumption's conditional proof sequence are no longer "available" to be used in justification of further lines in the proof.

The proviso in the above box deserves emphasis. It is important to understand that when the discharge line has been drawn and the conditional derived, the indented sentences enclosed within the discharge line cannot be used in the derivation of future steps in the proof. This is because in a conditional proof, we have no reason to suppose the indented steps follow validly from the premises. If we have no reason to suppose that the indented lines follow from the premises, we have no reason to suppose that lines that follow from the indented lines follow from the premises either. However, for reasons given above, the conditional derived as a result of a conditional proof sequence does validly follow from the premises, and this conditional may be used in the derivation of future steps if needed.

Let us apply this rule. In the following proof, we need to derive H ⊃ G. Employing the conditional proof strategy, we first indent and assume H. We then derive G. From this, we infer H ⊃ G according to the CP rule.

1. (H v S) ⊃ (~E & B)

2. E v G / H ⊃ G

3. | H   AP

4. | H v S   Add 3

5. | ~ E & B   MP 1, 4

6. | ~E   Simp 5

7. | G   DS 2, 6

8. H ⊃ G   CP 3–7

Notice that we began by assuming the antecedent of the conditional we were seeking to prove. We then applied Addition and added S in order to derive the antecedent of line 1. This allowed us to break up line 1 and derive ~E & B. We then applied Simp and derived ~E, which allowed us to infer the consequent of the conditional we were trying to prove. At step 8, we discharged our assumption and inferred the conditional whose antecedent was our assumption and whose consequent was the last line reached within the conditional proof sequence.

Understand that in the above proof, lines 3 through 7 do not follow from the premises; they merely follow from the assumption, which itself does *not* follow from the premises. However, for reasons given earlier, line 8 does follow from the premises—that's what the conditional proof sequence establishes.

Here is another example:

1. A ⊃ J

2. A ⊃ (J ⊃ T) / A ⊃ T

3. | A   AP

4. | J ⊃ T   MP 2, 3

5.		A ⊃ T	HS 1, 4
6.		T	MP 5, 3
7. A ⊃ T		CP 3–6	

Here is a more complicated CP:

1. (W & M) ⊃ (W ⊃ H)			
2. H ⊃ D			
3. M / (W & K) ⊃ D			
4.		W & K	AP
5.		W	Simp 4
6.		W & M	Conj 3, 5
7.		W ⊃ H	MP 1, 6
8.		H	MP 5, 7
9.		D	MP 2, 8
10. (W & K) ⊃ D		CP 4–9	

> The Conditional Proof rule stated above may be abbreviated as follows:
>
> To prove a sentence of the form **P ⊃ Q**: Indent, assume **P**, derive **Q**, end the indent, and assert **P ⊃ Q**.

The Conditional Proof rule has a variety of uses. For instance, the assumption for a conditional proof sequence does not have to be the first line after the premises. In the following proof, the conditional proof sequence begins in the middle of the proof:

1. B v (B ⊃ C)			
2. (B ⊃ C) ⊃ H			
3. H ⊃ A			
4. A ⊃ S / ~ B ⊃ S			
5. (B ⊃ C) ⊃ A	HS 2, 3		
6. (B ⊃ C) ⊃ S	HS 4, 5		
7.		~ B	AP
8.		B ⊃ C	DS 1, 7
9.		S	MP 8, 6
10. ~ B ⊃ S		CP 7–9	

Furthermore, the conditional established at the end of a conditional proof sequence need not be the last line of the proof. In the following proof, the conclusion is not a conditional and the conditional proof sequence in the argument's proof is not used to derive the conclusion of the argument. Rather, the conditional proof is used to derive a formula in the middle of the proof, and this formula is used to derive the conclusion of the argument.

1. G v B
2. (~ A ⊃ E) ⊃ ~ G
3. A v D
4. D ⊃ E
5. B ⊃ W   / W
6. | ~ A   AP
7. | D   DS 3, 6
8. | E   MP 4, 7
9. ~ A ⊃ E   CP 6–8
10. ~ G   MP 2, 9
11. B   DS 1, 10
12. W   MP 5, 11

Notice that in the above argument, we used the conditional proof method to derive ~ A ⊃ E. When ~ A ⊃ E was derived, we used it—~ A ⊃ E—to bring down ~ G. We then used the sentence ~ G to bring down B, by DS applied to lines 1 and 10. Finally, from B, the conclusion followed by Modus Ponens.

Remember that when an assumed premise is discharged and the conditional proof sequence has been concluded, the lines within that sequence may not be appealed to if further steps in the proof are inferred. So, in the above proof, after step 9, the lines within the conditional proof sequence (lines 6–8) may not be referred to as we derive further steps. Thus, the following proof is an incorrect proof:

1. A ⊃ (B & G)
2. (B v H) ⊃ F
3. F ⊃ R   / R
4. | A   AP
5. | B & G   MP 1, 4
6. | B   Simp 5
7. | B v H   Add 6
8. | F   MP 2, 7

9. A ⊃ F   CP   4–8   (Proof is correct so far)

10. F   MP 4, 9   INCORRECT. Lines 4–8 cannot be used after line 9.

11. R   MP 3, 10   INCORRECT. Line 10 is not allowed.

The proof is incorrect because step 10 violates the Conditional Proof rule.

Conditional proof may be used twice in a proof, as the following example illustrates:

1. (A v B) ⊃ E

2. (E v S) ⊃ A / (A ⊃ E) & (E ⊃ A)

3.	A	AP
4.	A v B	Add 3
5.	E	MP 1, 4

6. A ⊃ E   CP 3–5

7.	E	AP
8.	E v S	Add 7
9.	A	MP 2, 8

10. E ⊃ A   CP 7–9

11. (A ⊃ E) & (E ⊃ A)   Conj 6, 10

It is extremely important that you understand the following point: You cannot end a proof on an indented line. That is, the assumption of a conditional or an indirect proof sequence must be discharged before a proof is ended. So, the last line of a proof can never be an indented conditional or indirect proof sequence line. If you could legally end a proof on an indented line, without discharging the assumption, you could derive any arbitrary conclusion from any premise. For example, consider the following English argument:

1. George Costanza is neurotic.

2. If Wyoming is a city, then Seattle is a state.

3. Therefore, Seattle is a state.

In TL, using obvious abbreviations, this is:

1. G

2. W ⊃ S / S

Now, the following would be a correct proof if you were allowed to end a conditional proof (or an indirect proof) without discharging your assumption:

1. G

2. W ⊃ S / S

3. | W   AP

4. | S ·   MP 2, 3

This argument is obviously invalid. We were able to construct a "proof" of it only because we ended on an indented line.

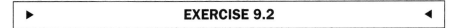

►   EXERCISE 9.2   ◄

Use the conditional proof method to prove each of the following:

(1)*1. A ⊃ (B & C) / A ⊃ C

(2) 1. (I v I) ⊃ I / I ⊃ I

(3)*1. J ⊃ (I ⊃ W)

   2. (I ⊃ W) ⊃ (I ⊃ S) / J ⊃ (I ⊃ S)

(4) 1. (A & B) ⊃ C

   2. A ⊃ B / A ⊃ C

(5)*1. ~ I v Z

   2. Z ⊃ A / ~ ~ I ⊃ (Z & A)

(6) 1. (J v I) ⊃ S

   2. A ⊃ J / A ⊃ (S v T)

(7) 1. A ⊃ ~ B

   2. B v (H & R) / A ⊃ (R v Z)

(8)*1. (J v I) ⊃ (A & B)

   2. (B v E) ⊃ (0 & S) / J ⊃ 0

(9) 1. H ⊃ (S & ~ L)

   2. L v ~ ~ I

   3. X ⊃ ~ I

   4. ~ X ⊃ B / H ⊃ B

(10) 1. A ⊃ J

   2. A ⊃ (J ⊃ B)

   3. J ⊃ (B ⊃ T) / A ⊃ T

(11) **1.** J ⊃ (I ⊃ W)

    2. S ⊃ (W ⊃ B)

    3. (I ⊃ B) ⊃ H / (J & S) ⊃ H

(12) **1.** A ⊃ B / A ⊃ [(B v S) v G]

(13) * **1.** E ⊃ A / (E & O) ⊃ A

(14) **1.** (E v G) ⊃ (Z & H)

    2. (Z v M) ⊃ W / E ⊃ W

(15) **1.** A ⊃ B

    2. B ⊃ R / A ⊃ R

(16) **1.** B ⊃ (S v R)

    2. S ⊃ P

    3. R ⊃ G

    4. B & H / ~P ⊃ G

(17) **1.** E ⊃ M

    2. E ⊃ W / E ⊃ (M & W)

(18) **1.** B ⊃ D

    2. C ⊃ E / (B & C) ⊃ (D & E)

(19) **1.** A ⊃ (B ⊃ G)

    2. A ⊃ (B ⊃ E)

    3. A ⊃ B / A ⊃ (G & E)

(20) **1.** A ⊃ (B & C)

    2. C ⊃ G

    3. G ⊃ N / A ⊃ (N v X)

(21) **1.** (A & B) ⊃ (C & D)

    2. B & (C ⊃ ~ H) / A ⊃ ~H

(22) **1.** A ⊃ (B & C)

    2. (B v D) ⊃ H

    3. H ⊃ ~ S

    4. S v J / A ⊃ J

NESTED PROOFS

It is permissible to construct a conditional proof or an indirect proof *inside* another conditional or indirect proof. That is, you may construct a conditional proof within a conditional proof, an indirect proof within an indirect proof, a conditional proof within an indirect proof, and an indirect proof within a conditional proof. Because such **nested proofs** can get complicated, we have rules to guide us. First, in order to make matters more precise, let us say that the scope of an assumption consists of the assumption itself along with all of its indented sentences. One conditional or indirect proof sequence lies inside the scope of another if its assumption is within the scope of the other proof sequence. When constructing nested proofs, you must obey the following rules:

1. You may not discharge an assumption unless all other assumptions within the assumption's scope have been discharged.

2. Every assumption must be discharged before a proof is completed.

Here is an example of a nested conditional proof:

1. G ⊃ (F ⊃ H) / F ⊃ (G ⊃ H)
2. | F   AP
3. | | G   AP
4. | | F ⊃ H   MP 1, 3
5. | | H   MP 2, 4
6. | G ⊃ H   CP 3–5
7. F ⊃ (G ⊃ H)   CP 2–6

Notice that the conclusion of this argument is a conditional whose antecedent is F and whose consequent is (G ⊃ H). Notice further that the consequent (G ⊃ H) is itself a conditional. At step 2 we assumed the antecedent of the conclusion. At step 3 we assumed the antecedent of the consequent of the conclusion, that is, we assumed the antecedent of (G ⊃ H). We then completed the inner conditional proof sequence and closed it off. Finally, we completed the outer conditional proof sequence and closed it off.

Concerning the nested proof above, the first of the two rules given above shows that you cannot end the first conditional proof sequence—the sequence whose assumption is F—and discharge its assumption until the conditional proof within has been completed and its assumption has been discharged. The second rule simply reminds you that you cannot end a proof on an indented line; all assumptions must be discharged before you derive the final conclusion.

Consider the following nested proof:

1. G ⊃ (B v ~A)
2. E v ~ B / ~ E ⊃ (G ⊃ ~ A)
3. | ~ E   AP
4. | ~ B   DS 2, 3
5. | G   AP
6. | B v ~ A   MP 1, 5
7. | ~ A   DS 4, 6
8. | G ⊃ ~ A   CP 5–7
9. ~ E ⊃ (G ⊃ ~ A)   CP 3–8

When we derive lines inside the inner conditional proof sequence, we can use lines within the outer conditional proof sequence, for its assumption has not yet been discharged. Thus at step 6 in the proof immediately above, we can appeal to steps 1 through 5 if we need to. However, once the inner proof sequence has been closed off and its assumption has been discharged, we cannot derive lines from the lines within its "region"—lines 5 through 7—as we continue on in the outer conditional proof sequence. Thus, after step 8, we cannot derive lines from lines 5–7 as we finish the rest of the proof. Similarly, after step 9, we cannot use the lines within the outer conditional proof—lines 3 through 8—to derive further lines in the proof.

Generally, you will need to nest two conditional proofs if the conclusion you seek to prove is a conditional whose consequent is itself a conditional. In such a case, the consequent of the conditional you are seeking to prove will itself contain an antecedent and consequent, which is to say that the conditional you are trying to derive is a conditional of the form **P** ⊃ (**Q** ⊃ **R**). To use a nested proof to derive a formula of this form, do the following:

1. indent and assume the antecedent, **P**, of the whole conditional;

2. indent again and assume the antecedent, **Q**, of the conditional constituting the consequent of the whole conditional.

The following illustrates the overall structure of the nested proof in such a case:

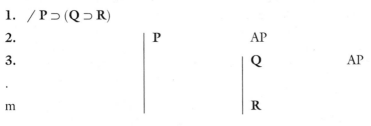

1. / **P** ⊃ (**Q** ⊃ **R**)
2. | **P**   AP
3. | **Q**   AP
  .
m | **R**

n | **Q ⊃ R**   CP 3-m

.   **P ⊃ (Q ⊃ R)**   CP 2-n

Some arguments require a triple-nested proof. Here is an example of the overall structure of such a proof:

1.		/ **P ⊃ [Q ⊃ (R ⊃ S)]**			
2.		**P**	AP		
3.			**Q**	AP	
4.				**R**	AP
.,					
.					
m				**S**	
n			**R ⊃ S**	CP 4-m	
.					
o		**Q ⊃ (R ⊃ S)**	CP 3-n		
.					

p. **P ⊃ [Q ⊃ (R ⊃ S)]**   CP 2-o

You may also nest an indirect proof sequence within a conditional proof sequence or a conditional proof sequence within an indirect proof sequence. The rules allow both types of proofs. Here is an indirect proof that is nested within a conditional proof:

1.	L ⊃ [~ M ⊃ (N & O)]			
2.	~ N & P / L ⊃ (M & P)			
3.		L	AP	
4.		~ M ⊃ (N & O)	MP 1, 3	
5.			~M	AP
6.			N & O	MP 4, 5
7.			N	Simp 6
8.			~N	Simp 2
9.			N & ~ N	Conj 7, 8
10.		M		IP 5-9

11.		P	Simp 2
12.		M & P	Conj 10, 11
13. L ⊃ (M & P)		CP 3–12	

> ► **EXERCISE 9.3** ◄

Nested proofs are recommended on the following problems.

(1)* 1. J ⊃ (I ⊃ W)

    2. W ⊃ (I ⊃ S) / J ⊃ (I ⊃ S)

(2) 1. A ⊃ (B ⊃ C)

    2. (C v H) ⊃ K / A ⊃ (B ⊃ K)

(3) 1. A ⊃ (B & E) / (J ⊃ A) ⊃ (J ⊃ E)

(4) 1. A ⊃ (B ⊃ J)

    2. A ⊃ (I ⊃ W)

    3. X ⊃ (B v I) / A ⊃ [X ⊃ (J v W)]

(5)* 1. H ⊃ (S & T)

    2. B ⊃ (A & G) / (T ⊃ B) ⊃ (H ⊃ G)

(6) 1. A ⊃ B

    2. (A & B) ⊃ I

    3. (H & I) ⊃ S / A ⊃ (H ⊃ S)

(7) 1. (A v B) ⊃ [(C v D) ⊃ (E & H)]

    2. (E & H) ⊃ J / A ⊃ (C ⊃ J)

(8) 1. A ⊃ [W ⊃ (E & H)]

    2. [A & (W & H)] ⊃ (D & ~ M) / A ⊃ (W ⊃ ~M)

PROVING SENTENCES TAUTOLOGICAL

In Chapter 5 we used truth-tables to prove sentences tautological. Recall that tautologies are also called **logical truths** because they can be shown true using only the procedures of logical theory, without investigating the physical world. This section presents one way to prove, using natural deduction, that a sentence is a tautology or logical truth.

Suppose we assume the antecedent of a particular conditional and proceed to construct, using no premises whatsoever, a conditional proof sequence that ends with the consequent of the conditional. This proves that *if* the antecedent of the conditional is true, the consequent of the conditional must be true as well. It follows that it is not possible for the antecedent to be true and the consequent false. Therefore, the conditional must be a tautology, because the only way a conditional can be false is if its antecedent is true when its consequent is false.

So, to prove a conditional sentence tautological with this method, construct a conditional proof sequence using no premises, with the antecedent of the conditional as the assumption. If you succeed in deriving the consequent, you will have proven that the conditional is tautological. Here is an extremely short example:

1. | J | AP
2. | J v B | Add 2
3. J ⊃ (J v B) | CP 1-2

This proves that J ⊃ (J v B) is a tautology. Here is another example:

1. | [(A v B) & ~A] | AP
2. | A v B | Simp 1
3. | ~A | Simp 1
4. | B | DS 2, 3
5. [(A v B) & ~A] ⊃ B CP 1– 4

Notice that we derived [(A v B) & ~A] ⊃ B without using any premises, which proves that the formula is tautological.

Let us prove that

$$[J \supset (B \supset R)] \supset [(J \supset B) \supset (J \supset R)]$$

is a tautology. This is a conditional whose antecedent is itself a conditional and whose consequent is itself a conditional as well. Indeed, the consequent of the antecedent is itself a conditional, and the antecedent and consequent of the consequent are each also conditionals. When you begin a proof such as this, proceed in two steps: (1) set up the overall structure of the proof; and (2) fill that structure in. In order to set up the overall structure, you must arrange the assumed premises in the proper order. Let us first assume the antecedent of the formula as a whole:

1. | J ⊃ (B ⊃ R) | AP

Next, assume within this sequence the next antecedent, that is, the antecedent of the whole formula's consequent:

1. | J ⊃ (B ⊃ R)   AP
2. | | (J ⊃ B)   AP

Within this nested sequence, assume the last antecedent:

1. | J ⊃ (B ⊃ R)   AP
2. | | (J ⊃ B)   AP
3. | | | J   AP

You now have the proof's overall structure worked out. The completed proof follows:

1. | J ⊃ (B ⊃ R) AP
2. | | (J ⊃ B) AP
3. | | | J   AP
4. | | | B ⊃ R   MP 1, 3
5. | | | B   MP 2, 3
6. | | | R   MP 4, 5
7. | | | (J ⊃ R) CP 3–6
8. | | (J ⊃ B) ⊃ (J ⊃ R) CP 2–7
9. [J ⊃ (B ⊃ R)] ⊃ [(J ⊃ B) ⊃ (J ⊃ R)] CP 1–8

A proof whose last line is **P** and that contains no premises is called a premise-free proof of **P**. If a premise-free proof of **P** can be constructed in TD, **P** is called a **theorem** of TD. In such a case, **P** is also said to be "truth-functionally true" because, using only the methods of truth-functional logic, **P** can be proven true. In this case, **P** is also called "logically true", because **P** can be proven true using only the methods of logic—without investigating the physical world.

A natural deduction proof demonstrates that the conclusion of the proof must be true if the premises are true. If a proof begins with no premises, and the assumption is discharged, this demonstrates that the conclusion must be true absolutely and not just if certain premises are true.

▶   **EXERCISE 9.4**   ◀

Construct premise-free proofs for the following tautologies.

1.*[(A ∨ B) & ~A] ⊃ B

2. [(J ⊃ I) & (I ⊃ I)] ⊃ (J ⊃ I)

3. A ⊃ [(A ⊃ B) ⊃ B]

4. (A ⊃ B) ⊃ [(A & E) ⊃ (B & E)]

5.*[(A ⊃ B) & (A ⊃ I)] ⊃ [A ⊃ (B & I)]

6. A ⊃ [(B & ~B) ⊃ K]

7. [(A ⊃ B) & ~B] ⊃ ~A

8. (A ⊃ B) ⊃ [A ⊃ (A & B)]

9. [(A ⊃ B) & (B ⊃ C)] ⊃ (A ⊃ C)

10. {[(A ⊃ B) & (E ⊃ S)] & (A v E) } ⊃ (B v S)

THE LAW OF NONCONTRADICTION

We have been presupposing that contradictions are logically impossible. Anything counts as logically possible except that which is contradictory. Recall that an explicit contradiction may be defined as any statement of the form **P** & ~ **P**. We used a truth-table in Chapter 6 to prove that any statement of the form **P** & ~ **P** is necessarily false, that is, can't possibly be true. The principle that contradictions are impossible is called the **Law of Noncontradiction** (or "LNC" for short). The LNC was formulated in ancient times by the world's first mathematicians and philosophers. Some logic students have a skeptical attitude toward the LNC. Now that you understand the nature of a formal proof, you have the background to understand one important consideration in support of the LNC: If a reasoning process begins with a contradictory premise, no matter what the contradiction is, by applying the obviously valid rules of Simplification, Addition, and Disjunctive Syllogism, *any and every* arbitrarily chosen conclusion validly follows. For example, in the following proof, A abbreviates *Ann is 16* and B abbreviates *The earth is perfectly flat*:

1. A & ~ A / B

2. A   Simp 1

3. A v B   Add 2

4. ~ A   Simp 1

5. B   DS 3,4

Assuming just the contradiction *Ann is 16 and Ann is not 16*, it validly follows that the earth is perfectly flat. Notice that this proof would be correct no matter what sentence B abbreviates.

In general, if someone accepts a contradiction **P** & ~**P**, then any and every sentence validly follows, as the next diagram demonstrates. Beginning with a contradiction, and using three obviously valid inference rules, the following line of reasoning proves **Q**, where **Q** is *any* sentence:

1. **P & ~ P / Q**

2. **P**   Simp 1

3. **P v Q**   Add 2

4. **~ P**   Simp 1

5. **Q**   DS 3, 4

Thus, if one contradiction is true, it validly follows that every statement is true: It is true that the total population of Japan is exactly 17, it is true that the moon is made entirely of paper, and so on. After reflecting on these considerations, perhaps you will agree that absurd consequences follow if we reject the Law of Noncontradiction.

CHAPTER 9 GLOSSARY

Conditional proof   A proof in which you indent, assume the antecedent of a conditional, deduce the consequent of the conditional, and end the indentation and assert the conditional.

Explicit contradiction   A statement of the form **P & ~ P**.

Law of Noncontradiction   The principle that contradictions are impossible.

Logical truth A statement that can be known on the basis of logical theory alone, without investigating the physical world.

Nested proof   A conditional or indirect proof inside another conditional or indirect proof.

Reductio ad absurdum proof   A proof in which you begin by assuming the logical opposite of the claim you wish to prove. You then demonstrate that this assumption implies a contradiction and deduce the opposite of the assumption, namely the claim you originally sought to prove. Reductio ad absurdum proofs are also called *indirect proofs* because in such a proof the conclusion is derived in an indirect way, by first deriving a contradiction from the denial or opposite of the conclusion and inferring the conclusion from that.

Theorem   A formula that can be proven true without the use of premises. A theorem of TD is a formula that can be proven true using a premise-free proof in TD.

10 REPLACEMENT RULES

T he rules we have used so far—inference rules—apply only to *whole* lines of proofs. For instance, if we are to apply Modus Ponens to two lines of a proof, one whole line must be of the form **P ⊃ Q** and another whole line must be simply the corresponding formula **P**. In contrast, the rules we will now consider—**replacement rules**—may be applied to whole lines or to *parts* of lines.

Before we introduce the first replacement rule, let us briefly consider the rationale underlying this new type of rule. Suppose we take a valid argument, remove a premise, and replace the premise with a logically equivalent formula. (Just as a mechanic might remove a car's carburetor and replace it with a mechanically equivalent carburetor.) Will the argument remain valid? (Will the car still run?) Certainly. Substituting one formula for another won't affect the argument's validity if the two *interchanged* formulas are equivalent. Now, suppose we take a valid argument, remove a subformula from within a premise, and replace the subformula with a logically equivalent formula. (A **subformula** is a well-formed formula that constitutes part of a bigger formula.) Will the argument remain valid? Clearly, it will remain valid. Substituting one subformula for another, when the two are logically equivalent, won't affect the argument's validity. This reasoning underlies the replacement rules.

Subformulas

In the following examples, the subformulas are underlined:

(<u>A v K</u>) & (<u>E ⊃ G</u>)   <u>~H</u> ⊃ (<u>J & J</u>)

However, in the following examples, the underlining does not mark off subformulas:

(G <u>v</u> S) ⊃ (R & H)   J <u>≡</u> (R v ~S)

THE COMMUTATIVE RULE

Any sentence **P** v **Q** is equivalent to the corresponding sentence **Q** v **P**. In a disjunction, the order of the two disjuncts doesn't affect the truth-value of the disjunction. And certainly any sentence **P** & **Q** will be equivalent to the corresponding sentence **Q** & **P**. The order of the conjuncts doesn't affect the truth-value of the conjunction. Our first replacement rule, the Commutative rule (Comm), reflects these two truths and is a combination of two rules, as shown in the box below:

The Commutative Rule

A formula **P** v **Q** may replace or be replaced by the corresponding formula **Q** v **P**.

A formula **P** & **Q** may replace or be replaced by the corresponding formula **Q** & **P**.

The replacement is made by rewriting the line and making the replacement as the line is being rewritten. Understand that this replacement operation may be performed on a whole line or on a subformula of a line, at any point in a proof. This should become clear after some examples.

We apply a replacement rule in two ways: to an entire line, or to a part of a line, that is, to a subformula within a line. If we apply a replacement rule to an entire line, we simply rewrite the whole line, writing the replacement in place of the original line. For example, in the following proof, as we go from line 2 to line 3, the Commutative rule is applied, and the entire line consisting of B v A is replaced with A v B:

1. (A v B) ⊃ E

2. B v A / E

3. A v B   Comm 2

4. E   MP 1, 3

After A v B was derived at line 3, we applied MP to lines 1 and 3 to derive E.

If we apply a replacement rule to a part of a line—to a subformula—we rewrite the whole line and make the replacement at the appropriate slot in the formula. In the following proof, as we go from line 1 to line 3, line 1 is rewritten in its entirety, but as it is being rewritten, the subformula A v B inside line 1 is replaced by the subformula B v A:

1. (A v B) ⊃ E

2. B v A / E

3. (B v A) ⊃ E   Comm 1

4. E   MP 2, 3

Notice that as we wrote in line 3, we rewrote *all* of line 1 and replaced A v B with B v A in the process. The inference was justified by the Commutative rule. Thus, line 3 is exactly like line 1 except that the subformula A v B was replaced with the formula B v A.

In each of the following, the Commutative rule is applied to a whole line. Specifically, the second formula is derived from the first by Comm:

1. A v B	**1.** A & B	**1.** (E ≡ F) v G	**1.** ~ (A & B) & ~ E
2. B v A	**2.** B & A	**2.** G v (E ≡ F)	**2.** ~ E & ~ (A & B)

However, in each of the following, the Commutative rule is applied to a *sub-formula* within the first line and the second formula is derived from the first by Comm:

1. (H & R) ⊃ G	**1.** (S v G) & F
2. (R & H) ⊃ G	**2.** (G v S) & F

So, Comm tells us that if the main operator is a wedge—or if it is an ampersand—we can reverse the order of the components without affecting the truth-value of the formula.

THE ASSOCIATIVE RULE

According to the Associative rule (Assoc), any sentence (**P** v **Q**) v **R** is equivalent to the corresponding sentence **P** v (**Q** v **R**), and any sentence (**P** & **Q**) & **R** is equivalent to the corresponding sentence **P** & (**Q** & **R**). In a triple disjunction or triple conjunction, the placement of the parentheses does not affect the truth-value of the whole. In other words, the parentheses can be shifted and the truth-value will not be affected. This can easily be confirmed on a truth-table. The following box summarizes the Associative rule.

The Associative Rule

A formula (**P** v **Q**) v **R** may replace or be replaced with the corresponding formula **P** v (**Q** v **R**).

A formula (**P** & **Q**) & **R** may replace or be replaced with the corresponding formula **P** & (**Q** & **R**).

In each case below, the second sentence was derived from the first by applying the Associative rule.

1. (A v B) v E	**1.** A & (B & E)	**1.** [(A & B) v (E & S)] v (G & H)
2. A v (B v E)	**2.** (A & B) & E	**2.** (A & B) v [(E & S) v (G & H)]

Here is a proof employing this new rule:

1. (A v B) v E
2. ~ A & ~ B / E
3. A v (B v E)   Assoc 1
4. ~ A   Simp 2
5. B v E   DS 3, 4
6. ~ B   Simp 2
7. E   DS 5, 6

Notice that at line 3 we applied the Associative rule to line 1. This allowed us to shift the brackets in order to set things up for the application of DS on line 5.

It is crucial that you remember this. If a pair of parentheses has a tilde attached, the Associative rule cannot be applied. For example, in the following case, you may not apply the Associative rule:

$$\sim (A \lor B) \lor E$$

So, Assoc tells us: if the operators are all ampersands or all wedges, we may shift the brackets without affecting the truth-value of the formula.

> No associative or commutative rules apply to the horseshoe. A truth-table will confirm that $(P \supset Q) \supset R$ is *not* equivalent to $P \supset (Q \supset R)$ and that $P \supset Q$ is *not* equivalent to $Q \supset P$.

THE DOUBLE NEGATION RULE

If Pat is not a non-Hindu, Pat is Hindu. If someone's tractor "ain't got no traction," that tractor has traction. In English, if one negation operator applies *directly* to another, the two cancel each other out and the result is as if neither of the two negation operators applied. The Double Negation rule (DNeg) in logic corresponds to ordinary English usage:

> *The Double Negation Rule*
>
> A formula $\sim \sim P$ may replace or be replaced with the corresponding formula **P**.

For example, we can apply Double Negation to replace

$$\sim \sim A \lor B$$

with

A v B

We also can replace ~ ~ A ⊃ B with A ⊃ B.

In each case below, the second sentence was derived from the first by an application of Double Negation:

1. G ≡ ~ ~ H	**1.** H v ~ ~ (R ⊃ G)	**1.** ~ ~ (C & E)	**1.** A
2. G ≡ H	**2.** H v (R ⊃ G)	**2.** (C & E)	**2.** ~ ~ A

Here are two short proofs that require Double Negation:

1. ~ ~A ⊃ S		**1.** ~ ~ B ⊃ K	
2. A` / S		**2.** B / K	
3. ~ ~A	DNeg 2	**3.** B ⊃ K	DNeg 1
4. S`	MP 1, 3	**4.** K	MP 2, 3

It is important that you understand the following: Two negatives cancel each other out only in case the first applies *directly* to the second, with no intervening parenthetical device. That is, DNeg may be applied to a formula that has two tildes only if one tilde applies directly to a second tilde, with no intervening parenthetical devices. Thus, in each of the following cases Double Negation is correctly applied to the first sentence in order to derive the second sentence:

1. ~ ~ A & ~ ~ B	**1.** ~ ~ (A v B)	**1.** J v ~ ~G
2. A & B	**2.** (A v B)	**2.** J v G

However, Double Negation may *not* be applied to the tildes in the following two cases:

~ (~ A & B)

A v ~ (~ B ⊃ G)

In the above two cases, the first tilde does not apply to the second tilde. Rather, the first tilde applies to the parenthesis that comes between the two tildes.

DEMORGAN'S RULE

In Chapters 3 and 5 you saw that any sentence ~ (**P** & **Q**) is equivalent to the corresponding sentence ~ **P** v ~ **Q**. (To say "Ann and Bob are not both home" is equivalent to saying "Either Ann is not home or Bob is not home.") You also saw that any sentence ~ (**P** v **Q**) is equivalent to the corresponding sentence (~ **P** & ~ **Q**). (To say "It's not the case that either Ann or Bob is home" is equivalent to saying "Ann is not home and Bob is not home.") Augustus DeMorgan (1806–71) was perhaps the first logician to draw attention to these logical equivalencies. Consequently, the following rule, which reflects these logical equivalencies, bears his name.

DeMorgan's Rule (DM)

A formula ~ (**P** & **Q**) may replace or be replaced with the corresponding formula ~ **P** v ~ **Q**.

A formula ~ (**P** v **Q**) may replace or be replaced with the corresponding formula ~ **P** & ~ **Q**.

A formula (**P** & **Q**) may replace or be replaced with the corresponding formula ~(~ **P** v ~ **Q**).

A formula (**P** v **Q**) may replace or be replaced with the corresponding formula ~(~ **P** & ~ **Q**).

Here is an algorithm that captures the essential meaning of DM. This algorithm, which will also count as an application of DM, allows the above replacements plus additional replacements as well.

If a formula or subformula has an ampersand or wedge as its main connective:

1.   Change the ampersand to a wedge or the wedge to an ampersand.

2.   Negate the formula to the immediate left of the ampersand or wedge and negate the formula to the immediate right of the ampersand or wedge.

3.   Negate the formula or subformula as a whole.

For example, let us use this algorithm to apply DeMorgan's rule to ~ (~ A v B) in three steps:

Given: ~ (~ A v B)

1.   First, we change the ampersand to a wedge or the wedge to an ampersand. In this case, we therefore change the wedge to an ampersand:

$$\sim (\sim A \ \& \ B)$$

2.   Next, we negate each side of the ampersand or wedge. Here, we negate the ~A and also negate the B:

$$\sim(\sim \sim A \ \& \sim B)$$

3.   Finally, we negate the formula as a whole:

$$\sim \sim (\sim \sim A \ \& \sim B)$$

Thus, the formula ~ (~ A v B) is equivalent to the formula ~ ~ (~ ~A & ~ B), and one may replace the other anywhere in a proof. (This equivalence can easily be confirmed on a truth-table.) However, the formula ~ ~ (~ ~A & ~ B) is a bit too top-heavy with tildes. In a real proof, we probably would dump the extra tildes as soon as possible by applying Double Negation. So, if we next apply

Double Negation to the formula, we can cancel out the double negatives, and this gives us

$$A \ \& \sim B$$

So, $\sim (\sim A \ v \ B)$ is actually equivalent to $A \ \& \sim B$.

Let's try another. If we use the algorithm to apply DM to $\sim (A \ \& \ B)$, the three steps are

Step 1. $\sim (A \ v \ B)$ (We changed the & to v)

Step 2. $\sim(\sim A \ v \sim B)$ (We negated each side of the v)

Step 3. $\sim \sim(\sim A \ v \sim B)$ (We negated the whole)

Thus, $\sim (A \ \& \ B)$ is equivalent to $\sim \sim (\sim A \ v \sim B)$. This, too, can easily be confirmed on a truth-table. And by applying Double Negation, $\sim \sim (\sim A \ v \sim B)$ may be replaced by $(\sim A \ v \sim B)$. So $\sim(A \ \& \ B)$ is equivalent to $\sim A \ v \sim B$. One may replace the other anywhere in a proof.

Here are additional examples. In each of the following, the second formula was derived from the first by applying the DM algorithm:

1. $\sim(A \ v \sim B)$ 1. $\sim (\sim A \ \& \ B)$ 1. $(A \ \& \sim B)$
2. $\sim \sim(\sim A \ \& \sim \sim B)$ 2. $\sim \sim(\sim \sim A \ v \sim B)$ 2. $\sim(\sim A \ v \sim \sim B)$

Notice that DeMorgan's rule applies only to conjunctions and disjunctions—it does not apply to horseshoes and triple bars.

Let us now employ DM within a proof. Consider the following argument:

1. $(A \ \& \ B) \ v \sim (E \ v \ F)$
2. $(\sim A \ v \sim B)$
3. $\sim F \supset (A \ v \ E) \ / \ E \ v \ (A \ v \ S)$

This argument is valid. However, using only the eight inference rules, it cannot be proven valid in TD. This is where the replacement rules are indispensable. Notice that line 2 displays a DM pattern. By applying DM to line 2, we can derive

4. $\sim (\sim \sim A \ \& \sim \sim B)$ DM 2

We next apply Double Negation twice to this formula:

5. $\sim (A \ \& \ B)$ DNeg 4

Notice that this is the negation of the left disjunct of line 1. This allows us to break line 1 down by DS:

6. $\sim (E \ v \ F)$ DS 1, 5

Because line 6 instantiates one of the DM forms, we apply DM to it:

7. ~ ~ (~ E & ~ F)   DM 6

Next, Double Negation allows us to derive

8. ~ E & ~ F   DNeg 7

This is fortunate, for now line 8 can be broken down by Simp:

9. ~ E   Simp 8

10. ~ F   Simp 8

Next, notice that lines 3 and 10 instantiate the premise section of MP. This gives us

11. A v E   MP 3, 10

Now we can use Comm to turn 11 into

12. E v A   Comm 11

Using Add, we move S into place:

13. (E v A) v S   Add 12

Finally, we use Association to shift the parentheses and derive our conclusion:

14. E v (A v S)   Assoc 13

THE DISTRIBUTION RULE

Do you remember the distributive principle of multiplication? One way to put that principle is

$$a \times (b + c) = (a \times b) + (a \times c)$$

For example,

$$3 \times (5 + 6) = (3 \times 5) + (3 \times 6)$$

We have a similar principle in logic. The Distribution (Dist) rule has two parts, as summarized in the following box:

The Distribution Rule

A formula **P** v (**Q** & **R**) may replace or be replaced with the corresponding formula (**P** v **Q**) & (**P** v **R**).

A formula **P** & (**Q** v **R**) may replace or be replaced with the corresponding formula (**P** & **Q**) v (**P** & **R**).

This rule reflects the fact that any sentence **P** v (**Q** & **R**) is equivalent to the corresponding sentence (**P** v **Q**) & (**P** v **R**) and any sentence **P** & (**Q** v **R**) is equivalent to the corresponding sentence (**P** & **Q**) v (**P** & **R**). (This can be confirmed on an eight-row truth-table.) The following proof employs Distribution at its first step:

1. A v (B & C)

2. (A v B) ⊃ E / E

3. (A v B) & (A v C)   Dist 1

4. A v B   Simp 3

5. E   MP 2, 4

This proof used Dist on line 3 to replace A v (B & C) with (A v B) & (A v C).

In the next proof, at step 3, a line of the form **P** & (**Q** v **R**) replaces a corresponding line of the form (**P** & **Q**) v (**P** & **R**):

1. (A & B) v (A & C)

2. A ⊃ E / E

3. A & (B v C)   Dist 1

4. A   Simp 3

5. E   MP 2, 4

Notice that when we apply the Distribution rule to a sentence, the main operator switches from the ampersand to the wedge or from the wedge to the ampersand.

▶ **EXERCISE 10.1** ◀

Using any of the eight inference rules plus any of the first five replacement rules, supply justifications for the following arguments.

(1)* **1.** A v (B & C)

2. (A v B) ⊃ S

3. ~ S v (Q v R) / R v Q

4. (A v B) & (A v C)

5. A v B

6. S

7. ~ ~S

8. Q v R

9. R v Q

(2)　**1.** (A v B) v C

　　2. ~ B & R

　　3. A ⊃ F

　　4. C ⊃ S / ~ (~ F & ~ S)

　　5. (B v A) v C

　　6. B v (A v C)

　　7. ~ B

　　8. A v C

　　9. F v S

　　10. ~ (~ F & ~ S)

(3)　**1.** ~ (A v B)

　　2. ~ B ⊃ Z

　　3. H ⊃ ~ Z

　　4. ~ H ⊃ (Q & R) / ~ (~ Q v ~ R)

　　5. ~ ~(~ A & ~ B)

　　6. ~ A & ~ B

　　7. ~ B

　　8. Z

　　9. ~ ~ Z

　　10. ~ H

　　11. Q & R

　　12. ~ (~ Q v ~ R)

(4)　**1.** H & (S v G)

　　2. (~ H v ~ S)

　　3. G ⊃ P / P

　　4. (H & S) v (H & G)

　　5. ~ (~ ~H & ~ ~ S)

　　6. ~ (H & S)

　　7. H & G

      8. G

      9. P

(5)*   **1.** A & B

      2. A ⊃ S

      3. S ⊃ Q

      4. (Q v P) ⊃ H

      5. ~ H v T / T

      6. A

      7. S

      8. Q

      9. Q v P

   10. H

   11. ~ ~ H

   12. T

▶   **EXERCISE 10.2**   ◀

Part I. Use the inference rules and the first five replacement rules to prove the following arguments valid.

(1) **1.** (A v B) ⊃ ~ E

     2. B v (G v A)

     3. ~ G / ~ E

(2) **1.** (H ⊃ S) v ~ R

     2. (A v B) v Q

     3. ~ B & ~ S

     4. (A v Q) ⊃ R / ~ H

(3) **1.** [(A v B) v C] ⊃ S

     2. B v (A v C)

     3. S ⊃ Q / Q

(4) **1.** (A v B) ⊃ G

2. R & H

3. ~ (~ A & ~ B) / G

(5)*1. ~ (A ∨ B)

2. ~ B ⊃ E

3. E ⊃ S / S

(6) 1. A ∨ (B & C)

2. (A ∨ C) ⊃ ~ S

3. H ⊃ S / ~ H

(7)* 1. (H & S) ∨ (H & P)

2. H ⊃ (G ∨ M) / G ∨ M

(8) 1. A & (B ∨ C)

2. (A & B) ⊃ H

3. (A & C) ⊃ O / H ∨ O

(9) 1. ~ (~ A ∨ ~ B)

2. B ⊃ H

3. H ⊃ S / S

(10)* 1. (R ⊃ S) ∨ ~ G

2. ~ ~ G

3. (R ⊃ S) ⊃ ~ ~ P

4. H ⊃ ~ P / ~ H

(11) 1. R ⊃ ~ (H ∨ B)

2. H ∨ (A & B)

3. S ⊃ R / ~ S

(12) 1. (G ≡ H) ∨ ~ F

2. ~ (A ∨ ~ F)

3. (G ≡ H) ⊃ P / P

(13)* 1. ~ (R & S)

2. ~ R ⊃ Q

3. ~ S ⊃ Z

4. ~ Q

    5. Z ⊃ (A ≡ B) / A ≡ B

(14) **1.** (P ⊃ Q) & (R ⊃ S)

    2. (Q v S) ⊃ ~ H

    3. P v (R & B) / ~ H

(15)***1.** ~ R ⊃ ~ S

    2. ~ ~ S

    3. ~ ~ R ⊃ H

    4. (H ⊃ A) & (Z ⊃ O) / A v O

(16) **1.** A v (B & C)

    2. ~ B

    3. ~ A v E / E

(17)***1.** S ⊃ ~ P

    2. (P & S) v (P & F) / F

(18) **1.** ~ (~ H & E)

    2. A & (B & E) / H v S

(19) **1.** ~ A

    2. (A & B) v (H & S) / S

(20)***1.** ~ ~ K

    2. ~(A & B) & ~ (E & F)

    3. (H & K) ⊃ [(A & B) v (E & F)] / ~H

(21) **1.** H ⊃ S

    2. K ⊃ H

    3. ~ (A v S) / ~ (K v A)

(22) **1.** ~ (A v B)

    2. (~ B v C) ⊃ (E & G)

    3. S ⊃ ~ G / ~ S v X

(23) **1.** A ⊃ (B v C)

    2. ~ (B & E)

    3. ~ (C v ~ E) / ~ A

(24) **1.** Q v R

2. [(P v Q) v R] ⊃ ~ S

3. H ⊃ S / ~ H

(25)*1. H ⊃ (E & P)

2. A v (B & ~ C)

3. A ⊃ ~ E

4. ~ C ⊃ ~ P / ~ H

(26) 1. (A v B) ⊃ ~ (H v E)

2. A v (B & H)

3. (A & ~ E) ⊃ (P & Q) / Q v X

(27)*1. ~ (A & B)

2. ~ A ⊃ (P & Q)

3. ~ B ⊃ (Q & S) / Q

(28) 1. R v S

2. ~ (A & R)

3. ~ (A & S) / ~ A

(29) 1. (A v B) ⊃ ~ E

2. (H & A) v (B & E) / A

Part II. Each of the following is a valid argument. In each problem, symbolize the argument and provide a proof of validity.

1.* It is not the case that both Ann and Bob are home. If Sue is home then Bob is home. Either Sue is home or James is not home. James is home. So, Ann is not home.

2. Neither Ann nor Bob is home. Either Bob is home or Ed is not home. If Jan is home then Ed is home. So, Jan is not home.

3. Either Ann is home or Bob or Dave is home. Neither Dave nor Cindy is home. Either it is not the case that Ann or Bob is home or Sue is home. Therefore, Sue is home.

4. Ann is home or Bob and Dave are home. If either Ann or Dave is home, then Cindy is not home. Either Jan is home or Cindy is home. Thus, Jan is home.

5. Ann is home and either Bob or Sue is home. If Joe is home then it is not the case that both Ann and Sue are home. If Joe is not home then Kriss is not home. Either Kriss is home or George is home. It is not the case that Ann and Bob are both home. So, George must be home.

FIVE MORE REPLACEMENT RULES
THE TRANSPOSITION RULE

Any sentence **P ⊃ Q** is equivalent to the corresponding sentence ~ **Q** ⊃ ~ **P**. (This can easily be confirmed on a four row truth-table.) This gives us our next replacement rule, the Transposition rule (Trans), summarized in the following box:

> ### The Transposition Rule
>
> A formula **P ⊃ Q** may replace or be replaced with the corresponding formula ~ **Q** ⊃ ~ **P.**

Here's one way to remember this rule: Do you remember the movie *Trading Places?* The character played by Eddie Murphy traded places with the character played by Dan Akroyd, and in the process, each became the opposite of what he was, in the sense that one went from being rich to being poor, and the other went from poor to rich. Now, when you apply Transposition to a formula **P ⊃ Q**, **P** and **Q** trade places and, in the process, each is also negated and becomes the opposite of what it was prior to the switch.

Here are some examples of inferences allowed by the Transposition rule:

1. A ⊃ B   1. ~ S ⊃ ~ G   1. (A & B) ⊃ ~ (H v E)   1. ~ A ⊃ B
2. ~ B ⊃ ~ A   2. ~ ~G ⊃ ~ ~S   2. ~ ~(H v E) ⊃ ~ (A & B)   2. ~ B ⊃ ~ ~A

In each case, the second sentence was derived from the first by an application of Transposition.

The following proof employs Transposition:

1. E ⊃ (A ⊃ B)
2. (~ B ⊃ ~ A) ⊃ S
3. (E ⊃ S) ⊃ H / H
4. (A ⊃ B) ⊃ S   Trans 2
5. E ⊃ S   HS 1, 4
6. H   MP 3, 5

THE IMPLICATION RULE

A sentence **P ⊃ Q** is equivalent to the corresponding sentence ~ **P** v **Q**. (This can easily be confirmed on a truth-table.) This is the basis of our next replacement rule, the Implication rule (Imp), as summarized in the following box:

> ### *The Implication Rule*
>
> A formula **P** ⊃ **Q** may replace or be replaced with the corresponding formula ~ **P** v **Q**.

The following algorithm, which you may use to apply Imp, captures the essential meaning of this rule:

   To replace a horseshoe with a wedge or a wedge with a horseshoe:

1. Trade the horseshoe for a wedge or the wedge for a horseshoe.

2. Negate the formula to the immediate left of the horseshoe or wedge.

In each example below, the second sentence was derived from the first sentence in accord with the Implication rule.

1. A ⊃ B	1. E v S	1. ~ H ⊃ ~ G	1. ~ F v ~ P
2. ~ A v B	2. ~ E ⊃ S	2. ~ ~H v ~ G	2. ~ ~F ⊃ ~ P

So, when the operator is either a horseshoe or a wedge, you may negate the left side and change the horseshoe to a wedge or the wedge to a horseshoe, provided that you leave the right side as it is.

   Here is a proof requiring Implication:

1. A ⊃ (G ⊃ E)

2. (E v ~ G) ⊃ S / A ⊃ S

3. A ⊃ (~ G v E)   Imp 1

4. A ⊃ (E v ~ G)   Comm 3

5. A ⊃ S   HS 2, 4

This proof could have been worked this way as well:

1. A ⊃ (G ⊃ E)

2. (E v ~ G) ⊃ S / A ⊃ S

3. (~ G v E) ⊃ S   Comm 2

4. (G ⊃ E) ⊃ S   Imp 3

5. A ⊃ S   HS 1, 4

THE EXPORTATION RULE

A simple truth-table test will show that any sentence (**P** & **Q**) ⊃ **R** is equivalent to the corresponding sentence **P** ⊃ (**Q** ⊃ **R**). In accordance with this, we have the Exportation rule (Exp), as summarized in the following box:

The Exportation Rule

A formula $(P \,\&\, Q) \supset R$ may replace or be replaced with the corresponding formula $P \supset (Q \supset R)$.

In each of the following examples, the Exportation rule has been applied to the first sentence in order to derive the second sentence:

1. $(A \,\&\, B) \supset E$		**1.** $[(H \vee S) \,\&\, G)] \supset M$
2. $A \supset (B \supset E)$		**2.** $(H \vee S) \supset (G \supset M)$

The following proof involves Exportation:

1. $A \supset (B \supset S) \,/\, B \supset (A \supset S)$

2. $(A \,\&\, B) \supset S$ Exp 1

3. $(B \,\&\, A) \supset S$ Comm 2

4. $B \supset (A \supset S)$ Exp 3

THE TAUTOLOGY RULE

Any sentence P is equivalent to the corresponding sentence $P \vee P$, and any sentence P is also equivalent to the corresponding sentence $P \,\&\, P$. Thus, the Tautology rule (Taut), summarized in the following box, is certainly valid:

The Tautology Rule

A formula P may replace or be replaced with the corresponding formula $P \vee P$.

A formula P may replace or be replaced with the corresponding formula $P \,\&\, P$.

Here is a proof employing this rule:

1. $A \supset (S \supset G)$

2. $A \vee A$

3. $S \vee S \,/\, G \,\&\, G$

4. A Taut 2

5. S Taut 3

6. $S \supset G$ MP 1, 4

7. G   MP 5, 6

8. G & G   Taut 7

At line 4, A v A was replaced with A, and A was later used to derive S ⊃ G. At line 5, S v S was replaced by S, and this was used to derive G. Tautology allowed us to replace G with G & G.

The following proof contains another application of the Tautology rule.

1. ~ (A & E)

2. ~ (A & B)

3. E v B / ~ A

4. ~ ~ (~ A v ~ E)   DM 1

5. ~A v ~E   DNeg 4

6. ~ ~(~ A v ~ B)   DM 2

7. ~A v ~B   DNeg 6

8. ~ E v ~ A   Comm 5

9. ~ B v ~ A   Comm 7

10. ~ ~E ⊃ ~ A   Imp 8

11. E ⊃ ~ A   DNeg 10

12. ~ ~B ⊃ ~ A   Imp 9

13. B ⊃ ~ A   DNeg 12

14. ~ A v ~ A   CD 3, 11, 13

15. ~ A   Taut 14

THE EQUIVALENCE RULE

Any sentence **P** ≡ **Q** is equivalent to the corresponding sentence (**P** ⊃ **Q**) & (**Q** ⊃ **P**). Also, a sentence **P** ≡ **Q** is equivalent to the corresponding sentence (**P** & **Q**) v (~ **P** & ~ **Q**). The Equivalence rule (Equiv), summarized in the following box, reflects these two principles:

The Equivalence Rule

A formula **P** ≡ **Q** may replace or be replaced with the corresponding formula (**P** ⊃ **Q**) & (**Q** ⊃ **P**).

A formula **P** ≡ **Q** may replace or be replaced with the corresponding formula (**P** & **Q**) v (~ **P** & ~ **Q**).

The following inferences are among those permitted by the Equivalence rule:

1. A ≡ B   1. A ≡ B   1. ~ D ≡ ~ K
2. (A ⊃ B) & (B ⊃ A)   2. (A & B) v (~ A & ~ B) 2. (~D ⊃ ~K) & (~ K ⊃ ~ D)

The following proof uses this rule.

1. A ≡ B

2. ~B / ~ A

3. (A ⊃ B) & (B ⊃ A)   Equiv 1

4. A ⊃ B   Simp 3

5. ~ A   MT 2, 4

ARE REPLACEMENT RULES WORTH THE BOTHER?

The replacement rules are sometimes difficult to work with, but, without them our proof system is limited. Many valid arguments can't be proven valid without the replacement rules. For example, using only the rules of inference, we could not construct a proof of the following simple and obviously valid argument:

1. (A v B) ⊃ C

2. (B v A) / C

Notice that MP does not apply directly to lines 1 and 2 of this argument. In order to prove this argument valid, we must employ the Commutative rule. Two alternative proofs are possible:

I. 1. (A v B) ⊃ C   II. 1. (A v B) ⊃ C
   2. (B v A) / C   2. (B v A) / C
   3. A v B   Comm 2   3. (B v A) ⊃ C   Comm 1
   4. C   MP 1, 3   4. C   MP 2, 3

The proofs below display some interesting patterns of reasoning that sometimes show up within longer proofs. Understanding them may help you plan strategy in cases of complex proofs.

(1)   1. I / J ⊃ I

      2. I v ~ J   Add 1

      3. ~ J v I   Comm 2

      4. ~ ~J ⊃ I   Imp 3

      5. J ⊃ I   DNeg 4

(2)   1. J v I

      2. ~ J v I / I

	3.	I v ~ J	Comm 2
	4.	~ I ⊃ ~ J	Imp 3
	5.	~ J ⊃ I	Imp 1
	6.	~ I ⊃ I	HS 4, 5
	7.	~ ~I v I	Imp 6
	8.	I v I	DNeg 7
	9.	I	Taut 8
(3)	1.	~ J / J ⊃ I	
	2.	~ J v I	Add 1
	3.	~ ~ J ⊃ I	Imp 2
	4.	J ⊃ I	DNeg 3
(4)	1.	J & ~ J / I	
	2.	J	Simp 1
	3.	~ J	Simp 1
	4.	J v I	Add 2
	5.	I	DS 3, 4
(5)	1.	J ⊃ I / J ⊃ (I v S)	
	2.	~ J v I	Imp 1
	3.	(~ J v I) v S	Add 2
	4.	~ J v (I v S)	Assoc 3
	5.	~ ~ J ⊃ (I v S)	Imp 4
	6.	J ⊃ (I v S)	DNeg 5
(6)	1.	J ⊃ I	
	2.	J ⊃ W / J ⊃ (I & W)	
	3.	~ J v I	Imp 1
	4.	~ J v W	Imp 2
	5.	(~ J v I) & (~ J v W)	Conj 3, 4
	6.	~ J v (I & W)	Dist 5
	7.	~ ~ J ⊃ (I & W)	Imp 6
	8.	J ⊃ (I & W)	DNeg 7

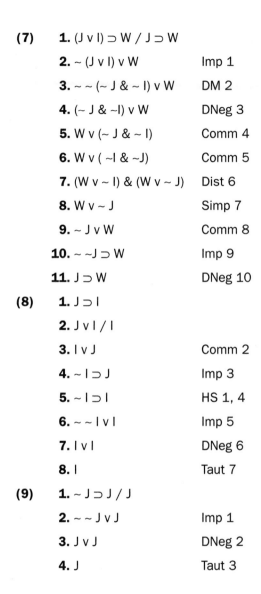

(7)   **1.** (J v I) ⊃ W / J ⊃ W

    2. ~ (J v I) v W   Imp 1

    3. ~ ~ (~ J & ~ I) v W   DM 2

    4. (~ J & ~I) v W   DNeg 3

    5. W v (~ J & ~ I)   Comm 4

    6. W v (~I & ~J)   Comm 5

    7. (W v ~ I) & (W v ~ J)   Dist 6

    8. W v ~ J   Simp 7

    9. ~ J v W   Comm 8

   10. ~ ~J ⊃ W   Imp 9

   11. J ⊃ W   DNeg 10

(8)   **1.** J ⊃ I

    2. J v I / I

    3. I v J   Comm 2

    4. ~ I ⊃ J   Imp 3

    5. ~ I ⊃ I   HS 1, 4

    6. ~ ~ I v I   Imp 5

    7. I v I   DNeg 6

    8. I   Taut 7

(9)   **1.** ~ J ⊃ J / J

    2. ~ ~ J v J   Imp 1

    3. J v J   DNeg 2

    4. J   Taut 3

►   **EXERCISE 10.3**   ◄

Using any of the inference and replacement rules, fill in the justifications for the following.

(1)*   **1.** J ⊃ I

    2. (~ I ⊃ ~ J) ⊃ S

3. (S v H) ⊃ [(A & B) ⊃ C]

4. A / ~ C ⊃ ~ B

5. ~ I ⊃ ~ J

6. S

7. S v H

8. (A & B) ⊃ C

9. A ⊃ (B ⊃ C)

10. B ⊃ C

11. ~ C ⊃ ~ B

(2)   **1.** J ≡ I

2. (J ⊃ I) ⊃ W

3. (~ J v ~ I)

4. W ⊃ S / S & (~ J & ~ I)

5. (J ⊃ I) & (I ⊃ J)

6. J ⊃ I

7. W

8. S

9. (J & I) v (~ J & ~ I)

10. ~ (~ ~J & ~ ~I)

11. ~ (J & I)

12. ~ J & ~ I

13. S & (~ J & ~ I)

(3)   **1.** A ⊃ B

2. E ⊃ B

3. ~ (~ A & ~ E)

4. B ⊃ G / G

5. ~ ~ (~ ~ A v ~ ~ E)

6. (~ ~A v ~ ~E)

7. A v E

8. B v B

   9. B

   10. G

(4)   **1.** ~ S ⊃ S

   2. ~ S v G

   3. G ⊃ ~ (J v I)

   4. ~ J ⊃ Z / Z

   5. ~ ~S v S

   6. S v S

   7. S

   8. ~ ~ S

   9. G

   10. ~ (J v I)

   11. ~ ~ (~ J & ~ I)

   12. ~ J & ~ I

   13. ~ J

   14. Z

(5)*   **1.** A ⊃ B

   2. (~ A v B) ⊃ J

   3. H ⊃ ~ J

   4. ~ H ⊃ I

   5. (I v I) ⊃ (S v S) / S

   6. ~ A v B

   7. J

   8. ~ ~ J

   9. ~ H

   10. I

   11. I v I

   12. S v S

   13. S

► **EXERCISE 10.4** ◄

Part I. Use any of the replacement and inference rules to prove the following.

(1) * **1.** H ⊃ ~ S
    2. (~ H v ~ S) ⊃ F
    3. F ⊃ B / B

(2) **1.** ~ A ⊃ B
    2. A ⊃ E
    3. B ⊃ S
    4. (E v S) ⊃ X / X

(3) **1.** J ⊃ I
    2. (~ I ⊃ ~ J) ⊃ S
    3. F ⊃ ~ S / ~ F

(4) **1.** H ⊃ S
    2. ~ J ⊃ ~ S / H ⊃ J

(5) * **1.** ~ (A & ~ B) ⊃ C
    2. A ⊃ B
    3. H v ~ C / H

(6) **1.** A ⊃ B
    2. A ⊃ ~ B
    3. ~ A ⊃ G / G

(7) * **1.** (J v J) ⊃ S
    2. J & F / S

(8) **1.** A ⊃ (B ⊃ C) / B ⊃ (A ⊃ C)

(9) **1.** H ⊃ (A ⊃ E)
    2. (E v ~ A) ⊃ S / H ⊃ S

(10) * **1.** (A & E) ⊃ S
    2. (E ⊃ S) ⊃ F / A ⊃ F

(11) **1.** ~ J v I
    2. ~ I v J / J ≡ I

(12) **1.** B ≡ G

    2. (B & G) ⊃ Z

    3. (~ B & ~ G) ⊃ Z / Z

(13)*1. J ≡ I

    2. ~ (J & I) / ~ J & ~ I

(14) **1.** A ⊃ (B & C)

    2. B ⊃ (C ⊃ S) / A ⊃ S

(15)*1. (A ⊃ B) & (E ⊃ B)

    2. ~ (~ A & ~ E) / B

(16) **1.** ~ [(J ⊃ I) & (I ⊃ J)]

    2. (A & S) ⊃ (J ≡ I)

    3. A / ~ S

(17) **1.** (A ⊃ B) ⊃ [(T v I) & (H ≡ J)]

    2. (T v I) ⊃ [(H ≡ J) ⊃ X] / (A ⊃ B) ⊃ X

(18) **1.** A ⊃ (B & E) / (A ⊃ B) & (A ⊃ E)

(19) **1.** H ⊃ ~ H

    2. ~ (H & S) ⊃ G / ~ (H v ~ G)

(20)*1. A ⊃ B

    2. A ⊃ ~ B / ~ A

(21) **1.** A v ~ I

    2. ~ A v G / I ⊃ G

(22) **1.** ~ A ⊃ S

    2. ~ E ⊃ S

    3. ~ (A & E) / S v H

(23)*1. (J ⊃ I) ⊃ W

    2. W ⊃ ~ W / J

(24) **1.** J v (I & S)

    2. J ⊃ S / S

(25)*1. J ⊃ (S & H)

    2. W ⊃ (~ S & ~ H)

    3. J v W / H ⊃ S

(26) 1. A ⊃ B / A ⊃ (B ∨ E)

(27) 1. A ⊃ ~ (B ⊃ G)

     2. A ∨ G / A ≡ ~ G

(28)*1. (A ⊃ B) ⊃ G

     2. (I ⊃ B) ⊃ ~ G / ~ B

(29) 1. (A ∨ B) ⊃ G

     2. (I ⊃ G) ⊃ S

     3. B / S

(30)*1. (J ⊃ I) ⊃ (I ⊃ J)

     2. (J ≡ I) ⊃ ~ (A & ~ B)

     3. I & A / A & B

(31) 1. A ≡ B

     2. A ∨ B

     3. A ⊃ (B ⊃ E) / E

(32) 1. A ⊃ (B & H)

     2. (H ∨ S) ⊃ Z / A ⊃ Z

Part II. Each of the following arguments is valid. In each problem, symbolize the argument and construct a proof.

1. If it is not the case that Ann swims, then it is not the case that Bob swims. If it is the case that either Bob does not swim or Ann does swim, then Sue swims. Therefore, Sue swims.

2. If Ann and Bob both swim then Ed will swim. Ann swims. Therefore, if Bob swims then Ed swims.

3. Ann jogs if and only if Ed jogs. Ed jogs if and only if Bob jogs. So, if Bob does not jog, then Ann does not jog.

4. Ann does not swim or Jan swims. If Jan swims, then Ed swims. So, if Ed does not swim, then Ann does not swim.

▶   **EXERCISE 10.5.**   ◀

Some interesting problems.

1. There has been a mix-up at the office. It is your job to determine in a logical way whether Pat did or did not go home three hours early. You have the following information.

It is not the case that Ann and Bob both worked overtime. Either Bob or Chris did not work overtime but it's not the case that both did not work overtime. Either Chris did not work overtime or Dave went home one hour early. If Ann did not work overtime then Edna did not go home two hours early. It is not the case that either Edna did not go home two hours early or Kari did not go home one hour early. If Dave went home one hour early then Pat went home three hours early.

Did Pat go home three hours early? Use natural deduction to *prove* that your answer follows from the information given above.

2. Use at least one replacement rule in each of the following problems.

    a. Show that MT is superfluous or redundant in our natural deduction system. That is, show how to derive a formula ~**P** from formulas **P** ⊃ **Q** and ~**Q** without using MT.

    b. Show that MP is superfluous.

    c. Show that HS is superfluous.

    d. Prove two other rules superfluous.

CHAPTER 10 GLOSSARY

Replacement rule A rule that allows you to replace one subformula with another logically equivalent formula or one formula with another logically equivalent formula. In contrast with *inference* rules, which must be applied to whole lines only, replacement rules may be applied to whole lines or to *parts* of lines.

Subformula A well-formed formula within a larger formula.

C h a p t e r

11

INDIRECT AND CONDITIONAL PROOFS WITH REPLACEMENT RULES

INDIRECT PROOFS WITH REPLACEMENT RULES

The replacement rules allow us to construct indirect proofs for many arguments that would not otherwise receive an indirect proof. First, let us review the Indirect Proof rule.

The Rule of Indirect Proof

Anywhere in a proof, you may do the following:

1. Indent and assume the negation of the conclusion you seek to prove, writing AP (for "assumed premise") as justification.

2. Using the assumption plus any previous lines that are available, derive a contradiction, that is, a whole line of the form **P** & ~**P**.

3. End the indentation and draw a vertical line in front of the indented steps (to mark them off from the other lines). On the next line assert the original conclusion that you sought to prove, that is, the assumption at step 1 above except with the negation removed. (This counts as "discharging" the assumption.) As a justification, write IP (for "indirect proof") and cite the indented lines.

Proviso: The indented lines produced are called an "indirect proof sequence." Once an assumption has been discharged and the conclusion has been derived, the lines in the indirect proof sequence may no longer be used to derive future lines in the proof. In other words, once the assumption has been discharged, the indented lines are no longer "available."

In short, to prove **P**: Indent, assume ~ **P**, derive a contradiction, end the indentation, assert **P**.

Here is an example of an indirect proof that employs replacement rules:

1. ~[(A & ~B) & ~C]

2. ~(B v C) / ~ A

3.	~ ~A	AP
4.	~ [A & (~B & ~C)]	Assoc 1
5.	~ [A & ~ (~ ~B v ~ ~C)]	DM 4
6.	~ [A & ~ (B v C)]	DNeg 5
7.	~ ~ [~ A v ~ ~ (B v C)]	DM 6
8.	~ A v ~ ~ (B v C)	DNeg 7
9.	~ ~ (B v C)	DS 3, 8
10.	B v C	DNeg 9
11.	(B v C) & ~ (B v C)	Conj 10, 2
12. ~ A		IP 3–11

We began the above proof by negating the conclusion. At the next step, line 4, we applied Assoc to line 1 and shifted the parentheses to the right one notch. At the next step, line 5, we applied DeMorgan's rule just to the subformula ~ B & ~ C in line 4. From there, the conclusion was derived using DNeg, DM, DS, and Conj. Here is a another sample:

1. A ⊃ B

2. B ⊃ E

3. A v B / E v H

4.	~(E v H)	AP
5.	~ ~(~E & ~ H)	DM 4
6.	(~E & ~ H)	DNeg 5
7.	~E	Simp 6
8.	~B	MT 2, 7
9.	~A	MT 1, 8
10.	B	DS 3, 9
11.	B & ~B	Conj 8, 10
12. E v H		IP 4–11

EXERCISE 11.1

Use the Indirect Proof rule to prove the following.

(1) 1. J ⊃ (A & B)

    2. (A v E) ⊃ R

    3. (E v J) / R

(2) 1. (A ⊃ A) ⊃ B

    2. (B v G) ⊃ C / C

(3) 1. A ⊃ (B ⊃ E)

    2. B ⊃ (E ⊃ ~ B) / ~ A v ~ B

(4)* 1. (H v S) ⊃ (A & B)

    2. (B v F) ⊃ K

    3. H v F / K

(5) 1. A ⊃ B

    2. I ⊃ ~ B

    3. ~ K ⊃ (A & I) / K v S

(6) 1. J ⊃ (I ⊃ R)

    2. J ⊃ I

    3. ~ K v (R v J) / ~ K v R

(7)* 1. (J & I) v E

    2. ~ E v I / J ⊃ I

(8) 1. (J & I) ⊃ (E v F)

    2. (J ⊃ E) ⊃ (I ⊃ F)

    3. I / F

(9) 1. (J v I) ⊃ (E & H)

    2. (H v A) ⊃ (B v ~ E)

    3. (B v K) ⊃ ~ (J & H) / ~ J

(10) 1. J ⊃ [(A ⊃ A) ⊃ (K v H)]

    2. H ⊃ ~ (X v ~ X) / J ⊃ K

(11) 1. A ⊃ (S ⊃ G)

    2. A ⊃ S

    3. I ⊃ (A v G) / G v ~ I

(12)*1. J ⊃ (~ I ⊃ S)

    2. (J ⊃ I) ⊃ S / S

(13) 1. ~[(A ⊃ B) v (B ⊃ A)]

    2. E v [S v (B ⊃ A)]

    3. ~H ⊃ ~(S v E) / H

(14) 1. A v B

    2. ~B v C / A v C

(15) 1. A ⊃ (B & C)

    2. (B v C) ⊃ W

    3. ~ C ⊃ (A & W) / W

(16) 1. A ⊃ B

    2. C ⊃ D

    3. (B v D) ⊃ H

    4. ~H / ~ (A v C)

(17) 1. A (B ⊃ C)

    2. ~C

    3. (A ⊃ ~B) ⊃ C / ~B

(18) 1. ~[(A & ~ B) & ~ C]

    2. ~(B v C) / ~A

(19) 1. A ⊃ B

    2. B ⊃ E

    3. A v B / E v Z

(20) 1. A & (H & B)

    2. [(J v H) & X] ⊃ E

    3. (C ⊃ ~E)

    4. X / ~C

(21) 1. ~ A v B

    2. B ⊃ C

    3. H ⊃ (A & ~C)

    4. H v D / D

```
┌────────────────────────────────────────────────────────────┐
│ ▶                    EXERCISE 11.2                        ◀  │
└────────────────────────────────────────────────────────────┘
```

Go back to the exercises in the previous chapter and prove five of the arguments valid using the Indirect Proof rule.

CONDITIONAL PROOF WITH REPLACEMENT RULES

The replacement rules allow us to construct conditional proofs for many arguments that would not otherwise be given a conditional proof. Before we begin, let us review the rule.

Rule of Conditional Proof

If, at any point in a proof, you wish to derive a conditional $P \supset Q$, you may do the following:

1. Indent and assume the antecedent of the conditional you seek to prove. (Write AP as justification.)

2. Using this assumption plus, if needed, any available lines occurring earlier, derive the consequent of the conditional.

3. Draw a vertical line in front of the indented steps (to mark them off from the other lines). The indented lines constitute a conditional proof sequence. The completion of this step is called discharging the assumption.

End the indentation and infer the conditional whose antecedent is the assumption with which you began the conditional proof sequence and whose consequent is the latest line in that sequence. As a justification, write CP (for "Conditional Proof") and cite the line numbers of the indented lines within the sequence.

Proviso: Once an assumption has been discharged and the conditional has been derived, the indented lines in the assumption's conditional proof sequence are no longer "available" to be used in justification of further lines in the proof.

In short: To prove a sentence of the form $P \supset Q$: Indent, assume P, derive Q, end the indent, and assert $P \supset Q$.

Examine the following proof:

1. $(A \supset B) \supset \sim (E \supset S)$

2. $(\sim E \vee S) \vee H$

3. H ⊃ J / B ⊃ J

4.	B	AP
5.	B v ~ A	Add 4
6.	~ A v B	Comm 5
7.	A ⊃ B	Imp 6
8.	~ (E ⊃ S)	MP 1, 7
9.	~ (~ E v S)	Imp 8
10.	H	DS 2, 9
11.	J	MP 3, 10
12. B ⊃ J	CP, 4–11	

Notice the derivation, between lines 4 and 7, that runs from B to A ⊃ B. That is a pretty tricky stretch of reasoning. In Chapter 10, we saw how to validly derive a sentence of the form **P ⊃ Q** from just the corresponding sentence **Q**. These lines constitute an instance of that (valid) derivation.

▶ **EXERCISE 11. 3** ◀

Use the Conditional Proof rule to prove each of the following.

(1)* **1.** A ⊃ (B ⊃ E)
  2. A ⊃ (H ⊃ E)
  3. ~ E / A ⊃ ~ (B v H)

(2) **1.** A ⊃ (B & C)
  2. B ⊃ (A & G) / A ≡ B

(3)* **1.** J ⊃ I
  2. (J & W) ⊃ E
  3. (I & E) ⊃ S / J ⊃ (W ⊃ S)

(4) **1.** J ⊃ (~I v W)
  2. (I ⊃ W) ⊃ (~I v S) / J ⊃ (I ⊃ S)

(5) **1.** (A & B) ⊃ C
  2. A ⊃ B / A ⊃ C

(6) **1.** I ⊃ Z
  2. Z ⊃ A / I ⊃ (Z & A)

(7)   **1.** (J v I) ⊃ S

    2. ~ J ⊃ ~ A / A ⊃ (~S ⊃ T)

(8)   **1.** H ⊃ (G ⊃ J)

    2. G ⊃ (J ⊃ W) / H ⊃ (~W ⊃ ~ G)
(Hint: a nested proof is recommended.)

(9)   **1.** A ⊃ [(J v I) ⊃ (M & N)]

    2. (N v S) ⊃ G / A ⊃ (I ⊃ G)
(Hint: a nested proof is recommended.)

(10)   **1.** A ⊃ (~B v C)

    2. ~H v ~C / H ⊃ (A ⊃ ~ B)
(Hint: a nested proof is recommended.)

(11)* **1.** A ⊃ (B v C)

    2. E ⊃ S

    3. B ⊃ C / A ⊃ C

►   **EXERCISE 11.4**   ◄

Go back to the exercises in the previous chapter and prove five of the arguments valid using the Conditional Proof rule.

PROVING TAUTOLOGIES

Suppose we begin an indirect proof by assuming the negation of a sentence **P**. Suppose we then proceed to construct, *using no premises whatsoever,* an indirect proof sequence. If we derive a contradiction, this proves that our initial assumption (~**P**) implies a contradiction. Our assumption is therefore contradictory, because only that which is contradictory implies a contradiction. It follows from this that the sentence **P** must be tautological, for in truth-functional logic, the negation of a contradiction is a tautology.

So, to prove that a particular sentence **P** is tautological with this method, construct an indirect proof sequence using no premises, with the negation of the sentence as the assumption. If you reach a contradiction, you've proven that the sentence **P** is tautological. For example:

1.	~ (A v ~A)	AP
2.	~ ~ (~ A & ~ ~A)	DM 1
3.	~ A & ~ ~A	DNeg 2

4.	~A & A	DNeg 3
5.	A & ~A	Comm 4
6. A v ~A	IP 1–5	

Notice that we derived a line of the form **P & ~ P** at line 5, which concluded the indirect proof sequence. We then discharged our assumption and derived the conclusion, A v ~A, without using any premises, which proves that A v ~ A is a tautology. Here is another example:

1.	~ ~ (A & ~ A)	AP
2.	(A & ~A)	DNeg 1
3. ~ (A & ~A)	IP 1–2	

This proves that ~ (A & ~A) is a tautology.

The following premise-free conditional proof shows that the sentence (A ⊃ B) ⊃ (~ B ⊃ ~ C) is tautological:

1.	A ⊃ B	AP
2.	~ B ⊃ ~ C	Trans 1
3. (A ⊃ B) ⊃ (~ B ⊃ ~ C)	CP 1–2	

Here is one more example. The following premise-free proof proves that (D v B) ⊃ ~ (~ D & ~ B) is tautological:

1.	~ [(D v B) ⊃ ~ (~ D & ~ B)]	AP
2.	~[~ (D v B) v ~ (~ D & ~ B)]	Imp 1
3.	~ ~[~ ~(D v B) & ~ ~ (~ D & ~ B)]	DM 2
4.	[~ ~(D v B) & ~ ~ (~ D & ~ B)]	DNeg 3
5.	(D v B) & (~ D & ~ B)	DNeg 4 (twice)
6.	(D v B)	Simp 5
7.	(~ D & ~ B)	Simp 5
8.	~ D	Simp 7
9.	B	DS 6, 8
10.	~ B	Simp 7
11.	B & ~ B	Conj 9, 10
12. (D v B) ⊃ ~ (~ D & ~ B)	IP 1–11	

Recall from Chapter 9 that if a sentence can be proven with a premise-free proof, the sentence is called a *logical truth* and is said to be "logically true" because it can be known to be true using the procedures of logical theory alone, without investigating the physical world. Also, recall that a sentence that can be proven with a premise-free proof in our natural deduction system TD is a *theorem of TD*.

▶   **EXERCISE 11.5**   ◀

Construct premise-free proofs for the following tautologies.

1. ~[(A ⊃ ~A) & (~A ⊃ A)]

2. ~(B & G) ⊃ (~B v ~G)

3.*~A v (B v A)

4. (A ⊃ B) ⊃ (~A v B)

5. (A ≡ B) ⊃ [(A ⊃ B) & (B ⊃ A)]

6. [(J & I) ⊃ I] ⊃ [J ⊃ (I ⊃ I)]

7. ~(A & ~A) v (B & ~B)

8.*~(J & I) v ~(~ J & ~ I)

9. ~J ⊃ (J ⊃ I)

10. ~{J & ~[(J ⊃ I) ⊃ I]}

11.*~J ⊃ ~ (I & J)

12. (J ⊃ I) v (I ⊃ J)

13. (J ⊃ ~J) v (~J ⊃ J)

14.*A ⊃ (I ⊃ A)

15. [~(A & ~B) & ~B] ⊃ ~A

16. [(J ⊃ I) ⊃ J] ⊃ J

17. (~A ⊃ B) v (A ⊃ E)

18. A ≡ [A & (S v ~S)]

19. (A ⊃ B) v (~B ⊃ A)

20. (A ⊃ B) ⊃ [(A ⊃ ~B) ⊃ ~A]

21. [(A ⊃ B) & (B ⊃ C)] ⊃ (~ C ⊃ ~ A)

22. [~ (A & B) & B] ⊃ ~A

23.*A ⊃ [(A ⊃ B) ⊃ B]

CHAPTER 11 GLOSSARY

Logical Truth   A sentence that can be known to be true using the procedures of logical theory alone, without investigating the physical world.

Tautology   In truth-functional logic, a sentence whose truth-table has all Ts in the final column. A tautology thus has the following feature: it cannot possibly be false.

TD   The name of the natural deduction system employed in Chapters 7–11.

Theorem of TD   A sentence that can be proven with a premise-free proof in the system TD.

Part

II

TWO INFORMAL TOPICS

Chapter

12   DEFINITION

Have you ever been in an argument with someone and suddenly realized that the other person does not understand what you are saying? As the prison guard said to Cool Hand Luke, "What we have here is a failure to communicate."

Communication is a difficult process. It takes work, it takes understanding, and often it ends in failure. When one person attaches one meaning to a word while the other person attaches a different meaning to it, neither really understands what the other is saying.

For instance, during the cold war, intellectuals from the Soviet Union argued that capitalism destroys freedom. However, proponents of capitalism argued that Soviet communism is the true enemy of freedom. How could two groups of intelligent people make apparently opposite claims? Part of the answer lay in the fact that proponents of capitalism attached one *definition* to the term "freedom" and Soviet communists attached an entirely different one. Today, we still cannot agree on the meaning of this important word.

When words aren't defined adequately, words are not understood; when words are not understood, reasoning breaks down, for you can't appreciate the reasoning within an argument if you don't understand the argument. And so logical theory is concerned with the nature of definition and with the process by which we come to understand the meanings of the words we employ.

A **definition** is an explanation of the meaning of a word or phrase. Definitions define words, not the entities we use words to refer to. For instance, a definition of *car* defines the word, it does not define the actual physical object you drive down the street. A definition can take several standard forms. Two of the most common are

- _____ means _____. For example: The word "triangle" means "three-sided closed figure."

- _____ is by definition _____. For example: A triangle is by definition a three-sided closed figure.

The word (or phrase) to be defined is the **definiendum,** and the word (or phrase) doing the defining is the **definiens.** So, in the above example, *triangle* is the definiendum, and *three-sided closed figure* is the definiens.

THE PURPOSES OF DEFINITION

Why do we construct definitions? Definitions serve several purposes. And, just as we use any tool better when we understand its purpose, when we understand the purposes definitions serve, we can better understand their nature, which in turn can guide us in constructing definitions of our own.

1. Describe a Common Meaning   Sometimes the purpose of a definition is simply to describe the word's commonly understood meaning. This common meaning is the **lexical meaning**—the type of meaning typically provided in a dictionary (a lexicon).

2. Remove Ambiguity   Sometimes we define words in order to remove *ambiguities* that arise. A word is **ambiguous** within a particular context if it can be interpreted as having two or more meanings in that context. Many words have two or more meanings, but the context usually tells us which meaning is intended. For instance, suppose Susan works in a town with no river, and needs to cash a check on her lunch break. "Do you know where the nearest bank is?" she asks. In such a context, it is obvious that she means a commercial bank rather than a river bank.

However, when the context does not indicate the meaning, a term is being used ambiguously. For instance, imagine that farmer John has a great big alfalfa *field* and is also an expert in the *field* of medieval literature. One day someone from the local Rotary club invites him to come and give a talk about his field. The term is ambiguous, and John needs more information before writing his speech.

> In Lewis Carroll's *Alice's Adventures in Wonderland,* Alice asks the Mouse to tell her his history, and the Mouse replies: "Mine is a long and sad tale!" Looking down "with wonder" at the Mouse's *tail,* Alice says, "It is a long tail, certainly, but why do you call it sad?"

Ambiguity can ruin communication. It can also lead to fallacious or ineffective argumentation, as we shall see in the next chapter.

3. Reduce Vagueness Sometimes a word's meaning is not precise enough for our purposes and we need to give it a more exact meaning. A definition is then put forward in hopes of making the meaning more precise. For instance, suppose a legislature imposes a tax on all "motor vehicles." It is obvious that the term *motor vehicles* applies to Chevrolet trucks, Ford Escorts, and such. However, does a snowmobile qualify as a motor vehicle? How about a jet-ski? It may not be clear whether *motor vehicle* was meant to apply in these cases. In this situation, the term *motor vehicle* is vague. The legislature may need to state a more precise definition of the term.

A word is **vague** if it has borderline cases. A **borderline case** for a word is a case where we are not sure whether the word applies or does not apply. In the above example, snowmobiles and jet-skis constitute borderline cases for the term *motor vehicle,* for we were not sure whether the term applies in those cases.

For another example, suppose a group is awarding a scholarship to a "needy" student. Joe lives on his own and earns only $12,000 per year. The question arises: Does the term *needy* apply in his case?

In most cases, a vague word or phrase will have various, perhaps many definite applications, and certain borderline cases where we aren't sure whether the word or phrase applies or not. With *motor vehicles*, cars and trucks are in the "definite" group, and jet-skis and snowmobiles are along the border. Thus, a word or phrase that has borderline cases is said to have "fuzzy boundaries of application."

The problem with a vague word is *not* that the word lacks a meaning, but that we are not sure of the *limits* of the word's application, that is, in some cases we are not sure if the word applies. To remedy vagueness, we need to clarify the meaning, and we do this by providing a more precise definition, one that will determine a decision in the questionable cases.

Many words are used in vague ways. Words such as *rich, poor, love, peace, happy, obscene,* and *nice* have numerous borderline cases. For instance, the word *rich* certainly applies in the case of Bill Gates, but does it apply in the case of Bill Clinton?

Incidentally, vagueness is not always a bad thing. Sometimes we want our words to be vague. For instance, suppose Joe goes to his first Chicago Bulls game and wants to identify Dennis Rodman. "Describe him for me," Joe says. So, his friend begins, "Well, he's tall." The word *tall* allows us to say something about someone's height without having to specify precisely what the height is.

4. Persuade People sometimes construct definitions to influence the attitudes or emotions of others, perhaps to persuade them to adopt a viewpoint. For instance, a socialist might define socialism as "democracy applied to the economy." The socialist hopes that the positive emotions aroused by the word *democracy* will help win approval for the socialist ideal. On the other hand, a libertarian might define socialism as "a state of affairs in which a powerful government dominates all aspects of life." The libertarian hopes that the negative emotions aroused by the term *dominates* will arouse negative feelings toward socialism.

5. Stipulate a New Meaning   Sometimes we need to give a word a new meaning, a meaning independent of actual usage. In such cases, we stipulate a new meaning. For example, as computers rose to prominence, we needed a word to describe those who break into computer systems, and the word *hacker* acquired a new meaning.

6. Provide a Theoretical Explanation   Some definitions are formulated in order to present a theory about the nature of something. For example, during the nineteenth century, James Clerk Maxwell theorized that light is an electromagnetic field. Maxwell's electromagnetic theory of light helped us understand the nature of light and at the same time provided a meaning for the word *light*. In the eighteenth century, Count Rumford discovered that heat is the motion (mean kinetic energy) of molecules. Rumford's discovery provided a theoretical picture of the nature of heat, and it also provided a meaning for the word *heat*.

FIVE TYPES OF DEFINITION

Because definitions can serve different purposes, we have different kinds of definitions. Closely related to the five purposes above are the following five types of definition.

1. Lexical Definitions   A **lexical definition** reports a word's commonly understood meaning. Lexical definitions are consequently true if they correctly report the common meaning and false if they do not. Dictionary definitions are usually lexical definitions.

2. Stipulative Definitions   A **stipulative definition** is a new proposal for the use of a word. Rather than report an existing standard meaning for a term, a stipulative definition institutes a meaning chosen for a purpose. Such a definition is neither true nor false, because it is not reporting a way a word is commonly understood.

The stipulative definition can be a first definition for a new term, as when *software* was given a brand new meaning, or it can be a new meaning given to a term that already has a meaning, as when *cool* began to mean something besides "low temperature."

In one episode of *Seinfeld*, someone gave Elaine a gift, and Elaine realized that the gift was actually something she had given the person the year before. Jerry invented a new word to describe such a person: the person was a "regifter." A regifter is someone who saves a gift and gives it to someone else as a gift later on (and thus avoids the trouble of buying a gift). Jerry "coined" a new word and stipulated its new meaning.

Science and mathematics often stipulate new meanings in order to simplify the expression of complex ideas. For example: mathematicians stipulate that an expression such as 7^5 means "$7 \times 7 \times 7 \times 7 \times 7$." A chemist stipulates that

2 H + O → H$_2$O means "Two hydrogen atoms combine with one oxygen atom to make a molecule."

3. Precising Definitions

A **precising definition** begins with a word that has a vague but established meaning and makes that meaning more precise so that the word can be applied in various borderline cases. Legislatures frequently construct precising definitions in order to clarify the meaning of a law. For instance, suppose a new building code is imposed on "dwellings" and the question arises: Does *dwelling* apply to a summer cabin? Does the term apply to a trailer in a trailer park? As a result of such questions, a bill might include a precising definition: "For the purposes of this law, the term *dwelling* shall mean 'any structure containing at least one bed and in which someone sleeps on a regular basis.' "

A precising definition aims to combat vagueness. However, in constructing a precising definition, we are not free to make up just any meaning we wish. The definition must remain in line with ordinary usage, yet it must add enough precision to help us decide borderline cases. Precising definitions thus differ from stipulative definitions in that a stipulative definition does not have to remain in line with established usage.

Our courts of law have had to formulate precising definitions for words such as *obscenity, free speech,* and *sexual harassment.* Currently, our society is attempting to make the meaning of the word *death* more precise. We are debating precisely when a person is medically dead; important issues concerning insurance liability, inheritance, and life support depend on the outcome of this debate.

4. Persuasive Definitions

A **persuasive definition** aims to influence attitudes. By defining a word in a certain way, we hope that the definition will call up a favorable or an unfavorable attitude toward what the word denotes. For example, someone opposed to abortion might define *abortion* as "the killing of a defenseless human being." On the other hand, a proponent of abortion might define *abortion* as "a woman choosing to have something done to her body."

An interesting example of a persuasive definition is the definition given by the World Health Organization (WHO) for the term *health:* "a state of complete mental, physical, and social well-being." Critics have raised several objections to this definition. First of all, this is not what we commonly mean by the word *health.* There have been many healthy Olympic athletes who nevertheless lived in unjust societies. For example, many of the German athletes at the 1936 Olympics, held in Germany under Hitler's watchful eye, did not live in a state of social well-being. Yet they were healthy athletes. The World Health Organization's definition of *health* is a persuasive definition because it is aimed at persuading us to accept a controversial political position: that U.N. efforts aimed at promoting world health must also promote a political agenda as well.

5. Theoretical Definitions

A **theoretical definition** characterizes the nature of something. It provides a theoretical picture of an entity, that is, a way of understanding the entity. For example, when the physicist defines light as an electromagnetic field, this provides a theory of the nature of light.

Definitions in science are often theoretical definitions because they draw on scientific theory to characterize things. For example, when the astronomer defines the *morning star* as "Venus when seen in the morning sky," we are given a new, theoretical way of understanding the morning star. When *heat* is defined as "the motion of molecules," we are given a theory as to the nature of heat. Theoretical definitions provide us with a theoretical picture of a part of our world, and this deepens our understanding of the world.

▶   **EXERCISE 12.1**   ◀

---

1. In one *Seinfeld* episode, Jerry's stereo wasn't working and Kramer came up with a scheme to acquire a new stereo for free: he would break the stereo into pieces, and Jerry would take it to the Post Office and tell them it had broken during shipment. It wouldn't cost the taxpayers anything, Kramer argued, because the Post Office would simply treat it as a "write-off." Jerry looked at Kramer and said, "What is a write-off?" When Kramer frowned and shrugged his shoulders, Jerry said, "You don't even know what a write-off is." What type of definition did Kramer need? Explain your answer.

2. *In the seventeenth century, Isaac Newton helped us understand the concept of physical force when he provided the following definition: "Force equals mass times acceleration" ($F = M \times A$). What type of definition was this?

3. Fred wants to go to dinner at Burger King for their tenth anniversary dinner, and Sue wants to go to Wendy's. In the middle of intense negotiations, Sue says to Fred, "Now you're just *quibbling*." Fred says, "I'm what?" What type of definition does Fred need?

4. Give an original example (that is, an example not discussed in this book) of each of the following kinds of definition.

   a. lexical

   b. theoretical

   c. precising

   d. stipulative

   e. persuasive

5. Give an original example of a vague word and explain why the word is vague.

6. Give an original example of an ambiguous use of a word and explain why the word is ambiguous in that context.

7. Make up a stipulative definition to introduce a new term and explain why your definition would be useful.

8. Write a precising definition for each of the following. If you think your definition is inadequate, explain why it is inadequate.

   a. old person

   b cool person

   c. rich person

   d. conservative

   e. liberal

   f. rock 'n' roll

   g. cold weather

   h. poverty

9. Write a persuasive definition for:

   a. socialism

   b. capitalism

   c. religion

   d. communism

   e. intellectual

   f. racism

   g. prejudice

   h. altruism

   i. selfishness

10. Give a theoretical definition for the following. (These may require some research.)

   a. soap

   b. detergent

   c. hormone

   d. lightning

   e. cloud

   f. smog

   g. atom

   h. electricity

11. Determine whether the following definitions are lexical, stipulative, precising, theoretical, or persuasive.

a.* *Decadent* means in a state of decline or decay.

b. *Cursory* means hasty, superficial, not thorough.

c.* A star is a large ball of hot, burning hydrogen gas.

d. *Cool* shall mean below sixty degrees and above forty degrees Fahrenheit.

e. Let's say a "plate scraper" is a person who finishes every bit of food on his plate and obnoxiously scrapes the plate with his fork to make sure he's gotten virtually every molecule of food.

f. *Jeopardy* means danger or risk of loss or injury.

g. A fanatic is a person possessed by an irrational zeal, especially for a ridiculous religious or political cause.

h.* A juggernaut is something that draws blind and destructive devotion, or to which people are ruthlessly sacrificed.

i. A gene is a portion of a DNA molecule.

j. By the term *capitalism* I shall mean a system in which people have the freedom to make their own economic decisions without the government telling them what to do.

k. An ingrate is an ungrateful person.

l. *Frank* means open and sincere in expression; undisguised, straightforward.

m. *Capacious* means "able to contain a large quantity."

n. An anomaly is a deviation from the normal order.

o.* *Ancillary* means "subordinate."

p. *Worker* means "person who is oppressed by a capitalist."

q. *Capitalist* means "person whose act of saving creates jobs for others and benefits society."

r. *Pulsar* means "dense neutron star with a high rate of spin."

TWO TYPES OF MEANING

As noted earlier, a definition is an explanation of the meaning of a word or phrase. Logicians draw a distinction between two general types of meaning—denotative meaning and connotative meaning—and this distinction helps illuminate the techniques we employ when we construct a definition.

> The **denotation** *of a word or expression, also called the* extension *of the word or expression, is all those entities to which the word or expression can truly be applied.*

So, for instance, the denotation of the word *city* includes Seattle, Los Angeles, Portland, Boston, and so on.

> The **connotation** *of a word or expression, also called the* intension *of the word or expression, is the features or attributes an entity must have for the word or expression to truly apply to it. That is, the connotation of a word is the properties that determine whether an entity is a member of the class of entities denoted by the word.*

For instance, the intension of the word *square* includes "figure with four equal angles," "figure with four equal sides," and so on. In short, the connotation of a word is the properties that determine whether the word applies or not.

Some words have an intension even though they have no extension. For example, words such as *unicorn, Leprechaun,* and *Santa Claus* have an intension, for these words have meanings; however they lack an extension, for they do not apply to any existing entities. (Such terms are said to be "empty" terms.)

Also, two different words can have the same intension. For example, *car* and *automobile* mean the same thing and thus have the same intension. Furthermore, if two words have the same intension, they have the same extension. For example, because *car* and *automobile* have the same intension, they obviously apply to exactly the same objects as well and thus have the same extension. The intension of a term determines the term's extension, for the intension serves as the criterion or guide for deciding which entities are or are not in the extension of the term.

The **extensional** or **denotative meaning** of a term consists of the class of objects to which the term may correctly be applied, that is, the members of the class that the term denotes. This collection is the **extension** or **denotation** of the term.

The **intensional** or **connotative meaning** of a term consists of the qualities or attributes the term connotes. In other words, the objects in a term's extension have common attributes or characteristics that lead us to apply the term to them, and this collection of attributes shared by all and only those objects in a term's extension is the **intension** or **connotation** of the term.

In some situations, it may be necessary to distinguish between the subjective, objective, and conventional connotations of a word. The **subjective connotation** of a word relative to a particular person consists of all the attributes that particular person *believes* to be possessed by the items in the word's extension. The subjective connotation of a word can vary from one speaker to another. For example, when Archie Bunker thinks of the word *welfare,* the connotation differs from that which arises in Michael ("Meathead") Stivik's mind. The **objective connotation** of a word consists of those attributes actually possessed by items in the extension, and the **conventional connotation** of a word consists of all the attributes the term commonly connotes.

Terms may be put into an **order of increasing intension**. In such a series, each term (after the first) connotes more attributes, which is to say that each term in the series is more specific than the one before. Here is a series arranged in order of increasing intension:

Animal, mammal, feline, house cat, calico cat

Terms may also be put into an **order of increasing extension**. In this type of series, each term (after the first) denotes a class having more members than the one before. For example:

Calico cat, house cat, feline, mammal, animal

Notice that the order of increasing extension equals the order of decreasing intension, and the order of increasing intension equals the order of decreasing extension.

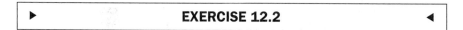

▶   **EXERCISE 12.2**   ◀

1. Choose a word and distinguish between your subjective connotation, the objective connotation, and the word's conventional connotation.

2. Arrange the following in order of increasing intension:

    a. Woodwinds, saxophones, alto-saxophones, musical instruments, old alto saxophones, orchestral instruments

    b.* Snare drums, drums, chrome snare drums, musical instruments, orchestral instruments, dented chrome snare drums

3. Arrange the following in order of increasing extension:

    a. alcoholic beverages, beverages, fine wines, wines, champagnes, cheap champagnes

    b.* athlete, ball player, catcher, baseball player

    c. cheese, dairy products, milk derivatives, American cheese, sliced American cheese

4. For each word, indicate: (a) the intension; (b) part of the extension.

    a. astronaut

    b. famous

    c.* college

    d. nutritious

    e. pope

    f. minister

    g. politician

5. State some of the attributes connoted by the following words:

    a.* Rambo

    b. communist

    c. snake

    d. liberal

    e. conservative

    f. religion

    g. science

CONSTRUCTING A DEFINITION: TECHNIQUES

In general, **extensional** or **denotative definitions** assign meaning to a word or phrase by giving *examples* of what the word or phrase denotes. This technique is accomplished in three common ways.

1. **Ostensive** (from the Latin word *ostens* for "display") or **demonstrative definitions** involve pointing or gesturing at an item belonging to the extension. You might say, "The word *circuit* means this" (as you point at a circuit). An ostensive definition is thus a nonverbal definition.

2. An **enumerative definition** assigns meaning by naming or listing ("enumerating") members of the extension. The listing can be partial or complete. For example, *comedian* means a person such as John Candy, Groucho Marx, or Bob Hope.

3. A **definition by subclass** assigns meaning by naming or listing subclasses of the class of entities denoted by a term. For example, *mammal* means cat, dog, monkey, shrew, aardvark.

An **intensional** or **connotative definition** assigns meaning by indicating the qualities or attributes a word or phrase connotes. That is, this definition lists the properties that an entity must have if the word or phrase is to apply correctly to it. This happens in three common ways.

1. A **synonymous definition** assigns meaning to a word by providing a synonym. In other words, the definiendum is defined in terms of one word or phrase that connotes the same attributes. For example:

physician means "medical doctor"

adage means "proverb"

bard means "minstrel"

2. An **operational definition** assigns meaning to a word or phrase by specifying an operation or set of procedures that determines whether the word or phrase is applied to an entity or not. For example, an operational definition of *acid* is: Dip a strip of blue litmus test paper in the substance. If the litmus paper turns pink, the substance is an acid.

3. An **analytical definition** attempts to explain the meaning of a word or phrase by specifying the characteristics possessed in common by those items to which the word or phrase applies. For example, a mathematician may define a *square* as "a closed figure with four equal sides and four equal angles." All squares have this set of features in common. Or a chemist might define an acid as "a substance that increases the hydrogen-ion concentration of water."

An important type of analytic definition is the **definition by genus and difference**. In this type of definition, a species or kind of entity is defined when we do two things: (a) we specify a general class or *genus* to which all the objects in the species belong, and (b) we narrow this down by indicating how this species differs from other species in the same genus.

A **class** is simply a group of entities having a specified characteristic in common. A **genus** is a general class divided into subclasses. A **species** is a subclass of a larger class. In the definition by genus and difference, when we specify the general class a species belongs to, we are giving the genus. When we state how the members of that species differ from others in the genus, we are stating the "difference."

The definition by genus and difference was favored by Aristotle and has always been a key part of Aristotelian logic. This definition is a type of analytic definition, for the goal is to specify the common attributes possessed by all the members of the species. Here are some examples:

Clock means	"an instrument	used to tell time."
Species ↑	genus ↑	difference ↑
Kitchen means	"a room	in which food is cooked."
Species ↑	genus ↑	difference ↑
Ice means	"water	that is frozen."
Species ↑	genus ↑	difference ↑
Daughter means	"an offspring	that is female."
Species ↑	genus ↑	difference ↑

Aristotle's definition of a human being is one of the most famous definitions in philosophy. The founder of logic defined a human being as a "rational animal." The genus (or general class) in this case is the class of animals, and what distinguishes the species (human beings) from the general class, according to Aristotle, is this difference: rationality. That is, the difference between the human species and the other species within the genus is rationality.

▶ **EXERCISE 12.3** ◀

In each of the following cases, determine whether the definition is an intensional definition or an extensional definition.

1. *A mother is a female parent.

2. Courage: Seattle School Superintendent John Stanford battling cancer.

3. *Sulky* means "sullen."

4. A substance is *soluble* if it dissolves when it is placed in water.

5. *A *politician* is someone such as Richard Nixon, Bill Clinton, Patty Murray.

6. Substance X is *harder than* substance Y: X can be used to scratch Y but Y cannot be used to scratch X.

7. *Classy: Buddy Love in the Jerry Lewis movie *The Nutty Professor.*

8. A groovy person is someone like Austin Powers, the international man of mystery.

9. Negative liberty is the absence of forceful interference in one's life.

10. *Sphere* means "set of points in three dimensional space equidistant from one point."

▶ **EXERCISE 12.4** ◀

For each definition below, identify the technique employed.

1. Political parties are organizations such as the Democratic Party, the Republican Party, the Socialist Workers Party.

2. *Freedom* means "liberty."

4. *Length* is that which is determined by placing a tape rule next to an object.

5. An acid is a substance that dissolves in water and turns blue litmus paper red.

6. A coxcomb is a conceited dandy, a fop.

7. *A coxcomb is a person such as Niles Crane.

8. A *curmudgeon,* as customarily used, means "a cantankerous person."

9. A curmudgeon is, for instance, Andy Rooney, Frank Costanza, Scrooge.

10. *Reptile* means lizard, snake, turtle, etc.

11.*A triangle is a closed, three-sided figure.

12. You want to know what sophisticated means? Okay. Dr. Frasier Crane. That's sophistication.

13. *Prime number* means "an integer whose only divisors are itself and 1."

14. A country bumpkin is someone such as Jethro Bodine or Jed Clampett.

▶   **EXERCISE 12.5**   ◀

---

1. Give a synonymous definition for each of the following.

  a. boring

  b. obnoxious

  c. funny

  d.* house

  e. punctilious

  f. pretentious

2. Give a denotative definition for each of (a) through (c) below.

  a. rock band

  b. politician

  c.* jet

3. Give a connotative definition for:

  a. spiritual

  b. respectful

  c. intellectual

  d.* athletic

  e. software

4. Give an operational definition for the following.

  a. funny

  b. loud

  c.* magnet

    d. buoyant

    e. electrically charged object

    f. hot

▶   **EXERCISE 12.6**   ◀

In each of the following, state the genus and the difference.

1. A tomcat is a male cat.

2. A dictator is an unelected head of state.

3.* A doe is a female deer.

4. A bachelor is an unmarried man.

5. An aardvark is an insect-eating mammal.

6. A soul is an immaterial substance.

7.* A brother is a male sibling.

8. A prime number is an integer whose only divisors are itself and 1.

RULES FOR INTENSIONAL DEFINITIONS

There exist a number of commonly accepted rules that guide the construction of adequate intensional definitions. Definitions in violation of these rules will generally be inadequate definitions.

Rule 1 The definition of a word or phrase should convey the essential properties connoted by the word or phrase. In this context, essential properties are properties that an entity cannot lack and still remain part of the extension of the word. For instance, suppose someone defines a clock as "a mechanism for telling time that contains gears, springs, and cogs, has twelve numbers on its face, and possesses an hour hand and a minute hand." The problem with this definition is that these properties (having gears, springs, hands, and such) are not *essential* features of a clock—digital clocks have no springs and yet they are still clocks. Because this definition cites nonessential features, it does not actually apply to many clocks. However, the following definition does not violate this rule: A triangle is a closed figure that has three sides and three angles. Definitions in mathematics typically cite only essential properties.

Rule 2 The definition should be neither too narrow nor too broad. If a definition is too narrow, it applies to too little; if a definition is too broad, it applies to too much. If the definition is neither too narrow nor too broad, it gives the necessary and sufficient conditions for applying the term.

For example, the following definition is too broad: "Pneumonia is a disease of the lungs." "Disease of the lungs" applies to more than just pneumonia. On the other hand, the following definition is too narrow: "Art is painted images." Other creations besides painted images count as art.

There is a story (perhaps apocryphal) that after Plato died, some students in the Academy, the school he founded, were trying to decide how to properly define a human being. They finally decided on this definition: "A human being is a featherless biped." This definition seemed accurate, until the day Diogenes threw the plucked chicken over the Academy's back fence.

Rule 3 The definition should prefer positive comparisons to negative ones. For example, suppose someone attempts to define a computer and begins by saying that a computer is *not* powered by water, is *not* made entirely of glass, and is *not* typically larger than a television set. This definition conveys little understanding of what a computer *is*. Many items besides computers fit this definition. For instance, an aardvark is not powered by water, is not made entirely of glass, and so on.

Rule 4 The definition should avoid vague, obscure, ambiguous, or figurative language. For example, suppose someone defines a television set as an Alka Seltzer for the mind. This figurative language provides little in the way of understanding for someone who does not already know what a television set is. Or imagine that someone defines love as a big, red rose in the garden of life. This image would apply to much more than just love. The German philosopher Martin Heidegger once defined truth as "the dissimulation of the dissimulated." This definition explains the obscure in terms of the more obscure, and fails to shed light on the notion of truth.

Rule 5 The definition should avoid language that plays on the emotions and attitudes. For example, someone who defines socialism as a system in which the government dominates all aspects of society is hoping that the proposed definition will cause negative feelings toward socialism, but the definition does not adequately explain what the word *socialism* actually means.

Rule 6 The definition should not be circular, with the definiendum used in the definiens. That is, a term should not be used to define itself. For example, the following definition is circular: "Freedom is a state of affairs in which people are free." One who does not already understand the meaning of *freedom* will not understand the full meaning of the phrase "a state of affairs in which people are free." Or, suppose someone defines a determinist as "one who believes in determinism." This too is a circular definition and sheds little light on the meaning of the term "determinist."

► **EXERCISE 12.7** ◄

Which of the six rules of definition is violated in each of the following?

a. A mind is a terrible thing to waste.

b. God is that which is not material, not limited, not visible.

c.* A protester is someone who protests.

d. A depressant is not a stimulant.

e. A soul is a nonmaterial substance that has no weight, physical size, or physical shape.

f.* Time is an old gypsy man ceaselessly moving from one stop to the next.

g. A human being is a carnivorous animal.

h. Rock music is music that is not jazz, not country, not classical, and not blues.

i. *Religious* means obsessively concerned with things that don't really matter.

j.* Peace means the absence of war.

k. *Atheist* means someone who sneers at religion and thumbs his nose at truth and goodness.

l. Fascism means a totalitarian government.

m.* A star is a stellar body.

n. A college student is a student attending a university.

o. A student is one who studies.

p. A square is a figure that has four equal sides.

q. "Beauty is eternity gazing at itself in the mirror." (Kahlil Gibran)

CHAPTER 12 GLOSSARY

Ambiguous   A word is ambiguous within a particular context if it can be interpreted as having two or more meanings in the given context.

Borderline case   A borderline case for a word is a case where we are not sure whether the word applies or does not apply.

Class   A group of entities having a specified characteristic in common.

Connotation (or "intension") of a word or expression   The features or attributes something must have for the word or expression to correctly apply to it.

- The *subjective connotation* of a word relative to a particular person consists of all the attributes that particular person believes to be possessed by the items in the word's extension.

- The *objective connotation* consists of those attributes actually possessed by items in the extension.

- The *conventional connotation* consists of all the attributes the term commonly connotes.

Definiendum   The word (or phrase) to be defined.

Definiens   The word (or phrase) doing the defining.

Definition   An explanation of the meaning of a word or phrase.

Definition by genus and difference   An important type of analytic definition in which a *species* or kind of entity is defined in two steps: a) we specify a general class or *genus* to which all the objects in the species belong; and (b) we narrow this down by indicating how this species differs from other species in the same genus.

Denotation (or "extension") of a word or expression   All those entities to which the word or expression can truly be applied.

Extensional (or "denotative") definition   A definition that assigns meaning to a word or phrase by giving *examples* of what the word or phrase denotes. There are three types:

- *Enumerative definitions* assign meaning by naming or listing members of the extension.

- *Ostensive (or "demonstrative") definitions* involve pointing or gesturing at an item belonging to the extension.

- A *definition by subclass* assigns meaning by naming or listing subclasses of the class of entities denoted by a term.

Extensional (or "denotative") meaning of a term   The class of objects to which the term may correctly be applied, that is, the members of the class that the term denotes.

Genus   A general class divided into subclasses.

Intensional (or "connotative") definition   A definition that assigns meaning by indicating the qualities or attributes a word or phrase connotes, that is, by listing the properties that a entity must have if the word or phrase is to apply to it. There are three common types:

- A *synonymous definition* assigns meaning to a word by providing a synonym.

- An *operational definition* assigns meaning to a word or phrase by specifying an operation or set of procedures that determines whether the word or phrase is applied to a entity or not.

- An *analytical definition* attempts to explain the meaning of a word or phrase by specifying the characteristics possessed in common by those items to which the word or phrase applies. (See also Definition by Genus and Difference.)

Intensional (or "connotative") meaning   The qualities or attributes the term connotes, that is, the common attributes or characteristics that lead us to apply the term.

Lexical meaning   The commonly understood meaning of a word or phrase.

Lexical Definition   A definition that reports a word's commonly understood meaning.

Order of increasing extension   A series in which each term (after the first) denotes a class having more members than the one before.

Order of increasing intension   A series in which each term (after the first) connotes more attributes, which is to say that each term in the series is more specific than the one before.

Persuasive Definition   A definition that aims to influence attitudes.

Precising Definition   A definition that provides a more precise meaning for a word that formerly had a vague but established meaning. The more precise meaning provides additional guidance as to how the word is to be applied in various borderline cases.

Species   A subclass of a larger class.

Stipulative Definition   A definition that constitutes a new meaning for a word or phrase.

Term   A word or phrase that could serve as the subject term of a declarative sentence.

Theoretical Definition   A definition that characterizes the nature of something. Such a definition provides a theoretical picture of an entity, that is, a way of understanding the entity.

Vague   A word is vague if it has borderline cases.

Chapter

13

INFORMAL
FALLACIES

Have you ever been in an argument in which the other person says to you, "No, that's a fallacy!" Sometimes people use the term *fallacy* to refer to a mistaken belief. More often, it is used to refer to an error in reasoning. However, in logic, a **fallacy** is usually defined more narrowly as a defective argument that nevertheless can appear to some to be a correct argument. Fallacies are thus deceptive because they can appear correct even though they are flawed. (The word *fallacy* actually stems from the Latin verb *fallere*, which means "to deceive.") Fallacies are interesting because they can be psychologically persuasive even though they are logically in error.

There are two general types of fallacy. **Formal fallacies** contain errors that may be identified by inspecting the form or abstract logical structure of the argument without reference to the content or subject matter of the reasoning. The flaw lies in the general form of the reasoning rather than in the content. (Formal fallacies were discussed in Chapter 6.) However, **informal fallacies** contain flaws that are not simply due to the logical form of the argument (hence "informal"). With informal fallacies, one must examine the actual content of the reasoning in order to find the error; for the erroneous reasoning is not a matter of pure logical form alone.

The first systematic study of logical fallacies, *On Sophistical Refutations,* was written by Aristotle in the fourth century B.C. "That some reasonings are genuine," Aristotle wrote, "while others seem to be so but are not, is evident." Aristotle catalogued a number of different types of informal fallacy. Since the time of Aristotle, many philosophers have attempted to classify the various informal argumentative fallacies according to a more general scheme. Many systems of classification have been proposed, but no scheme has won widespread support among logicians. We shall follow the proposal of logician Gary Jason by grouping the fallacies into the following three categories: fallacies of no evidence, fallacies of little evidence, and fallacies of language.

Fallacies of no evidence present no evidence whatsoever. That is, the premises provide no logical support for the conclusion. **Fallacies of little evidence** present some evidence, but the evidence is too little or is flawed in some way. **Fallacies of language** involve a misuse of language.

▶ A NOTE ON TERMINOLOGY

During the Middle Ages, logic was one of the core subjects taught in the universities of Europe. In those days, Latin was the language of scholars, and numerous fallacies were given Latin names, many of which remain in use. For most of the fallacies below, we will give both the English name and the traditional Latin name.

FALLACIES OF NO EVIDENCE

▶ THE ARGUMENT AGAINST THE PERSON
  (*ARGUMENTUM AD HOMINEM*)

Senator Jones has just finished giving a speech in favor of a new government program, and Pat and Jan have been watching the senator's speech on C-SPAN. Pat thinks the senator has a good argument, but Jan disagrees:

Jan: That's all a bunch of baloney.

Pat: Why?

Jan: Why? Well . . . he . . . uh. . . . Didn't you know that he takes illegal campaign contributions, he's been divorced twice, and he's an alcoholic? Besides that, he has been known to kick his dog. He is a rotten person.

Jan is attacking the senator with an **argument against the person (ad hominem argument)**. In this type of fallacy, someone attacks a person's character or circumstances in order to oppose or discredit the person's argument or viewpoint. In other words, one attacks the person rather than the person's argument.

This type of "argument" is a fallacy because such personal matters are, by themselves, irrelevant to the truth or falsity of an argument's conclusion. A bad person can still give a good argument, and someone in a bad circumstance can still be right. The mere fact that someone has been bad does not mean his argument is bad. These considerations suggest that an argument must be judged on its logical merits, apart from the arguer's character or circumstances.

Ad hominem arguments fall into two general categories: **abusive ad hominem** arguments attack the person's character, and **circumstantial ad hominems** attack the person's circumstances. Within each category are several common variations. Let us briefly look at each category.

▶ THE ABUSIVE AD HOMINEM

When someone is losing an argument or has no good case to present, he or she may fall back on the personal attack. The first argument above, Jan's argument against the senator, was a standard **abusive ad hominem.** However, arguers have several ways to verbally attack a person, and thus several common types of ad hominem argument.

Tu Quoque ("You're Another") The fallacy known as the **tu quoque** (pronounced too-KWOH-kway) is one variation. Imagine a parent lecturing her teenage son on the dangers of drugs, and the son says, "Why should I listen to you; you drink booze and that's a drug." The implication is that the mother's reasoning is no good for even she doesn't live up to it. Or picture this: a doctor who smokes gives a speech on the evils of smoking. He even lights up a cigarette during the question and answer session. People might be tempted to think that his argument is worthless because he fails to follow his own advice.

The tu quoque attempts to discredit a person's argument by charging the person with hypocrisy or inconsistency. The tu quoque sometimes appears as, "Look who's talking." Essentially, the charge is "We don't need to take his argument seriously because he doesn't believe his own argument, since he doesn't practice what he preaches." This is a fallacy because what a person does with her personal life is logically irrelevant to the logical connection between the premises and the conclusion of her argument. A hypocrite can still give a good argument. If in 1942 Hitler had given a speech in which he argued that genocide and anti-Semitism are immoral, his argument would have been a good one (assuming he used good reasoning), even though he was at the time practicing the very immoralities he condemned. Of course, although the issue of whether someone practices what he preaches is logically irrelevant, it can nevertheless be psychologically persuasive.

Guilt by Association The **guilt by association** fallacy is another type of abusive ad hominem. In this fallacy, one person attacks a second person's associates in order to discredit the person and thereby her argument. For instance, imagine that Professor Jones is giving a lecture on the Constitution and someone says, "Her argument is a bunch of nonsense, didn't you know she associates with suspected communists?"

By attacking the person's associates, the speaker hopes that the audience will reject the person's argument. This fallacy owes some of its psychological punch to our tendency to sometimes judge people by the company they keep. However, even someone who has some "bad" associates might still give a good argument. A person's argument needs to be judged on its logical merits, not on the personal relationships the arguer has or has not made.

▶ THE CIRCUMSTANTIAL AD HOMINEM

Suppose Joe Blow, a fairly rich industrialist, gives an argument in favor of lower taxes. In response, someone opposes Joe's argument with, "His argument is no good. After all, he stands to gain if tax rates are lowered." This is a **circumstantial ad hominem** fallacy. Joe's argument needs to be judged in terms of its logical merits and not in terms of the benefits that someone in his circumstances may receive. A person who stands to gain if the conclusion of his argument is accepted might nevertheless be giving a good argument. If you disagree with this last point, consider a similar situation. At a City Council meeting, Ed the environmentalist is arguing against a policy that would greatly increase the amount of water pollution. An opponent argues that Ed goes swimming, so he stands to benefit from preventing water pollution. Therefore, the opponent concludes, Ed's argument is no good. Is that a good argument against Ed? When someone's argument is attacked on the grounds that the person stands to gain if his conclusion is accepted, the fallacy is sometimes called the "vested interest" fallacy.

In sum, the key feature of all forms of ad hominem argumentation is that the arguer attempts to discredit the argument by personally attacking the arguer rather than the argument. Although this can be psychologically persuasive, it is logically faulty, for even a bad person can sometimes give a good argument. Again, an argument needs to be considered on its own logical merits, apart from the personal characteristics or circumstances of the arguer.

Let us consider one important exception. In a court of law, if the reliability of a witness is at issue, the witness's character or circumstances can be logically relevant. For if a person has a questionable character, or is a known liar, or has a strong motive to lie, or is being coerced by corrupt associates, the jury has good reason to doubt her truthfulness. The attack, in a court of law, on the witness's character or circumstances in such a situation is not fallacious at all.

▶ THE APPEAL TO FORCE (*ARGUMENTUM AD BACULUM*, "ARGUMENT FROM THE STICK")

The ad baculum—**appeal to force**—fallacy is committed when an arguer appeals to force or the threat of force to make someone accept a conclusion. This move is sometimes made when rational argument has failed. The arguer basically says, "Accept my conclusion or you may be harmed."

For example, suppose everyone in the office wants to stay late and finish the project except Ed. Ed thinks everyone should go home at the regular time. The others say to Ed, "You'd better decide to stay or we'll make things miserable for you tomorrow." Ed has been hit with an ad baculum argument. Or imagine an ambassador at a U.N. meeting saying this to a delegate from a small country: "We have the world's biggest bomb, so vote for our proposal."

Of course, the premises of an ad baculum argument are logically irrelevant to the conclusion, for the fact that harm may come to you if you don't accept an argument's conclusion is no reason to think the argument's conclusion is *true*.

Sometimes it can be costly or dangerous to oppose falsehood and seek the truth. However, an ad baculum may be psychologically persuasive—if someone is frightened into accepting a conclusion.

Here is another example: "You really shouldn't join that protest march, for if you do we're likely to beat you up." The problem here is that a threat of force is logically irrelevant to the issue, which is, is it right or wrong to join the protest march? The possibility of harm if one draws a particular conclusion is not a good reason to suppose that the conclusion is false or wrong.

▶ THE APPEAL TO PITY *(ARGUMENTUM AD MISERICORDIAM)*

The **appeal to pity** fallacy attempts to evoke pity from the audience (the reader or listener) and then to use that pity to force the audience to accept the conclusion. Attorneys have been known to use this strategy. For instance, an attorney whose client is charged with armed robbery might play up his client's unfortunate childhood in hopes that the jurors will feel so sorry for the defendant that they will be sympathetic and lenient.

Students sometimes go into their teacher's office and argue along the following lines: "If I don't earn a 2.0 in this class, my parents won't pay for my school, and I won't be able to afford to go to school next year. Please let me do an extra-credit project to boost my grade." This is an appeal to pity.

The ad misericordiam argument is a fallacy because it treats pity as the only relevant factor to take into account in reaching the conclusion and ignores logical factors, that is, reasons in support of the conclusion. Of course, we do sometimes take pity into account, but only along with other logically relevant factors.

▶ THE APPEAL TO THE PEOPLE *(ARGUMENTUM AD POPULUM)*

People sometimes think differently when they are in a crowd. When alone or with a few friends they are generally rational, but in a crowd they may become excited and give in to irrational urges and ideas. The **appeal to the people** fallacy attempts to arouse and use the emotions of a group or crowd in order to win acceptance for a conclusion. It relies on the psychological fact that individuals often want to join in the enthusiasm of a crowd, and it encourages people to give in to such feelings rather than to think logically. Because most of us want peer approval, the idea of being a part of the group can motivate us to accept a group conclusion. This strategy is a fallacy, for the mere fact that the group supports a conclusion is not, by itself, a logically relevant reason to accept the conclusion.

For example, a television ad shows a group of teenagers all wearing a particular brand of sunglasses. In the background a voice says, "Everyone is wearing Brand X sunglasses. Don't you want to be part of the action? Don't you think it is time you bought some?" The implicit argument here is: Because the group is wearing these, you ought to wear them.

Or an ad says: "Polls show that Smith, the Democrat, will be elected by a landslide, so join the Democratic Party and be on the winning side." This type of ad populum argument is also known as the "bandwagon argument" because it (figuratively speaking) presents a bandwagon full of people and asks the listener to jump on with everyone else.

The ad populum is also known as the "appeal to the gallery" because of the way some politicians have been known to use the technique during legislative debates. Instead of speaking to and reasoning with their fellow legislators in the legislative chamber, they appeal to the emotions of the visitors watching the proceeding from the gallery of the legislative building. They hope that if their fellow legislators hear the people in the gallery cheering, this, rather than reason, will decide their vote. Politicians who use this technique are called demagogues.

The various versions of the ad populum share a common structure: Join in with everyone else; accept my conclusion and be accepted and valued by the group.

▶ Snob Appeal

A related fallacy is the **snob appeal fallacy.** Here the arguer claims that if you will adopt a particular conclusion, you will be a member of a special, elite group of people who are better than everyone else. Because many people have a desire to be special, some give in to the temptation and draw the conclusion. For example, the voice on an advertisement says, "Only people with the most sophisticated taste buy XYX brand pens. Buy one today and be one of the few who own a superior pen. Only $200." If Joe buys the pen, and if his reason is the advertisement and not facts about the actual quality of the pen, Joe has been persuaded by a fallacious argument.

▶ The Fallacy of Irrelevant Conclusion (*Ignoratio Elenchi*, "Ignorance of the Proof")

In the **fallacy of irrelevant conclusion,** someone puts forward premises in support of a stated conclusion, but the premises actually support a different conclusion instead. For instance, suppose the senate is debating a bill that would place laptop computers in every public school classroom. Senator Smith is supposed to present an argument for the bill.

"Our children are our most precious investment," the senator begins. "The public schools help prepare the next generation for responsible adulthood. Only a scrooge would oppose children. We must pass this bill."

The senator is supposed to be arguing in favor of a bill that would place laptops in schools, but his premises are actually directed at a completely different conclusion: namely, the generally agreed conclusion that public schools are a good thing. His premises do nothing to show that schools will actually do a better job if classrooms have laptops. The senator is "arguing beside the point."

An ignoratio elenchi succeeds in being psychologically persuasive by evoking the audience's approval and favorable emotions for the conclusion the speaker's premises actually support. It then attempts to transfer this approval over to the conclusion the speaker was supposed to be arguing for, which is a different conclusion altogether. In the above case, the senator's speech evokes positive emotions in support of children and public schools and then attempts to subtly slip that positive approval over to the laptop bill—*without actually giving an argument for the laptop bill*. If everyone is vigorously nodding their heads in approval at the mention of public education and the welfare of children, they may still be nodding their heads when the senator mentions the laptop bill.

Sometimes, an ignoratio elenchi is simply the lazy way out. For instance, suppose Representative Jones is scheduled to argue in favor of a bill that would increase federal funding for housing. However, Jones hasn't done her homework and doesn't have any facts at hand. What does she do? She talks about the general human need for decent housing, and tells an emotional story about her own poverty-stricken childhood in a house with no central heating. This is an ignoratio elenchi, for it completely ignores the key issue: Will this particular bill actually provide decent housing? Is there a better way to help the poor, or is this the most effective use of scarce resources? Presumably, everyone agrees that decent housing is needed and that it is a good thing—that is not the point at issue. The point at issue is whether or not this particular bill is justified. The congresswoman's speech is supposed to prove one thing (that this particular bill is needed) but is instead directed at a different thing (that decent housing is a good thing). Her argument "misses the point."

Here are a few additional examples of the ignoratio elenchi fallacy:

Robbery, theft, and kidnapping are on the rise, so we need to impose the death penalty more frequently.

We need to increase welfare spending because people are going hungry.

Welfare cheating is rampant, so we must abolish welfare.

▶ BEGGING THE QUESTION (*PETITIO PRINCIPII*, "POSTULATION OF THE BEGINNING")

The **begging the question** fallacy appears in several ways. The most straightforward way is simply to employ the conclusion of the argument as a premise in support of that same conclusion. For example:

Joe: God exists.

Fred: Why suppose that's true?

Joe: Because God exists.

However, seldom is anyone going to simply place the conclusion word-for-word into the premises like this. Rather, an arguer might use phraseology that conceals the fact that the conclusion is masquerading as a premise. The conclusion is rephrased to look different and is then placed in the premises. Here are two examples:

It is immoral to be a communist, for communism is an immoral doctrine.

He is angry right now. How do I know? Because right now he is mad.

In these cases, because the premise is a rephrased version of the conclusion, the conclusion is being used as a premise. As a result, the real question—is the conclusion true?—is being begged. That is, the listener is being "begged" to accept a conclusion without being given an independent reason for the conclusion.

Incidentally, this fallacy could be committed unintentionally—an arguer might not realize that his premise is simply a rephrased version of his conclusion. For instance, a person might innocently argue:

Free trade is a good thing, for the unimpeded flow of products is a good thing.

This person may not realize that free trade is "an unimpeded flow of products." Or someone who doesn't realize that "liberty" and "freedom" are synonyms might argue:

Liberty is good, for freedom is a good thing.

The most common form of this fallacy occurs when someone gives an argument that contains a premise that *presupposes* the conclusion. In order to explain this, we must first clarify the nature of a *presupposition*. Suppose someone says, "The King of the United States has issued a proclamation." This sentence cannot be true unless there is actually a king of the United States. So anyone asserting or accepting the sentence must take it for granted that the United States has a king. The sentence therefore *presupposes* that the United States has a king. In general, one sentence **S** presupposes a sentence **P** if it is the case that anyone asserting or accepting **S** must take it for granted that **P** is true. Granting the truth of **S** is in effect to also grant the truth of **P.** Now, consider the following argument for God's existence:

1. The Bible says God exists.
2. The Bible is inspired by God.
3. Anything inspired by God is true.
4. Therefore, God exists.

Notice that the second premise cannot be true unless God exists. To grant the truth of this premise is *already* to suppose God exists. The premises therefore presuppose the conclusion, and the argument "begs the question." Because nobody would accept the premises unless they already accepted the conclusion, the argument begs us to take for granted what it is supposed to prove.

Here is another argument that begs the question:

Ann: There must be life after death.

Rita: Why? What evidence do you have?

Ann: Well, the Ouija board is proof. It transmits messages from those who have died and entered the next life.

One variation on this fallacy is the long circular argument. Here is an example.

Ed: Why do you believe Pat is trustworthy?

Ned: Because Sue is trustworthy and Sue told me Pat is trustworthy.

Ed: But why do you believe what Sue says?

Ned: Because Rita is trustworthy and she told me that Sue is trustworthy.

Ed: But why do you believe what Rita says?

Ned: Because Fred is trustworthy and Fred says that Rita is trustworthy.

Ed: But why do you believe what Fred says?

Ned: Because Pat is trustworthy and Pat says that Fred is trustworthy.

The premises form a chain that circles back on itself and ends where it began. This argument begs the point at issue, which is: Why believe Pat is trustworthy? In the end, we are asked to believe Pat is trustworthy because Pat is trustworthy. In this type of argument, if the circle is long enough, the listener might not notice that the argument is circular.

▶ THE APPEAL TO IGNORANCE (*ARGUMENTUM AD IGNORANTIUM*)

In the **appeal to ignorance** fallacy, someone argues that a proposition is true simply on the grounds that it has not been proven false. For example, suppose a believer in UFOs argues, "There must be UFOs because nobody has proven there aren't any." Or, "I believe in astrology; after all, it has not been disproved." This fallacy is also committed if someone argues that a proposition must be false because it has not been proven true. For example, an atheist might argue: "God doesn't exist, for nobody has proven God exists."

We must here state a qualification. In some cases, if something were true, evidence of its truth would exist. In such a case, the absence of evidence for the proposition *is* evidence of its falsity. For instance, a teacher might argue, "If an elephant were in this classroom, we would be very much aware of it. So, because we have no evidence of an elephant in this classroom, one probably isn't here." This is not a fallacious argument.

Courts of law present another typical exception. In a court of law, the defendant is considered innocent until proven guilty. In the absence of evidence of the defendant's guilt, he or she is considered not guilty. This inference, that the defendant is not guilty because he has not been proven guilty, is *not* a fallacious argument—assuming the principle that a person is legally innocent until proven guilty in a court of law.

▶ THE RED HERRING FALLACY

In the **red herring** fallacy, the arguer tries to divert attention from her opponent's argument by changing the subject and drawing a conclusion about the new issue. For instance, suppose a company has been accused of polluting a stream, and the company spokeswoman is being grilled by the press. In the middle of the heated press conference, she says, "Our company is the largest donor to the city's Boys and Girls Clubs. Why, last year we gave them a combined total of $25,000." The astonished reporters now begin questioning the official about the company's charitable giving, and the argument is now *completely off track*. By switching the subject to the donations, the spokeswoman has introduced a "red herring" into the argument.

To introduce a red herring into an argument is to throw the argument off track; it is to divert attention from the real point at issue. The fallacy possibly got its name from the tactic for training hunting dogs of dragging a strong-scented red herring across a trail to cause the dogs to lose the scent. Most of the dogs would follow the herring's scent and be led completely off track; but the best dogs would stay with the original scent. In the context of an argument, a "red herring" throws us off the scent, that is, throws the argument off track.

Arguers introduce a red herring into an argument for any number of reasons. Perhaps someone knows they are losing an argument and is trying to change the subject. Perhaps unconsciously the arguer is uncomfortable with the point at issue and wants to change the subject. Perhaps someone is afraid of an issue, wants to avoid it, and changes the topic. Here's an example:

- Bud's daughter has her friend Veronica over for dinner. Bud is a butcher at the local Safeway. Because he is also a big beef-eater, Bud cooks beef for dinner. Veronica is a veterinarian and a vegan. At the dinner table, Veronica argues that eating meat is immoral. Bud listens for a while and then fires back, "Eating meat is legal, so are you vegans going to have all meat-eaters put in jail? You'd have to arrest millions of people. We'd have to have concentration camps, we would all lose our freedom. That is something only a communist would favor." Bud the beef-eating butcher has just slapped a big fat red herring on the plate of Veronica the vegan veterinarian. Veronica's point concerned the *morality,* not the *legality* of eating meat.

An arguer may innocently interject a red herring into an argument thinking (mistakenly) that it is a pertinent point; however, in most cases, the herring is presented in an attempt to avoid an argument's conclusion. In some cases, if people are divided as to the real point at issue, debate can arise over whether or not a point is or is not a red herring. One more example:

Jim: Fred, I need to talk with you about the hundred dollars you owe me. I think you should pay your debt. You promised.

Fred: Let's say you and me play some pool tonight.

Jim: Fred, how about that money?

Fred: How about them Seahawks?

▶ THE GENETIC FALLACY

When we explain the origin (the "genesis") of a thing, we have given a genetic explanation. The **genetic fallacy** is committed when someone attacks a view by disparaging the view's origin. Often this involves attacking the manner in which the view was acquired. In short, the origin of the view is attacked rather than the evidence for the view, and this is offered as a reason to reject the view.

For example, during the heyday of Soviet communism, communists would argue against modern economic theory by pointing out that it originated in the minds of "bourgeois" economists, that is, economists who taught at major universities in the capitalist world. These universities and their highly paid professors, Soviet communists claimed, were "hired prize fighters" of the ruling capitalist class. So, because it originated in minds of the ruling capitalist class, and because the ruling capitalist class is evil, it was argued, modern economic theory is false. This is fallacious. Modern economic theory, as taught in any university, should be evaluated on the basis of the evidence, and on the basis of the logic and arguments offered for and against various theories. To simply dismiss it on the basis of its origin is to dismiss it for an illogical reason.

For another example, some psychologists have argued against the truth of religious belief by arguing that (a) religious beliefs originate in a fear of the unknown, and (b) this fear produces a desire to have some higher power protect us from unknown forces. It is argued that if this psychological condition is the *source* of religious belief, belief in God is unmasked as irrational or false. Of course, the premises of such an argument are logically irrelevant to the conclusion, for even if religious belief did originate in a fear of the unknown, this fact would not show at all that religious belief is false.

▶ POISONING THE WELL

When the arguer uses emotionally charged language to bash an argument or position before arguing against it, that constitutes **poisoning the well.** For instance, at the start of a lecture critical of capitalism, suppose the speaker begins, "What can we say about this selfish, dog-eat-dog system known as capitalism?" Right away, before any evidence has been given, the audience is given a negative image of capitalism. Or, at the start of a lecture on communism, suppose a speaker begins, "What can be said in defense of these bloodthirsty communists?" For the rest of the lecture, people will have in mind this negative image.

Here is an especially tricky version of this fallacy: In a debate, one debater says, "My opponent cannot accept the truth, so after I make my next point, she will firmly protest it." This poisons the well, for after the next point is made, the opponent is going to be in a real bind. If she protests the point, she confirms her opponent's prediction and appears unwilling to accept the truth. If she doesn't protest the point, she seems to concede the point. Either way, she loses. (The best response in such a case may be to point out the nature of the fallacy before giving a response to it.)

▶ **EXERCISE 13.1** ◀

For each fallacy listed below, make up an argument that commits the particular fallacy. Provide some context for each argument.

a. The petitio principii fallacy

b. The red herring fallacy

c. The argumentum ad misericordiam fallacy

d. The abusive ad hominem fallacy

e. The circumstantial ad hominem fallacy

f. The tu quoque fallacy

g. The genetic fallacy

h. The ad populum fallacy

i. The argumentum ad baculum

j. Poisoning the well

k. The argumentum ad ignorantium

l. The ignoratio elenchi fallacy

▶ **EXERCISE 13.2** ◀

Identify the fallacy in each of the following.

1. *A mom tells her child why the child should not take drugs, and this is the reply: "But you took drugs as a kid, so why should I listen to your argument?"

2. That book was written by a radical, atheist, communist sympathizer. Don't buy it.

3. A scientist who works at the Tobacco Institute, an industry-funded research think tank, argues that cigarettes are actually good for you. A critic counters, "That's a bunch of baloney. Do you know where he works? The Tobacco Institute. Look who is paying the salaries of their researchers. They probably got a raise for saying cigarettes are healthy."

4. *Joe: Hydrogen burns.

Pete: No it doesn't. Give me an argument.

Joe: Okay. Hydrogen is combustible, therefore it burns.

5. The world is good because it was made by a good God. How, you ask, do we know God is good? Look at this world—it is such a good place that the God who made it must be good.

6. If you are a true-blue, loyal American, you will support this candidate.

7. If I don't earn at least B in this class, my parents won't pay for school and I'll be kicked out of the house. I think I should be allowed to do extra credit.

8. Logic is absolutely necessary for organized thinking. Anyone who wants to think in an organized way must study logic.

9. *Mafia goon: It would be smart to buy our protection policy. It's only two hundred dollars a month, and it will save you thousands of dollars a month of damage to your facility.

10. I really think Joe should get the job. He lost all his money investing in a fraudulent land deal, his kids are hungry, and his wife will leave him if he doesn't find work soon.

11. Fred says we should increase spending on welfare. But he is a welfare case worker who makes his living dispensing welfare money to poor people. Can't you see he's advocating this because he will benefit? His argument should be rejected.

12. You should read the new novel, *Love on Main Street*. It's a bestseller, and everyone will be talking about it at lunch next week.

13. *Our company stands accused of polluting the river, but the Army Corps of Engineers has been causing all sorts of damage to the river for decades.

14. Nobody has ever proven UFOs don't exist, so we must suppose the sightings are legitimate.

15. I am sure he respects me, because he told me so, and he wouldn't lie to someone he respects.

16. You say your bird can sing, but no bird can sing. So, your bird doesn't sing.

17. We criticize the Soviets for repressing freedom of speech, but we don't allow people to yell "fire" in a crowded theater.

18. Dr. Smith says smoking is bad, but he chain-smokes, so his arguments are no darned good.

19. *Professor Levy's argument for increasing teacher pay is not worth listening to. She's a teacher, so she'll benefit from such a policy.

20. The African diplomat has no business criticizing America for its race relations. Africa has intertribal conflicts so severe that literally millions of Africans have been killed by other Africans just in the past few decades.

21. That man's argument against rent control is faulty—he's a landlord who will benefit if rent control is not instituted.

22. That man's argument for rent control is faulty—he's a renter who will benefit if rent control is instituted.

23. *Nobody has ever proven that a fetus has rights. So, a fetus does not have rights.

24. Nobody has ever proven that there is no such thing as ESP. So ESP is a real phenomenon.

FALLACIES OF LITTLE EVIDENCE

▶ THE FALLACY OF ACCIDENT

In the **fallacy of accident,** a general rule is applied to a specific case, even though unforeseen and accidental features of the case make it an *exception* to the general rule. For example, suppose a completely intoxicated patron stumbles out of Bud's Brontosaurus Burger Bar and asks the valet parking attendant to retrieve his car. When the valet refuses, a bystander argues: "Property should be returned to its legal owner. That totally inebriated man who just staggered out of the restaurant is the legal owner of that car. You should return it to him immediately." This is a fallacy of accident.

In the middle of a movie, in a dark, crowded theater, Joe the practical joker decides to yell, "Someone's got a machine gun," in order to see if anyone is crushed to death in the ensuing stampede. Later, he argues to the police, "I should not be arrested—I have free speech." This is also the fallacy of accident. The argument is fallacious, for the principle of free speech was not intended to protect this *type* of vocal activity. The fallacy in this case misapplies a general rule. Here are two more examples:

- A group of anti-Chinese agitators incites a riot in a city's Chinatown district and, as a result, six businesses are burned down and hundreds of people hurt. The agitators commit the fallacy of accident when they argue, "We have free speech, so we should not be arrested for inciting this riot." (Such an argument wouldn't fly far in any reasonable courtroom.)

- Fred, the physical fitness expert argues, "Exercise is good for you, so Herman should exercise even though he has a severe case of pneumonia."

The fallacy of accident arises because a general rule can rarely be formulated so as to apply to every possible individual case within a predetermined range. As a result, general rules typically have exceptions. The arguer who commits this fallacy neglects this important fact.

▶ THE STRAW MAN FALLACY

Sometimes, when person A criticizes person B's argument, person A first summarizes B's argument, then criticizes the summarized version of B's argument, and concludes that B's actual argument has been refuted. The **straw man fallacy** is committed when (1) A's summary of B's argument is a weakened, exaggerated, or distorted version of B's original argument, and (2) A attacks this summarized argument (instead of B's actual argument) but then concludes that B's *original* argument has been refuted.

The straw man strategy is to exaggerate or simplify your opponent's argument to make it appear ridiculous or weak, refute the weakened argument, and conclude that you have refuted your opponent's actual argument. You have (metaphorically) set up a straw man in order to knock him down. You have attacked a weakened or distorted version of your opponent's argument rather than the real thing, yet you have concluded that his argument has been refuted.

Here is a straw man argument that purports to refute Einstein's theory:

- Einstein's theory of relativity states that everything is relative. If everything is relative, there is no truth, and if there is no truth, his theory is not true. So, if his theory is true it is not true, which means it is self-contradictory. So, the theory of relativity is false.

This argument sets up a straw man, for Einstein's position is not simply the claim that everything is relative. Such a portrayal of Einstein's view is too simplistic. (For instance, according to Einstein, the speed of light is not relative, and Einstein also did *not* hold that truth is relative.) A simplified and distorted version of Einstein's theory was attacked and refuted, but the arguer concluded that Einstein's actual theory had been refuted.

▶ THE APPEAL TO QUESTIONABLE AUTHORITY (*ARGUMENTUM AD VERECUNDIAM*)

Sometimes we base a conclusion on the testimony of an authority. For instance, "That chemical is a poison, for Dr. Brown says so, and he's an expert on poisons." This type of reasoning can produce a strong argument *if the authority is trustworthy and if the authority knows a lot about the matter at issue*. However, if someone attempts to support a claim by appealing to an authority when the authority is not trustworthy, or when the authority is ignorant or unqualified, or prejudiced, or has a motive to lie, or when the issue lies outside the authority's field of competence, the ad verecundiam fallacy is committed. In short, the **ad verecundiam fallacy** involves an improper appeal to an authority.

Advertising frequently involves the ad verecundiam fallacy. For example:

- Clint Eastwood says I should buy a Chevrolet truck. So, I should buy a Chevrolet truck.

- Michael Jordan says Wheaties cereal is good for me. So, I am going to buy Wheaties.

These are ad verecundiam fallacies, for the famous personalities mentioned above are not experts or authorities in the area of trucks or breakfast cereals.

However, the following two arguments avoid the ad verecundiam fallacy because Mr. Eastwood knows a lot about acting and Mr. Jordan knows a lot about basketballs:

- Clint Eastwood endorses the Ace Acting School. So, that would be a good school to attend.

- Michael Jordan says Wilson makes good basketballs. So, their basketballs are probably high quality.

▶ THE FALLACY OF HASTY GENERALIZATION

In the **fallacy of hasty generalization,** someone draws a generalization about a group on the basis of observing an unrepresentative sample of the group. A sample can be unrepresentative in several ways: the sample might be too small to be representative, as when someone meets three construction workers, finds all three to be crude, and concludes that all construction workers are crude. A sample might be unrepresentative because it is exceptional or unusual in some way, as when someone goes to a Metallica concert, observes the behavior of the fans, and concludes that fans behave that way at all rock concerts.

Archie Bunker:	You can't trust any of them type of people.
Michael (Meathead):	How do you know, Archie?
Archie Bunker:	When I was a kid, a family of them lived down the street and the whole family was no good.

Here is an example of a type of hasty generalization that is sometimes made in the field of health. Joe Blow falls sick. As he is laying in bed, he happens to develop a craving for prunes. For three days, he eats nothing but prunes. On the third day, Joe feels great. In addition, he has lost twenty pounds. A new cure has been discovered: the "Prune Cure." Joe begins telling his friends about the amazing curative properties of the common prune. He writes a book and starts a new health fad known as "Prunitarianism." Joe's argument for prunitarianism is based on one single case: his own case. His argument commits the fallacy of hasty generalization.

▶ THE FALSE CAUSE FALLACY

False cause fallacies are grouped into two important types, both known by their Latin names. In a **post hoc ergo propter hoc** ("after this, therefore because of this") fallacy, someone concludes that A caused B simply on the grounds that A preceded B in time. For example, suppose a tribe in the Amazon rain forest has a tradition of beating a drum every time there is a lunar or a solar eclipse. Each time they perform this ritual, the moon or sun reappears. The tribe concludes that their drumming *causes* the moon or sun to reappear. This is a fallacy, because temporal succession does not prove causation, that is, the mere fact that A occurred before B does not prove that A caused B.

The following argument is another example: "Beginning in the 1930s, a large number of laws were added to the law books, and look what we have now: more crime than ever before. So, having lots of laws *causes* crime. We need to eliminate all those extra laws, to eliminate most of our crime." In this argument, the link between the premises and the conclusion depends on a cause-effect connection that probably doesn't exist.

In a **non causa pro causa** ("not the cause for the cause") fallacy, someone claims that A is the cause of B, when in fact (1) A is not the cause of B, but (2) the mistake is not based merely on one thing coming after another thing. One version of this fallacy is the "fallacy of accidental correlation." Suppose someone argued that every major war during the twentieth century has happened under a Democratic president, and therefore, Democratic administrations have caused war. The problem with this argument is that the mere *correlation* of the two phenomena (Democratic administrations and wars) does not prove that they are causally related. For example, at ocean beaches and lakes there is a correlation between ice-cream sales and drownings. Generally, the months with the highest ice cream sales also have the highest numbers of deaths by drowning, and the lowest numbers for drownings coincide with the lowest ice cream sales. However, we should not conclude from this correlation that ice cream sales cause drownings (nor that drownings cause ice cream sales). In fact, ice cream sales and drownings both relate to a third factor, a common cause of both—the presence of sunshine. Sunshine encourages people to swim and it also encourages people to eat ice cream.

In other cases, the non causa pro causa fallacy may involve an oversimplified claim of causation. Here, someone selects and focuses on one cause out of many and treats this as the sole cause. For example, suppose a riot occurs on a college campus and several fraternity members are involved. The police chief, let us imagine, ignores other factors and focuses blame only on the fraternity brothers. This would be a fallacy if other factors and individuals were ignored in a "rush to judgment."

▶ THE SLIPPERY SLOPE FALLACY

In the **slippery slope fallacy,** also known as the "domino argument," someone objects to a position P on the grounds that P will set off a chain reaction leading to trouble, but no reason is given for supposing the chain reaction will occur. Metaphorically, if we adopt a certain position, we will start sliding down a slippery slope and we won't be able to stop until we slide all the way to the bottom (where some horrible result lies in wait). Here is an example: At a meeting of the Social Sciences division of Harmony Community College, the secretaries ask if they can have their own coffee-break room. But the economics professor, Professor McScrooge, mounts an objection. "If we give them their own room, the next thing you know, they'll be asking for their own exercise room, then they'll want a sauna, and a hot-tub, and a tanning salon, and before you know it, they'll be working about four hours per day." McScrooge's position: If we give them an inch, they'll want a mile, and we can't afford to give them a mile, so we can't give them even an inch.

The slippery slope arguer believes that we must not adopt position P because what justifies adopting P also justifies adopting Q, and what justifies adopting Q will also justify adopting R, and so on to disaster. The slippery slope fallacy is committed when the argument's conclusion depends on a chain reaction but no reason is given to think the chain reaction will really occur.

The slippery slope fallacy has been called the "camel's back" because it reminds us of the old story of the straw that broke the camel's back. (We cannot place even one straw on the camel's back, for if we are justified in adding one straw to the load on his back, we are justified in placing a straw on top of the first straw, and if we are justified in adding one straw, we are justified in adding a straw on top of that, and so on, and we will eventually be justified when we place the straw that breaks the camel's back. But it is wrong to break his back. So, not even one straw can be placed on his back.)

▶ THE FALLACY OF WEAK ANALOGY

In the **fallacy of weak analogy,** an analogical argument is presented but the analogy is not strong enough to support the conclusion. (Chapter 25 will discuss analogical argumentation.) Consider the following example:

- We must force people to believe in the true religion just as we must force a suicidal person away from the edge of a cliff.

Now, most of us agree that a suicidal person may rightly be forced away from the edge of a cliff. However, this argument is fallacious, for there is little similarity—analogy—between forcing a suicidal person away from the edge of a cliff and forcing someone to believe the true religion. In the face of such weak analogies, we sometimes say, "You're mixing apples with oranges."

During the Vietnam War, when student radicals protested both the war and capitalism, I once heard a Progressive Labor Party (a pro-Mao group) speaker argue that we shouldn't criticize Mao tse Tung for having killed millions of political opponents while building his communist society in China, for "you have to break a few eggs in order to make an omelet." The argument was that just as there's nothing wrong with breaking a few eggs to make an omelet, there's nothing wrong with killing a few million political opponents in order to build a communist society. This argument was fallacious because killing human beings isn't at all like breaking a few eggs for your omelet.

▶ THE FALLACY OF FALSE DILEMMA

In the **fallacy of false dilemma,** someone assumes only two alternatives exist, eliminates one of these two, and concludes in favor of the second, when in fact more alternatives exist but the additional alternatives haven't been considered. Here are some examples:

- "You either hate America or you love her. You obviously don't love America, so you hate America."

- "Buy me that new coat, or else I'll be miserable. If I'm miserable, I'll make you miserable, too, so buy me the coat."

- "You are either for me or against me. You don't want to be against me, so you had better be for me."

- "Either we see the movie or we stay home. You obviously don't want to stay home. So let's see the movie."

Why would someone present two options but leave out or ignore an alternative? Perhaps they are afraid to face the alternative. Perhaps they don't want to spend time thinking about alternatives. Perhaps they are unaware of the alternative.

▶ THE FALLACY OF SUPPRESSED EVIDENCE

When the arguer leaves out or covers up evidence that would count heavily against the conclusion, the **fallacy of suppressed evidence is committed**. For example:

Car salesman: This car was driven by a little old lady, and she only drove it on Sundays to church. It's a great buy. (Unstated fact: Her reckless nephew took it joy-riding every Friday night and totaled it three times.)

Politician: I received no money from the XYZ Company, therefore I am innocent of the charges that I was bribed. (Unstated fact: He received thousands of dollars from its foundation in return for favorable votes.)

▶ THE FALLACY OF SPECIAL PLEADING

In the **fallacy of special pleading,** the arguer applies a principle to someone else's case but makes a special exception to the principle in his own case. In other words, the arguer allows the exception in his own case but does not allow the exception in the case at hand. For example, an official of political party X argues, "The other party takes special-interest money, so join our party instead." But the official doesn't mention that party X also takes special-interest money.

▶ **EXERCISE 13.3** ◀

For each fallacy below, make up an argument that commits the particular fallacy listed. Provide some context for each argument.

a. The ad verecundiam fallacy

b. The fallacy of false dilemma

c. The fallacy of weak analogy

d. The fallacy of accident

e. The fallacy of hasty generalization

f. The fallacy of post hoc ergo propter hoc

g. The fallacy of non causa pro causa

h. The fallacy of special pleading

i. The straw man fallacy

j. The fallacy of suppressed evidence

k. The slippery slope fallacy

▶ **EXERCISE 13.4** ◀

Identify the fallacy in each of the following.

1.*Doctor Smith is a hypocrite, for he kills rats in his laboratory, and doctors take an oath to protect life.

2. It is apparently okay to speed, for ambulances can speed.

3. Professor McOrnery: I do not allow any questions in class. Here is why: If I allow one student to ask a question, someone else will have a question; if someone else asks a question, another will, and pretty soon I'll be doing nothing but answering questions.

4. We should not have a student government at this school. After all, at home, we do not allow teenagers to run the family.

5.*Actor Joe Blow says nuclear power is dangerous. So, it is dangerous.

6. Watching television cannot be harmful to children. After all, it keeps them out of mischief—when they are watching it, they aren't out on the street.

7. I always rub my lucky quarter before a game. Most of the time I win after I do that. It brings me good luck.

8. Drinking milk with bourbon cures the flu. Every time I've had the flu, I drink milk mixed with bourbon and my condition improves within a day or two.

9. It doesn't matter that you had to stop on the way to rescue someone who was drowning. You promised to meet me for coffee at six, you're late, and one should not break one's promises.

10.*America. Love it or leave it. You won't love it, so, you should leave it.

11. There are only two types of people in this world: Those who work and those who sponge off those who work. So, people who don't work are sponges.

12. You are either for us or agin' us. You aren't for us, so, you're agin' us.

13.*Because it is right to speak the truth, it is right to tell Martha that she looks ridiculous in that dress, even if this is her 80th birthday.

14. We have to buy a car. We'll buy either an expensive one or a cheap one. We can't afford to buy an expensive one, so we'll buy that cheap one for five hundred dollars.

15. The Constitution guarantees freedom of speech. Therefore, it was okay that we broke into the meeting and began screaming obscenities at the top of our lungs.

16.*I'll never go to another doctor. I've been to two, and they didn't help me at all.

17. Tests on Mr. Brown have not found any physical cause of his illness, so it must be psychological.

18. It's against the law to cut someone with a sharp instrument. Dr. Verrier, the heart surgeon, should be arrested.

19. The Constitution guarantees the free exercise of religion. Our religion, which is a revival of the Aztec religion and practices mass human sacrifice, should therefore be protected by law.

20.*I ate your sandwich at noon and I got sick at three, so your sandwich made me sick.

21. It is illegal to go through a red light, so that ambulance should get a ticket for going through that light.

22. Either you support welfare or you want the poor to die. You wouldn't want the poor to die, so, you must support welfare.

23. Four local teenagers were arrested for selling drugs. Teenagers are nothing but a bunch of drug-crazed thugs.

24. Animal-rights activists say dogs and cats have rights. But if we grant them that premise, the next thing they'll argue is that birds, trees, fleas and mosquitoes have rights, and mosquito spray will become an illegal substance. We therefore must not agree that dogs and cats have rights.

25. The larger the city, the more churches. The larger the city, the more prostitution. So, churches cause prostitution.

26.*Humans are similar in many ways to cows and chickens. Similar hearts, similar lungs, and so on. Humans have a right to life. So, cows and chickens do, too.

27. The Constitution guarantees freedom of speech, so I was within my rights telling this little old lady I would sell her the Brooklyn Bridge for one thousand dollars.

FALLACIES OF LANGUAGE

The terms *ambiguity* and *equivocation* are used to refer to multiplicity of meaning. A word is ambiguous in a context if it can be interpreted in more than one way. A word is used equivocally in a passage if it has one meaning in one place but a different meaning in another place. The first two fallacies in this category involve arguments in which a straight path of reasoning collides with a roadblock involving multiplicity of meaning.

▶ THE FALLACY OF EQUIVOCATION

The **fallacy of equivocation** begins when the arguer uses a particular word or phrase with one meaning in one place and uses the same word or phrase with another meaning in another place. The arguer then commits the fallacy by regarding what she has established on the basis of the one meaning as equally established with respect to the other meaning. As a result, the conclusion depends on a word (or phrase) being used in two different senses in the argument. The premises are true on one interpretation of the word, but the conclusion follows only from a different interpretation. We do not normally allow the meaning of a word or phrase to shift within an argument. If the listener fails to notice the shift in meaning, he might unwittingly accept the conclusion when he actually should not. For example, the following argument commits the fallacy of equivocation:

- Sugar is an important chemical constituent of your body, so, eating candy and food that is loaded with sugar can't harm you.

This is a fallacy because the form of sugar that is a key chemical constituent of your body is different from the form of sugar in a candy bar. The premise is talking about one kind of sugar while the conclusion is talking about a different kind, although the listener may not realize it. The premise is true if *sugar* means one thing, the conclusion is true if *sugar* means something else. The meaning of *sugar* must shift if the argument is to have both a true premise and a true conclusion. Without such a shift, either the argument has false a premise or it is invalid.

Now, when someone gives an argument, we normally suppose their words retain a constant meaning throughout the whole process. In the case of equivocation, if someone thinks that the conclusion is actually proven by the premise, the person has not noticed the shift in meaning.

Consider this argument:

A person ought to do what is right.

I have a right to eat unhealthy foods.

So, I ought to eat unhealthy foods.

The shift in meaning occurs with the word *right*. That which is "right" is one thing, but a "right" possessed by a person is another thing entirely. If *right* is given a constant meaning throughout the argument, the argument is faulty. Someone would consider the conclusion proven only if she did not notice that the meaning has shifted. Having noted the meaning shift, she would also note that the premise and the conclusion are really talking about two different things. If the meaning of *right* is not allowed to shift during the argument, either the argument has false a premise or it is invalid. Notice that in a fallacy of equivocation, the two meanings must be closely related if the shift in meaning is to fool anyone. Here are two additional examples.

People shouldn't protest over a mere word.

Racism is a mere word.

So people shouldn't protest over racism.

In the second premise of this argument, the expression "racism" refers to the word itself. However, in the conclusion, the word *racism* refers to the actual social phenomenon.

A plane is a woodworking tool.

A Boeing 747 is a plane.

So, a Boeing 747 is a woodworking tool.

This argument is valid if the word *plane* is given an unchanging meaning. But then the premises are not both true.

Here are a few more examples of the fallacy of equivocation.

Only man is rational.

No woman is a man.

So, no women are rational.

Pat Buchanan is a politician whose views are on the right, so he must be speaking the truth, because *right* means "true."

The legislature can revoke laws.

The law of gravity is a law.

So the legislature can revoke gravity and we'll all be able to float through the air.

Liberal means generous.

Senator Jones is a liberal.

So he is generous.

You believe in the miracles of modern science.

So you do believe in miracles.

You should therefore believe in the miracles of the Bible.

▶ THE FALLACY OF AMPHIBOLY

A *sentence* is **amphibolous** if (a) it can be interpreted in more than one way and (b) because of the grammatical structure of the sentence, it is unclear which meaning is intended. An example of amphiboly is the following sentence, which appeared at the start of a newspaper article: "Police went to the home where they shot the entire family." Did the police kill everyone? Here is another example. Captain Spaulding, the great African explorer, once described an encounter with an elephant thus: "The other morning I got up and shot an elephant in my pajamas."

In an amphibolous *argument,* a statement is ambiguous because of its grammatical construction. One interpretation makes the statement true, the other makes it false. If the ambiguous statement is interpreted one way, the premise is true but the conclusion is false; but if the ambiguous statement is interpreted the other way, the premise is false. The meaning must *shift* if the argument is to go from a true premise to a true conclusion. If the meaning is not allowed to shift during the argument, either the argument has a false premise or it is invalid. Now, when someone gives an argument, we normally suppose their words retain a constant meaning throughout the whole process. Either way we interpret the ambiguous sentence, the argument is defective.

Here is an example of amphiboly. Imagine that a person's legal will states: "I leave my house and my parakeet to Ann and to Bob." The attorney concludes that Ann keeps the house and Bob keeps the parakeet. Or, Jethro Bodine is on his way to Greenwich Village to have a look at the Empire State Building. Why does he expect to find it there? He gives this argument: "The tour guide said that standing in Greenwich Village, the Empire State Building could easily be seen. The Empire State Building therefore must be in Greenwich Village."

Here are two admittedly silly examples of amphiboly.

I saw a newspaper headline that read: "Elderly Often Burn Victims."

So, old people must be really cruel.

The report says that Brenda is a pretty tall young lady.

Therefore, Brenda must be attractive.

A famous example of amphiboly is found in Herodotus' *Histories*. Croesus, the King of Lydia, was thinking of going to war against Cyrus, the king of the Persians. Because the Oracle of Delphi was known to be wise, Croesus asked the oracle, "Should I go to war?" The oracle replied: "If Croesus goes to war with Cyrus, he will destroy a mighty kingdom." Croesus then confidently went to war, expecting certain victory. However, Croesus was defeated by Cyrus. After the battle, Cyrus spared Croesus's life, and Croesus wrote to the Oracle for an explanation. The reply was as brief as the original revelation: "In going to war, you did indeed destroy a mighty kingdom, your own. . . ."

Notice how amphiboly differs from equivocation. The fallacy of equivocation traces to an ambiguity in one or more *words*, whereas the fallacy of amphiboly traces the ambiguity to the grammatical structure of an entire *sentence*.

Here's a combination of equivocation and amphiboly. A recipe says: "Cook the prunes in a brandy and tapioca mixture and serve them when thoroughly stewed." The cook, upon reading this, breaks out the whiskey, having concluded that the dish is best served when he is totally drunk.

▶ THE FALLACY OF COMPOSITION

In the **fallacy of composition,** someone uncritically assumes that what is true of a *part* of a whole is also true of the whole. For example, after meeting the individual members of an orchestra, someone reasons as follows:

- Each member of the orchestra is a good musician, so, the orchestra must be a good orchestra.

The problem is that although each *individual* member of the orchestra might be an accomplished musician, it might be that the group *as a whole* makes a lousy orchestra.

Here is another example. Suppose A, B, and C are reporters who have never met each other. The news editor reasons as follows:

- A is a happy person. B is a happy person. C is a happy person. So, if we put them all together on assignment overseas, they will be a happy group of people.

The problem here, of course, is that each person might individually be happy, but when combined in a group, the individuals might not be happy at all. Similarly, someone might argue the following:

- Each of these basketball players is an excellent athlete, so, the team must be an excellent team.

The problem with this reasoning is that each player might be excellent, but they might function poorly as a group.

Advertisers sometimes encourage people to commit the fallacy of composition when they divide a large price into many small monthly payments and focus only on the *size* (rather than the number) of the monthly payments. For instance, after watching an add on TV, Pat thinks: "That huge exercise set is only $19.95 per month. That's not very much for such a big piece of equipment. Maybe we'd better buy it."

Here are a few further examples of the fallacy of composition:

- Because each scene in the play was well done, the whole play was well done.

- Sodium is poison. Chlorine is poison. Salt is made of sodium and chlorine. So salt is a poison.

- Each man dies, so, some day humanity will die.

- I can buy this new car for only ninety-nine dollars per month. Each monthly payment is small, so, the whole price must be small.

- This piece of pie is triangular, so, it must have come from a triangular pie.

The fallacy of composition involves a logically fallacious transfer of a property from the part to the whole, based on the idea that because the parts have an attribute, the whole has the attribute. In many cases, this fallacy involves an abuse or misuse of language, for a term that was meant to be applied only to individuals or to parts of a whole is mistakenly transferred to a whole. For instance, in the following fallacy, the expression "has a mother" is meant to be applied only in individual cases, but it is transferred to a whole:

- Because each human being has a mother, there must be some woman who is the mother of the whole human race.

▶ THE FALLACY OF DIVISION

The **fallacy of division** is the reverse of the composition fallacy. In this fallacy, someone uncritically supposes that what is true of the whole must be true of the parts. Here are three examples.

- Wilbur is on vacation and sees the biggest hotel he has ever seen in his life: "Wow. The hotel is huge, so the rooms must be big rooms."

- Fonebone buys a new car and accidentally breaks the door handle: "Oh, oh. This car is expensive, so each part must be expensive."

- The average student owns 1.6 coats. So that student over there must own 1.6 coats.

In these cases, someone is assuming that because the whole has a particular property, the parts must have that property as well. In many cases, this fallacy involves an abuse or misuse of language, for a term that was meant to be applied only to a whole is mistakenly transferred to the parts of a whole.

▶   **EXERCISE 13.5**   ◀

For each fallacy below, make up an argument that commits the particular fallacy listed. Provide some context for each argument.

   a. The fallacy of equivocation

   b. The composition fallacy

   c. The fallacy of amphiboly

   d. The fallacy of division

▶   **EXERCISE 13.6**   ◀

Name the fallacy in each of the following.

1. Modern societies have conflicts. So, the people within them must have conflicts.

2. *Our team is strong. Pat is on our team, so, Pat must be strong.

3. This house must have a huge living room, for look how big the house is!

4. The company has performed poorly, so, the employees must have performed poorly.

5. Chemists have studied free radicals. John is a chemist, so, John has studied free radicals.

6.*Captain Spaulding, the great African explorer, got up in the morning and saw an elephant in his pajamas. So, elephants wear pajamas.

7. The Bullitt foundation is generous, so, the people working there are generous.

8. The party was wild. Joe was at the party. So Joe was wild.

9. The consumer price index rose last month, so, that stereo you are going to buy has gone up in price over the past month.

10. College grads earn 30 percent more than high school grads, so, Pat must earn 30 percent more than Jan, for Pat went to college and Jan only went through high school.

11.*Each person in that mob is a decent person who normally wouldn't hurt anyone, so, the mob won't hurt anyone.

12. This first chapter is long, so, the book must be long.

13. A slice of stale bread is better than nothing. Nothing is better than God. A slice of stale bread is better than God.

14. Pete removed his computer from the shipping crate and threw it in the garbage. He must not have liked the computer, because he threw it away.

15. Encouraging people to take the bus won't cut down on energy, for a bus uses more energy than a car.

16. Because salt is not poisonous, the constituents of salt, sodium and chlorine, are not poisonous.

17.*Father to son: "It's wrong to steal." Son: "But Dad, you watch baseball and they steal bases."

18. If we really believed in liberty, we would tear down all the prisons, for prisoners don't have liberty.

19. It is morally right to give food to starving families. Our family is always starving by the time we sit down at the table. Therefore, people should give our family food.

20. Every item of clothing she is wearing is in style. Therefore, her outfit is in style.

21.*Nobody can read all of Shakespeare in a day, so, nobody can read *Hamlet* in a day, because it is part of Shakespeare's works.

22. America is a wealthy nation, so, Jim, who is an American, must be wealthy.

23. I can lift each individual part of my car, so, I can lift my whole car.

24. America is 10 percent atheist. So, your family of ten must contain one atheist.

25. Half of America is Protestant, half is Catholic. So, Sue must be half Protestant and half Catholic.

26. An atom bomb kills more people than a conventional bomb. Therefore, atom bombs killed more people during World War II than conventional bombs.

27. *Every member of the Young Democrats Club is under twenty. So, the club must be under twenty years old, and it was founded less than twenty years ago.

28. Each brick in that house weighs less than ten pounds, so, the house weighs less than ten pounds.

29. There are more cats than dogs, so, cats consume more food than dogs, so, Felix the cat eats more than Abner the dog.

30. Water will quench your thirst. Water is composed of hydrogen and oxygen. So, hydrogen will quench your thirst.

31. The average family has 2.3 children. So the Smith family must have 2.3 kids.

CHAPTER 13 GLOSSARY

Ambiguity   A word is ambiguous in a context if it can be interpreted in more than one way.

Amphiboly   A sentence is an amphiboly if (a) it can be interpreted in more than one way and (b) because of the grammatical structure of the sentence, it is not clear which meaning is intended.

Equivocation   The act of using a word in one sense in one part of an argument and in another sense in another part of the argument. A word is used *equivocally* in a passage if it has one meaning in one place but a different meaning in another place.

Fallacy   A defective argument that nevertheless can appear to some to be a correct argument.

- *Formal* fallacies are fallacies that contain errors that may be identified by inspecting the form or abstract logical structure of the argument without reference to the content or subject matter of the reasoning.

- *Informal* fallacies are fallacies that contain flaws that are not simply due to the logical form of the argument.

- *Fallacies of no evidence* are informal fallacies that present no evidence whatsoever.

- *Fallacies of little evidence* are informal fallacies that present some evidence, but the evidence is too little or is flawed in some way.

- *Fallacies of language* are informal fallacies that involve a misuse of language.

A SUMMARY OF THE FALLACIES

► FALLACIES OF NO EVIDENCE

Argument Against the Person Fallacy (*argumentum ad hominem*)   An argument that attacks a person's character or circumstances in order to oppose or discredit the person's viewpoint. There are two types.

- *Abusive* ad hominem arguments attack the person's character.

- *Circumstantial* ad hominems attack the person's circumstances.

Tu Quoque Fallacy ("you're one, too")   A type of abusive ad hominem that attempts to discredit a person's position by charging the person with hypocrisy or inconsistency. Essentially, the charge is, "We don't need to take his argument seriously because he doesn't practice what he preaches," or "He practices that which he condemns so we don't need to accept his argument."

Guilt by association Fallacy   A type of abusive ad hominem in which one person attacks a second person's associates in order to discredit the person and thereby his argument.

Appeal to Force (*argumentum ad baculum*, literally "argument from the stick")   A fallacy committed when an arguer appeals to force or to the threat of force to make someone accept a conclusion. (Sometimes made when rational argument has failed.)

Appeal to Pity *(argumentum ad misericordiam)*   A fallacy committed when the arguer attempts to evoke pity from the audience and tries to use that pity to make the audience accept the conclusion.

Appeal to the People *(argumentum ad populum)*   A fallacy committed when an arguer attempts to arouse and use the emotions of a group or crowd to win acceptance for a conclusion.

Snob Appeal Fallacy   A fallacy committed when the arguer claims that if you will adopt a particular conclusion, you will be a member of a special, elite group that is better than everyone else.

Fallacy of Irrelevant Conclusion (*ignoratio elenchi*, meaning "ignorance of the proof")   A fallacy in which someone puts forward premises in support of a stated conclusion, but the premises actually support a different conclusion.

Begging the Question Fallacy (*petitio principii*, meaning "postulation of the beginning")   A fallacy committed when someone employs the conclusion in some form as a premise in support of that same conclusion.

Appeal to Ignorance   *(argumentum ad ignorantium)* In this fallacy, someone argues that a proposition is true simply on the grounds that it has not been proven false (or that a proposition must be false because it has not been proven true).

Red Herring Fallacy   A fallacy committed when the arguer tries to divert attention from his opponent's argument by changing the subject and then drawing a conclusion about the new subject.

Genetic Fallacy   A fallacy committed when someone attacks a view by disparaging the view's origin or the manner in which the view was acquired. The origin of the view is attacked rather than the evidence for the view, and this is offered as a reason to reject the view.

Poisoning the Well   The use of emotionally charged language to discredit or bash an argument or position before arguing against it.

▶ FALLACIES OF LITTLE EVIDENCE

Fallacy of Accident   A fallacy committed when a general rule is applied to a specific case, but because of unforeseen and accidental features of the case, the general rule should not be applied to the case. The argument ignores the fact that the accidental features of the case make it an *exception* to the general rule.

Straw Man Fallacy   A fallacy committed when an arguer (a) summarizes his opponent's argument; (b) the summary is an exaggerated, ridiculous, or oversimplified representation of the opponent's argument that makes the opposing argument appear illogical or weak; (c) the arguer refutes the weakened, summarized argument; and (d) the arguer concludes that the opponent's actual argument has been refuted.

Appeal to Questionable Authority Fallacy   *(argumentum ad verecundiam)* When someone attempts to support a claim by appealing to an authority that is untrustworthy, or when the authority is ignorant or unqualified, or is prejudiced, or has a motive to lie, or when the issue lies outside the authority's field of competence.

Fallacy of Hasty Generalization   A fallacy committed when someone draws a generalization about a group on the basis of observing an unrepresentative sample of the group, that is, a sample that is either too small to be representative or that is unrepresentative because it is exceptional or unusual in some way.

False Cause Fallacy   A fallacy involving faulty reasoning about causality. There are two important types of this fallacy.

- In a *Post Hoc Ergo Propter Hoc* fallacy ("after this, therefore, because of this") someone concludes that A is the cause of B simply on the grounds that A preceded B in time.

- In a *non causa pro causa* fallacy ("not the cause for the cause") someone claims that A is the cause of B, when in fact (1) A is not the cause of B, but (2) the mistake is not based merely on one thing coming after another thing.

One version of this fallacy is the **fallacy of accidental correlation.** In this fallacy, the arguer concludes that one thing is the cause of another thing from the mere fact that the two phenomena are correlated.

Slippery Slope Fallacy (or "domino argument")   In this fallacy, someone objects to a position P on the grounds that P will set off a chain reaction leading to trouble; but no reason is given for supposing the chain reaction will actually occur. Metaphorically, if we adopt a certain position, we will start sliding down a slippery slope and we won't be able to stop until we slide all the way to the bottom (where some horrible result lies in wait).

Fallacy of Weak Analogy   A fallacy committed when an analogical argument is presented but the analogy is too weak to support the conclusion.

Fallacy of False Dilemma   A fallacy committed when someone assumes there are only two alternatives, eliminates one of these two, and concludes in favor of the second, when more than the two stated alternatives exist, but have not been considered.

Fallacy of Suppressed Evidence   In this fallacy, evidence that would count heavily against the conclusion is left out of the argument or is covered up.

Fallacy of Special Pleading   In this fallacy, the arguer applies a principle to someone else's case but makes a special exception to the principle in his own case.

► FALLACIES OF LANGUAGE

Fallacy of Equivocation   In this fallacy, a particular word or phrase is used with one meaning in one place, that word or phrase is used with another meaning in another place, and what has been established on the basis of the one meaning is regarded as established with respect to the other meaning. As a result, the conclusion depends on a word (or phrase) being used in two different senses in the argument. The premises are true on one interpretation of the word, but the conclusion follows only from a different interpretation.

Fallacy of Amphiboly   A fallacy containing a statement that is ambiguous because of its grammatical construction. One interpretation makes the statement true, the other makes it false. If the ambiguous statement is interpreted one way, the premise is true but the conclusion is false; but if the ambiguous statement is interpreted the other way, the premise is false. The meaning must *shift* if the argument is going to go from a true premise to a true conclusion. If the meaning is not allowed to shift during the argument, either the argument has false a premise or it is invalid.

Fallacy of Composition   A fallacy in which someone uncritically assumes that what is true of a *part* of a whole is also true of the whole.

Fallacy of Division   A fallacy in which someone uncritically assumes that what is true of the whole must be true of the parts.

Part

III

ARISTOTELIAN
CATEGORICAL LOGIC

THE LOGIC OF CATEGORICAL STATEMENTS

14

More than 2,300 years ago, the Greek philosopher Aristotle (384–322 B.C.) developed the first comprehensive system of logical theory. In his treatises on logic, Aristotle laid the systematic foundations of logical theory when he articulated the concept of a deductive argument and the concept of a counterexample to an argument, identified logical fallacies, distinguished validity from soundness, and specified validity in terms of abstract *forms* of argumentation rather than in terms of specific content or subject matter.

The logical theory developed by Aristotle was taught in the school he founded in Athens, the Lyceum. Later, in the Middle Ages, Aristotle's logical theory was a core requirement in Europe's first universities. Logic has been considered a valuable part of a university education ever since. It is fair to say that the theory developed by Aristotle has played a significant role in the history of human thought.

After 2,300 years, Aristotle's logical theory is still useful for analyzing many different types of argumentation. Indeed, admission tests for most fields of graduate study, including graduate school in law, medicine, and business, usually test the student's abstract reasoning ability with problems drawn from the logical theory first systematized by Aristotle.

Until the twentieth century, Aristotle's logical theory was the only formal system of logic in the entire world; and until approximately the 1930s, the formal logic taught in just about every college and university consisted almost entirely of Aristotelian logic. However, Aristotelian logic has a limitation: it was designed to deal primarily with a particular type of argument, an argument composed of "categorical" statements and known as a "categorical syllogism." During the twentieth century, logicians began working with new types of argumentation that required new logical theory, and thus twentieth-century logical theory has come to differ in many ways from the classical logic of the ancient

Greek philosopher who is nevertheless still known as the Father of Logic. Consequently, it is now customary to identify the logical theory developed in the twentieth century as modern logic and to refer to Aristotle's logical system as traditional logic, Aristotelian logic, syllogistic logic, or categorical logic. Our entry into syllogistic or categorical logic begins with a look at categorical sentences.

CATEGORICAL SENTENCES

Consider the sentence "All aardvarks are mammals." This sentence expresses what logicians call a "categorical proposition." The standard interpretation is that the subject term *(aardvarks)* denotes a class or category of entities—the class of aardvarks—and the predicate term *(mammals)* also denotes a class of entities—the class of mammals. A **class** is a collection of objects having a specified characteristic in common. The sentence "All aardvarks are mammals" asserts that all of the subject class is *included* in the predicate class, which is to say that every aardvark is included in the category of "mammal." In general, class A is included or contained within class B if every member of A is a member of B.

Next, consider the categorical sentence "No dogs are reptiles." The subject class is the class of dogs and the predicate class is the class of reptiles, and this sentence asserts that *no* member of the subject class (the class of dogs) is a member of the predicate class (the class of reptiles), which is to say that all of the subject class is *excluded* from the predicate class. In general, class A is excluded from class B if no member of A is a member of B.

In contrast, the categorical sentence "Some Democrats are Lutherans" asserts that some members of the subject class (Democrats) are included in the predicate class (Lutherans). Furthermore, "Some Democrats are *not* Lutherans" asserts that some members of the subject class (Democrats) are not included in the predicate class (Lutherans).

In general, a **categorical sentence** is any sentence asserting either that all or some of the class denoted by the subject term is included or excluded from the class denoted by the predicate term. Thus, we have precisely four general kinds of categorical sentences:

1. Those that assert that the whole subject class is included in the predicate class, that is, every member of the subject class is a member of the predicate class. For example, "All raccoons are nocturnal." The standard logical *form* of this kind of categorical sentence is written schematically as

All so and so's are such and such.

We can abbreviate this format with the expression "All S are P," where S stands in for the subject term (*raccoons* in our example above) and P stands in for the predicate term (*nocturnal* in the example above).

2. Those that assert that the whole subject class is excluded from the predicate class, that is, no member of the subject class is a member of the predicate class. For example, "No insects are mammals." The standard logical form of this kind of categorical sentence is

No so and so's are such and such.

We can abbreviate this format with "No S are P," where S stands in for the subject term (*insects* in our example) and P stands in for the predicate term (*mammals*).

3. Those that assert that part of the subject class is included in the predicate class, that is, some members of the subject class are members of the predicate class. For example: "Some dogs are brown." The standard form of this kind of categorical sentence is

Some so and so's are such and such.

We can abbreviate this format with "Some S are P" where S stands in for the subject term (*dogs* in our latest example) and P stands in for the predicate term (*brown*).

4. Those that assert that part of the subject class is excluded from the predicate class, that is, some members of the subject class are not members of the predicate class. For example: "Some dogs are *not* brown." The standard form of this kind of categorical sentence is

Some so and so's are not such and such.

We can abbreviate this format with "Some S are not P" where S stands in for the subject term (*dogs*) and P stands in for the predicate term (*brown*).

Before we go further, we need to clarify the meaning of the word *some*. In everyday English, *some* can mean "one or more but not all," or it can mean simply "one or more." For the sake of definiteness, in logic the quantifier *some* is given its simplest meaning, "one or more," which means the same as "at least one." Thus, as we shall use it, *some* means "one or more, possibly all."

So, "Some S are P" means simply *one or more* members of S are members of P. You might make a mental note that "Some S are P" does *not* imply that some S are *not* P. Given only the information that "Some S are P," it might be that all S are P, and it might be that some S are and some S are not P.

Now, a bit more terminology. Within logical theory, *all, no,* and *some* are **quantifiers**. The words *are* and *are not* are **copulas**. In each of these categorical sentences, the copula serves to link the subject term with the predicate term. In the categorical sentence "All raccoons are nocturnal," *all* is the quantifier, *raccoons* is the subject term, *are* is the copula, and *nocturnal* is the predicate term.

QUALITY AND QUANTITY

Every categorical sentence has a "quality" and a "quantity." The quality is **affirmative** if the sentence affirms class membership and the quality is **negative** if class membership is denied. The negative quality is expressed by a negating word, *no* or *not,* and the affirmative quality is expressed by the absence of a negating word. Thus: A categorical sentence of the form "All S are P" is affirmative, and one of the form "Some S are P" is also affirmative, but sentences of the forms "No S are P" and "Some S are not P" are negative in quality.

Quantity is an entirely different matter. The quantity is **universal** if the sentence makes a claim about every member of the class denoted by the subject term, and the quantity is **particular** if the sentence makes a claim about some of the class denoted by the subject term. The universal quantity is typically expressed by the quantifier *all* or the quantifier *no* while the particular quantity is typically expressed by the quantifier *some.* Thus, "All S are P" and "No S are P" are universal, and "Some S are P" and "Some S are not P" are particular. If you question why "No S are P" is considered universal, consider that the statement makes a claim about *all* the members of the subject class, for it says that each and every member of the subject class is excluded from the predicate class. That is, "No S are P" says that each and every S is not a P, which makes a universal claim about *all* S.

As far as logical form is concerned, all categorical sentences differ in terms of just these two respects—quality and quantity—and there are therefore only four possible combinations of quality and quantity: universal + affirmative, universal + negative, particular + affirmative, particular + negative. Thus, there are four possible forms of categorical sentence, and these forms are given the following names:

- Sentences of the form All S are P are **universal affirmative** sentences.

- Sentences of the form No S are P are **universal negative** sentences.

- Sentences of the form Some S are P are **particular affirmative** sentences.

- Sentences of the form Some S are not P are **particular negative** sentences.

It has been customary since the Middle Ages to label the two affirmative forms A and I (from the first two vowels of the Latin word *affirmo,* which means "I affirm"), and the two negative forms E and O (from the two vowels in the Latin word *nego,* which means "I deny"). Thus:

A: Universal Affirmative   (All S are P)
E: Universal Negative   (No S are P)
I: Particular Affirmative   (Some S are P)
O: Particular Negative   (Some S are not P)

> ► **EXERCISE 14.1** ◄

For each sentence below, answer the following:
(a) Is the sentence an A, E, I, or O sentence?
(b) Name the quantity and quality.

1.* Some cats are not pets.

2. No pets are dangerous animals.

3. Some witnesses are persons of interest.

4.* All adults are persons who are over the age of twenty-one.

5. Some bulls are animals you wouldn't want to anger.

THE TRADITIONAL SQUARE OF OPPOSITION

Two categorical statements with the same subject and predicate terms are said to be **in opposition** if they differ in quality or quantity or both. For example, "All submarines are green" stands in opposition to "No submarines are green." During the Middle Ages, in the universities of Europe, the table on the following page was developed in order to display the logical relations among opposing categorical statements. This table, known as the *traditional square of opposition*, summarizes some of the key parts of Aristotelian logic.

However, before we turn to the table, an assumption concerning existence must be made explicit. The square of opposition will concern logical relations between the four abstract sentence forms, *All S are P, No S are P, Some S are P,* and *Some S are not P.* So, when we inspect the table, we will be considering abstract logical forms, which means we will be disregarding questions of content. Thus, we will not be concerned with the actual subject classes and actual predicate classes S and P might stand for. We will be abstracting from questions of content. However, a question about existence arises. As we think about the logical relations among categorical propositions, should we presuppose that the subject terms refer to actually existing things? Or not? In other words, when we consider a sentence form such as *All S are P* or *No S are P*, should we assume that S's actually exist? In everyday life, we sometimes utter a universal affirmative or universal negative sentence but do *not* presuppose that the subject term refers to actually existing entities. For instance, a farmer who says "All trespassers are subject to prosecution" does not presuppose there actually are trespassers on his property. A teacher who says "No late students are eligible for extra credit" does not presuppose that there actually are late students. (Perhaps this semester all students will arrive on time.)

So, when we think about the logical relations among the four types of categorical sentence, should we presuppose that the subject terms apply to actually existing entities? The answer we give to this question, as you will see later, makes a big difference. If we do presuppose that the subject terms designate or apply to actually existing things, we are taking the **existential viewpoint**. If we think about the logical relations among categorical statements from the viewpoint that does not presuppose that the subject terms designate or apply to actually existing things, we are taking the **hypothetical viewpoint**. The traditional square of opposition presupposes the existential viewpoint, that is, it presupposes that there do exist things to which the subject terms apply. As we shall see later, the square of opposition must be drastically altered if we take the hypothetical rather than the existential viewpoint. Thus, in the table below, it is assumed that for each form, the subject class (S) has at least one member.

One more preliminary point. A categorical statement is said to have **existential import** if the class referred to by its subject term has at least one member. So, for example, if we assume existential import for the statement "All illie pies are gentle creatures," we are assuming that there does exist at least one illie pie. Since the traditional square of opposition (below) presupposes the existential viewpoint, it rests on the *assumption* that the categorical statements under consideration all have existential import. Aristotle made this assumption when he catalogued the logical relations among categorical statements, and so the assumption of existential import is sometimes called the "Aristotelian presupposition."

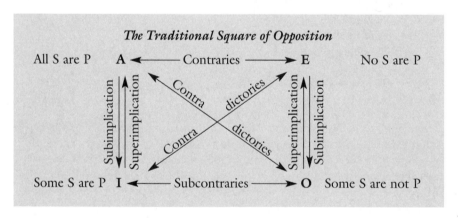

The Traditional Square of Opposition

Five important relations of opposition appear on this table: contradiction, contrariety, subcontrariety, subimplication, and superimplication. The first three of these relations can be explained in quick succession. The fourth and fifth will require a separate treatment.

- Two statements are **contradictories** if and only if they cannot both be true and they cannot both be false. So, in all possible situations, if one is true, the other is false. In other words, the two will always have opposite truth-values.

For instance, the A statement "All Swedes are Lutherans" is the contradictory of the opposing O statement "Some Swedes are not Lutherans." If "All Swedes are Lutherans" is true, then "Some Swedes are not Lutherans" is false, and if "Some Swedes are not Lutherans" is true, then "All Swedes are Lutherans" is false.

- Two statements are **contraries** if they cannot both be true, but might both be false. So, if two statements are contraries, at least one is false.

For instance, the A statement "All Swedes are Lutherans" and the E statement "No Swedes are Lutherans" are contraries. They cannot both be true but both might be false.

- Two statements are **subcontraries** if they cannot both be false, but might both be true. So, if two statements are subcontraries, at least one is true.

For instance, the I statement "Some Swedes are Lutherans" and the opposing O statement "Some Swedes are not Lutherans" are subcontraries. They cannot both be false but both might be true.

On the traditional square of opposition, the arrows for contradiction indicate the following:

- If an A statement is true, the opposing O statement is false and if an A statement is false, the opposing O statement is true. Also, if an O statement is true, the opposing A statement is false and if an O statement is false, the opposing A statement is true.

- If an E statement is true, the opposing I statement is false and if an E statement is false, the opposing I statement is true. Also, if an I statement is true, the opposing E statement is false and if an I statement is false, the opposing E statement is true.

The arrows for contrariety indicate:

- Opposing A and E statements cannot both be true but may both be false. So, if an A statement is true, the opposing E statement is false; and if an E statement is true, the opposing A statement is false.

The arrows for subcontrariety tell us:

- Opposing I and O statements cannot both be false but both may be true. So, if an I statement is false, the opposing O statement is true; and if an O statement is false, the opposing I statement is true.

The remaining two relations on the traditional square of opposition are superimplication and subimplication. We interpret the two subimplication arrows as follows: The downward arrow only takes in and transmits truth. This relation is called *subimplication*; it is also called *subalternation*. Thus, the two subimplication arrows indicate:

- If an A statement is true, then the opposing I statement is true; and if an E statement is true, then the opposing O statement is true.

- However, if an A statement is known to be false, nothing can be concluded as to the truth-value of the opposing I statement; and if an E statement is known to be false, nothing can be concluded as to the truth-value of the opposing O statement.

We interpret the superimplication arrows as follows: the upward arrow only takes in and transmits falsity. This relation is called *superimplication*; it is also called *superalternation*. The two superimplication arrows tell us:

- If an I statement is false, then the opposing A statement is false; and if an O statement is false, then the opposing E statement is false.

- However, if an I statement is known to be true, nothing can be concluded as to the truth-value of the opposing A statement; and if an O statement is known to be true, nothing can be concluded as to the truth-value of the opposing E statement.

In sum, the traditional square of opposition indicates the following logical relations between pairs of categorical statements with the same subject and predicate terms:

1. If A is assumed true, then E is false, I is true, O is false.

2. If E is assumed true, then A is false, I is false, O is true.

3. If I is assumed true, then E is false, A is undetermined, O is undetermined.

4. If O is assumed true, then A is false, E is undetermined, I is undetermined.

5. If A is assumed false, then E is undetermined, I is undetermined, O is true.

6. If E is assumed false, then A is undetermined, I is true, O is undetermined.

7. If I is assumed false, then A is false, E is true, O is true.

8. If O is assumed false, then A is true, E is false, I is true.

In the list above, "undetermined" means that given just the assumption at hand, we cannot, from that alone, determine the truth-value of the statement.

▶   **EXERCISE 14.2**   ◀

Use the square of opposition to answer the following questions.

1.* Assuming "All aardvarks are mammals" is true, what can be inferred about the truth-values of the following? True, false, or undetermined?

    a. No aardvarks are mammals.

    b. Some aardvarks are mammals.

    c. Some aardvarks are not mammals.

2. Assuming "All aardvarks are mammals" is false, what can be inferred about the truth-values of the following? True, false, or undetermined?

    a. No aardvarks are mammals.

    b. Some aardvarks are mammals.

    c. Some aardvarks are not mammals.

3. Assuming "No aardvarks are mammals" is true, what can be inferred about the truth-values of the following? True, false, or undetermined?

    a. All aardvarks are mammals.

    b. Some aardvarks are mammals.

    c. Some aardvarks are not mammals.

4. Assuming "No aardvarks are mammals" is false, what can be inferred about the truth-values of the following? True, false, or undetermined?

    a. All aardvarks are mammals.

    b. Some aardvarks are mammals.

    c. Some aardvarks are not mammals.

5.* Assuming "Some aardvarks are mammals" is true, what can be inferred about the truth-values of the following? True, false, or undetermined?

    a. All aardvarks are mammals.

    b. No aardvarks are mammals.

    c. Some aardvarks are not mammals.

6. Assuming "Some aardvarks are mammals" is false, what can be inferred about the truth-values of the following? True, false, or undetermined?

    a. All aardvarks are mammals.

    b. No aardvarks are mammals.

    c. Some aardvarks are not mammals.

7. Assuming "Some aardvarks are not mammals" is true, what can be inferred about the truth-values of the following? True, false, or undetermined?

    a. All aardvarks are mammals.

    b. No aardvarks are mammals.

    c. Some aardvarks are mammals.

8. Assuming "Some aardvarks are not mammals" is false, what can be inferred about the truth-values of the following? True, false, or undetermined?

    a. All aardvarks are mammals.

    b. No aardvarks are mammals.

    c. Some aardvarks are mammals.

▶   **EXERCISE 14.3**   ◀

Use the traditional square of opposition to decide if the following one-premise arguments are valid or invalid. (Definition: "illie pies" are strange beasts said to inhabit the vast uncharted wilds of the Pacific Northwest.)

a.* It is false that no illie pies are blue. So, some illie pies are blue.

b. All illie pies are blue. So, it is false that no illie pies are blue.

c. No illie pies are green. So, some illie pies are not green.

d. It is false that some illie pies are green. So, no illie pies are green.

e.* All illie pies are orange. So, some illie pies are orange.

f. It is false that some illie pies are green. So, it is false that all illie pies are green.

g. It is false that some illie pies are not round. So, it is false that no illie pies are round.

h.* It is false that all illie pies are hairy. So, it is true that some illie pies are hairy.

i. Some illie pies are not green. So, no illie pies are green.

j. It is not the case that no illie pies are green. Therefore, some illie pies are green.

k. Some illie pies are not green. Thus, not all illie pies are green.

l. No illie pies are green. So, some illie pies are green.

m. No illie pies are green. So, all illie pies are green.

n. All illie pies are green. Consequently, some illie pies are not green.

o. Some illie pies are green. So, it is false that no illie pies are green.

p. Some illie pies are green. So, all illie pies are green.

| ► | **EXERCISE 14.4** | ◄ |

Part I. For each of the following pairs *of statements, specify whether the two statements are contradictories, contraries, or subcontraries.*

1.* All snowboarders are young.

No snowboarders are young.

2. Some snowboarders are young.

Some snowboarders are not young.

3. All snowboarders are young.

Some snowboarders are not young.

4. No snowboarders are young.

Some snowboarders are young.

Part II.

1.* What is the contradictory of "All extraterrestrials are green"?

2. What is the contradictory of "No extraterrestrials are green"?

3. What is the contrary of "All extraterrestrials are green"?

4. What is the subcontrary of "Some extraterrestrials are green"?

TRANSLATING ENGLISH SENTENCES INTO STANDARD CATEGORICAL FORMS

You have probably noticed that the categorical sentences that we have been working with all fit into standardized formats. They don't sound quite like sentences used in everyday life. There is a reason for this. Our flexible language allows us a great deal of linguistic freedom and many different ways to say a given thing. But this raises problems for logical theory. If we are to formulate a workable system of logic with precise rules and principles, the rules must be stated for sentences and arguments that fit a limited number of certain standardized forms. For, given the incredible variety of different types of English sentence, rules that would apply to any English sentence of any type would be rules so incredibly complex as to be practically useless. Another virtue of standardized forms is that sentences in standardized forms are clearer, have a more precise meaning, and are freer of emotional overtones. Such sentences are easier to reason about.

In everyday English, sentences that express categorical statements appear in many different forms. To permit the precise application of Aristotle's logical techniques, English sentences expressing categorical statements must be rewritten in standardized form. A categorical sentence is in **standard form** if

1. It begins with a quantifier, and the quantifier is either *all, no,* or *some.*

2. The subject term, which designates the subject class, appears next. This term is a noun, or a qualified noun (for example, *blue car*), or a noun clause (for example, *increase in temperature*), and it designates the subject class.

3. The principal verb—the copula—appears next and is a present-tense form of the verb *to be,* which is to say that it is either *are* or *are not.*

4. The predicate term, which designates the predicate class, appears last. This term is a noun, or a qualified noun, or a noun clause, and it designates the predicate class.

The following principles should help you translate ordinary English sentences into the standard form of a categorical sentence.

▶ Translating Nonstandard Subject and Predicate Terms

The subject and predicate terms of a standard categorical statement each must contain a noun or other term that denotes a class of things. So, for instance, the sentence "Some roses are red" is not in standard form, for its predicate term is an adjective ("red") and adjectives do not by themselves denote classes (they denote characteristics of things). If a predicate term is an adjective only, we need to supply a noun or a noun-like expression so that the term denotes a class. Thus "Some roses are red" becomes "Some roses are red flowers." Now, *red flowers* denotes a class of things. In general, if a statement has an adjectival predicate, replace this predicate with a term designating the class of all objects of which the adjective may truly be predicated. Here are further examples:

Nonstandard:	Standard:
All tigers are carnivorous.	All tigers are carnivorous animals.
All deer are fleeing the fire.	All deer are things that are fleeing the fire.
All millwrights are striking.	All millwrights are persons who are striking.
Some aardvarks are cute.	Some aardvarks are cute animals.
All cars are metallic.	All cars are metallic things.

▶ Translating Unexpressed Quantifiers

Consider a sentence such as "Mammals have lungs." The sentence has no quantifier. Yet someone who states this sentence surely means, *"All* mammals have lungs," so, we can rewrite the sentence and place the universal quantifier *All* at the front of the sentence. Similarly, the sentence "Fish are not warm-blooded" lacks a quantifier. But someone who says, "Fish are not warm-blooded" surely means, "No fish are warm-blooded creatures," and we can rewrite the sentence and add the quantifier *No*. If someone says, "A dog is a warm-blooded creature," their sentence lacks a quantifier. However, they most likely mean to assert that *all* dogs are warm-blooded, and their sentence can be rewritten "All dogs are warm-blooded creatures." When an English sentence lacks a quantifier, we must interpret the sentence as best we can and supply the most appropriate quantifier. Here are some further examples:

Nonstandard:	Standard:
A tiger is a mammal.	All tigers are mammals
A fish is not a mammal.	No fish are mammals.
Emeralds are green.	All emeralds are green things.
A whale is a beautiful creature.	All whales are beautiful creatures.

Suppose someone says, "Dogs are friendly." The sentence lacks a quantifier, and furthermore it is unclear what quantifier the person intends. However, if the person has common sense and knows that at least some dogs are unfriendly, he probably intends the sentence to be interpreted "Some dogs are friendly animals," and we can rewrite the sentence and supply the quantifier *some*.

Sentences in which the subject term is preceded by nonstandard quantifiers such as *most, few,* and *at least one*, should be rewritten and assigned the quantifier *some*, for any quantity less than *all* will be translated as *some*. For example:

Nonstandard Quantifier:	Standard Interpretation:
Most cats are cute.	Some cats are cute animals.
Few aardvarks are handsome.	Some aardvarks are handsome animals.
At least one platypus is cute.	Some platypuses are cute animals.
Several pigs are smart.	Some pigs are smart animals.
Many bears are timid.	Some bears are timid animals.
There are bears in the woods.	Some bears are animals in the woods.
Republicans live next door.	Some Republicans are persons who live next door.
A tiger roared.	Some tigers are animals that roared.

▶ FIXING MISSING OR NONSTANDARD COPULAS

Recall that the copula connects the subject and predicate terms. The only copulas allowed are *are* and *are not*. Consider the sentence "All mice eat cheese." This sentence expresses a universal affirmative statement. However, the sentence does not contain *are* or *are not*, and so we need to supply a proper copula. In order to do this, we will have to rewrite the predicate term to preserve the sense of the original sentence but in such a way as to also allow for the copula *are*. When we do this, the sentence becomes "All mice are cheese-eaters." Similarly, "All hens lay eggs" needs the copula *are:* "All hens are egg-laying animals."

English sentences often use other forms of the verb *to be*. Such sentences are not in standard form and also need to be rewritten so that they contain *are* or *are not*. In general, we do this by moving the verb into the predicate, and adding *are* or *are not*. For example, "Some dogs shake hands" should be rewritten "Some dogs are animals that shake hands. And "Some persons who go to college will become educated" can be rewritten as "Some persons who go to college *are persons* who will become educated." Similarly, "Some dogs bite" can be rewritten as "Some dogs are animals that bite." The nonstandard "All ducks swim" translates into "All ducks are animals that swim." Finally, "Some birds fly south for the winter" may be rewritten as "Some birds are animals that fly south during the winter."

When we supply a proper copula, we must always supply the present tense of the verb. "Some Democrats will be elected" should be rewritten as "Some Democrats are persons who will be elected." "Some Democrats were technocrats" becomes the present tense "Some Democrats are technocrats." Also, when we add a copula, we must supply the plural *are*, so "No dog is a cat" should be rewritten "No dogs are cats."

▶ TRANSLATING SINGULAR STATEMENTS

We often reason about a specifically identified individual thing, and in such cases we express our reasoning with singular statements. A **singular statement** makes an assertion about a specifically identified entity or thing. For instance, "Aristotle is a logician," or "Hillary Clinton is an attorney." Each of these two sentences is about one specifically identified individual. Can a logic that was designed for categorical statements handle singular sentences? The answer, surprisingly, is yes. The strategy is to treat the singular term in a singular sentence as if it designates a one-member class. That is, a singular term is understood as a term that refers to all the members of a class that has only one member. The sentence "Aristotle is a logician" is translated "All persons identical to Aristotle are persons who are logicians." Aristotle is in a class by himself! The singular sentence "Hillary Clinton is an attorney" becomes "All persons identical to Hillary Clinton are attorneys." For one more example, the singular sentence "The moon has craters" is translated "All things identical to the moon are things that have craters."

▶ DEALING WITH ADVERBS

Temporal adverbs (*when, whenever, anytime, always,* and so on) need to be translated into times. Spatial adverbs (*where, everywhere, anywhere,* and so on) need to be translated into places. For example, "He always wears a bright tie to work" translates into "All times he goes to work are times he wears a bright tie." Similarly, "She never is nostalgic" becomes "No times of her life are times she is nostalgic." And "He is always upbeat" needs to be rewritten as "All times of his life are times he is upbeat."

Spatial adverbs are handled similarly. "Nowhere on earth are there unicorns" translates to "No places on earth are places where there is a unicorn."

Likewise, "Nowhere on Earth is there a perfect person" may be rewritten "No places on earth are places where there is a perfect person." A sentence such as "Illie pies live in the woods of the Pacific Northwest" becomes "Some places in the woods of the Pacific Northwest are places in which illie pies live." The sentence "Love is everywhere" translates as "All places in the world are places where there is love."

► TRANSLATING CONDITIONAL SENTENCES

A conditional (if-then) sentence, such as "If it's a mouse then it's a mammal," is interpreted as a universal sentence that makes a claim about *all* the items referred to by the antecedent. Specifically, the conditional sentence is interpreted as attributing the property expressed in the consequent to each and every item denoted by the antecedent. Therefore, "If it's a mouse, then it's a mammal" is interpreted as attributing the property of being a mammal to each and every mouse. Accordingly, "If it's a mouse, then it's a mammal" is translated "All mice are mammals." Similarly, "If it is a mouse, then it is *not* a reptile" becomes "No mice are reptiles." For another example, "If it's an IBM clone, then it's a PC" is translated "All IBM clones are PCs" and "If it is an IBM clone, then it is not a Mac" becomes "No IBM clones are Macs."

► EXCLUSIVE STATEMENTS

Consider the sentence "None but the brave are free." This is called an *exclusive* sentence because the sentence's wording indicates that the predicate term applies exclusively to the group designated by the subject term. This sentence translates as "All free persons are brave persons." Similarly, a sentence such as "None except the brave are free" is also translated "All free persons are brave persons," for *none but* and *none except* are used in analogous ways in these two sentences. Notice that when we translate exclusive sentences into standard universal affirmatives, we reverse the subject and predicate.

Consider "Only students are invited to the reception." This too is exclusive and translates as "All persons invited to the reception are students." A sentence such as "Only elected officials will attend the meeting" indicates that those who will attend the meeting will all be elected officials. Consequently, "Only elected officials will attend the meeting" is translated "All persons who will attend the meeting are elected officials." Again, notice that when we translate exclusive sentences into standard universal affirmatives, we reverse the subject and predicate.

A sentence beginning with "The only" is typically *not* an exclusive sentence. For instance, the sentence "The only available meals are TV dinners" translates as "All available meals are TV dinners." Likewise, the sentence "The only people invited were anarchists" becomes "All persons invited are anarchists." Similarly, "The only creatures that live in the woods are trolls" translates as "All creatures that live in the woods are trolls." When it appears at the start of a sentence, the expression "The only" typically translates as "All."

▸ EXCEPTIVE STATEMENTS

Consider this sentence: "All except truckers are happy with the new regulations." How shall we translate this? The sentence indicates two things: 1. All non-truckers are happy with the new regulations. 2. No truckers are happy with the new regulations. Consequently, the sentence may be translated "All non-truckers are happy with the new regulations and no truckers are happy with the new regulations." Generally, a sentence of the form "All except A's are B" translates as a conjunction of the form "All non-A's are B and no A's are B."

▸   **EXERCISE 14.5**   ◀

---

1. Translate the following statements into A statements.

   a. *A drop of seawater tastes salty.

   b. Iron accepts a magnetic charge.

   c. Teenagers are naturally full of energy.

   d. If it's a Jeep, then it has four wheel drive.

   e. *Whoever uses heroin is self-destructive.

   f. Truly religious people are charitable.

   g. A velvet Elvis painting is a beautiful thing.

   h. Anytime Joe drives the car, the car breaks down.

   i. Hamburgers are a good source of cholesterol.

   j. *Only the rational are truly free.

   k. *Wherever he goes, there is a cloud of gloom over him.

   l. None but paid guests were admitted to the party.

   m. It's always raining.

   n. Human beings are naturally selfish.

2. Translate the following statements into E statements.

   a. *A drop of seawater does not taste salty.

   b. Iron does not accept a magnetic charge.

   c. Teenagers are not full of energy.

   d. Whoever uses heroin is not self-destructive.

   e. *Truly religious people are not charitable.

f. A velvet Elvis painting is not beautiful.

g. When Joe drives the car, the car does not break down.

h. Hamburgers are not a good source of cholesterol.

i.*Only the knowledgeable are not truly free.

j. If it's an antique, then it won't have a computer-designed frame.

k.*It never rains.

l. Human beings are not selfish.

m. No one who is a student is rich.

3.   Translate the following statements into I statements.

    a.*There are energetic teenagers.

    b. Most socialists are egalitarians.

    c. Some Democrats are from the South.

    d. Many Republicans are religious.

    e.*There are several contented voters.

    f. A few communists play Scrabble.

    g. There have been some old skateboarders.

    h.* Several people saw smoke.

    i. Sometimes it rains.

    j.*There are some sacred places.

4.   Translate the following statements into O statements.

    a.*There are dishonest politicians.

    b. There are teenagers who aren't energetic.

    c. Most cats are not mean.

    d.*There are inactive seniors.

    e. Many homes don't have VCRs.

    f. Most teenagers aren't rich.

    g.*There have been musicians who cannot read music.

    h. Sometimes it doesn't rain.

    i. Some places are not sacred.

    j. Some of Pink's hot dogs have no chili on them.

EQUIVALENCE RULES FOR ARISTOTELIAN LOGIC

Three logical operations can be performed on categorical sentences, and once these operations are specified we can state equivalence rules for categorical statements.

▶ PRODUCING THE CONVERSE OF A CATEGORICAL STATEMENT

If we switch the subject and predicate terms in a categorical statement, the resulting sentence is the **converse** of the original sentence. For example, if we begin with "All sharks are hungry things" and switch the terms, we produce the converse "All hungry things are sharks." Here are further examples:

Statement:	Converse:
A: All aardvarks are mammals.	All mammals are aardvarks.
E: No dogs are cats.	No cats are dogs.
I: Some politicians are poets.	Some poets are politicians.
O: Some politicians are not poets.	Some poets are not politicians.

Recall that two statements are **logically equivalent** if they imply each other, which is to say that in all situations they have matching truth-values. In a case of two equivalent statements, if one is true, the other will be true, and if one is false, the other will be false. Notice that the E statement "No dogs are cats" is logically equivalent to its converse, "No cats are dogs." And the I statement "Some politicians are poets" is logically equivalent to its converse "Some poets are politicians." In general, *converting an E or an I statement produces a new statement that is logically equivalent to the original statement.* Thus, the converse of an E or an I statement always has same truth-value as the original statement.

However, converting an A or an O statement does not always result in a sentence with the same truth-value as the original. For instance, the A statement "All aardvarks are mammals" is true, but its converse "All mammals are aardvarks" is false. This A statement and its converse are therefore not equivalent. In general, converting an A or an O statement does not produce an equivalent statement.

▶ PRODUCING THE OBVERSE OF A CATEGORICAL STATEMENT

If you perform the following two operations on a categorical statement, the result is the **obverse** of the original statement:

1. Change the quality (without changing the quantity) from affirmative to negative or negative to affirmative.

2. Replace the predicate term with its term complement.

The **complement of a class C** is the class consisting of all those things outside the class C. For example, the complement of the class of cars is the class of all non-cars, the complement of the class of mammals is the class consisting of all non-mammals, that is, all things that are not mammals. The **term complement** is the term denoting the class complement. In general, if T is a term designating the class C, the term complement of T is the term designating the complement of C. The term complement is formed by adding or subtracting the prefix *non* or some equivalent to the term. In effect, we form the term complement by "negating" the term. The term complement of *cars* is *non-cars,* the term complement of *cats* is *non-cats,* and so on.

A class and its complement are mutually exclusive classes (nothing is a member of both classes), and they are also jointly exhaustive classes (anything whatsoever is a member of one or the other). A term and its complement are thus sometimes called "contradictory terms."

Now, let us obvert the A statement "All Germans are Lutherans." The two steps required to transform a statement into its obverse are

1. We change the quality without changing the quantity. So we change the statement from the affirmative to the negative: All Germans are not Lutherans, that is, no Germans are Lutherans.

2. Next, we replace the predicate with its term complement: No Germans are non-Lutherans.

Let's try another. What is the obverse of "Some teenagers are smokers"? First, we change the statement from a particular affirmative to a particular negative, and then we negate the predicate. This gives us "Some teenagers are not nonsmokers." Notice that *not* is part of the copula while *non* is part of the predicate. Here are further examples:

Statement:	Obverse:
No Greeks are Persians.	All Greeks are non-Persians.
Some bullets are projectiles.	Some bullets are not non-projectiles.

Notice that if one of these statements is true, its obverse is true, and if one is false, its obverse is false. These examples illustrate the point that *the obverse of an A, E, I, or O statement is always equivalent to the original statement.* Thus, the obverse of a statement always has the same truth-value as the original statement. Notice that if you obvert a categorical statement twice and cancel out any double negatives, you will end up back where you started, with your original statement.

► **Producing the Contrapositive of a Categorical Statement**

If you perform the following two operations on a categorical statement, the result is the **contrapositive.**

1. Switch the subject and predicate.

2. Replace each term with its term complement.

For example, if we begin with "All apples are crispy things, " the two steps are

Step 1   All crispy things are apples.

Step 2   All noncrispy things are non-apples.
  In general, *the contrapositive of an A statement is equivalent to the original statement and thus has the same truth-value. Similarly, an O statement and its contrapositive are also equivalent and thus have matching truth-values.* For example, consider the O statement "Some apples are not crispy things." The contrapositive is formed in two steps:

Step 1   Some crispy things are not apples.

Step 2   Some noncrispy things are not non apples.
  However, E and I statements are not always equivalent to their contrapositives.

Statement:	Converse:	
A: All S are P	All P are S	
E: No S are P	No P are S	(equivalent)
I: Some S are P	Some P are S	(equivalent)
O: Some S are not P	Some P are not S	
Statement:	Obverse:	
A: All S are P	No S are non-P	(equivalent)
E.:No S are P	All S are non-P	(equivalent)
I: Some S are P	Some S are not non-P	(equivalent)
O: Some S are not P	Some S are non-P	(equivalent)
Statement:	Contrapositive:	
A: All S are P	All non-P are non-S	(equivalent)
E: No S are P	No non-P are non-S	
I: Some S are P	Some non-P are non-S	
O: Some S are not P	Some non-P are not non-S	(equivalent)

▶   **EXERCISE 14.6**   ◀

---

1.   For each statement below, construct the converse and state whether the converse is equivalent or not to the original statement.

   a.*All spiders are creepy things.

   b. No spiders are creepy things.

    c. Some spiders are creepy things.

    d.*Some spiders are not creepy things.

2. For each statement below, construct the obverse and state whether the obverse is equivalent or not to the original statement.

    a.*All Corvairs are safe vehicles.

    b. No Corvairs are safe vehicles.

    c. Some Corvairs are safe vehicles.

    d. Some Corvairs are not safe vehicles.

3. For each statement below, construct the contrapositive and state whether the contrapositive is equivalent or not to the original statement.

    a. *All politicians are honest persons.

    b. No politicians are honest persons.

    c. Some politicians are honest persons.

    d. Some politicians are not honest persons.

DROPPING THE ASSUMPTION
OF EXISTENTIAL IMPORT

Some classes are **empty classes** and some classes are said to be "nonempty" or "not empty." A class is empty if there do not actually exist entities of that category. For example, the class of unicorns is an empty class. A class is *not empty* if there are actually existing things of that category. For example, the class of cats is nonempty.

Recall that a categorical statement is said to have existential import if its subject term refers to actually existing entities. For example, "All dogs are mammals" has existential import (because there are dogs). So, if a statement has existential import, its subject term designates a nonempty class.

Recall also that a statement *lacks* existential import if its subject term does *not* refer to actually existing entities. For example, "All unicorns have horns" lacks existential import (because there are no unicorns). So, if a statement lacks existential import, its subject term designates an empty class.

As noted earlier, the logic developed by Aristotle was built on the assumption of existential import—the assumption that the subject terms of the categorical statements under consideration refer to actually existing entities. In other words, traditional Aristotelian logic was built on the assumption that the subject terms of categorical statements under consideration designate nonempty classes. Consequently, traditional Aristotelian logic restricts itself to categorical statements whose subject terms refer to actually existing things. The discussion up to this point was therefore built on the presupposition that the subject terms of the categorical sentences under consideration refer to actually existing things.[1]

However, in everyday matters we sometimes wish to reason *without* assuming existential import. For instance, "Let he who is without sin cast the first stone" is not meant to imply the actual existence of a sinless human being. Existential import is not assumed when this statement is made. And "All trespassers will be shot" is not supposed to imply the existence of trespassers. The property owner who puts up a "No trespassing" sign hopes there will be no trespassers. In everyday life we do not always assume existential import.

In science, scientists sometimes reason without assuming existential import. For example, Einstein reasoned about what would happen *if* a human observer were to travel at the speed of light, even though no human observer can actually travel at the speed of light. Three hundred years earlier, Galileo made mathematical calculations dealing with perfectly frictionless surfaces, even though no such thing exists.

And so it came to pass that during the nineteenth century, logicians began to ask: Must a system of logic assume existential import? What changes must we make in logical theory if we drop the assumption? As a result of these and related speculations, new developments in logical theory sprang up during this period from theorists such as Augustus DeMorgan (1806–71), George Boole (1815–64) and John Venn (1834–23). Incidentally, some of these developments in logic helped pave the way for the logical languages later used in computer science.

What changes must we make in logical theory if we drop the blanket assumption of existential import? One important change was suggested by the logician George Boole. Boole suggested that if we drop the existential import assumption and do not suppose that every subject term refers to actually existing entities, we shall have to specify a new interpretation of the A and E categorical statements. Specifically, we should interpret an A statement, "All S's are P," as "Nothing is an S and not a P." Furthermore, we should interpret the E statement "No S's are P's" as "Nothing is an S and also a P." Notice that on Boole's interpretation, the A and E statements neither assert nor deny the actual existence of any S's. That is, from the "Boolean standpoint" the universal statements are taken as "mum" on the question of whether or not any S's actually exist.

However, Boole gave the I and O statements essentially the same interpretation Aristotle had given them: "Some S are P" means "At least one S exists and is a P" and "Some S are not P" means "At least one S exists and is not P." Boole's interpretation is known as the **Boolean Interpretation.** If we adopt the Boolean interpretation of categorical sentences, we are taking the "hypothetical" rather than the "existential" viewpoint.

---

[1]The entities designated by a term are said to "fall under" the term. So, for example, all (and only) dogs fall under the term *dogs*. A term is an "empty term" if nothing falls under it. Terms such as *unicorn*, *round square*, and *gryphon* are thus empty terms. A term under which everything in the universe falls is a "universal" term. For example, *thing that is either a mouse or not a mouse* is such a term. Neither empty nor universal terms are allowed into Aristotelian syllogistic logic.

Aristotelian Interpretation:	Boolean Interpretation:
A: All S are P	Nothing is S and not P
E: No S are P	Nothing is S and P
I: Some S are P	Something is S and P
O: Some S are not P	Something is S and not P

THE MODERN SQUARE OF OPPOSITION

If we drop the blanket assumption of existential import and give categorical statements the Boolean interpretation, we must drop most of the arrows from the square of opposition. For instance, according to the traditional square of opposition, opposing A and E statements are contraries. However, if we do not assume existential import, then opposing A and E statements can both be true, for if there are no S's, then the A statement on the Boolean interpretation ("Nothing is an S and not a P") is clearly true and the E statement on the Boolean interpretation ("Nothing is an S and a P") is also true. Thus, on the Boolean interpretation, A and E statements are not contraries in the traditional sense.

For another instance, according to the traditional square of opposition, opposing I and O statements are subcontraries. However, if we do not assume existential import, then opposing I and O statements can both be false, for if there are no S's, then it is false that some S are P and it is also false that some S are not P. So, on the Boolean interpretation, I and O statements are not subcontraries in the traditional sense. Indeed, when drop the assumption of existential import from the square of opposition, the two arrows of contradiction are all that remain. The revised square, shown in the box below, is called the Modern Square of Opposition.

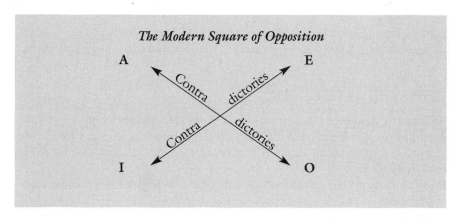

The Modern Square of Opposition drops the blanket assumption of existential import and consequently contains no arrows for contrariety, subcontrariety, superimplication or subimplication. If it seems illogical to suppose that "All S are P" does *not* imply "Some S are P," consider the following example. A grocer announces that all shoplifters will be shot. As a result, nobody shoplifts. So, the grocer's statement, "All shoplifters will be shot," remains true, but "Some shoplifters are shot" is false. In other words, in this case, the A statement does not imply the opposing I statement.

Also, if it seems illogical to suppose that "All S are P" and " No S are P" are *not* contraries, consider the previous example. The grocer announces, "All shoplifters will be shot" and as a result, nobody shoplifts. In this case, it will remain true that all shoplifters will be shot but it will also be true that no shoplifters will be shot. In other words, in this case, the opposing A and E statements are *not* contraries (since they might both be true).

Let us call a categorical statement that is interpreted from the Boolean standpoint a **Boolean categorical.** The modern square of opposition sums up logical relations among opposing Boolean categoricals. The general forms of the converses, obverses, and contrapositives of the Boolean categorical statements are summarized on the following table.

Boolean categorical:	Converse:	
A: Nothing is S and not P	Nothing is P and not S	
E: Nothing is S and P	Nothing is P and S	(equivalent)
I: Something is S and P	Something is P and S	(equivalent)
O: Something is S and not P	Something is P and not S	

Boolean categorical:	Obverse:	
A: Nothing is S and not P	Nothing is S and non-P	(equivalent)
E: Nothing is S and P	Nothing is S and not non-P	(equivalent)
I: Something is S and P	Something is S and not non-P	(equivalent)
O: Something is S and not P	Something is S and non-P	(equivalent)

Boolean categorical:	Contrapositive:	
A: Nothing is S and not P	Nothing is non-P and not non-S	(equivalent)
E: Nothing is S and P	Nothing is non-P and non-S	
I: Something is S and P	Something is non-P and non-S	
O: Something is S and not P	Something is non-P and not non-S	(equivalent)

CHAPTER 14 GLOSSARY

Affirmative sentence   A sentence that affirms class membership.

Assumption of Existential Import   The assumption, made within Aristotelian logic, that the categorical statements under consideration have existential import.

Boolean categorical   A categorical statement that is given the Boolean interpretation.

Boolean interpretation   The interpretation of categorical sentences that was developed by George Boole and that dispensed with the assumption of existential import for universal sentences.

Categorical sentence   A sentence whose subject term denotes a *class* or *category* of entities and whose predicate term also denotes a class of entities and that asserts that all or some of the class denoted by the subject term is included or excluded from the class denoted by the predicate term. We examine four types in this chapter:

- **Particular affirmative sentence**   Categorical sentence asserting that part of the subject class is included in the predicate class, that is, some members of the subject class are members of the predicate class.

- **Particular negative sentence**   Categorical sentence asserting that part of the subject class is excluded from the predicate class, that is, some members of the subject class are not members of the predicate class.

- **Universal affirmative sentence**   Categorical sentence asserting that the whole subject class is included in the predicate class, that is, every member of the subject class is a member of the predicate class.

- **Universal negative sentence**   Categorical sentence asserting that the whole subject class is excluded from the predicate class, that is, no member of the subject class is a member of the predicate class.

Class   A collection of objects having a specified characteristic in common.

Complement of a class C   The class consisting of all those things outside the class C.

Contradictories   Two statements are contradictories if and only if they cannot both be true and they cannot both be false. In all possible situations, if one is true, the other is false.

Contrapositive of a sentence   The sentence that results if you perform the following two operations on a categorical sentence: 1. Switch the subject and predicate. 2. Replace each term with its term complement.

Contraries   Two statements are contraries if they cannot both be true, but might both be false. So, if two statements are contraries, at least one is false.

Converse of a sentence   The sentence that results if we switch the subject and predicate terms in a categorical sentence.

Copula   A word that serves to link the subject term with the predicate term.

Empty class   A class is empty if no entities of that category exist. (And a class is nonempty if entities of that category do exist.)

Existential import   A categorical statement is said to have existential import if its subject term refers to a class that is not empty, that is, the subject term refers to actually existing entities. (A statement lacks existential import if its subject term refers to a class that is empty.)

Existential viewpoint   If we presuppose that the subject terms of categorical sentences under consideration designate actually existing things, we are taking the existential viewpoint.

Hypothetical viewpoint   If we think about the logical relations among categorical statements from the viewpoint that does not presuppose that the subject terms designate actually existing things, we are taking the hypothetical viewpoint.

Logical equivalence   Two statements are logically equivalent if they imply each other, which is to say that in all situations they have matching truth-values.

Negative sentence   Sentence in which class membership is denied.

Obverse of a sentence   The sentence that results if you perform the following two operations on a categorical sentence: 1. Change the quality (without changing the quantity) from affirmative to negative or negative to affirmative. 2. Replace the predicate term with its term complement.

Opposition   Two statements with the same subject and predicate terms are said to be in opposition if they differ in quality or quantity or both.

Particular sentence   A categorical sentence that makes a claim about some of the class denoted by the subject term.

Quantifier   A word such as *all* or *some* that specifies a quantity for the subject term of the sentence.

Singular sentence   A sentence that makes an assertion about a specifically identified entity or thing.

Some   One or more, that is, at least one. Another way to put this: one or more, possibly all.

Square of Opposition   A table representing the logical relations between opposing categorical statements.

Standard form (of a categorical sentence)   A categorical sentence that satisfies the following criteria: 1. It begins with a quantifier, and the quantifier is either *all, no,* or *some.* 2. The subject term, which appears next, is a noun, or a qualified noun, or a noun clause, and it designates the subject class. 3. The copula appears next and is a present-tense form of the verb *to be,* which is to say that it is either *are* or *are not.* 4. The predicate term appears last. This term is a noun, or a qualified noun, or a noun clause, and it designates the predicate class.

Subcontraries   Two statements are subcontraries if they cannot both be false, but might both be true. If two statements are subcontraries, at least one is true.

Term complement   The term denoting the class complement.

Universal sentence   A categorical sentence that makes a claim about every member of the class denoted by the subject term.

15

CATEGORICAL
SYLLOGISMS

T he heart of Aristotelian logic is the study of the "categorical syllogism."
In general, a syllogism is a deductive argument. A *categorical syllogism* is
an argument that contains three categorical statements, the statements contain
three different terms altogether, each statement contains two different terms,
and no two statements contain the same two terms. So, each term appears in
only two of the sentences. It is assumed that the terms are used in the same
sense throughout the argument. The term appearing in both premises (and thus
not in the conclusion) is called the **middle term**. The conclusion's predicate
term is called the **major term**. The term appearing as the subject of the conclusion is the **minor term**. The **major premise** is the premise containing the major
term, and the **minor premise** is the premise containing the minor term. A categorical syllogism is in **standard form** if the major premise is written first, the
minor premise is written second, and the conclusion is last.

LOGICAL FORM

The logical form of a categorical syllogism is specified in terms of the mood and
figure of the syllogism. A categorical syllogism's **figure** is determined by the
placement of its middle term. As the table below shows, the middle term has
four possible patterns of placement, which produce four different figures. In the
table, S stands for the minor term, P stands for the major term, and M stands
for the middle term. (The minor term is assigned S (subject) because it is the
subject of the conclusion, the major term gets P (predicate) because it is the
predicate of the conclusion, and the middle term gets M for *m*iddle.)

Figure 1	Figure 2	Figure 3	Figure 4
M - P	P - M	M - P	P - M
S - M	S - M	M - S	M - S
S - P	S - P	S - P	S - P

The **mood** of a categorical syllogism is simply a listing of the type—A, E, I, or O—of each of its categorical statements. For example, consider the following syllogism:

No aardvarks are writers.

Some aardvarks are happy.

Therefore, some happy things are not writers.

Its mood is EIO, because its major premise is an E statement, its minor premise is an I statement, and its conclusion is an O statement. The mood will therefore be specified by three letters, with the first letter indicating the type of major premise, the second letter the type of minor premise, and the third letter indicating the type of conclusion.

Because the middle term of the syllogism above *(aardvarks)* is positioned according to Figure 3, this syllogism is of the mood *EIO in the third figure,* or, more briefly, EIO-3. This specification of mood and figure gives the **logical form** of the syllogism.

If a categorical syllogism in standard form is valid, any syllogism of the same logical form will also be valid, and if a standard form syllogism is invalid, any syllogism of the same logical form will be invalid. To assess a syllogism, we need only check out its mood and figure—its logical form—to see if that is a valid form. The actual terms in the syllogism—the subject matter—make no logical difference at this level of abstraction.

There are 256 possible syllogistic forms in all, and ancient and medieval logicians studied all 256. Counterexamples were found for all but twenty-four of the forms, and they concluded that these twenty-four were the only valid forms of categorical syllogism.

The following are valid unconditionally, without the assumption of existential import:

Figure 1	Figure 2	Figure 3	Figure 4
AAA	EAE	IAI	AEE
EAE	AEE	AII	IAI
AII	EIO	OAO	EIO
EIO	AOO	EIO	

If we assume existential import, we must add to this the following:

Figure 1	Figure 2	Figure 3	Figure 4	Required Assumption:
AAI	AEO		AEO	S exists
EAO	EAO			S exists
		AAI	EAO	M exists
		EAO		M exists
			AAI	P exists

▶   **EXERCISE 15.1**   ◀

---

1. Rewrite each categorical syllogism in standard form and name the mood and figure.

    a. *All dogs are mammals. No mammals are reptiles. So, no dogs are reptiles.

    b. Some mammals are primates. Some primates are small animals. So, some mammals are small animals.

    c. No mammals are fish. Some fish are cold-blooded. So, some mammals are not cold-blooded.

    d. Some aardvarks are not old things. All old things are wrinkled. So, no aardvarks are wrinkled.

    e. *No Pink's hot dogs are hamburgers. Some hamburgers are fattening things. So, no Pink's hot dogs are fattening things. (Pink's Hot Dogs has been an L.A. "institution" since 1939.)

    f. Some Pink's hot dogs are things covered with chili. All things covered with chili are things that taste good. So all Pink's hot dogs are things that taste good.

    g. All Pink's hot dogs are things that are delicious. All things that are delicious are things that are nutritious. So, all Pink's hot dogs are things that are nutritious.

    h. Some Pink's hot dogs are not things that are covered with mustard. All things that are covered with mustard are things that are messy. So, some Pink's hot dogs are not things that are messy.

2. In the problems above, identify the major, minor, and middle terms.

3. In the problems above, determine which syllogisms are valid by checking the mood and figure against the two lists of valid syllogistic forms. If a syllogism is valid only if an assumption is made, state the assumption.

4. Construct a syllogism in the mood EIO, figure 2.

5. Construct a syllogism in the mood EAE, figure 1.

6. Construct a syllogism in the mood AII, figure 1.

7. Construct a syllogism in the mood AEE, figure 3.

8. Construct a syllogism in the mood IAI, figure 4.

9. Construct a syllogism in the mood OAO, figure 3.

10. Construct a syllogism in the mood AOO, figure 3.

VENN DIAGRAMS

The British logician John Venn developed a way to diagram and visually display the information contained in a categorical statement. A **Venn diagram** is a drawing of overlapping circles where each circle represents a class denoted by a term and each circle is labeled for the term it denotes. Because a single categorical statement has two terms, a Venn diagram for a single categorical statement will have two overlapping circles, one for the subject term and one for the predicate term. For example, here are the two circles for the Venn diagram for the A statement "All aardvarks are mammals":

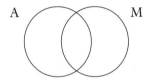

You may find it helpful at first to think of members of the class denoted by the term as being in the corresponding circle. We imagine that all the aardvarks have been rounded up and are inside the first circle, snorting and rooting around for ants. And all the mammals have been rounded up by a modern-day Noah and are resting inside the second circle. (The lions are lying down next to the lambs, and so on.)

Next, in order to enter information into these diagrams, we shall employ two methods: shading an area, and placing an X into an area. Shading an area indicates the area is empty. Putting an X in an area indicates the area is not empty. That is, an X indicates that at least one thing exists in the area. If no marks appear in an area, this indicates that nothing is known of the area—the area may be empty, or it may have members. Consequently, when we enter the information for the sentence "All aardvarks are mammals," the diagram looks like this, assuming existential import:

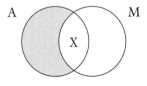

All A are M.

The Venn diagrams for the other three categorical sentence forms follow, assuming existential import.

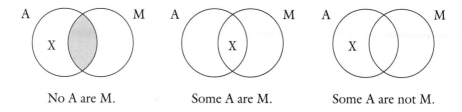

No A are M.   Some A are M.   Some A are not M.

Now, let us examine the Venn diagram for the A statement when it is given a Boolean interpretation. Recall that on the Boolean interpretation, an A statement such as "All aardvarks are mammals" is understood as "Nothing is an aardvark and not a mammal." If we drop the assumption of existential import and assume the Boolean standpoint, the diagram for the A statement "Nothing is an aardvark and not a mammal" looks like this:

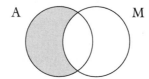

Notice that the diagram no longer has an X. The diagram no longer explicitly asserts the existence of any entity. It only tells us where there are *not* entities. This Venn diagram tells us there is nothing in the part of the A circle outside the M circle—no member of A is outside M. However, the diagram does not say anything *is* A and M. The overlapping area may have members or it may be empty. In other words, the diagram tells us where there are *no* members of A, but it does not say any A actually exist. It is neutral—noncommittal—on the existence of A's. *If* there is an A, we can tell from the diagram where it will be; but we can't tell, from the diagram, whether or not there actually are any A's. The Boolean interpretation is sometimes called the "hypothetical viewpoint" because of its hypothetical or "if-then" nature.

On the Boolean interpretation, the opposing universal negative statement in this case is "Nothing is an aardvark and a mammal." The Venn diagram for this universal negative statement, using the Boolean interpretation, looks like this:

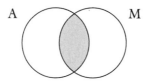

Notice that according to the Boolean interpretation, the A and E statements tell us where there are *no* aardvarks, but they do not say any aardvarks actually exist.

Because the two particular statements, I and O statements, have the same meaning in both Aristotelian and Boolean logic, that is, I and O both presuppose existential import, they have the same diagrams in both modern and classical systems.

In sum, on the Boolean interpretation, the Venn diagrams for the four categorical forms indicate the following:

A: No members of S are outside P

E: No members of S are inside P

I: At least one S exists and it is a member of P

O: At least one S exists and it is not a member of P

TESTING A CATEGORICAL SYLLOGISM FOR VALIDITY WITH VENN DIAGRAMS

A set of Venn diagrams can be drawn for an entire categorical syllogism. Furthermore, once the diagram is drawn, we can determine, from a visual inspection of the circles, whether or not the argument is valid. The first step is to set up the circles correctly.

In the diagram and example below, S stands for the minor term, P stands for the major term, and M stands for the middle term. The circle labeled S represents the category of things referred to by the minor term, circle P represents the category of things referred to by the major term, and circle M represents the category of things referred to by the middle term.

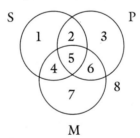

In order to represent all possibilities, the circles must overlap to form the eight different areas numbered above. Each of these areas represents a different class of things:

Area 1: Anything here is an S, but not an M, and not a P.

Area 2: Anything here is an S, a P, and not an M.

Area 3: Anything here is a P, but not an S, and not an M.

Area 4: Anything here is an S, an M, but not a P.

Area 5: Anything here is an S, a P, and an M.

Area 6: Anything here is a P, an M, but not an S.

Area 7: Anything here is an M, but not an S, and not a P.

Area 8: Things out here would be neither P, S, nor M.

The following procedure allows us to test, somewhat formally, with Venn diagrams, a categorical syllogism for validity.

1. Draw three overlapping circles, one for each term, in such a way that eight areas are formed.

2. Label the circles for the three terms.

3. Enter the information for the premises only.

4. If the argument contains a universal premise, enter its information first. If the argument contains two universal premises, either premise can be entered first.

5. When placing an X in an area, if one part of the area has been shaded, place the X in an unshaded part.

6. When placing an X in an area, if a circle's line runs through the area, place the X squarely on the line separating the area into parts. An X on a line means that the individual represented by the X may be on either side of the line, which is to say that it is not known which side of the line the X is actually on.

7. An X should never be placed on the intersection of two lines.

8. To leave an area blank is to assert that we have no information about the area.

9. The test for validity is as follows:

> *A categorical syllogism is valid if, when the information from just the premises has been entered into the diagram, visual inspection of the diagram reveals that the information contained in the conclusion has been diagrammed as well. In diagramming only the premises, we have also diagrammed the conclusion.*

This shows that the conclusion contains no more information than is already presented in the diagram of the premises. In a sense, the diagram for the premises "contains" the diagram for the conclusion. If the diagram for the premises contains all the information presented in the conclusion, it would be impossible for the premises to be true and the conclusion false, and the argument is there-fore valid. The test for invalidity follows:

> *A categorical syllogism is invalid if, when we have diagrammed the prem-ises, information must be added to the diagram in order to represent the conclusion.*

If the diagram for the premises does not contain all the information in the conclu-sion, it is possible for the premises to be true and the conclusion false, that is, the conclusion could be false even though the premises are true, and the argument is therefore invalid. Now, let us begin with Aristotelian categorical syllogisms.

DIAGRAMMING ARISTOTELIAN CATEGORICAL SYLLOGISMS

In this section, we shall use the Venn diagram method to test Aristotelian cat-egorical syllogisms for validity. Each argument will therefore carry a presuppo-sition, namely, the assumption that the subject term of each categorical state-ment refers to at least one actually existing thing. Let us begin with an AAA-1 syllogism:

All mammals are warm-blooded.

All aardvarks are mammals.

Therefore, all aardvarks are warm-blooded.

Our first step is to draw the three overlapping circles and label them A, M, and W as follows:

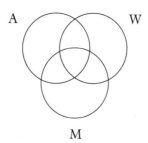

Because both premises are universal, we can enter them into the diagram in either order. Let us enter the shading for the major premise first. This premise tells us that all mammals are warm-blooded. That is, no mammals are not warm-blooded. In terms of the diagram, this indicates that all of the M circle that is outside the W circle is empty. Therefore, the part of the M circle that is outside the W circle must be shaded to indicate that it is empty:

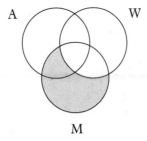

We are assuming here that at least one mammal exists; however, we must wait to enter the X for this assumption until we diagram the shading for the next premise.

Next, we enter into the diagram the information from the minor premise, "All aardvarks are mammals." This premise asserts that all aardvarks belong to the class of mammals. That is, all of the members of the A circle are inside the M circle. To enter this information into the diagram, we must shade the part of the A circle that lies outside the M circle to indicate that this area is empty:

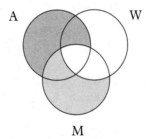

Because we are making the Aristotelian assumption, we are assuming there are aardvarks and there are mammals. With the shading completed, we can now enter the X's to indicate the existence of at least one aardvark and at least one mammal. Notice that one X placed in the only unshaded part of the A circle is also automatically in the M circle, and so this one X will serve to represent both existential presuppositions:

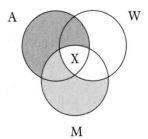

This figure now represents the information contained in the premises. We purposely do *not* enter the information from the conclusion.

Now we inspect the diagram to see if the information in the conclusion is already represented in the diagram of the premises. If it is, then the conclusion cannot possibly be false if the premises are true, and the argument is valid. The conclusion, "All aardvarks are warm-blooded," indicates that all members of the class of aardvarks are also members of the class of warm-blooded creatures. That is, nothing in the A circle is not also in the W circle. And that is indeed exactly what our diagram indicates, for all sections of the A circle outside the W circle have been shaded. Nothing more would need to be added to our diagram in order to have it also represent the conclusion. In a sense, the information contained in the conclusion is already contained in the premises. The conclusion would have to be true if the premises are true, and the argument is therefore valid.

Let us test another syllogism.

Some farmers are not happy individuals.

No happy individuals are greedy individuals.

So, some greedy individuals are farmers.

We begin by entering the information from the second premise, because it is universal in nature. In terms of our diagram, the premise tells us that no member of the H circle is a member of the G circle. Consequently, we shade the part of the H circle that lies inside the G circle:

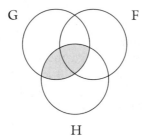

Next, the first premise indicates that some farmers are not happy. To record this information, we must place an X inside the F circle but outside the H circle. However, we must avoid any shaded areas, for these areas are supposed to be empty. Now, where do we place our X? The remaining region is divided into two parts, and the premises do not indicate which of the two parts contains the X. In this case, we place the X *on the line* between the two parts. We "straddle the fence," so to speak. This indicates that, given the information in the premises, we do not know to which of the two areas X belongs.

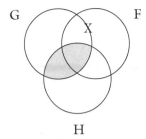

Next, the presuppositions. We must only place an X inside the unshaded part of the H circle, to indicate the existence of at least one happy person, because the first X took care of the assumption of at least one farmer. However the remaining part of the H circle is divided into two parts. We therefore place this X on the line dividing the parts. We must also place an X inside the G circle to indicate the existence of a greedy person. Since the premises do not indicate the side, the X goes on the line:

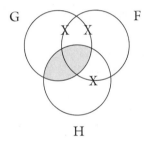

Now, the conclusion indicates that some greedy individuals are farmers, but the diagram of the premises does not guarantee this at all. The diagram of the conclusion is not "contained" within the diagram of the premises. It is therefore possible that the premises are true and the conclusion is false. The syllogism is invalid.

Let us diagram another syllogism. The following syllogism has already been symbolized:

All C are G.

Some G are S.

Therefore, some S are C.

We diagram the universal premise first. It indicates that all members of the C class also belong to the G class. In terms of our diagram, this is to say that all members of the C circle are also inside the G circle; the area of the C circle that lies outside the G circle is therefore empty and is shaded. Thus:

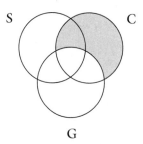

Next, the remaining premise indicates that at least one member of the G class is a member of the S class. We therefore place an X in a spot within the G circle that is also within the S circle. However, the region we are interested in is divided into two parts, and our premises do not tell us which part to place our X in. In such a case, we place the X on the line between the two areas. This indicates that we do not know to which of the two areas the X belongs. We also place an X within the C circle to indicate the existence of a C.

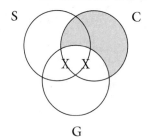

The conclusion, that some S are C, is not represented in the diagram. Given just the information in the premises, the conclusion is not guaranteed. This argument is therefore also shown to be invalid.

One more example. Consider the following EIO-3 syllogism:

No S are F.

Some S are R.

Therefore, some R are not F.

First, we diagram the universal premise. In terms of our diagram, this premise tells us that no members of the S circle are also inside the F circle. We therefore shade the part of the S circle that is inside the F circle.

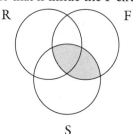

To diagram the next premise, we must place an X inside a part of the S circle that is also a part of the R circle:

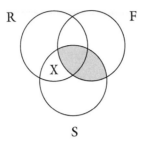

The Aristotelian existential assumption is represented adequately by this single X. The conclusion, that some R are not F, is indeed represented by this diagram of the premises. The argument is thus valid.

DIAGRAMMING FROM THE BOOLEAN STANDPOINT

In this section, we drop the blanket existential presupposition and therefore do not assume that the subject terms of universal statements refer to actually existing things. Let us begin with the following EAO-1 syllogism:

Nothing is a mammal and a bird.

Nothing is an aardvark and not a mammal.

Something is an aardvark and not a bird.

Our first step is to enter the information for the first premise. We shade the area representing individuals that belong to both the M circle and the B circle:

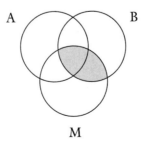

Next, to enter the information for the other premise, we shade the area representing individuals belonging to the A circle but not the M circle:

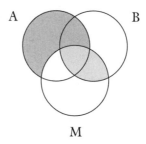

A ⬤ B

M

The conclusion, that something is inside the A circle but not inside the B circle, is not represented in our diagram. That is, the information from the conclusion is not contained in the diagram of the premises. The argument is therefore invalid.

Incidentally, if we add the Aristotelian presupposition here and assume that the subject terms refer to actually existing things, the Venn diagram for this EAO-1 syllogism looks like this:

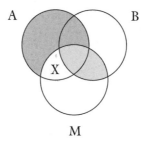

A ⬤ B

M

The X represents the Aristotelian presupposition that at least one mammal and at least one aardvark exist. Notice that with this presupposition, the syllogism is valid. In this case, the Boolean syllogism is invalid, but the corresponding Aristotelian syllogism is valid.

Let us diagram the following syllogism assuming the Boolean standpoint:

Nothing is a mammal and not warm-blooded.

Nothing is an aardvark and not a mammal.

Therefore, nothing is an aardvark and not warm-blooded.

Our first step is to draw the three overlapping circles and label them A, M, and W as follows:

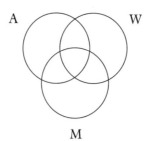

A ⬤ W

M

Let us begin with the major premise first. This premise tells us that all of the M circle that is outside the W circle is empty. Therefore, the part of the M circle that is outside the W circle must be shaded to indicate that it is empty:

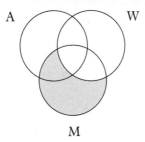

Next, we add the information from the minor premise by shading the part of the A circle that lies outside the M circle to indicate that this area is empty:

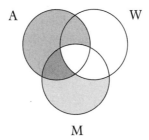

Because we are not making the Aristotelian assumption, we are not assuming the existence of aardvarks nor of mammals. With the shading completed, we examine the figure to see if it represents the information contained in the conclusion. The conclusion indicates only that nothing is in the A circle that is not also in the W circle. And that is indeed exactly what our diagram indicates, for all sections of the A circle outside the W circle have been shaded. We need add nothing more to our diagram to have it also represent the conclusion. In a sense, the information contained in the conclusion is already contained in the premises. The conclusion would have to be true if the premises are true. The argument is thus valid. One more example:

All astronauts are brave.

Some people are not astronauts.

So some people are not brave.

From the Boolean standpoint, the first premise tells us only that the part of the A circle that lies outside the B circle is empty:

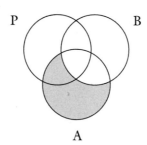

The next premise indicates that at least one person is not an astronaut. We must place an X in a part of the P circle that lies outside the A circle. Because the region is divided in two, the information we have does not tell us which side of the line to place the X on, and so we place the X squarely on the line:

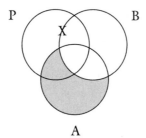

The conclusion, that some people are not brave, is not represented by this diagram, and the argument is therefore invalid.

The existential or Aristotelian presupposition makes a difference in some cases, and in other cases it makes no difference. For instance, the following syllogism is valid from both the Boolean and Aristotelian standpoints:

No P are M.

All S are M.

No S are P.

The Aristotelian and Boolean diagrams look like this:

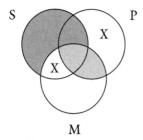
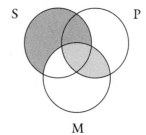

Notice that the two diagrams are the same except for the placement of the X. The following is invalid on both the Boolean and the Aristotelian standpoints:

All M are P.

No S are M.

No S are P.

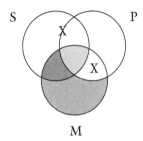

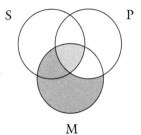

Aristotelian assumption   Boolean assumption

▶   **EXERCISE 15.2**   ◀

Assume the Aristotelian standpoint and test the following syllogisms for validity with Venn diagrams. For each syllogism, also list the mood and figure.

1. *Some politicians are lawyers. All lawyers are ambidextrous individuals. So, some ambidextrous individuals are politicians.

2. All truck drivers are rugged individuals. Some farmers are rugged individuals. Therefore, some farmers are truck drivers.

3. All truckers are happy persons. No prisoners are happy persons. Thus, no prisoners are truckers.

4. *All hedgehogs are cute. All small mammals are cute. So, all small mammals are hedgehogs.

5. Some musicians are not skateboarders. All musicians are happy persons. Therefore, some happy persons are not skateboarders.

6. No snowboarders are rich. No rich persons are MTV fans. So, no snowboarders are MTV fans.

7. No truck drivers are musicians. All musicians are artistic persons. Consequently, no artistic persons are truck drivers.

8. *Some political zealots are idealistic persons. No idealistic persons are sourpusses. So, some sourpusses are not political zealots.

9. No scientists are poets. Some scientists are ambidextrous individuals. Therefore, some ambidextrous individuals are not poets.

10. Some actors are romantics. Some pilots are not actors. So, some pilots are not romantics.

11. No A are B. All C are A. So, no C are B.

12. *All A are B. All A are C. So, some C are B.

13. Some A are B. All C are B. So, some C are A.

14. All A are B. No B are C. Therefore, some C are A.

15. All A are B. All A are C. Thus, all C are B.

16. All A are B, because all A are C and all B are C.

17. *No A are B, therefore some C are B, because no A are C.

18. Some A are B. Some A are C. So, some B are C.

19. Some A are not B, because some A are not C, and all C's are B.

20. No A are B, so no C are A because all B are C.

21. All M's are P. No S's are M's. So, no S are P's.

22. All P are M. Some S are M. So, some S are P.

23. *All P are M. All S are M. Consequently, all S are P.

24. Some M are not P. Some S are M. Therefore, some S are not P.

25. No M are P. All M are S. So, some S are not P.

26. No A are B. No C are B. So, no C are A.

27. No H are G. All F are G. So, no F are H.

28. No H are G. Some F are G. So, some F are not H.

29. Some G are not H. No F are G. So, some F are not H.

30. Some H are G. Some F are G. Therefore, some F are H.

▶   **EXERCISE 15.3**   ◀

Assume the Boolean standpoint and use Venn diagrams to decide which of the above syllogisms are valid and which are invalid.

THE SORITES

A **sorites** (from the Greek word *soros* for "pile") is a series of four or more categorical statements with one of the statements designated as the conclusion and the rest designated as premises. Here is an example of a sorites:

All hedgehogs are mammals.

All mammals are animals.

No plants are animals.

Therefore, no plants are hedgehogs.

This argument is composed of categorical statements, but it is not a standard-form, two-premise syllogism, and so the method of the previous section cannot be directly applied. However, we can reason through this sorites as follows. From the first two premises:

All hedgehogs are mammals.

All mammals are animals.

it follows that

All hedgehogs are animals.

We have deduced an "intermediate conclusion" from the first two premises. Now, if we combine this intermediate conclusion with the sorites' third premise, namely,

No plants are animals.

it follows that

No plants are hedgehogs.

This is the conclusion of the sorites.

A sorites is actually a chain of interlocking categorical syllogisms in which the intermediate conclusions have been left out. The sorites gets its name from the picturesque idea that it is a pile (soros) of conclusions. When we string together syllogisms to form a sorites, with intermediate conclusions omitted, we produce an interconnected chain of reasoning. And just as a chain is only as strong as its weakest link, a sorites is only as strong as its weakest logical link. If one syllogism within the chain is invalid, the whole sorites is invalid.

Let us say that a sorites is in standard form if it meets the following conditions:

1. All statements are standard form categorical statements.

2. Each term occurs twice.

3. The predicate term of the conclusion appears in the first premise.

4. Every statement (except the last) has a term in common with the statement immediately following.

We can reason through a sorites by employing the following procedure:

1. We pair together two premises that have a common term and from these derive a conclusion. This conclusion should have a term in common with one of the further statements in the sorities.

2. We pair together these two statements and draw a conclusion from this second pair.

3. We repeat this procedure until all the premises have been used. The resulting conclusion should be the conclusion of the sorites.

Evaluation of a sorites is direct: If each two-premise syllogism in the chain is valid, the sorities is valid. If even one syllogism in the chain is invalid, the whole sorities is invalid.

TESTING A SORITES WITH VENN DIAGRAMS

A sorites can be tested with Venn diagrams by the following procedure.

1. Place the sorites in standard form.

2. Deduce the first intermediate conclusion from the first two premises, and test this syllogism with a Venn diagram.

3. Deduce the next intermediate conclusion and test the resulting syllogism again.

4. Test each syllogism within the sorites. Each must be valid for the whole sorites to be valid. If even one component syllogism is invalid, the sorites is invalid.

Let us apply this to the following symbolized sorites:

No B are C.

Some D are C.

All A are B.

Therefore, some D are not A.

First, we place the sorites into standard form:

All A are B.

No B are C.

Some D are C.

Therefore, some D are not A.

The first two premises imply the following intermediate conclusion:

No A are C.

This sorites thus breaks down into the following two syllogisms:

All A are B.

No B are C.

Therefore, no A are C.

And:

No A are C.

Some D are C.

Therefore, some D are not A.

Now, to test this sorites, we draw the Venn diagrams for these two syllogisms, and if both syllogisms are valid, the sorites is valid. If one or more of the constituent syllogisms is invalid, the sorites is invalid. The two diagrams follow, assuming the Boolean standpoint:

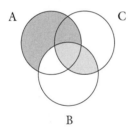

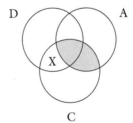

Because both are valid, the sorites is valid.

One more example of a sorites follows.

All C are D.

All A are B.

All B are C.

Therefore, all D are A.

In standard form this becomes

All A are B.

All B are C.

All C are D.

Therefore, all D are A.

Now, from the first two premises, we can deduce the following intermediate conclusion:

All A are C.

This sorites therefore breaks down into the following two syllogisms:

All A are B.

All B are C.

So, all A are C.

And:

All A are C.

All C are D.

So, all D are A.

The Venn diagrams reveal that although the first syllogism is valid, the second is invalid; this sorites is therefore invalid.

| ► | **EXERCISE 15.4** | ◄ |

Rewrite each sorites below in standard form. Test each with Venn diagrams. Assume the Aristotelian standpoint.

1. *All sane persons are persons who can do logic. All persons fit to serve on a jury are sane persons. No sons of yours are persons who can do logic. So, no sons of yours are persons fit to serve on a jury.

(This sorites is from a logic text written by Lewis Carroll, the nineteenth-century logician and author of *Alice's Adventures in Wonderland* and *Through the Looking Glass*.)

2. Some government officials are ambassadors. All ambassadors are tactful individuals. All government officials are people in public life. So, some people in public life are tactful individuals.

3. Athletes are energetic. No student of psychology is happy. Energetic people are happy. Therefore, no athletes are students of psychology.

4. *No chess players are nonathletic. Some doctors are chess players. Athletes are not lethargic. Therefore some doctors are not lethargic.

5. All A are B. Some C are A. All C are D. So, some D are B.

6. Some S are D. All D are A. All S are G. Therefore, some G are A.

7.*All A are B. Some C are A. All C are D. So, some D are B.

8. All A are B. Some C are A. All C are D. So, some D are B

9. All A are B. No C are B. All D are C. So, no D are A.

10. All A are B. Some C are not D. No B are D. So, no A are C.

11.*All A are B. All C are D. No D are B. So, no C are A.

12. All A are B. No C is B. All D are C. So, no D are A.

ENTHEMEMES

When we present a line of reasoning, we often leave out obvious premises. We assume that others possess a large body of common knowledge and suppose they will fill in the gaps. Sometimes we forget an assumption and inadvertently leave it out. Other times, we leave the conclusion unstated and let the reader draw the conclusion for himself or herself. An **enthememe** (from the Greek word for "have in mind") is an argument that is missing one or more of the premises or the conclusion, or both.

Premises left unstated are called **suppressed premises**. Incidentally, premises left unstated often are left unstated for a legitimate reason: leaving out trivial or obvious premises can help focus the reader's attention on the important parts of the argument. Also, stating extremely obvious points can both bore and insult your listener. However, when attempting to *prove* an enthememic argument valid, the suppressed parts must be added.

For example, consider the following argument:

All aardvarks are mammals.

So, all aardvarks are warm-blooded.

This argument is invalid as it stands, but only because the arguer has left a premise unstated. If the following premise is added, the argument is valid: "All mammals are warm-blooded."

How do we go about supplying a suppressed premise if we are presented with an enthememe? Many philosophers advocate the **principle of charity**: assume the other person is trying to be logical, and attempt to find a premise that, when added to the enthememe, makes the argument as reasonable as possible. If possible, look for a true premise that would also make the argument valid (if it claims to be valid) or strong (if it claims to be strong). In the case of an enthememic categorical syllogism, look for a premise that relates any terms not yet related and that also produces a valid syllogism that has true premises.

For example, suppose Joe's safety-conscious sister is trying to decide between buying a 1987 Chevrolet and a 1998 Ford. Joe argues as follows:

Cars with airbags are safer than cars without them.

So, the 1998 car is safer than the 1987 car.

Joe's argument clearly has a suppressed premise. Now, if we reject the principle of charity, we might suppose that the missing premise is something entirely unreasonable, such as

All 1998 cars are made entirely of green cheese.

If we were to impute this premise to the argument, we could say that the completed argument is both invalid and unsound. And someone could ridicule Joe for giving such a ridiculous argument. However, the principle of charity advises us to suppose instead that Joe is a reasonable guy who would leave unstated only a reasonable premise. The following premise, if added, would make Joe's argument valid:

The 1987 car has no airbag, and the 1998 car has an airbag.

If we add this, the argument is now valid and arguably sound as well.

► **EXERCISE 15.5** ◄

In the following enthememes, supply missing premises or missing conclusions. In each case, try to supply missing elements in such a way that the result is a valid argument.

1.*All dictatorships trample on human rights. Therefore, the government of Ruritania tramples on human rights.

2. Men are from Mars, and women are from Venus. So, Pat is from Venus.

3. Anyone who sympathizes with Kramer is unorthodox. Susan is unorthodox.

4.*Some mammals live in the water, for whales do.

5. You owe taxes only if you earn a profit. You owe no taxes.

6. Only gas-powered vehicles are allowed in the race. Therefore, Joe's truck is not allowed in the race.

7.*Because all cats are felines, tigers are felines.

8. Many adults are drug users, because caffeine is a drug.

9. Nobody who eats fatty food is healthy. So, people who eat hamburgers aren't healthy.

10. Anyone who watches television gets a superficial view of the world. So, you waste your time if you watch television.

11.*Jones must be healthy, for he runs a mile every day.

12. All shrews are mammals. So, all shrews have hair.

13. Some cats are domesticated. So, some mammals are domesticated.

REFUTATION BY LOGICAL ANALOGY

As we have seen, if an argument displays a valid form, any argument of the same form will also be valid, for all substitution instances of a valid argument form will be valid arguments. (Indeed, a valid form is defined as a form *all* of whose instances are valid arguments.) Also, if an argument has true premises and a false conclusion, it is automatically an invalid argument, for an argument is invalid if it is even *possible* that the argument has true premises and a false conclusion. This suggests a way to show that a particular argument is **formally invalid**, that is, that it displays an invalid form: Produce an argument that has exactly the same logical form as the original argument but that also has obviously true premises and a false conclusion. Because the second argument is obviously invalid, and because it displays the same logical form as that displayed by the first argument, the form in question must not be a valid form. This way of showing an argument formally invalid is sometimes called **refutation by logical analogy** since it involves producing a logically analogous argument that has true premises and a false conclusion. This strategy is also called **refutation by counterexample** since the logically analogous yet invalid argument is a counterexample to the original argument.

For example, suppose someone argues the following:

All whales are mammals, and all whales are swimmers, so, all mammals are swimmers.

Now, to show that this argument displays an invalid form, we must produce a similar argument that has the same form but that also has true premises and a false conclusion. First, the logical form of this argument is

All A's are B's, and all A's are C's, so, all B's are C's.

Now, here is another argument with this same form:

All cats are mammals, and all cats are feline, so, all mammals are feline.

This analogous argument obviously has the same form as the first argument, yet this argument has obviously true premises and a false conclusion, which shows that this second argument is invalid. Thus, the form in common is an invalid form (because one of its instances is invalid). This second argument is a counterexample to the first argument's form, and shows the first argument to be formally invalid.

Thus, if someone argued, "All mammals are swimmers, for whales are all mammals and all whales are swimmers," you might say in response, "That is like arguing, 'all mammal are feline, for all cats are mammals, and all cats are feline.' " This shows the person that their argument has an invalid form.

However, a word of caution is in order. This procedure shows that an argument's *form* is invalid. It does not show that the particular argument *must* itself be invalid, for in some cases, a valid argument can fit an invalid form. For example, consider this argument:

Necessarily, the number two is even. Therefore, the number two is even.

Suppose we characterize the form as

P. Therefore Q.

The argument does fit this form, for if we replace P with "Necessarily, the number two is even," and if we replace Q with "The number two is even," we generate the argument. Furthermore, this form is an invalid form. However, the argument is itself obviously a valid argument. Thus, a valid argument can fit an invalid form. Of course, this argument also fits the following form:

Necessarily P. Therefore P.

This is a valid form. Indeed, the argument is valid in virtue of fitting *this* form.

So, if an argument's *form* is shown invalid, this does not in every case show that the argument is itself invalid. However, if an argument's form is shown invalid, this is evidence against the claim that the argument is valid. Furthermore, if the argument fails to fit any known valid form, this creates a strong presumption against the argument. Suspicion arises that the argument is invalid, and a defender of the argument must step up to the plate and defend the argument with further reasoning.

▶   **EXERCISE 15.6**   ◀

Provide a refutation by logical analogy for each of the following arguments and argument forms.

1.*Some people are selfish, and some people are spiritual, so, some people are selfish and spiritual.

2. All cheetahs are fast runners. Some horses are fast runners. So, some horses are cheetahs.

3. All angels can swing dance on the point of a ballpoint pen. Some things that can swing dance are clumsy. Therefore, some clumsy things are angels.

4.*Some Democrats are classical liberals. No classical liberals are socialists. So, some socialists are not Democrats.

5. No poor people are socialists. No socialists are Catholics. Therefore, no Catholics are poor people.

6. Some truck drivers are not philatelists. All Republicans are truck drivers. Therefore, some Republicans are not philatelists.

7.*All aardvarks are mammals. Some pets are mammals. So, some pets are aardvarks.

8. No books are conscious. No trees are books. Therefore, no trees are conscious.

9. Some A are B. All C are A. So, some C are B.

10. All A are B. Some C are not A. Therefore, all C are not B.

11.*No A are B. All A are C. Therefore, no C are B.

12. All A are B. Some B are C. So, some C are A.

13. Some A's are B. Some B are C. So, some A are C.

CHAPTER 15 APPENDIX:
Rules For Evaluating Categorical Syllogisms

We have been testing categorical syllogisms for validity by drawing and interpreting Venn diagrams. Another method of testing validity employs rules rather than visual diagrams. First, some terminology.

▶ DISTRIBUTION

The following rules require an understanding of "distribution." Within a categorical statement, a term is said to be **"distributed"** if the statement makes an assertion about every member of the class denoted by the term, i.e., it assigns an attribute to every member of the class denoted by the term. For example, to say that the term S is distributed is to say that the statement asserts something about every member of the S class.

In a universal affirmative sentence *All S are P*, only S is distributed. In a universal negative sentence *No S are P*, S and P are both distributed. In a particular affirmative sentence *Some S are P*, neither S nor P is distributed. And in a particular negative sentence *Some S are not P*, P is distributed.

Sentence type	Terms distributed
A	S
E	S P
I	none
O	P

▶ DETERMINING VALIDITY BY RULES

Assuming a syllogism is in standard form, it is a valid syllogism if it violates none of the following rules:

1.   The middle term must be distributed in at least one premise.

If a syllogism violates this rule, the syllogism is said to commit the **fallacy of undistributed middle**. Here is an example:

All poets are romantics.

All musicians are romantics.

So, all musicians are poets.

In this argument, the middle term is *romantics*, and since this term is not distributed in either premise, the argument violates the first rule.

2.   If either term is distributed in the conclusion, it must be distributed in the premises.

If a syllogism violates this rule, the fallacy is the **fallacy of illicit major** if the major term is distributed in the conclusion but is not distributed in the major premise, and the mistake is the **fallacy of illicit minor** if the minor term is distributed in the conclusion but is undistributed in the minor premise.
    Here is a syllogism that commits the fallacy of illicit major:

All shrews are mammals.

All hedgehogs are shrews.

So, no hedgehogs are mammals.

This syllogism violates the rule because the major term *mammals* is distributed in the conclusion but not in the major premise.
    And this syllogism commits the illicit minor fallacy:

All artists are dreamers.

All dreamers are thoughtful individuals.

So, no thoughtful individuals are artists.

This syllogism violates the rule because the minor term *thoughtful individuals* is distributed in the conclusion but not in the minor premise.

3. The syllogism cannot have two negative premises.

If a syllogism violates this rule, the fallacy is the **fallacy of exclusive premises.** Here is an example:

No truck drivers are skaters.

No philosophers are skaters.

So, no philosophers are truck drivers.

Here is another example:

No comedians are punctilious persons.

No punctilious persons are easily deceived persons.

Therefore, no comedians are easily deceived persons.

4. If one premise is negative, the conclusion must be negative, and if the conclusion is negative, one premise must be negative.

A syllogism that violates this rule is said to commit either the **fallacy of drawing a negative conclusion from affirmative premises** or else the **fallacy of drawing an affirmative conclusion from a negative premise**. Here is an example of each:

All mammals are warm-blooded creatures.

All dogs are warm-blooded creatures.

So, no dogs are mammals.

And:

No reptiles are mammals.

No birds are mammals.

Therefore, some birds are reptiles.

5. The argument cannot have two universal premises and a particular conclusion.

In other words, if both premises are universal, the syllogism cannot have a particular conclusion. If a syllogism violates this rule, the fallacy is the **existential fallacy**. For example, consider the following syllogism:

All aardvarks are mammals.

All hedgehogs are mammals.

So, some hedgehogs are aardvarks.

A categorical syllogism that violates *only* rule 5 is valid from the Aristotelian standpoint but is invalid from the Boolean standpoint. This rule therefore applies only to the modern or Boolean interpretation of the categorical syllogism.

```
►                    EXERCISE 15.7                    ◄
```

Go back to exercises 15.1, 15.2, and 15.3 and identify the invalid categorical syllogisms. For each invalid syllogism, state the rule that it violates.

CHAPTER 15 GLOSSARY

Categorical syllogism   An argument that contains three categorical statements; the statements contain three different terms altogether, each statement contains two different terms, and no two statements contain the same two terms.

Distributed term   A term within a categorical statement is said to be "distributed" if the statement makes an assertion about every member of the class denoted by the term, that is, it assigns an attribute to every member of the class denoted by the term.

Enthememe   An argument that is missing one or more of the premises or the conclusion, or both.

Figure of a categorical syllogism   A specification of the pattern of placement, inside the syllogism, of the syllogism's middle term. Four possible patterns of placement exist, and therefore four figures.

Formally invalid   An argument is formally invalid if it displays an invalid form.

Formally valid   An argument is formally valid if it displays a valid form.

Logical form of a categorical syllogism   The syllogism's general logical structure, expressed by listing the syllogism's mood and figure.

Major premise   The premise containing the major term.

Minor premise   The premise containing the minor term.

Major term   The conclusion's predicate term.

Middle term   The term appearing in both premises (and thus not in the conclusion).

Minor term   The term appearing as the subject of the conclusion.

Mood of a categorical syllogism   Something that is specified by listing the type—A, E, I, or O—of each of the syllogism's categorical statements.

Principle of Charity (for interpreting enthememes)   If a person presents an enthememic argument, assume the person is trying to be logical, and attempt to find a premise or conclusion that, when added to the enthememe, makes the argument as reasonable as possible.

Refutation by logical analogy   To show that a particular argument is formally invalid: produce an argument that has exactly the same logical form as the original argument but that also has obviously true premises and a false conclusion. Because the second argument is obviously invalid, and because it displays the same logical form as that displayed by the first argument, the form in question must not be a valid form. This way of showing an argument formally invalid is also sometimes called refutation by counterexample since the logically analogous yet invalid argument is a counterexample to the original argument.

Sorites A series of four or more categorical statements with one of the statements designated as the conclusion and the rest designated as premises.

Standard form A categorical syllogism is in standard form if the major premise is written first, the minor premise is written second, and the conclusion is last.

Suppressed premise Premise that is left unstated.

Valid argument form An argument form all of whose instances are valid.

Venn diagram A drawing of overlapping circles designed to represent the information contained within one or more categorical statements, with each circle representing a class denoted by a term and each circle labeled for the class it denotes.

IV

MODERN QUANTIFICATIONAL LOGIC

Chapter

16

QUANTIFICATIONAL LOGIC I: THE LANGUAGE QL

onsider the following argument:

1. All aardvarks are cute.

2. Some aardvarks are filthy.

3. Therefore, some cute things are filthy.

Does this argument seem valid to you? It is. If the premises are true, the conclusion must be true. However, this argument cannot be *proven* valid using truth-functional techniques alone, because the argument contains no truth-functional operators such as "and," "if then," or "or," and so it cannot be symbolized with truth-functional connectives such as the ampersand and wedge. Therefore, within TL, our language for truth-functional logic, the only way we can abbreviate the argument is by assigning a single sentence constant to each premise and one to the conclusion. For instance:

1. A

2. B / C

(The premises are numbered and the symbolized conclusion is place after the slash.)

Although the English argument is valid, this abbreviation in TL is an instance of an invalid form. The argument instantiates an invalid form when translated into TL because the logical features of the argument that make it valid are not represented within TL. Consequently, if we wish to apply modern natural deduction techniques to this type of argument, we must move to a new symbolic language that represents these additional, nontruth-functional, logical features.

The opening argument is a **quantificational argument**. This is an argument built out of quantifiers and the units to which these attach. In order to prove the opening argument valid with a modern natural deduction system, we must employ a second branch of modern logic, quantificational logic. The first step in the process will be to introduce a second formal logical language that can handle quantificational arguments. After that, we will examine proof techniques that apply to arguments symbolized in this formal logical language.

In this chapter, we will identify the various parts of a quantificational argument and introduce logical symbols that abbreviate those parts. Then we will put all this together in the form of a second formal logical language, which will be named "QL" for "quantificational language."

TWO TYPES OF SENTENCES

A clear distinction exists between *singular* sentences and *general* sentences. We shall begin with the former.

▶ SINGULAR SENTENCES

Singular sentences are composed of *singular* terms and *general* terms. A **singular term** refers to or describes one specifically identified thing. Two types of singular terms will concern us: (1) proper names and (2) definite descriptions. We use proper names such as "John F. Kennedy" and "the Space Needle" to refer by name to one specific thing. Definite descriptions such as "The fourth husband of Zsa Zsa Gabor" or "the first person to walk on the moon" refer by unique description to one specific thing. In general, we use a singular term to single out a specific thing so that we may then say something about it.

Terms such as *curved, tall,* and *blue* do not refer to or uniquely describe one specifically identified thing. Rather, they represent properties (qualities, attributes, or characteristics) that a number of entities, or things, might have in common. Terms such as *curved, tall,* and *blue* are called **general terms** because, instead of singling out one unique thing for discussion (as singular terms do), they characterize a general category of things having a property in common.

The simplest type of singular sentence we will consider contains just two parts:

• A singular subject expression in which one specific thing is singled out for discussion by a singular term;

• A predicate expression that contains a general term attributing a property to the specific thing designated by the subject expression.

Here are several examples:

1. Herman Snodgrass is tall.

2. Rita's favorite bike is blue.

3. The tallest building in Chicago has more than ninety floors.

The subject expressions in these sentences are, in order:

1. Herman Snodgrass

2. Rita's favorite bike

3. The tallest building in Chicago

Note that the subject clause of 1 is a proper name, and the two other subject clauses are definite descriptions.

The predicate expressions in the three sentences are, in order,

1. _____ is tall.

2. _____ is blue.

3. _____ has more than ninety floors.

The blank placed at the beginning of each predicate expression represents the place where the singular term of a subject expression goes.

In each of the three complete singular sentences, the subject expression singles out one thing for discussion and the predicate expression says something about the thing that has been singled out. Each of the predicate expressions says something about the individual designated by the subject expression by attributing to the individual thing a property—the property represented by the predicate's general term. Thus, for instance, sentence 1 says something about the subject, Herman Snodgrass, by attributing or ascribing to him the property of being tall. A singular sentence such as "George Harrison is a musician" attributes a property (the property of being a musician) to the individual designated by the subject clause (that is, the individual named George Harrison).

In short, singular subject expressions refer to or designate entities, or things, for discussion and predicate expressions specify properties things may have. We use singular subject expressions to single things out for discussion, and we use predicate expressions to say something about them, that is, to attribute properties to them. (Some predicate expressions represent relations between things, a topic we'll examine in the next chapter.)

The next step is to specify the symbolic elements that will abbreviate singular sentences. Let us begin with the predicate expressions from above, which are repeated here:

1. _____ is tall.

2. _____ is blue.

3. _____ has more than ninety floors.

These three predicate expressions are not complete sentences, for each lacks a subject expression. In a sense, they are incomplete sentences with holes where a subject term would go. We call them **open sentences** because they have an open space in them where a subject expression may be placed. The absence of a subject expression also means that these open sentences do not designate a specific individual and so they do not make an assertion about anything in particular. Each is therefore neither true nor false. Open sentences do not have truth-values.

However, an open sentence becomes a complete sentence if you add an appropriate subject expression. The result is known as a *substitution instance* of the open sentence. The substitution instance is a complete sentence that has one of the two truth-values.

We will use the lower case letters *x, y,* and *z,* known as **individual variables,** to represent the open spaces—the blanks—in these predicate expressions. Within predicate expressions, variables will serve as placeholders for singular subject expressions. The three predicate expressions may now be abbreviated:

1.   x is tall.

2.   x is blue.

3.   y has more than ninety floors.

Notice that in each open sentence, the variable marks where a singular subject expression may be added to produce a complete sentence. In effect, the variable serves as a placeholder for whatever the subject expression might be. In the case of each of these open sentences, it makes no difference whether we use as a variable x, or y, or z—the choice of variable is completely arbitrary.

So far, we have used variables to represent the blanks in the predicate expressions. Next, we will abbreviate the remainder of those expressions using capital letters A, B . . . Z. These letters, when used to abbreviate predicates, are **predicate constants.** When symbolizing sentences, it is the custom in quantificational logic to place the predicate constant to the left of the variable. Thus, if we abbreviate the predicate expression "is tall" with the constant T, "is blue" with B, and "has more than ninety floors" with O, the predicate expressions may be completely abbreviated as

1.   Tx

2.   Bx

3.   Ox

Notice that in each case we placed the predicate constant to the left of the variable. We also chose a predicate constant that reminds us of the predicate it abbreviates.

Next, we shall use the lower case letters *a, b, c . . . t,* known as **individual constants,** to abbreviate singular terms. If we abbreviate "Herman Snodgrass" with h, "Rita's favorite car" with r, and "the tallest building in Chicago" with t, the three subject terms become

1. h

2. r

3. t

We now have all the elements necessary to symbolize singular sentences. If we go back to the symbolized open sentences—the predicate expressions—and replace the variables in those open sentences with the individual constants abbreviating the subject expressions, we get the complete abbreviations of our three singular sentences:

1. Th (This abbreviates "Herman Snodgrass is tall.")

2. Br (This abbreviates "Rita's favorite bike is blue.")

3. Ot (This abbreviates "The tallest building in Chicago has more than ninety floors.")

Again, notice that the predicate constants are placed to the left.

The formula "Th" may be understood as indicating that the individual represented by h has the property represented by T. That is, Herman Snodgrass has the property of being tall, which is to say that Herman Snodgrass is tall. Incidentally, we say properties are "shared" by individuals. For instance, this property, represented in English by "is tall," and represented in our symbols by the open sentence Tx, is shared by a number of individuals, namely, by all those who are tall.

Truth-functional combinations of singular sentences—singular sentences joined by truth-functional operators—will also count as singular sentences. The abbreviation of such sentences should be fairly obvious. Here are some examples:

English sentence:	Symbolic equivalent:
McCoy is Irish or Scotty is Irish.	Im v Is
If Spock is logical then Spock is not emotional.	Ls ⊃ ~ Es
It's not the case that Kramer is employed.	~ Ek
It's not true that both Niles and Frasier are dandies.	~ (Dn & Df)

▶   **EXERCISE 16.1**   ◀

Using our new symbols, symbolize the following.

1.* Grandpa Munster banks at the blood bank.

2. Either Wimpy eats a hamburger or Popeye eats some spinach.

3. The starship *Enterprise* has 21 decks, and it also has a complement of 72 officers and 428 enlisted crew members.

4. Jethro Bodine wants to become a "double-nought" spy and Granny likes to cook possum stew.

5.* If Moe sings, then Curly will make a face and Larry will poke Curly in the eye.

6. Either Jerry won't swim or Kramer won't swim.

When a variable in an open sentence is replaced by a singular term, the result is called a **substitution instance** of the open sentence. For example:

Open sentence: Three substitution instances:

1. Cx   Cj, Ck, Cl
2. Lx   La, Lb, Lc
3. Wy   Wa, Wb, Wc

A specific individual item is said to "satisfy" an open sentence just in case replacement of the variable in the open sentence with a singular term designating that item results in a substitution instance that expresses a true proposition. For example, if we let Tx abbreviate "x is tall," Michael Jordan satisfies this open sentence, because the substitution instance Tj, where j abbreviates "Michael Jordan," expresses a truth. However, Danny DeVito does not satisfy this open sentence, because he is not tall.

GENERAL SENTENCES

Recall that a singular sentence attributes a property to one specifically identified individual item or thing. In contrast, a **general sentence** makes a claim not about a specific individual but about some or all members of a group of individuals. Two types of general sentences will concern us here. **Universal general sentences** make a claim about *all* of the members of a group. **Existential general sentences** make a claim about *some* members of a group, where *some* is understood as meaning "at least one" or (equivalently) "one or more." (Existential general sentences are also called "particular" general sentences.) The subject phrase of a universal general sentence typically contains a quantity word indicating universality—either *all, no, every,* or some related term. The subject phrase of an existential general sentence also typically contains a quantity word—usually this will be *some* or the phrase *at least one* or a related expression.

► UNIVERSAL GENERAL SENTENCES

Let us begin by symbolizing the doctrine known as **materialism**. This is the view that absolutely everything in existence is made of nothing but matter.

(Matter may be specified as the topic that physics studies, namely particles such as photons, protons, neutrons, electrons, quarks, and so on.) So, according to materialism, no nonmaterial entities such as spirits, ghosts, angels, God, or souls, exist. Now, in English, the doctrine of materialism may be put this way:

Everything is made of matter.

How should this claim be represented in symbols? One idea that first comes to mind is this. Let Mx abbreviate "x is made of matter," let e abbreviate "everything," and then put the two together, substituting e for x:

Me

However, this won't do. The letter *e* is an individual constant, and individual constants may be used only to abbreviate singular terms, terms that refer by name or definite description to one specifically identified item. Because *everything* is definitely not a singular term, we cannot use *e* to abbreviate it. Of course, we could use *e* to abbreviate a singular term such as *Ernest's bowling ball*. In this case, the formula Me would abbreviate "Ernest's bowling ball is made of matter." Anyway, because *e* can't be used for the term *everything*, we are back to where we started. How shall we abbreviate the materialist thesis? We could try

Mx

However, this is just the open sentence "x is made of matter." The materialist thesis means more than this. The sentence "Everything is made of matter" is used to claim that *all* things satisfy the open sentence Mx, that is, all things have the property of being made of matter.

In order to symbolize the thesis of materialism, we will proceed by a series of paraphrases. First, let us paraphrase the materialist thesis

Everything is made of matter.

into

Every thing is such that it is made of matter.

Next, let us paraphrase this by replacing *thing* with the variable x:

Every x is such that it is made of matter.

Because the pronoun *it* relates back to the variable x in the subject clause, let us also replace *it* with x:

Every x is such that x is made of matter.

Next, the predicate expression "x is made of matter" may be abbreviated Mx. (We have already seen the rationale for this abbreviation.) Finally, the quantifier word *every* will be abbreviated using the symbol "(x)". This symbol, known as a universal quantifier, shall abbreviate "Every x is such that." Putting this all together, the materialist's thesis "every x is such that x is made of matter" may be abbreviated as follows:

$$(x)\ Mx$$

This may be read as, "Every x is such that x is M" or "For all x, x is M."

This formula may actually be understood in various equivalent ways:

1. Every x is such that it is material.

2. For every x, it is the case that x is material.

3. Every x satisfies "x is material."

4. Every x satisfies "Mx."

5. For all x: x is material.

6. For all x: Mx.

7. All x's are material.

8. Everything is material.

9. All things are material.

We will also use y or z as individual variables. So "Everything is material" might just as well be symbolized with the y or with the z variable. Thus, either

$$(y)\ My$$

or

$$(z)\ Mz$$

would suffice to symbolize the claim of materialism. The symbols (x), (y), and (z) are all universal quantifiers.

The next universal sentence has a more complex logical structure. Consider the sentence "All dogs have fleas." How should this be symbolized? One suggestion that comes to mind at first is this. Let Fx abbreviate "x has fleas," let d abbreviate "*all* dogs," and substitute d for x:

$$Fd$$

However, this won't work. The individual constant d may only be used to abbreviate a singular term—either a proper name or a definite description. Remember that a singular term singles out one specific thing for discussion. Because *all dogs* is neither the proper name of one specific thing nor a definite description of one specific thing either, we cannot abbreviate *all dogs* using the individual constant d. Of course, if a specific dog named Dave had fleas, we could abbreviate "Dave has fleas" with Fd; but we can't abbreviate "all dogs have fleas" this way.

We shall again work through a series of paraphrases to get to the proper symbolization. Let us first paraphrase "All dogs have fleas" into

Every dog is such that it has fleas.

Next, this is equivalent to

Every thing such that it is a dog is also such that it has fleas.

This in turn may be rewritten as

Every x such that x is a dog is such that x has fleas.

This may be abbreviated

Every x such that Dx is such that Fx.

where Dx abbreviates "x is a dog" and Fx abbreviates "x has fleas."

We are not finished symbolizing the sentence under consideration. However, before we look at the next paraphrase, it will be helpful if we perform a brief thought-experiment. Afterward, the next paraphrase will make more sense. Suppose you are a computer technician whose job is to upgrade all the modems in the building. That's twenty-seven floors' worth of modems. The problem is that some of the computers in the building have modems, and some do not. You have an assistant, Rocky, who just retired from the military and who likes precise, direct orders. Fortunately, the building has only two types of computers, IBM computers and Compaq computers, and all the IBM computers have modems and none of the Compaqs has a modem. Now, you want to give Rocky a precise instruction that will tell him that *all* the IBM computers have modems. How about this: "Pick any computer, *if* it is an IBM, *then* it has a modem." Now, doesn't that if-then construction indicate that all IBMs (in the building) have modems?

After reflecting on this, consider the next step in this series of paraphrases. We earlier paraphrased

Every thing that is a dog is such that it has fleas.

as

Every x such that Dx is such that Fx.

Now, we will say instead

For any and every x you might pick, if it is a dog, then it has fleas.

That is,

For every x: if Dx then Fx.

In other words, for anything you might pick, if it happens to be a dog, it has fleas. That is one way to indicate that all dogs have fleas.

Finally, this may be abbreviated:

$$(x)\,(Dx \supset Fx)$$

This expression may be understood in a number of equivalent ways:

1. For every x, if x is a dog, then x has fleas.

2. For any x, if x is a dog, then x has fleas.

3. For every x, if x satisfies "x is a dog" then x satisfies "x has fleas."

4. For every x, if x satisfies Dx then x satisfies Fx.

5. For all x, if x is a dog, then x has fleas.

► EXISTENTIAL GENERAL SENTENCES

In quantified logic, *some* is given the same meaning as the phrase *at least one.* "Some things are green" is equivalent to "At least one thing is green." How shall we symbolize "Some things are green"? One suggestion that comes to mind is to let s abbreviate *some* or *at least one* and to let G abbreviate *is green:*

$$Gs$$

However, this symbolization won't work. The letter *s* is an individual constant. Therefore, it can only be used to abbreviate a singular term. Because *some* and *at least one* are not singular terms—they do not single out one specifically identified item for discussion—they cannot be represented by an individual constant such as *s.*

We are back to our original question: How shall we abbreviate "Some things are green"? Again, we shall proceed through a series of paraphrases. Because *some* means *at least one,* our sentence may be written "At least one thing is green." This may be paraphrased as

At least one thing is such that it is green.

Substituting a variable for *thing,* we can rewrite this as

At least one x is such that it is green.

If we replace the pronoun *it* with the variable it relates to, we produce

At least one x is such that x is green.

Now, we already know how to abbreviate "x is green." This is symbolized Gx. We'll abbreviate the quantifier phrase *at least one* with the symbol "(∃x)," known as an existential quantifier. Putting this together, we produce

$$(\exists x)\,Gx$$

which may be read

There is at least one x such that Gx.

or more simply,

There is an x such that Gx.

We can understand this formula in various alternative but equivalent ways.

1. There exists at least one x such that "x is green" applies to x.
2. For at least one x: x is green.
3. There exists at least one x such that x satisfies "x is green."
4. Some x are such that they satisfy "x is green."
5. Something is green.
6. At least one thing is green.
7. There exists at least one green thing.

Consider next a more complicated existentially quantified sentence, "Some cars are noisy." How shall we put this into symbols? First, let us paraphrase:

There is at least one noisy car.

This is equivalent to

There exists at least one thing such that it is a car and it is noisy.

That is,

There exists at least one x such that x is a car and x is noisy.

Letting Cx abbreviate "x is a car," and letting Nx abbreviate "x is noisy," this sentence goes into symbols easily:

$$(\exists x)\,(Cx\ \&\ Nx)$$

We now have identified two types of general subject clauses:

a. Those containing the quantifier *all* or one of its cognate terms.
b. Those containing the quantifier *some* or one of its cognate terms.

Sentences whose subject phrase is of the first type are **universal quantifications**. Sentences whose subject phrase is of the second type are **existential quantifications**.

A SYNTAX FOR OUR NEW LANGUAGE

We have been speaking in somewhat informal terms so far. However, it is time to specify a formal language for the logical theory we are exploring. Let us call the language **QL** (for "Quantificational Logic"). In this section we will set out the syntax for QL; the semantics for QL will be developed in Chapter 19.

THE VOCABULARY OF QL

Sentence constants: A . . . Z

Predicate constants: A′ B′ . . .

Individual constants: a, b, c . . . w

Individual variables: x, y, z , x′, y′, z′ (each with or without a prime mark)

John Doe names: u, u′ . . . v, v′. . . .

Sentence operators: ~ v ⊃ & ≡

Quantifier symbols: (x) (y) (z) (∃x) (∃y) (∃z)

Grouping indicators: ( ) [ ] { }

The prime marks on the predicate constants distinguish predicate constants from sentence constants. However, when symbolizing predicates, we will normally omit the primes on the predicate constants and simply use the capital letter alone as an abbreviation of the predicate constant. (In other words, a predicate constant without a prime is technically an abbreviation of a predicate constant.) In most contexts it will be easy to distinguish the two types of constant because a sentence constant stands alone and a predicate constant is paired with one or more individual constants. Also, we will normally use only x, y, and z as variables, but in a context where more than three different variables are needed, we can add prime marks to x, y, and z to generate up to six different variables. The symbols known as "John Doe names" will be explained in Chapter 18.

► FORMATION RULES OF QL

In the following, **c** is used as a metalinguistic variable ranging over individual constants of QL and **v** is used as a metalinguistic variable ranging over variables and John Doe names of QL. **P** and **Q** serve as metalinguistic variables ranging over wffs of QL.

Q1. Any sentence constant is a wff.

Q2. Any n-place predicate followed by n individual constants is a wff.

Q3. If **P** is a wff, ~ **P** is a wff.

Q4. If **P**, **Q** are wffs, then (**P** & **Q**), (**P** v **Q**), (**P** ⊃ **Q**), (**P** ≡ **Q**) are wffs.

Q5. If **P** is a wff containing a constant **c**, and **v** is a variable that does not appear in **P**, then (**v**) [**P** with **c/v**] and (∃**v**) [**P** with **c/v**] are wffs, where [**P** with **c/v**] abbreviates sentence **P** with one or more occurrences of **c** replaced uniformly by **v** and the whole surrounded by a pair of grouping indicators.

Any formula that can be constructed by a finite number of applications of these rules is a sentence or well-formed formula of QL. Nothing else is a sentence or well-formed formula of QL.

▶ **ABBREVIATORY CONVENTION**

In formulas that contain no dyadic operators, formulas such as (x) (Fx), (∃x) (Hx), and (x) (~ Fx), you may omit the second pair of parentheses because these parentheses are actually redundant—they add no new information to the formula. The three formulas above thus may be abbreviated as (x) Fx, (∃x) Hx, and (x) ~ Fx. However, the second pair of parentheses is needed in formulas that contain dyadic operators, formulas such as (x) (Hx ⊃ Gx) and (∃x) (Jx v Bx). We will discuss the reason these parentheses are needed later in this chapter.

If you compare the syntax for QL with the syntax for TL, you will notice that any wff of TL will count as a wff of QL, although not every wff of QL will count as a wff of TL. One more point. Recall the difference between an object language and a metalanguage. In the present context, English is the metalanguage and QL is the object language.

SYMBOLIZING GENERAL SENTENCES

The rest of this chapter will primarily address the symbolization of various types of general sentences. We begin with a type of sentence that was near and dear to Aristotle's heart, the categorical sentence.

CATEGORICAL SENTENCES

A general sentence that asserts that all or some members of a group have or lack a specified property, is a **categorical sentence**. As noted in Chapter 14, Aristotle (384–322 B.C.) was the first logician to study and systematize the logic associated with categorical sentences. Because categorical sentences are still of importance in modern logic, it will be instructive to examine the symbolizations of such sentences.

A categorical sentence is **universal** if it asserts something about *all* the members of a group, and it is **existential** if it asserts something about *some* of the members of a group. (Existential categorical sentences are also called "particular" categorical sentences.) In addition, a categorical sentence is **affirmative** if it asserts that things have a certain property and it is **negative** if it denies that things have a certain property. This gives us four different types of categorical sentences: universal affirmative, universal negative, existential affirmative, and existential negative.

Universal Affirmative   The standard form for a universal affirmative is the following:

All ___ are ___.

Modern logicians interpret this standard form as follows. In a well-formed universal affirmative sentence, the first blank is filled in with an expression that refers to a group of entities, or things, and the second blank is filled in with an expression that attributes a property to the things belonging to that group. If we take the sentence "All dogs have fleas" and rewrite it as "All dogs are flea-infested critters," we can see that it fits the standard form of a universal affirmative sentence. Here are a number of additional examples of this type of categorical sentence:

1. All whales are mammals.

2. All maples are trees.

3. All mosquitoes are annoying pests.

4. All persons are mortal beings.

Sentence 1 attributes (to each and every whale) the property of being a mammal, sentence 3 attributes (to each and every mosquito) the property of being an annoying pest, and so on.

We have already seen how to symbolize sentences such as these. However, because a bit of repetition can be helpful when learning a new language, let us work through the steps involved in the symbolization of sentence 1. First,

All whales are mammals.

is equivalent to

Every thing that is a whale is such that it is a mammal.

This may be paraphrased

Every x such that x is a whale is such that x is a mammal.

In other words,

For every x, if x is a whale, then x is a mammal.

If we now abbreviate "x is a whale" with Wx, and "x is a mammal" with Mx, sentence 1 goes into symbols as

$$(x)\ (Wx \supset Mx)$$

This translates as, For any x, if x is a whale, then x is a mammal. That is how we shall express, within logic, the claim that all whales are mammals.

Of course, when symbolizing universal affirmatives, working through the above four steps is unnecessary. When we understand the rationale behind the symbolization process, universal affirmatives can be translated into symbols in one quick step. The remaining three universal affirmative sentences, sentences 2 through 4, are symbolized thus:

2. All maples are trees. $(x)(Mx \supset Tx)$
3. All mosquitoes are annoying pests. $(x)(Mx \supset Ax)$
4. All persons are mortal beings. $(x)(Px \supset Mx)$

A word of caution is in order. Some persons, when they are learning to symbolize universal affirmative sentences, try using an ampersand in place of a horseshoe. For example, when symbolizing

All maples are trees.

they try:

$$(x)(Mx \ \& \ Tx)$$

This symbolization is inaccurate, and it is important that you understand why. To assert $(x)(Mx \ \& \ Tx)$ is to assert that

For every x, x is a maple *and* x is a tree.

In other words, $(x)(Mx \ \& \ Tx)$ asserts that each and every thing in the entire universe is both a maple and a tree. That is to say, *you* are a maple tree, the chair you are sitting on is a maple tree, the moon is a maple tree, each and every atom in this room is a maple tree, and so on. Generally, then, although there are some exceptions, the main connective of the open sentence attached to the right of the universal quantifier expression should not be an ampersand. In most cases, the symbolization of a universal categorical statement will employ a horseshoe.

Various alternative but equivalent ways exist by which to express a universal affirmative sentence in English. For example, instead of saying, "All whales are mammals," we could just as well say any of the following:

1. Whales are mammals.

2. Any whale is a mammal.

3. A whale is a mammal.

These are all symbolized as:

$$(x)(Wx \supset Mx)$$

Universal Negative Sentences   A universal negative sentence is a sentence that is used to deny that any of the members of a group have a certain property. To deny that any of the members of a group have a certain property is to assert that *none* of the members of the group *have* the property. Therefore, the standard form for a universal negative may be represented as

No _____ are _____ .

As in the case of a universal affirmative, the first blank is to be filled in with an expression that refers to a group of things and the second blank holds an expression representing the property the things in the group lack. Examples of this type of sentence include:

No birds are reptiles.

No rocks are living things.

No horses are winged creatures.

No Vulcans are illogical creatures.

Let us examine the symbolization of the first example. Notice first that:

No birds are reptiles.

is equivalent to

All birds are nonreptiles.

That is, "No bird has the property of being reptilian" is equivalent to "All birds have the property of being nonreptilian." Thus, we could also represent the standard form of a universal negative as

All _____ are not _____ .

Let us therefore paraphrase the first example:

For every x, if x is a bird, then x is *not* reptilian.

In symbols, this is simply

$$(x) \, (Bx \supset \sim Rx)$$

Notice that universal negatives are symbolized exactly as we symbolized universal affirmatives except for the addition of a suitably placed negation operator.

Existential Affirmative Sentences   An existential affirmative sentence is used to assert that some (that is, at least one) of the members of a group have a certain property or characteristic. The standard form of such a sentence is:

Some _____ are _____.

where the first blank is filled in with an expression referring to a group of entities, and the second blank contains an expression attributing a property to some of the members of the group. One of the sentences we examined earlier, "Some cars are noisy," is an example of an existential affirmative. Other examples include

Some trees are poplars.

Some horses are pets.

Some comedians are funny people.

In order to translate the second example into symbols, we can again think through a series of paraphrases. First, "Some horses are pets" is equivalent to

At least one thing is such that it is a horse and it is a pet.

That is,

At least one x is such that x is a horse and x is a pet.

Letting Hx abbreviate "x is a horse" and letting Px abbreviate "x is a pet," this translates into symbols as

$$(\exists x)\,(Hx \ \& \ Px)$$

The first of the three existential affirmatives above goes into symbols as follows:

Some trees are poplars.

$$(\exists x)\,(Tx \ \& \ Px)$$

Another word of caution is in order. Some persons, after using a horseshoe in the symbolization of a universal quantification, attempt to use a horseshoe in the symbolization of an existential affirmative. For instance, they might try to symbolize "Some trees are maples" with $(\exists x)\,(Tx \supset Mx)$. This is *not* an accurate symbolization. Let us consider why. The formula $(\exists x)\,(Tx \supset Mx)$ asserts that

There is at least one x such that *if* x is a tree then x is a maple.

If you will reflect on the meaning of this sentence, you will notice that this is not equivalent to "Some trees are maples." For one reason, "Some trees are maples" asserts the existence of at least one maple tree, while the suggested translation, with its if-then or hypothetical construction, does not assert the existence of any maple tree. The sentence "There is an x such that *if* x is a tree then x is a maple" does not assert the actual existence of a maple tree for the sentence could be true even though no maple trees exist. So, (∃x) (Tx ⊃ Mx) does not accurately symbolize "Some trees are maples." Although there are exceptions, the main connective of an existential sentence is typically not a horseshoe.

Existential Negative Sentences An existential negative sentence is used to claim that some of the members of a group do *not* have a certain property. The standard form of such a sentence is

Some ____ are not ____ .

For example:

1. Some students are not Republicans.

2. Some basketball players are not stamp collectors.

3. Some trees are not maples.

These existential negatives are symbolized exactly as we symbolized existential affirmatives, except for the addition of an appropriately placed negation operator:

1. Some students are not Republicans.   (∃x) (Sx & ~ Rx)
2. Some basketball players are not stamp collectors.   (∃ x) (Bx & ~ Sx)
3. Some trees are not maples.   (∃x) (Tx & ~ Mx)

In the table below, which sums up this discussion, S represents the subject term and P represents the predicate term. For ease of reference, the four categorical forms are traditionally labeled A, E, I, and O, and we shall follow this tradition.

A: Universal affirmative	All S's are P	(x) (Sx ⊃ Px)
E: Universal negative	No S's are P	(x) (Sx ⊃ ~ Px)
I: Existential affirmative	Some S's are P	(∃x) (Sx & Px)
O: Existential negative	Some S's are not P	(∃x) (Sx & ~ Px)

THE OLD "QUANTIFIER SWITCH" TRICK

Sentences that contain a universal quantifier can be translated into sentences containing an existential quantifier, and vice versa. The rules that govern the translations are based on the following general truths:

• If *all* individuals have a certain property, then it is *not* the case that *some* individual does *not* have the property.

- If it is *false* that *all* individuals have a certain property, then it must be that *some* individual does *not* have the property.

- If *some* individuals have a certain property, then it must be *false* that *all* individual do *not* have the property.

- If it is *false* that *some* individuals have a certain property, then it must be that *all* individuals do *not* have the property.

  With these principles in mind, let us begin with a simple example, "Everything is material." This is equivalent to

Not even one thing is not material.

Now, "Everything is material" is symbolized with (x) Mx, and "Not even one thing is not material" is symbolized as ~ (∃x) ~ Mx. Thus, (x) Mx may be expressed equivalently as ~ (∃x) ~ Mx .
  Next, notice that

Something is spiritual.

which is symbolized as (∃x) Sx, is equivalent to

It is not the case that everything is not spiritual.

which is symbolized as ~ (x) ~ Sx. Thus, (∃x) Sx may be expressed equivalently as: ~ (x) ~ Sx .
  Similarly,

It is not the case that at least one thing is spiritual.

is equivalent to

All things are nonspiritual.

That is:

$$\sim (\exists x) \, Sx \text{ is equivalent to } (x) \sim Sx.$$

And

It is not the case that all things are spiritual.

is equivalent to

At least one thing is nonspiritual.

That is,

~ (x) Sx is equivalent to (∃x) ~ Sx.

When we rewrite a sentence and switch one quantifier for the other, we do three things:

(a) switch one quantifier for the other

(b) negate each side of the quantifier

(c) cancel out any double negatives that result.

For example, if we apply this procedure to (x) Fx, we produce ~ (∃x) ~ Fx. And if we apply this to (∃x) Fx, we produce ~ (x) ~ Fx.

Let us apply this procedure to a negated quantifier. We begin with ~ (x) Fx. At step (a), we produce ~ (∃x) Fx; at step (b) we produce: ~ ~ (∃x) ~ Fx ; and at step (c), when we cancel out the double negative, we produce: (∃x) ~ Fx. One more example: If we start with ~ (∃x) Fx, the three steps are

(a) ~ (x) Fx

(b) ~ ~ (x) ~ Fx

(c) (x) ~ Fx

SWITCHING QUANTIFIERS ON CATEGORICALS

These switcheroo operations can be applied to categorical sentences. That is, a categorical sentence that contains a universal quantifier can be translated into an equivalent sentence that contains an existential quantifier and vice versa. However, when we switch quantifiers on a categorical sentence, we must make some adjustments in the structure of the whole sentence. The rules that guide the switches are based upon four general principles:

1.   "All A's are B's" is equivalent to "There is no A that is a non-B." For example, "All aardvarks are brown" is equivalent to "There is no aardvark that is not brown." In symbols:

$$(x) (Ax \supset Bx) \text{ is equivalent to } \sim (\exists x)(Ax \,\&\, \sim Bx)$$

2.   "No A's are B's" is equivalent to "There is no A that is a B." For example, "No aardvarks are brown" is equivalent to "There is no aardvark that is brown." In symbols:

$$(x) (Ax \supset \sim Bx) \text{ is equivalent to } \sim (\exists x)(Ax \,\&\, Bx)$$

3.   "Some A's are B" is equivalent to "It's not the case that all A's are non-B." For example, "Some aardvarks are brown" is equivalent to "It is not the case that all aardvarks are nonbrown." In symbols:

$$(\exists x) (Ax \,\&\, Bx) \text{ is equivalent to } \sim (x) (Ax \supset \sim Bx)$$

4. "Some A's are non-B" is equivalent to "It's not the case that all A's are B." For example, "Some aardvarks are not brown" is equivalent to "It is not the case that all aardvarks are brown." In symbols

$$(\exists x)(Ax \And \sim Bx) \text{ is equivalent to } \sim (x)(Ax \supset Bx)$$

In sum, A, E, I, and O categorical sentences may be symbolized using either universal or existential quantifiers. If S represents the subject term and P the predicate term,

A: $(x)(Sx \supset Px)$ or equivalently $\sim (\exists x)(Sx \And \sim Px)$
E: $(x)(Sx \supset \sim Px)$ or equivalently $\sim (\exists x)(Sx \And Px)$
I: $(\exists x)(Sx \And Px)$ or equivalently $\sim (x)(Sx \supset \sim Px)$
O: $(\exists x)(Sx \And \sim Px)$ or equivalently $\sim (x)(Sx \supset Px)$

▶   **EXERCISE 16.2**   ◀

Symbolize each sentence below twice, first with a universal quantifier, second with an existential quantifier. Example: "All bats are cute critters":

$(x) (Bx \supset Cx)$ $\sim (\exists x) (Bx \And \sim Cx)$

1. *All things are good.

2. Something is right.

3. Something is not right.

4. *It's not the case that all things are good.

5. It is not the case that something is right.

6. It is not the case that something is not right.

7. *No biographies are novels.

8. Some fish are not hungry.

9. *Every Vulcan is logical.

10. Some opossums are cuddly.

11. All members of the Martin van Buren gang are toughies. (The van Burens are a New York City street gang made up of guys who refuse to go to architecture school and such.)

12. It is not the case that all opossums are cuddly.

13. It is not the case that some rats are not cute animals.

14. It is not the case that all Vulcans are not passionate.

SYMBOLIZING COMPLICATED
GENERAL SENTENCES

Consider the following sentence:

Some dogs are either noisy or obnoxious.

This is similar in form to an existential affirmative sentence. First, let us paraphrase it as

There is at least one x such that x is a dog and x is either noisy or x is obnoxious.

This goes into symbols easily:

$$(\exists x)\,[Dx\,\&\,(Nx\text{ v }Ox)]$$

where Dx abbreviates "x is a dog," Nx abbreviates "x is noisy," and Ox abbreviates "x is obnoxious."

Consider the following sentence:

Any person who likes the Blues Brothers movie has a sense of humor.

This is similar in form to a universal affirmative sentence. First, let us paraphrase this:

Every x such that x is a person and x likes the Blues brothers movie is such that x has a sense of humor.

This is equivalent to

For every x, if x is a person and x likes the Blues Brothers movie, then x has a sense of humor.

If we let Px abbreviate "x is a person," Lx abbreviate "x likes the Blues Brothers," and Hx abbreviate "x has a sense of humor," this becomes

$$(x)[(Px\,\&\,Lx)\supset Hx]$$

Notice that a relative clause such as "who likes the Blues Brothers" translates as a combination of a predicate constant and a variable.

Adjectives typically translate into symbols as predicate constants. For instance, "All old werewolves are gruesome" is symbolized as

$$(x)[(Ox\,\&\,Wx)\supset Gx]$$

where Ox abbreviates "x is old," Wx abbreviates "x is a werewolf," and Gx abbreviates "x is gruesome." This formula reads "For all x, if x is O and x is W then x is G."

Consider the following sentence:

All Vulcans are logical and in control of their emotions.

If we let Vx abbreviate "x is a Vulcan," Lx abbreviate "x is logical," and Cx abbreviate "x is in control of his or her emotions," this translates thus:

$$(x) [Vx \supset (Lx \ \& \ Cx)]$$

(For any x, if x is a Vulcan, then x is logical and x is in control of his or her emotions.)

Here are a few more examples:

All friendly old cats purr.

$$(x)[(Fx \ \& \ (Ox \ \& \ Cx)) \supset Px]$$

Some fish are blue.

$$(\exists x)(Fx \ \& \ Bx)$$

Some tall, sleepy giraffes are cute.

$$(\exists x)[(Tx \ \& \ (Sx \ \& \ Gx)) \ \& \ Cx]$$

As the sentences above demonstrate, we usually choose predicate constants that remind us of the English predicates they abbreviate.

Consider this sentence:

If everyone cheers, then someone will boo.

This is actually a compound consisting of two separate quantified sentences joined by an if-then connective. The two component sentences are

Everyone cheers.

Someone will boo.

Symbolized, "Everyone cheers" becomes

$$(x)(Px \supset Cx) \qquad \text{(For any x, if x is a person, then x cheers.)}$$

And "Someone will boo" becomes

$$(\exists x) \ (Px \ \& \ Bx) \quad \text{(There is at least one x such that x is a person and x boos.)}$$

Joining these two by the if-then connective produces the proper symbolization:

$$(x)(Px \supset Cx) \supset (\exists x) \ (Px \ \& \ Bx)$$

DENYING EXISTENCE

Sentences that deny the existence of something are easily symbolized. For example, consider the *Cadborosaurus.* If a real Loch Ness "monster" is ever discovered, the scientific name for the beast will apparently be "*Cadborosaurus.*" Anyway, suppose someone asserts the following:

There exists at least one *Cadborosaurus.*

This easily goes into symbols as $(\exists x)$ Cx. Of course, if we wish to use the variable y, the symbolization is $(\exists y)$ Cy. Now, suppose someone else says:

Cadborosauri do not exist.

This person means the following:

There are no *Cadborosauri.*

This is symbolized as

$$\sim (\exists x)\, Cx$$

This formula says, "It is not the case that there is at least one x such that x is a *Cadborosaurus.*

If we switch quantifiers according to the translation scheme introduced earlier, this may also be expressed as

$$(x) \sim Cx$$

which abbreviates "Each and every thing is not a cadborosaurus."

Suppose Cap has been searching for a Cadborosaurus for fifty years. If one is finally discovered, Cap the Cadborosuarus hunter will surely be happy. In English, this is:

If at least one Cadborosaurus exists, then Cap will be happy.

Now, how shall we symbolize this sentence? First, the English sentence is itself a compound of two shorter sentences. More specifically, the English sentence is a conditional sentence with an antecedent and a consequent. So when we translate the sentence into a formula, the formula's main connective should be a horseshoe. Second, the antecedent of the conditional is a quantified sentence ("At least one Cadborosaurus exists") and the consequent is a singular sentence ("Cap will be happy"). In symbols, the sentence thus becomes:

$$(\exists x)\, (Cx) \supset Hc$$

THE ONLY WAY TO GO

It is sometimes difficult to figure out the proper symbolization of a sentence that contains *only.* Suppose we wish to abbreviate "Only politicians are honest."

Notice that this is *not* equivalent to "All politicians are honest." Rather, this is to say:

The only persons who are honest are politicians.

Another way to put this is

All honest persons are politicians.

Paraphrased, this becomes

For any x, if x is honest, then x is a politician.

In symbols, this is:

$$(x) (Hx \supset Px)$$

However, to claim that all honest persons are politicians is *not* to claim that all politicians are honest. (Compare: "All bachelors are men" is not equivalent to "All men are bachelors.") Thus, to claim that only politicians are honest is *not* to claim that all politicians are honest. In symbols, "All politicians are honest" reads:

$$(x) (Px \supset Hx)$$

Notice also that "All honest persons are politicians" is equivalent to "All nonpoliticians are not honest." Thus, (x)(Hx $\supset$ Px) is equivalent to (x) (~ Px $\supset$ ~ Hx).

A sentence of the form "Only so and so's are such and such" is called an **exclusive sentence** because it indicates that the predicate term applies exclusively to the things designated by the subject term. Notice that when we symbolized the exclusive sentence above ("Only politicians are honest"), we reversed the positions of the subject and predicate terms.

Here is another example. The exclusive sentence "Only boaters are happy with the new regulations" is equivalent to "The only people happy with the new regulations are boaters." In symbols this becomes:

$$(x) (Hx \supset Bx)$$

where Hx abbreviates x is a person who is happy with the new regulations and Bx abbreviates x is a boater. Notice that the expression "The only," when it appears at the start of a sentence, translates as "All."

Here is a more complicated case:

Only teachers with certification were hired.

That is,

All who were hired were teachers who had certification.

We paraphrase this:

For any x, if x was hired, then x was a teacher and x had certification.

In symbols:

$$(x)[H x \supset (Tx \& Cx)]$$

WHAT IS A CAT-DOG?

Consider the sentence "Cats and dogs are mammals." At first glance, the proper symbolization would appear to be (x) [(Cx & Dx) ⊃ Mx].
However, this formula literally means: For any x, if x is both a cat *and* a dog at the same time, then x is a mammal. In other words, all 'cat-dogs' are mammals. Of course, this is an incorrect symbolization. The correct formula is

$$(x) [(Cx \lor Dx) \supset Mx]$$

This says, for any x, if x is *either a cat or a dog*, then x is a mammal. Notice that we used a *wedge* to capture the idea that cats *and* dogs are all mammals. This sentence could also have been symbolized

$$(x) (Cx \supset Mx) \& (x)(Dx \supset Mx)$$

This sentence is a compound of two component sentences, namely, "All cats are mammals" and "All dogs are mammals," and it translates as "All cats are mammals and all dogs are mammals."

Let us consider two final types of sentences. In most cases, a sentence that begins with "any" is a universal sentence. Consider "Any person that sings is happy." This is equivalent to "All persons who sing are happy." If we let Px abbreviate "x is a person," Sx abbreviate "x sings," and Hx abbreviate "x is happy," the sentence goes into symbols as

$$(x) [(Px \& Sx) \supset Hx]$$

This says, for any x, if x is a person and x sings, then x is happy.

An exceptive sentence is generally symbolized as a conjunction of two sentences. For example, consider "All except truckers are happy with the new express lanes." This is equivalent to the following conjunction: "All nontruckers are happy with the new express lanes, and no truckers are happy with the new express lanes." In symbols, with obvious abbreviations, this conjunction is:

$$(x) (\sim Tx \supset Hx) \& (x) (Tx \supset \sim Hx)$$

▶   **EXERCISE 16.3**   ◀

Symbolize the following.

1. *If Elaine dances, then everyone will laugh.

2. No old books are interesting.

3. If everybody is happy, then Newman will be sad.

4. Some books are long and dull.

5. *Either everything is material or everything is nonmaterial.

6. Everything is either material or is nonmaterial.

7. Anything that harms someone and is intentional is against the law.

8. If someone slips on a banana peel, then someone will laugh.

9. *Some movies are philosophical and some are not.

10. There are no flying dragons.

11. All aardvarks and tapirs are mammals.

12. At least one nondetectable strain of the AIDS virus exists.

13. Only the brave are free.

14. *There is a green car in the garage.

15. Not all Klingons are bad.

16. Every contestant who is under twenty-one will be given a prize.

17. *Every human being is either "right brained" or "left brained."

18. A hot dog with peanut butter on it is a tasty treat.

19. Some people are either too busy or too preoccupied.

20. Not every long book is a good book.

21. Whoever eats a diet of nothing but hamburgers has a high cholesterol level.

22. Anyone who likes the Marx Brothers has good taste in humor.

23. *If Dr. Frasier Crane is broadcasting, then everyone in Seattle is listening.

24. Everything is temporary and perishable.

25. If everything is temporary and perishable, then life has no objective meaning.

26. Everybody is happy except John.

27. All persons except the demonstrators are happy with the changes.

28. The only people in attendance were people who are fans of the original *Star Trek*.

29. Only rockers were partying.

30. Only a *Star Trek* fan would like *Lost in Space*.

CHAPTER 16 GLOSSARY

Categorical sentence A general sentence that asserts that all or some members of a group have or lack a specified property. A categorical sentence is *universal* if it asserts something about *all* the members of a group, and it is *existential* if it asserts something about *some* of the members of a group. A categorical sentence is *affirmative* if it asserts that things have a certain property and it is *negative* if it denies that things have a certain property. Thus, there are four different types of categorical sentences: universal affirmative, universal negative, existential affirmative, and existential negative.

Constant A symbol that stands for a specifically identified thing. In QL, *individual* constants such as a, b, and c are symbols that abbreviate singular terms and *predicate* constants are symbols such as A, B, and C that abbreviate predicates.

General sentence A sentence that makes a claim about some or all members of a group. This text considers two types of general sentences.

- *Universal* general sentences make a claim about *all* of the members of a group.

- *Existential* general sentences make a claim about *some* members of a group, where *some* is understood as meaning "at least one."

General term A term that may be used to characterize a general category of entities, or things, that have a property in common. Unlike a singular term, a general term may be applied to many things and it represents a property that many things could have in common.

Materialism The view that absolutely everything in existence is composed of nothing but matter.

Open sentence An incomplete sentence that is lacking a subject expression and that has a space where a subject expression may be placed.

QL An artificial symbolic language for quantificational logic.

Quantificational argument An argument built out of quantifiers and the units to which these attach.

Singular sentence A sentence that is made up of two parts: (1) a singular subject expression in which one specific thing is singled out for discussion by a singular term; (2) a predicate expression that contains a general term attributing a property to the specific thing designated by the subject expression.

Singular term An expression that refers to or describes one specifically identified thing. In general, a singular term is used to single out a specific thing so that we may say something about it. This text addresses two types of singular term.

- *Proper names* are capitalized names that refer by name to one specific thing.

- *Definite descriptions* are descriptions that refer by unique description to one specific thing.

Variable In QL, a symbol such as x, y, or z that serves as a placeholder for individual constants. A variable in QL performs the function that is performed by the word "thing" within English.

Chapter

17

THE LANGUAGE OF QUANTIFICATIONAL LOGIC II: RELATIONS

We earlier divided singular sentences into two components: subject phrases and predicate phrases. A predicate phrase looks like a sentence with a hole where its subject phrase was or should be. For instance, the following is a predicate phrase: _____ *is happy.* The subject phrase has been knocked out and the hole is marked by a blank. We used open sentences to symbolize predicate phrases. For instance, the predicate phrase _____ *is happy* is symbolized by Hx. In the formula, the x marks the hole where a subject term should go, and the capital letter—the predicate constant—abbreviates the predicate phrase. Recall that if the blank in a predicate phrase is filled in with a singular term, the result is a completed sentence and if the variable in Hx is replaced with an individual constant, the result, say, Hp, is a substitution instance of the open sentence and represents a complete sentence.

All of the predicate phrases we have examined so far have been **monadic,** or "one-place" predicate phrases. A monadic predicate phrase contains just one blank and is used to attribute a property to just one thing. In contrast, a **dyadic,** or two-place predicate expression contains two blanks. Here are some examples of dyadic predicate expressions:

_____ knows _____

_____ is older than _____

_____ is taller than _____

In order to turn one of these dyadic predicate phrases into a complete sentence, we must insert two singular terms, one in each blank. Below we have inserted *Pat* in the left blank and *Chris* in the right blank of each of the preceding dyadic predicates:

Pat knows Chris.

Pat is older than Chris.

Pat is taller than Chris.

The dyadic predicate phrases listed above are symbolized just as we symbolized the monadic predicates, except for the addition of a second variable representing the second blank. Thus:

____ knows ___ is abbreviated as Kxy.
____ is older than ___ is abbreviated as Oxy.
____ is taller than ___ is abbreviated as Txy.

In each case, we used the variable x in place of the left blank and the variable y in place of the right blank. If we next abbreviate *Pat* with p and *Chris* with c, we produce the following three substitution instances of these open sentences:

Kpc　(This abbreviates "Pat knows Chris.")
Opc　(This abbreviates "Pat is older than Chris.")
Tpc　(This abbreviates "Pat is taller than Chris.")

Notice that we place the dyadic predicate constant to the left and the two singular terms to the right.

　　A dyadic predicate phrase represents a two-place property or, as it is more often called, a *relation* between two things. We use a dyadic predicate to assert the existence of a relation between two individual things.

SENTENCES WITH A QUANTIFIER–DYADIC PREDICATE COMBO

Now, let us consider a sentence that combines a single quantifier with a dyadic predicate. For example, "Jan knows everybody." First, this is equivalent to

For all x such that x is a person, Jan knows x.

This may be paraphrased as

For all x, if x is a person, then Jan knows x.

If we now abbreviate *Jan* with j, *Jan knows x* with Kjx, and *x is a person* with Px, our sentence may be symbolized as

$$(x)\ (Px \supset Kjx)$$

　　Consider next the sentence, "Fred knows somebody." We may paraphrase this as

There exists at least one x such that x is a person and Fred knows x.

In symbols this is

$$(\exists x)(Px \ \& \ Kfx)$$

where Kfx abbreviates the predicate *Fred knows x* and Px abbreviates *x is a person*.

Now, compare "Fred knows somebody" with "Somebody knows Fred." The sentence "Somebody knows Fred" is equivalent to

There exists at least one x such that x is a person and x knows Fred.

In symbols this becomes

$$(\exists x) \ (Px \ \& \ Kxf)$$

where Kxf abbreviates *x knows Fred*. Notice the difference in meaning between Kfx and Kxf in the last two QL sentences.

In the following examples, Txy abbreviates *x is taller than y*, and the other abbreviations are obvious:

All brontosauri are taller than Pat. $(x)(Bx \supset Txp)$

The formula reads, For any x, if x is a brontosaurus, then x is taller than Pat.

Pat is taller than all brontosauri. $(x)(Bx \supset Tpx)$

The formula reads, For any x, if x is a brontosaurus, then Pat is taller than x.

Not all brontosauri are taller than Sam. $\sim (x)(Bx \supset Txs)$

The formula reads, It is not the case that for any x, if x is a brontosaurus, then x is taller than Sam.

ANY AND *EVERY*

The terms *any* and *every* function differently in certain kinds of complex sentences. In the following example, the two words function in unison:

Wimpy likes every hamburger. $(x)[Hx \supset Lwx]$
Wimpy likes any hamburger. $(x)[Hx \supset Lwx]$

Here Hx abbreviates *x is a hamburger,* w refers to Popeye's friend Wimpy, and Lwx abbreviates *Wimpy likes x*.

However, here *every* and *any* function in opposing ways:

Wimpy does not like every hamburger. $\sim (x)[Hx \supset Lwx]$
Wimpy does not like any hamburger. $(x)[Hx \supset \sim Lwx]$

Examine the following additional symbolizations:

Everybody loves Raymond. $(x)(Px \supset Lxr)$

The formula reads, For every x, if x is a person, then x loves Raymond.

Not everybody knows Fred. $\sim (x)(Px \supset Kxf)$

This reads, It is not the case that for every x, if x is a person, then x knows Fred.

Fred doesn't know anybody.   ~ (∃x)(Px & Kfx)

The formula reads, It is not the case that there is an x such that x is a person and Fred knows x.

Fred doesn't know everybody.   ~ (x)(Px ⊃ Kfx)

This reads, It is not the case that for every x, if x is a person, then Fred knows x.

Nobody knows Fred.   (x) (Px ⊃ ~Kxf)

This reads, For any x, if x is a person, then x does *not* know Fred. Notice that this could also be symbolized:

$$\sim (\exists x) (Px \ \& \ Kxf)$$

This formula reads, It is not the case that there is an x such that x is a person and x knows Fred.

REFLEXIVE SENTENCES

Here are examples of reflexive sentences and their proper symbolizations. To symbolize

Susan is proud of herself.

we may write

Pss

with the understanding that Pxy abbreviates *x is proud of y* and s abbreviates *Susan*. Pss indicates that Susan is proud of Susan.

Some reflexive sentences are more general in nature. Consider the following examples and their symbolizations:

Somebody knows himself.	(∃x)(Px & Kxx)
Everybody knows herself.	(x)(Px ⊃ Kxx)
Everybody likes himself.	(x) (Px ⊃ Lxx)

In the above three sentences, Kxx abbreviates "x knows x" and Lxx abbreviates "x likes x."

▶   **EXERCISE 17.1**   ◀

Symbolize the following English sentences.

1.*Pam is taller than Sue, but Sue is older than Pam.

2. Archie Bunker does not like any "liberal."

3. Archie Bunker does not like every liberal.

4. Wimpy respects himself.

5.*Someone knows himself.

6. Everybody knows Bill Gates.

7. All elephants are larger than Nathan.

8. Lorraine likes any horse.

9. Somebody knows Katie.

10.*Johnny is a friend of Elliot's.

11. Everything with caffeine in it is liked by Elmer.

12. Sam dislikes somebody.

13.*Sam dislikes everybody.

14. Everybody loves Raymond, but nobody loves Ned.

15. If everybody loves Raymond, then somebody loves George.

16. If someone likes Jan, then Jan will be pleased.

17. If nobody likes Jan, then Jan will be sad.

18. If every person loves himself or herself, then every person is selfish.

19. Nobody cares about Joe.

20. Joe doesn't care about anybody.

21.*It is not the case that everybody loves Raymond.

22. It is not the case that: if everybody loves Raymond, then somebody loves George.

SENTENCES WITH OVERLAPPING QUANTIFIERS

In a sentence with "overlapping" quantifiers, one quantifier appears within the scope of another quantifier. We can illustrate the basic techniques needed to translate sentences that contain overlapping quantifiers by examining four cases involving cause-and-effect relationships.

Case 1 According to the philosophical principle known as the principle of universal causation, everything has a cause. This is taken to mean just that each thing has some cause *or other*. The *or other* records the idea that although each thing has a cause, the cause of one thing might be different from the cause of another thing. Not all things need have one and the same cause. The principle may be put this way:

> *Everything is caused by something or other.*

This sentence is saying something about everything, and should therefore begin with a universal quantifier. Let us paraphrase the sentence:

For every x, there exists at least one y such that y is the cause of x.

In other words, for every thing x, there is something y such that y is x's cause. Now, if we let Cyx abbreviate *y causes x* or *y is the cause of x*, the principle of universal causation may be symbolized as

$$(x) \, (\exists y) \, Cyx$$

Every individual thing has a cause, although it is not necessarily the same cause in each case. Note that it does not follow that some one thing (such as God) exists that is the sole cause of everything.

Case 2 Suppose now that we consider "Something causes everything." This sentence is about a something, and so it should begin with an existential quantifier. The sentence can be paraphrased:

There exists a y such that for every x in the universe, y is the cause of x.

In other words, there exists some one thing y that is itself the cause of everything. In this case, the sentence asserts the existence of a cause of the universe, a creator of the cosmos. In symbols, this translates as

$$(\exists y) \, (x) \, Cyx$$

Notice that the two formulas we have just formed, $(x) \, (\exists y) \, (Cyx)$ and $(\exists y) \, (x) \, (Cyx)$, match except for the order of the quantifiers. And as we have seen, these two sentences have two distinct meanings. In general, when universal and existential quantifiers appear next to each other in a formula, the *order* of the two quantifiers affects the meaning of the sentence. We'll come back to this point later on.

Case 3 Consider the claim, "Everything causes something or other." This sentence is again about everything, and so its symbolization should begin with a universal quantifier. This sentence says that for every x, there is at least one y such that x causes y. In other words, for anything whatsoever, there exists something or other that it causes. But this need not be the same y in each case. (Hence the *or other* in the sentence.) That is, each thing causes something, but it doesn't follow from this that one thing exists that is caused by everything. If we let Cxy abbreviate *x causes y*, the sentence goes into symbols as $(x) \, (\exists y) \, Cxy$.

Case 4   Now let us rearrange the order of the quantifiers in the previous formula:

$$(\exists y)\,(x)\,Cxy$$

This has an entirely different meaning. According to this, something is caused by everything. There is at least one y such that for every x, x causes y. In other words, one thing is caused by everything. Again, when dealing with "mixed" quantifier sentences, the order of the quantifiers matters.

It will be helpful to remember this: In most cases, when an existential quantifier appears first, followed by a universal quantifier, the sentence asserts that at least one thing stands in a relation to all things. When a universal quantifier appears first, followed by an existential quantifier, the sentence asserts that for each and every thing, some entity stands in a relation to that thing.

- (x) (∃y) Axy: For every x, there is a y such that x has relation A to y. (Every x has relation A to some y.)

- (∃x) (y) Axy: There is at least one x such that for every y, x has relation A to y. (At least one x is such that it has relation A to every y.)

Quantifier Scope and the Meaning of "Overlapping"

We now have six quantifiers to work with: (∃x), (x), (∃y), (y), (∃z), and (z). In a sentence containing just one quantifier, the scope of the quantifier is the quantifier itself plus the expression enclosed in parentheses to the quantifier's immediate right. In the following, the scope of each quantifier is underlined:

(x)(Fx ⊃ Gx)	(x)(Fx) ⊃ Sa	(x)(Fx) ⊃ (∃x)(Hx)
(∃y)(Fy v Gy)	(∃y)(Hy) v Gb	(x)(Hx) v Gx

If a quantifier appears immediately to the right of another quantifier, the scope of the first quantifier is that quantifier itself plus the scope of the second quantifier. In the following sentence, the second quantifier lies within the scope of the first quantifier:

$$(x)(\exists y)Fxy$$

In such a case, the quantifiers are *overlapping quantifiers.*

Notice that as far as scope is concerned, a quantifier functions exactly as if it were a tilde. In other words, the scope of a quantifier is just what the scope would be if the quantifier were to be removed and replaced with a tilde.

A quantifier is said to "bind" the variable it contains plus any occurrences of that variable that lie within the quantifier's scope. Thus, in the following sentence, the quantifier binds all three variables:

$$(x)\,(Fx \supset Gx)$$

(Continued)

If a variable is not bound by a quantifier, it is a *free variable*. In each of the following, the variable x is bound by the universal quantifier and the variable y is a free variable.

$$(x) (Fx) \supset Gy \qquad (x) (Fx \supset Gx) \lor Sy$$

In the following, the third occurrence of the variable x is a free occurrence of that variable:

$$(x) (Fx) \supset Gx$$

Any sentence that contains one or more free variables is an open sentence. Open sentences, expressions such as Fx or Hxy, contain no subject terms and so do not express claims about things. It therefore makes no sense to ask, concerning an open sentence, "Is it true or false?" Consequently, open sentences do not have truth-values.

A sentence that is not open is said to be closed. To turn an open sentence into a closed sentence, one or the other of two things must be done: (a) the variables must be replaced by constants; or (b) each variable must be bound by an appropriate quantifier, that is, by a quantifier that contains the same variable. Unlike open sentences, closed sentences represent claims that are either true or false, and so every closed sentence has a truth-value.

Consider the sentence (x) (Hx ⊃ Fy). The parentheses to the right of the quantifier indicate that the scope of that quantifier covers the whole sentence. This puts Fy within the scope of (x). However, the variable y within Fy is not bound by (x), because the variable y does not occur within the quantifier (x). In the sentence (x) (Hx ⊃ Fy), then, y occurs as a free variable, even though it is within the scope of the quantifier.

Consider next the sentence (x) (y) Kxy. The second quantifier sits within the scope of the first quantifier. However, the variable y in Kxy is bound only by the second quantifier, and the variable x is bound only by the first quantifier.

"WHAT ARE YOU TALKING ABOUT?" THE UNIVERSE OF DISCOURSE

In logic, a variable is said to "range" over a domain. The **domain** of the variable is the set of things the variable can take as values. The values of a variable are just the things represented by the singular terms that can replace the variable in a sentence. The domain of a variable is also known as the **universe of discourse** for the sentence containing the variable. When we specify the universe of discourse for a sentence, we specify what the sentence is making its claim about. When the domain of a variable is everything in the universe, the variable has an **unrestricted** or **universal domain**.

We have been presupposing an unrestricted universe of discourse in the symbolizations so far specified. That is, we have been assuming that our variables range over all things in the universe. However, in this section, it will simplify things greatly if we stipulate a restricted domain or restricted universe of discourse. In order to see how a restricted domain simplifies things, compare the following two symbolizations of the same English sentence. First, assuming an unrestricted universe of discourse, that is, the universal domain, the sentence

All persons have rights.

goes into symbols as

$$(x)(Px \supset Rx)$$

This says, for any x in the entire universe, if x is a person, then x has rights.

However, if we now stipulate that the variable x ranges *only over persons*, "All persons have rights" may be symbolized as

$$(x) \, Rx$$

with the understanding that it is only persons that are under discussion. Because the variable now ranges only over persons, the universal quantifier (x) in effect asserts "All persons" and the sentence in effect says, For all persons, each has rights.

Of course, (x) Rx by itself—assuming an unrestricted domain—simply says that all things in the universe have rights. But because the universe of discourse is now restricted to persons and nothing but persons, the "things" referred to are all and only persons, and so (x) Rx asserts only that all persons have rights.

Similarly, if we stipulate that our domain is all and only Vulcans, "All Vulcans have hidden emotions" may be symbolized (x) Hx with the understanding x ranges only over *Vulcans* and Hx abbreviates *x has hidden emotions*.

Let us now stipulate that the variables in the following examples range over persons and nothing but persons. Consider the sentence

Someone knows someone.

In order to symbolize this, let us paraphrase it into an appropriate form. First, "someone knows someone" is equivalent to

For some x and for some y, x knows y.

In symbols, this is abbreviated thus:

$$(\exists x) \, (\exists y) \, Kxy$$

where Kxy abbreviates *x knows y*. (Remember that we are assuming the variables range only over persons.) Next, consider the following:

Everyone knows everyone.

Another way to put this is

For every x, and for every y, x knows y.

In QL, this becomes:

$$(x) \, (y) \, Kxy$$

If a sentence contains two universal quantifiers in a row or two existential quantifiers in a row, we can switch the order of the adjacent quantifiers without altering the meaning of the sentence. For example, "Someone knows someone" may be symbolized by either

$$(\exists x)(\exists y)(Kxy)$$

or

$$(\exists y)(\exists x)(Kxy)$$

And "Everyone knows everyone" may be symbolized by either

$$(y)(x)(Kxy)$$

or

$$(x)(y)(Kxy)$$

In general, if a sentence contains two universal quantifiers—each with a different variable attached—or two existential quantifiers with different variables, the order of the adjacent quantifiers makes no difference. However, as we saw earlier, when two quantifiers appear next to each other in a sentence, and one is universal and the other is existential, the order of the two quantifiers affects the meaning of the sentence. Changing the order of the quantifiers changes the meaning of the sentence.

DEAN MARTIN, UNIVERSAL LOVE, AND A SUMMARY OF LOGIC RELATIONS

Let us now sum this up by symbolizing some sentences about people, assuming a domain limited to human beings. Consider first a principle that was made famous in a 1965 song by that former rat-packer, the one and only Dean Martin:

Everybody loves somebody (sometime).

This great truth amounts to the following claim: Pick any person in the world. No matter who you pick, there is someone, somewhere, whom he or she loves. Everyone is such that he or she loves someone. Now, this goes into symbols through the following series of paraphrases:

Every x is such that x loves someone.

Every x is such that there is someone such that x loves this someone.

Every x is such that there is some y such that x loves y.

$$(x) \, (\exists y) \, Lxy$$

Consider next, "Someone loves everyone." (This would have been a good title for a follow-up song to Dean Martin's big 1965 hit.) Anyway, this sentence asserts that there is someone who is such that no matter whom you pick, this someone loves the one you picked. Paraphrasing, this is "There is at least one x such that for any y, x loves y." In QL:

$$(\exists x) (y) \text{ Lxy}$$

Notice the difference between the following two sentences.

1. Everyone is loved by someone (or other).

2. Someone loves everyone.

The two have different meanings. Sentence 1 says that no matter whom we pick, someone loves that person. But sentence 2 says that someone—a very special someone—exists who, no matter whom we pick, loves the person we pick. This someone is a saint.

Sentence 1 is a universal statement, because it is about all persons. But sentence 2 is an existential statement, because it is about this one special someone. Consequently, sentence 1 begins with a universal quantifier and sentence 2 begins with an existential quantifier:

1. $(x)(\exists y) \text{ Lyx}$

2. $(\exists y)(x) \text{ Lyx}$

Study the following examples carefully. Remember that the domain is limited to people.

English:	Symbolization:
Somebody is known by everybody.	$(\exists y) (x) \text{ Kxy}$
Everybody knows somebody or other.	$(x) (\exists y) \text{ Kxy}$
Everybody is known by somebody or other.	$(y)(\exists x) \text{ Kxy}$
Everyone is either nice or phony.	$(x) (\text{Nx v Px})$
Either everyone is nice or everyone is phony.	$(x) \text{ Nx v } (x) \text{ Px}$

This discussion can be summed up in a table[1]

English:	Formula:	Image:
Everyone loves everyone.	$(x) (y) \text{ Lxy}$	The '60s
Everyone loves someone.	$(x)(\exists y) \text{ Lxy}$	Dean Martin
Everyone loves himself.	$(x) \text{ Lxx}$	Egoism
Someone loves everyone.	$(\exists x) (y) \text{ Lxy}$	Mother Teresa
Someone loves someone.	$(\exists x) (\exists y) \text{ Lxy}$	Romeo and Juliet
Someone loves himself.	$(\exists x) \text{ Lxx}$	Narcissus
Everyone is loved by someone.	$(x) (\exists y) \text{ Lyx}$	A soap opera
Someone is loved by everyone.	$(\exists x) (y) \text{ Lyx}$	Mickey Mouse

[1]I owe the idea for this table to Professor Karl Hillstrom, Green River Community College.

▶   **EXERCISE 17.2**   ◀

Symbolize the following in QL.

1. *If everybody knows somebody, then somebody knows everybody.

2. There is a person who is universally respected.

3. There is a person who knows everybody.

4. Everybody respects someone or other.

5. *There's a person whom everyone respects.

6. Some people do not know anybody.

7. *If someone helps someone, then God is pleased.

8. Nobody who loves someone is all bad.

9. A lawyer who represents himself or herself has a fool for a client.

10. *Anyone who loves no one is to be pitied.

11. Pat is taller than any aardvark and is shorter than any elephant.

12. Everything that happens has a purpose.

13. It's not true that everything that happens has a purpose.

14. Nothing that happens has a purpose.

15. Some things happen for no purpose.

16. For each person, there is someone who cares for them.

17. *Either everybody was happy or everybody was acting.

18. Everybody is either happy or faking it.

TO BE OR NOT TO BE: THE LOGIC OF IDENTITY

"It depends on what the meaning of is is."—President Bill Clinton

The verb *to be* can be used in three different ways. First, it can be used to assert the existence of something, as in the following sentence:

There is a divine being.

Here, *is*, a form of the verb *to be*, is used to assert the existence of a deity. This *is* is called the "is of existence" or the "existential is." If we let Dx abbreviate "x is a divine being," this sentence is abbreviated

$$(\exists x)\; Dx$$

This is the existential use of the verb *to be*.

Second, *is* may be used to assert that a predicate expression applies to an object, as in the following sentence:

Fred is happy.

This is the *is* of predication. In symbols, this is

$$Hf$$

This is the predicative use of the verb *to be*.

In this section, we shall not be concerned with either of these two uses of *is*. Rather, we shall focus our attention on a third use of this verb. Consider the following sentence:

Robert Zimmerman is Bob Dylan.

Here *is* is used to assert that the individual named Robert Zimmerman is the same individual as the one named Bob Dylan. In other words, Robert Zimmerman, it is claimed, is Bob Dylan. This is the *is* of identity. Call this the identificational use of *to be*.

An interesting example of the *is* of identity concerns the heavenly object known to ancient astronomers as the Evening Star. In ancient times, the term *the Morning Star* had one meaning, and the term *the Evening Star* carried another meaning. It was thought that the two names referred to two different stars. Astronomical observations eventually proved that the two terms refer to the same individual object, namely, the planet Venus. That is, through observation, it was discovered that

The Morning Star is the Evening Star.

expresses a true proposition. In other words, the morning star and the evening star are not two different entities, they are one and the same entity.

Here are additional examples of the identificational use of *is*:

Taiwan is Formosa.

Muhammad Ali is Cassius Clay.

Olympia is the capital of Washington state.

Richard Starkey is Ringo Starr.

Do not confuse the concept of identity with the concept of similarity. In everyday speech, we might point to two hamburgers and say that they are "identical," meaning only that the two are very similar. However, according to the concept of identity used in logic, two different hamburgers, no matter how similar, could never be identical, that is, they could never be one and the same hamburger.

THE IDENTITY SIGN

Let us now introduce a symbol for the *is* of identity. We will abbreviate

___ is identical with ____

as

$$x = y$$

and we will abbreviate

___ is not identical with ____

as

$$\sim(x = y)$$

In order to incorporate the identity sign into QL, we must return to QL's syntax list, which we started on page 348, and add

Q6. If **c** and **d** are individual constants, then $(\textbf{c} = \textbf{d})$ is a wff.

In addition, we must add = to QL's vocabulary.

▶ SYMBOLIZING WITH THE IDENTITY SIGN

Using this notation, we can now abbreviate sentences that could not previously have been accurately symbolized. In the following, assume the domain is restricted to persons. Consider:

There are at least two musicians.

Without the identity sign, one might *try* symbolizing this as

$$(\exists x)\,(\exists y)\,(Mx \ \& \ My)$$

That is, for some x and for some y, x is a musician and y is a musician. However, this won't do, for it is possible that in this case x and y stand in for the same person, in which case it is possible, given the symbolism under consideration, that only one musician exists. That is, if only one musician were to exist, $(\exists x)\,(\exists y)$ $(Mx \ \& \ My)$ would nonetheless be true. This attempted symbolization fails to capture the meaning of the sentence at hand. Similarly, suppose we know that

a.   Someone in the house is an electrician.

and

b.   Someone in the house is a truck driver.

It does not necessarily follow that two persons are in the house; perhaps the two occurrences of the variable word *someone* refer to one and the same person.

The solution is to symbolize "There are at least two musicians" as follows:

$$(\exists x)\,(\exists y)\,[(Mx \,\&\, My) \,\&\, \sim (x = y)]$$

This reads, There is an x and there is a y and x is a musician, and y is a musician, and it is not the case that x is the same individual as y. Now, if it is not the case that x is the same individual as y, then x and y must be two different individuals, which guarantees that there are at least two musicians.

Next, consider:

There are at least three musicians.

In order to avoid unnecessary complications, we shall leave out some of the parentheses in the long conjunctive clause:

$$(\exists x)\,(\exists y)\,(\exists z)\,[(Mx \,\&\, My \,\&\, Mz) \,\&\, \sim (x = y) \,\&\, \sim (y = z) \,\&\, \sim (x = z)]$$

This reads, There exists at least one x, there exists at least one y, there exists at least one z such that x is a musician, y is a musician, and z is a musician, and x is not identical with y, y is not identical with z, and x is not identical with z. Now, if x is not the same individual as y, then x and y must be two different individuals; if y is not the same individual as z, then y and z must be two different individuals; and if x is not the same individual as z, then x and z must be two different individuals. This guarantees that there exists at least three musicians.

Suppose we wish to abbreviate the following:

There exists at most one Creator.

The symbolization (assuming an unrestricted domain) is

$$(x)\,(y)\,[(Cx \,\&\, Cy) \supset (x = y)]$$

According to this, for any x and any y, if x is a creator and y is a creator, then x is identical with y. This guarantees that there do not exist two different creators, which is to say that at most one Creator exists. And to symbolize

There exist at most two Creators.

we need

$$(x)\,(y)\,(z)\,\{[(Cx \,\&\, Cy \,\&\, Cz)] \supset [(x = y) \lor (x = z) \lor (y = z)]\}$$

This reads, For any x, for any y, and for any z, if x is a creator, y is a creator, and z is a creator, then either x is the same individual as y or x is identical with z or y and z are one and the same individual. This guarantees that at most two Creators exist.

Next, a slightly different construction:

There exists exactly one Creator.

This becomes

$$(\exists x)\,(y)\,[Cx \,\&\, (Cy \supset (y = x))]$$

And

There exist exactly two Creators.

abbreviates as

$$(\exists x)\, (\exists y)\, (z) \{[Cx\ \&\ Cy\ \&\ \sim (x = y)]\ \&\ [Cz \supset (z = x)\ v\ (z = y)]\}$$

This reads, There exists at least one x, there exists at least one y, and for any z, x is a creator, y is a creator, and it is not the case that x is identical with y, (so x and y are two different entities), and if z is a creator then z is identical with x or z is identical with y. This guarantees that exactly two creators exist.

We can also use the identity sign to abbreviate exceptive statements. Here is the first of several examples:

Rita is happier than anyone else except Joe.

In QL, this may be rendered

$$(x)\ \{[Px\ \&\ \sim(x = r)\ \&\ \sim(x = j)] \supset Hrx\}$$

This says that for any x, if x is a person and x is not Rita and x is not Joe, then Rita is happier than x. Next:

Everyone is happy except me and my monkey.

In QL, this is:

$$(x)\ [P\,x\ \&\ \sim(x = j)\ \&\ \sim(x = m)] \supset Hx]\ \&\ (\sim Hm\ \&\ \sim Hj)$$

where j abbreviates *my monkey* and m abbreviates *me*. Note that we are simplifying things a bit by leaving out a pair of brackets within the triple conjunction. Here is another example:

Rodney Dangerfield is the only person who respects Rodney Dangerfield.

If the constant r designates *Rodney Dangerfield* and Rxy abbreviates *x respects y*, this goes into QL as:

$$Rrr\ \&\ (x)\ [\sim (x = r) \supset \sim Rxr]$$

The identity sign also comes into play when we symbolize superlative statements. For instance:

Rita is the loveliest meter maid.

In QL, this may be put as

$$(x)\ [(Mx\ \&\ \sim (x = r)) \supset L\,r\,x]\ \&\ Mr$$

where Mx abbreviates *x is a meter maid*, Lxy abbreviates *x is lovelier than y*, and r designates *Rita*. This QL sentence says that for each x, if x is a meter maid and x is not Rita, then Rita is lovelier than x, and it adds that Rita is a meter maid.

▶   **EXERCISE 17.3**   ◀

Symbolize the following.

1.*Betty is the new winner.

2. Charlie is not the richest person in the world.

3. Rita has at least two jobs.

4. There is at most one president.

5.*There is no greatest number.

6. Washington state has more than one senator.

7.*Nobody except citizens may vote.

8. Someone is older than everyone else.

9.*There is only one deity.

10. There are exactly two deities.

11. There is no deity.

12. God is greater than all other beings.

13. The Beatles had two different drummers.

14. Ringo was the best drummer the Beatles ever had.

15. Only John Lennon was a member of both the Beatles and the Plastic Ono Band.

16.*Pat loved everybody except himself.

17. All were hired except Chris.

18. Raegina is the oldest member of the team.

19. Everyone is happy except Pat and Chris.

20. Only two slide rules remain in existence.

21.*Jupiter's moon, Io, is the only solar system body other than Earth known to have erupting volcanoes.

22. No more than three deities exist.

23. Millard is the only person more long-winded than Rob.

24. There are at most two surviving dinosaurs.

25. Any star other than the sun lies outside our solar system.

26. Virgo A is the largest known galaxy.

27. No planet is larger than Jupiter.

28. Everybody (except old Jed) is younger than old Jed.

CHAPTER 17 APPENDIX: *Properties of Relations*

The claim that Julie is older than Peter is abbreviated in QL as

$$Ojp$$

where O represents the two-place relation of "being older than." Relations have properties of their own and important philosophical arguments often depend on which properties it is correct to suppose a particular relation actually has. We shall look at three especially important properties of relations.

▶ TRANSITIVITY

A relation is a **transitive relation** just in case if one thing bears the relation to a second thing, and the second thing bears the relation to a third thing, the first must bear the relation to the third. In other words, if A has the relation to B, and if B has the relation to C, then A has the relation to C. The relation of "being older than" thus qualifies as transitive, for if A is older than B, and B is older than C, A must be older than C.

> *In general, for any two-place relation R, R is transitive if and only if:*
>
> $$(x) \, (y) \, (z) \, [(Rxy \, \& \, Ryz) \supset Rxz]$$

A relation qualifies as intransitive just in case if one thing bears the relation to a second and the second bears the relation to a third, the first cannot possibly bear the relation to the third. The relation of "being the grandmother of" thus qualifies as intransitive, for if one person is the grandmother of a second person, and the second is the grandmother of a third person, the first person cannot possibly be the grandmother of the third.

> *In general, for any two-place relation R, R is intransitive just in case:*
>
> $$(x) \, (y) \, (z) \, [(Rxy \, \& \, Ryz) \supset \sim Rxz]$$

Relations that are neither transitive nor intransitive are nontransitive. The relation *is a friend of* is an example of a nontransitive relation, for if Betty is a friend of John and John is a friend of Beth, it does not follow that Betty and Beth are friends and it does not follow that they are not friends. Other examples of nontransitive relations are *admires, is jealous of, loves,* and *gave a present to.*

Examples of transitive relations are *is the same weight as, owns more pencils than, has a larger bank account than,* and *received more votes than.* Examples of intransitive relations are *is exactly ten years older than, ate exactly twice as many hamburgers as,* and *owns one more car than.*

▶ SYMMETRY

A relation is a **symmetrical relation** just in case if one thing has the relation to a second thing, the second thing has that relation to the first thing. In other words, if A has the relation to B, then B has the relation to A. For example, the relation *is the same age as* is symmetrical, for if one thing is the same age as a second thing, the second thing is the same age as the first.

> *In general, for any two-place relation R, R is symmetrical if and only if:*
>
> $$(x)\,(y)\,(Rxy \supset Ryx)$$

Other examples of symmetric relations are *is married to, lives in the same city as, is a cousin of, is the same height as,* and *is a friend of.*

A relation qualifies as asymmetrical just in case if one thing has that relation to a second, the second cannot possibly have that relation to the first. For example, *is taller than* is asymmetrical, for if one person is taller than a second, the second must not be taller than the first.

> *In general, a relation R is asymmetrical if and only if:*
>
> $$(x)\,(y)\,(Rxy \supset \sim Ryx)$$

Other asymmetrical relations include *is older than, is father of, is mother of,* and *is east of.*

If a relation is neither symmetrical nor asymmetrical, it is nonsymmetrical. Thus, *knows the name of* is nonsymmetrical, for if one person knows the name of a second person, it doesn't follow that the second knows the name of the first, and it also doesn't follow that the second does not know the name of the first. Other examples of nonsymmetric relations are *is a sister of, painted a picture of, admires,* and *greeted.*

▶ REFLEXIVITY

A **reflexive relation** is one in which, if one thing has the relation to something or something has the relation to the first thing, then the first thing has that relation to itself. For example, the relation "belongs to the same church as" is reflexive, for if Clyde belongs to the same church as Bonnie, Clyde belongs to the same church as himself.

> *In general, a relation R is reflexive just in case:*
>
> $$(x) \, (\exists y) \, [(Rxy \lor Ryx) \supset Rxx]$$

Other examples of reflexive relations are *weighs as much as, went to the same school as, is as tall as,* and *has the same parents as.*

A relation is irreflexive just in case nothing has the relation to itself. Thus, *is older than* is irreflexive, for nothing could possibly be older than itself. In general, a relation R is irreflexive just in case

$$(x) \sim Rxx$$

Other examples of irreflexive relations are *is a brother of, is older than, weighs more than,* and *owns more pencils than.*

If a relation is neither reflexive nor irreflexive, it is nonreflexive. The relations *gave a present to, looked at, sang to,* and *likes* are examples of nonreflexive relations.

▶ EXERCISE 17.4 ◀

1. List additional examples of relations that are

    a.* transitive   **f.** nonsymmetric

    b. intransitive   **g.** reflexive

    c. nontransitive   **h.** irreflexive

    d. symmetric   **i.** nonreflexive

    e. asymmetric

2. Classify the following as transitive, symmetric, and so on. (Note: each relation will possess more than one of the properties covered in this section.)

a. is south of

b.*is the sister of

c. is at least ten pounds heavier than

d. is proud of

e. is a relative of

f. is the boss of

g. has more stamps than

h. borrowed a book from

i. is not taller than

j. went to the circus with

k. is acquainted with

l.* is identical with

CHAPTER 17 GLOSSARY

Domain of a variable   The set of things the variable can take as values. The values of a variable are just the entities represented by the singular terms that can replace the variable. The domain of a variable is also known as the *universe of discourse* for the sentence containing the variable.

Dyadic (or "two-place") predicate expression   A predicate phrase that contains two blanks and is typically used to represent a relation between two things.

Monadic (or "one-place") predicate phrase   A predicate phrase that contains one blank and is used to attribute a property to just one thing.

Overlapping quantifiers   In a sentence with overlapping quantifiers, one quantifier appears within the scope of another quantifier.

Reflexive relation   A relation that possesses the following feature: if A has the relation to B or B has the relation to A, A has that relation to itself.

Symmetrical relation   A relation that possesses the following feature: if one thing has the relation to a second thing, the second thing has that relation to the first thing.

Transitive relation   A relation that possesses the following feature: if one thing bears the relation to a second thing and the second thing bears the relation to a third thing, the first must bear the relation to the third.

Universal domain   Everything in the universe. When the domain is everything in the universe, the domain is unrestricted or universal.

Universe of discourse (for a sentence or group of sentences)   The collection of things that the sentence or sentences are making a claim about. Essentially, the collection of things we are talking about on an occasion is our universe of discourse on that occasion.

Chapter

18

Proofs with Monadic Predicates

I n this chapter, we will combine the truth-functional rules with four quantifier inference rules and one quantifier replacement rule. With this expanded set of rules, we can prove valid many valid arguments that contain quantifiers and predicates. Our first rule governs inferences that run from a universal general sentence to an instantiation of the general sentence.

THE UNIVERSAL INSTANTIATION RULE

Before we can state this new inference rule, some terminology is necessary. First, a wff of QL is called a **universal quantification** if and only if it begins with a universal quantifier whose scope covers the entire wff. Examples include the following:

   (x)Fx   (y)Fy   (z)Fz   (x) (Fx ⊃ Gx)   (y) (Fy v Gy)

If we remove the quantifier from a universal quantification and uniformly replace each occurrence of the variable it binds with a constant, we produce an **instantiation (substitution instance) of the universal quantification.** For example, suppose we begin with the universal quantification

$$(x) (Wx \supset Mx)$$

If we remove the quantifier and in place of each occurrence of the variable we uniformly place the constant n into the formula, we get:

$$Wn \supset Mn$$

If we instantiate with the constant f we get:

$$Wf \supset Mf$$

Similarly, if we begin with the universal quantification

$$(x) Gx$$

the following are instantiations:

Gm

Gp

Gs

The distinction just drawn between a universal quantification and its substitution instances has its roots in ordinary English discourse. We sometimes draw a generalization and, when challenged or asked for an example, we say, "For instance" and give a specific instance of the general point. For instance, suppose Pat makes the broad generalization, "All pigs are smart." When you ask for an example, Pat might say, "For instance, my Grandpa had a pig named Wilbur and Wilbur was very smart."

Before we formulate our first rule, one technicality: a little further down the road, after our first two rules have been formulated, our system is going to need to use a new type of symbol called a **John Doe name.** John Doe names will be special symbols that function in our system just as the name "John Doe" functions as a temporary name assigned to an unnamed, unidentified mystery male. ("John Doe is approximately thirty years old, Caucasian, six feet tall.") Our first two quantifier rules each need a clause that mentions John Doe names, but we will wait until the third rule before actually explaining these special symbols and putting them to work in proofs.

Now we can turn to our first quantificational inference rule.[1] A universal quantification makes a claim about all things in the universe of discourse. For instance, (x) Gx might be used to assert that all things are good. If a universal quantification is true, its claim is true of each and every thing in the universe of discourse. Therefore, if a universal quantification is true, each and every instantiation of it is true as well. This logic is reflected in our first quantificational inference rule:

Universal Instantiation (UI)

From a universal quantification, one may infer any instantiation, provided that the instantiation was produced by uniformly replacing each occurrence of the variable that was bound by the quantifier with a constant or John Doe name.

In each case below, the second line was derived from the first by an application of UI:

1. (x)Fx   **1.** (x)(Fx ⊃ Gx)   **1.** (x) ~ Fx   **1.** (x) ~ (Fx ⊃ Gx)
2. Fa   **2.** Fc ⊃ Gc   **2.** ~ Fs   **2.** ~ (Fb ⊃ Gb)

---

[1] The deduction rules in this chapter are based on rules presented in William Gustason and Dolph Ulrich, *Elementary Symbolic Logic* (Prospect Heights, Ill.: Waaveland Press, 1989)

In each of these examples, the first sentence is a universal quantification and the second sentence is an instantiation of the universal quantification. In each example, the second sentence was derived from the first by performing two steps: first the quantifier was removed from the first sentence, and then each occurrence of the variable was uniformly replaced by a constant. The replacement must be uniform: Notice, for example, that in the second and fourth instantiations above, the same constant is used all the way across the formula. Thus, the following inference, from 1 to 2, is an incorrect application of UI because the variable has not been *uniformly* replaced by a constant:

1. (x)(Fx ⊃ Gx)

2. Fa ⊃ Gb   (Illegal)

Here is natural deduction proof that employs UI:

1. (x)(Fx ⊃ Gx)

2. (x)(Gx ⊃ Hx) / Fs ⊃ Hs

3. Fs ⊃ Gs   UI 1

4. Gs ⊃ Hs   UI 2

5. Fs ⊃ Hs   HS 3, 4

The first step in this proof was the application of UI at line 3. The quantifier was stripped off line 1, the variable was uniformly replaced by the constant s, and the result was written on line 3. Then, at line 4, UI was applied to the formula on line 2. Again, the quantifier from line 2 was removed and the variable was replaced by s in order to produce line 4. One application of UI can employ the same constant used in a previous application, if needed. In this case, we had to use the same constant on lines 3 and 4 in order to set up the application of HS on line 5. If we had instantiated with s on line 3 and then with, say, b on line 4, we would have derived two formulas that would not fit the HS rule, namely:

3. Fs ⊃ Gs   UI 1

4. Gb ⊃ Hb   UI 2

Remember that UI may be applied only to a universal quantification. In a universal quantification, the universal quantifier appears at the start of the sentence and has the entire sentence as its scope. Thus, UI may *not* be applied to sentences such as the following:

$$\sim(x)(Fx \supset Gx)$$

$$(x)(Fx) \supset (x)(Gx)$$

UI cannot be applied to these sentences because in each case the scope of the universal quantifier fails to cover the entire sentence. Accordingly, the following inferences from 1 to 2 are not permissible:

1. ~ (x) (Fx ⊃ Gx)   **1.** (x) (Fx) ⊃ (x) (Gx)
2. ~ (Fa ⊃ Ga)   **2.** Fa ⊃ Ga

EXISTENTIAL GENERALIZATION

A sentence of QL is a **singular sentence of QL** if and only if it contains no quantifiers and no variables. If we begin with a singular sentence containing one or more constants or John Doe names, and if we prefix to this an existential quantifier and uniformly replace one or more occurrences of a constant (or John Doe name) with the variable occurring in the existential quantifier, we produce the **existential generalization** of that singular sentence. In each example below, the first sentence is a singular sentence and the sentence below it is an existential generalization of the singular sentence:

1. Fa   **1.** Hb & Rb   **1.** Gs v Fs
2. (∃x) Fx   **2.** (∃x) (Hx & Rx)   **2.** (∃y) (Gy v Fy)

A singular sentence containing one constant asserts that one particular thing—the thing designated by the constant—has a certain property. The existential generalization of such a sentence asserts that something or other has the property. Certainly if a specific individual designated by a singular term has a certain property, it logically follows that *something* has that property. If a singular sentence is true, the existential generalization of that sentence must be true as well.

For example, from "George is sad," it certainly follows that something is sad. And from "The Space Needle is yellow" it certainly follows that something is yellow. This simple and obviously valid pattern of reasoning is reflected in the next rule.

> ### Existential Generalization (EG)
>
> *From a sentence that contains a constant or a John Doe name, you may infer any corresponding existential generalization, provided that (a) the variable used in the generalization does not already occur in the sentence generalized upon and (b) the generalization results by replacing at least one occurrence of the constant or John Doe name with the variable, and no other changes are made.*

In each example below, the first sentence is a singular sentence and the second sentence was derived by applying EG to the singular sentence.

1. Fa   **1.** Fb   **1.** ~ Fc   **1.** Fd v Hd
2. (∃x) Fx   **2.** (∃x) Fx   **2.** (∃x) ~ Fx   **2.** (∃x) (Fx v Hx)

The following proof employs EG:

1. $(\exists x) (Fx) \supset (x) (Hx \supset Gx)$

2. Fa

3. Hs / Gs

4. $(\exists x) Fx$ EG 2

5. $(x) (Hx \supset Gx)$ MP 1, 4

6. $Hs \supset Gs$ UI 5

7. Gs   MP 3,6

At line 4 in this proof, we applied EG to line 2. The existential quantifier was added, and the constant a (from line 2) was replaced by the variable x. The formula that was thus produced at line 4, namely $(\exists x)Fx$, is the antecedent of line 1. This fact allowed us to apply Modus Ponens to lines 1 and 4 to derive line 5. At the next line, we applied UI to line 5 and removed the quantifiers. Notice that we employed the constant s on the applications of UI. This allowed us to make the MP move on line 7.

EXISTENTIAL INSTANTIATION

Sometimes, in everyday reasoning, people reason about an unidentified, unnamed individual—a "mystery individual." For example, the police might learn that someone robbed a bank Tuesday, and they might have a lot of information about the robber even though they don't know the robber's name. They may assign a temporary name—perhaps John, or Jane, Doe—to the unknown person to simplify referring to and reasoning about him or her. In the next inference rule, we will employ this practice of assigning a temporary name to an unnamed individual for the purpose of reasoning about that individual.

In the upcoming proofs, when we need to refer to and reason about an entity of a certain type, but don't know which particular thing of that type this unnamed entity is, we will assign to it a John Doe name. We will use the letters u and v as our John Doe names, and these letters will be used only in such a situation. (If we need more than two John Doe names, we can prime the letters and use u', u'', v', v'', and so on.) Notice that a John Doe name is not a constant, for a constant refers to something specifically identified by name or by definite description.

Now, if a wff begins with an existential quantifier whose scope is the entire wff, the wff is called an **existential quantification**. If the existential quantifier is removed from an existential quantification and each of the variables bound by that quantifier is uniformly replaced by a constant or a John Doe name, the result is an **instantiation or substitution instance of the existential quantification.** The inference rule Existential Instantiation may now be formulated as follows:

> ### *Existential Instantiation (EI)*
>
> *From an existential quantification, you may infer an instantia-*
> *tion, provided that (a) each occurrence of the variable bound by*
> *the quantifier in the existential quantification is uniformly*
> *replaced with a John Doe name and no other changes are made*
> *and (b) the John Doe name does not appear in any earlier line of*
> *the deduction.*

Here are some applications of Existential Instantiation. In each case, the second sentence was derived from the first by applying EI.

1. $(\exists x)$ Fx	**1.** $(\exists x)$ (Hx v Gx)
2. Fv	**2.** Hv v Gv

In each case above, the first sentence is an existential quantification and the second sentence is an allowed instantiation of the existential quantification. Remember that each occurrence of the variable must be replaced uniformly when the instantiation is written.

It is crucial that you instantiate only with a new John Doe name when applying EI. In order to see why we must impose this restriction, consider what we could do if the restriction were dropped. Suppose the restriction has been dropped and we can apply EI using any constant or John Doe name we wish, including one already in the premises, one already in the conclusion, and one used in a previous instance of EI. In such a case, the three following invalid arguments could easily be proven valid:

Argument 1

1. There is a millionaire.

2. So, Pat is a millionaire.

In QL, this is

$$(\exists x) \text{ Mx } / \text{ Mp}$$

If we could apply EI with any constant or John Doe name, we could use EI to prove this argument valid in one step:

1. $(\exists x)$ Mx / Mp

2. Mp   EI 1 (Incorrect)

Argument 2

1. There exists an octogenarian.

2. There exists a teenager.

3. So, there exists someone who is both a teenager and an octogenarian.

In QL, this is

    1. (∃x) Ox

    2. (∃x) Tx / (∃ x) (Ox & Tx)

If we could apply EI using any constant or John Doe name—with no restrictions—we could instantiate twice with the same constant and prove this argument valid:

1. (∃x) Ox

2. (∃x) Tx / (∃x) (Ox & Tx)

3. Oa   EI 1 (Incorrect)

4. Ta   EI 1 (Incorrect)

5. Oa & Ta   Conj 3, 4

6. (∃x)(Ox & Tx)   EG 5

Argument 3

1. Bob is an octogenarian.

2. Something is a teenager.

3. So, Bob is a teenager.

In QL:

    1. Ob

    2. (∃x) Tx / Tb

Again, if we could use any constant or John Doe name when applying EI, we could construct the following proof:

1. Ob

2. (∃x) Tx / Tb

3. Tb   EI 2 (Incorrect)

Each of these three arguments is invalid, of course. And each proof involves a violation of EI's restrictions. Thus, in the first, we instantiated the existential quantification using a constant appearing in the conclusion. In the second, we instantiated using a constant that had been used in a previous existential instantiation. In the third, we instantiated using a constant that appears in the premises. In order to avoid invalid inferences such as the three above, we must make sure that when we apply EI, we use a John Doe name that is new to the proof.

    The following proof employs a correct application of EI:

1. (x)(Fx ⊃ Gx)

2. (∃x)Fx / (∃x) Gx

3. Fv   EI 2

4. Fv ⊃ Gv   UI 1

5. Gv   MP 3, 4

6. (∃x)Gx   EG 5

Notice that in this proof we applied EI before we applied UI. When you apply EI you must use a John Doe name new to the proof, whereas you may universally instantiate using any constant or John Doe name. For this reason, it is generally a good practice to apply EI before applying UI if possible.

MEMORIES OF GEOMETRY CLASS: UNIVERSAL GENERALIZATION

The next rule will make better sense if we first consider a logical technique employed frequently in mathematics. Recall the Pythagorean theorem. The square of the hypotenuse of a right triangle is equal to the sum of the squares of the legs. (A right triangle is a triangle that has one right angle. In a right triangle, the sides that include the right angle are called the legs of the triangle, and the side opposite the right angle is the hypotenuse.) Throughout the centuries, many different proofs of the theorem have been presented, but the first is attributed to Pythagoras, for whom it is named, in about 525 B.C.

The proofs of this theorem usually begin with a diagram and a statement similar to, "Let figure ABC be any right triangle." A mathematical chain of reasoning is then applied to triangle ABC and it is proved that the square of the hypotenuse is equal to the sum of the squares of the legs. This result is then asserted for *all* right triangles—for right triangles in general—and not just for the particular triangle ABC. Now, why is this reasoning valid? It is valid because no special assumptions are made in the proof concerning triangle ABC, assumptions that would differentiate ABC from any other right triangle. Therefore, if something is proven about ABC, this will be true of *any* right triangle in general. However, if the proof had depended on special assumptions about ABC, such as an actual measurement of its angles or sides, the proof would only establish something about ABC, and not about right triangles in general. The idea here is that if we had assumed something special about triangle ABC, what we concluded about *it* could not be applied to all other triangles. But because we knew nothing about ABC that would distinguish it from any other right triangles, what we proved about it would apply also to all other right triangles. This basic logic is reflected in the next rule.

If we take a singular sentence that contains one or more constants or John Doe names and prefix to it a universal quantifier term, and uniformly replace each occurrence of a constant or John Doe name in the sentence with the variable contained in the quantifier, the result is the **universal generalization** of the singular sentence. Here are some examples:

1. Fu	**1.** Fa ⊃ Ha	**1.** ~ (Fa & Ga)
2. (x)Fx	**2.** (x)(Fx ⊃ Hx)	**2.** (x) ~ (Fx & Gx)

In each case above, the first sentence is a singular sentence and the second sentence is a universal generalization of the singular sentence.

We may now formulate our fourth rule:

Universal Generalization (UG)

From a sentence containing a John Doe name, one may infer the corresponding universal generalization, provided that

(a) *The John Doe name that is replaced by a variable does not occur in any preceding line derived by EI;*
(b) *The generalization results by replacing each occurrence of the John Doe name with the variable and no other changes are made;*
(c) *The variable you use in the generalization does not already appear in the sentence from which you are generalizing; and*
(d) *The John Doe name does not appear in any assumed premise that has not already been discharged.*

Consider the following proof:
1. (x)(Gx)

2. (x)(Cx) / (x)(Gx & Cx)

3. Gu   UI 1

4. Cu   UI 2

5. Gu & Cu   Conj 3, 4

6. (x)(Gx & Cx)   UG 5

Let us emphasize the rationale for this rule. If we correctly apply UG to a singular statement containing a John Doe name, what is true in the case of this John Doe name is true not only in the case represented by the John Doe name but in all like cases. That is, what is true of that which the John Doe name designates is true of all like things. Therefore, after we have finished reasoning about what this name represents, we may generalize our results and assert that what is true of that designated by the John Doe name is true of all like things.

The following proof also employs UG:

1. (x)(Fx ⊃ Gx)

2. (x)(Gx ⊃ Sx)

3. (x)(Sx ⊃ Rx) / (x)(Fx ⊃ Rx)

4. Fu ⊃ Gu   UI 1

5.	Gu ⊃ Su	UI 2
6.	Su ⊃ Ru	UI 3
7.	Fu ⊃ Su	HS 4, 5
8.	Fu ⊃ Ru	HS 6, 7
9.	(x)(Fx ⊃ Rx)	UG 8

Notice that we are allowed to universally generalize line 8 because the John Doe name—u—does not occur on a previous line derived by EI.

This last point brings up an important issue. We are not allowed to apply UG to a John Doe name if that name appears on a line derived by EI. In order to avoid mix-ups and mistakes, it will be helpful if we apply UG only to John Doe names u, u′, u″, and so on, and we shall use only John Doe names v, v′, v″, and so on when we apply EI. That is, from here on, we shall reserve John Doe names from the "u" group for UG, and John Doe names from the "v" group will be set aside for EI.

Here are further examples of our new rules. In the next problem, we begin by removing the universal quantifiers. We then apply a truth-functional rule and end by attaching the existential quantifier:

1.	(x) (Hx ⊃ Bx)	
2.	Hg / (∃x) Bx	
3.	Hg ⊃ Bg	UI 1
4.	Bg	MP 2, 3
5.	(∃x) Bx	EG 4

Above, we instantiated premise 1 using the constant g. It is not a coincidence that this is the same constant that appears in line 2. We could have used any constant when we instantiated line 1, but if we had instantiated line 1 using a constant other than g, then we could not have applied Modus Ponens to lines 2 and 3.

Next, examine the first line of the following argument:

1.	(∃x) (Jx) ⊃ (x) (Kx ⊃ Lx)	
2.	Jb & Kb / Lb	

Because line 1 does not begin with a quantifier whose scope extends over the entire line, we cannot instantiate that line. Neither UI nor EI applies in such a case. Therefore, our only hope is to break 1 up into its two quantified components:

3.	Jb	Simp 2
4.	(∃x) Jx	EG 3
5.	(x) (Kx ⊃ Lx)	MP 1, 4

In the above steps, we derived the antecedent of line 1 and brought down 1's consequent by Modus Ponens. Now we can instantiate line 5 and get rid of its quantifier:

6. Kb ⊃ Lb   UI 5

7. Kb   Simp 2

8. Lb   MP 6, 7

We instantiated line 6 with the constant b so that we could apply Modus Ponens later in the proof.

   In the previous proof we used something derived from line 2 to break up line 1. The following argument calls for the same strategy:

1. (x) (Bx ⊃ Gx) ⊃ (∃x) (Ax & Hx)

2. (x) (Bx ⊃ Hx) & (x) (Hx ⊃ Gx) / (∃x) Hx

Because 1 contains two quantified formulas joined by a horseshoe, we cannot apply to it either UI or EI. Remember: You cannot apply UI to a line unless the universal quantifier sits at the start of the line and also has the *entire* rest of the line as its scope. Our first job is to somehow break line 1 up into parts that can be instantiated. We therefore begin with the only move available—we simplify line 2:

3. (x) (Bx ⊃ Hx)   Simp 2

4. (x) (Hx ⊃ Gx)   Simp 2

Next, we remove the quantifiers:

5. Bu ⊃ Hu   UI 3

6. Hu ⊃ Gu   UI 4

Notice that these two lines fit the requirements for the Hypothetical Syllogism rule. Thus:

7. Bu ⊃ Gu   HS 5, 6

From this, we derive the antecedent of 1:

8. (x) (Bx ⊃ Gx)   UG 7

Now we can bring down the consequent of 1:

9. (∃x) (Ax & Hx)   MP 1, 8

Finally:

10. Av & Hv   EI 9

11. Hv   Simp 10

12. (∃x) Hx   EG 11

▶ **EXERCISE 18.1** ◀

Using truth-functional inference rules and the four new quantifier rules, supply justifications for the following natural deduction proofs.

(1)* **1.** (x) [Fx ⊃ (Gx & Sx)]

    2. (x)[(Gx & Sx) ⊃ (Hx v Rx)] / (x)[Fx ⊃ (Hx v Rx)]

    3. Fu ⊃ (Gu & Su)

    4. (Gu & Su) ⊃ (Hu v Ru)

    5. Fu ⊃ (Hu v Ru)

    6. (x)[Fx ⊃ (H x v Rx)]

(2) **1.** (x) [Px ⊃ (Sx v Hx)]

    2. (∃x)(Px & ~ Sx) / (∃x) (Px & Hx)

    3. Pv & ~ Sv

    4. Pv

    5. Pv ⊃ (Sv v Hv)

    6. Sv v Hv

    7. ~ Sv

    8. Hv

    9. Pv & Hv

    10. (∃x) (Px & Hx)

(3) **1.** (∃x) (Fx) ⊃ (x) (Ax ⊃ Bx)

    2. Fc

    3. Ag / (∃y) (By)

    4. (∃x) (Fx)

    5. (x) (Ax ⊃ Bx)

    6. Ag ⊃ Bg

    7. Bg

    8. (∃y) (By)

(4) **1.** (x) (Hx) ⊃ (x) (Sx ⊃ Px)

    2. ~ (x) (Sx ⊃ Px)

    3. ~ (x) (Hx) ⊃ (x) (Mx) / (x) (Mx)

4. ~ (x) (Hx)

5. (x) (Mx)

▶   **EXERCISE 18.2**   ◀

Using truth-functional inference rules and the four new quantifier rules, supply proofs for the following valid arguments.

(1)* 1. (x) (Wx ⊃ Sx)

   2. (x)(Sx ⊃ Px) / (x) (Wx ⊃ Px)

(2) 1. (x) (Hx ⊃ Jx)

   2. (∃x) (Hx) / (∃x) (Jx)

(3) 1. (x) (Sx ⊃ Gx)

   2. Sa / Ga

(4) 1. (x) (Ax ⊃ Bx)

   2. ~ Bc / ~ Ac

(5)* 1. (x) (Mx)

   2. Hg / Hg & Mg

(6) 1. (∃x) (Fx & ~ Mx)

   2. (x) (Fx ⊃ Hx) / (∃x) (Hx & ~ Mx)

(7)* 1. Sa ⊃ (x) (Fx)

   2. Ha v Sa

   3. ~ Ha

   4. (x) (Fx ⊃ Gx) / (x) (Gx)

(8) 1. (x) (Hx & Sx)

   2. (∃x) (Hx) ⊃ (∃x) (Gx) / (∃x) (Sx) & (∃x) (Gx)

(9) 1. (x)(Hx) ⊃ (∃x)(Sx)

   2. (x) (Fx) v (x)(Hx)

   3. ~ (x) (Fx) / (∃x) (Sx)

(10)* 1. (x) (Ax ⊃ Bx)

   2. (x) (Jx ⊃ Fx)

   3. (x) (Ax v Jx) / (x) (Bx v Fx)

(11) **1.** (x) (Ax ⊃ Bx)

     2. (x) (Ax) / (x) (Bx)

(12)* **1.** (x) (Hx ⊃ Gx)

     2. (x) (~ Gx) / (x) (~ Hx)

(13) **1.** (x) (Sx) ⊃(x) (Gx)

     2. ~(x) (Gx) / ~ (x) (Sx)

(14) **1.** (∃x) (Bx) ⊃ (x) (Hx ⊃ Gx)

     2. Bb & Hb / Gb

(15) **1.** (x) (Fx ⊃ Sx)

     2. Fa & Fb / Sa & Sb

(16)* **1.** (x) (Hx ⊃ Qx)

     2. Ha v Hb / Qa v Qb

(17) **1.** (∃x) (Sx) ⊃ (x) (Hx ⊃ Gx)

     2. Sp & Hp / Gp

(18)* **1.** (∃x) (Fx) ⊃ (x) (Sx)

     2. (∃x) (Hx) ⊃ (x) (Gx)

     3. Fs & Hg / (x) (Sx & Gx)

►   **EXERCISE 18.3**   ◄

Using truth-functional inference rules, truth-functional replacement rules, and the four new quantifier rules, supply proofs for the following valid arguments.

(1)* **1.** (x) {Sx ⊃ [Px ⊃ (Bx & Qx)]}

     2. (∃x) (Sx) / (∃x)(Px ⊃ Qx)

(2) **1.** (x) (Ax v ~ Bx)

     2. Hs / (∃x) [(Hx & Ax) v (Hx & ~ Bx)]

(3) **1.** (∃x) (Jx) ⊃ (x) (Hx v Sx)

     2. (x) (Fx v ~ Sx)

     3. (x) (Jx) / (∃x)(Hx v Fx)

(4)* **1.** (x) [Hx ⊃ (Sx v Gx)]

     2. ~ Sb & ~ Gb / ~ Hb

(5) 1. (∃y) (Sy & Hy)

    2. (x) [(Gx v Rx) ⊃ ~ Sx] / (∃x)(~ Gx)

(6)* 1. (x) [Hx ⊃ (Bx & Wx)]

    2. (∃y) (~ By) / (∃z) (~ Hz)

(7) 1. (x) [Hx ⊃ (Sx v Gx)]

    2. (∃y) (~ Sy & ~ Gy) / (∃x)(~ Hx)

.(8) 1. (∃x) (Gx) ⊃ (x) (Hx ⊃ Sx)

    2. (∃x) (Gx v Sx)

    3. (x)(Sx ⊃ Gx) / (x)(~ G x ⊃ ~ Hx)

ONE NEW REPLACEMENT RULE: QUANTIFIER EXCHANGE

The previous four new rules are all inference rules. The next rule is a replacement rule. Recall the difference: inference rules may only be applied to whole lines of a proof, whereas replacement rules may be applied to a designated part of a line. In Chapter 16, we saw how to exchange one quantifier for another. That discussion is the basis for the following replacement rule:

> ### *Quantifier Exchange (QE)*
>
> *If **P** contains either a universal or an existential quantifier, **P** may be replaced by or may replace a sentence that is exactly like **P** except that one quantifier has been switched for the other in accord with the following steps:*
>
> *(a) Switch one quantifier for the other.*
> *(b) Negate each side of the quantifier.*
> *(c) Cancel out any double negatives that result.*

For example, if we apply QE to (x) Fx we get this: ~ (∃x) ~ Fx. And if we apply QE to (∃x) Fx we get this: ~ (x) ~ Fx.

    In each of the following five examples, the second sentence was derived from the first by applying QE:

1. ~ (x) Fx
2. (∃x) ~ Fx

1. ~(∃x) Fx
2. (x) ~ Fx

1. (x) [~(Fx ⊃ Gx)]
2. ~ (∃x) (Fx ⊃ Gx)

1. ~ (∃x) (Fx & Gx)
2. (x) [(~ (Fx & Gx)]

1. (x) (Fx ⊃ Gx)
2. ~ (∃x)[~ (Fx ⊃ Gx)]

The following proofs illustrate the use of this new rule.

1. (x) (~ Hx) ⊃ (x) (Fx ⊃ Gx)

2. ~ (∃x) Hx

3. Fa / (∃x) Gx

4. (x) ~ Hx   QE 2

5. (x) (Fx ⊃ Gx)   MP 1, 4

6. Fa ⊃ Ga   UI 5

7. Ga   MP 3, 6

8. (∃x) Gx   EG 7

On line 4 of this proof, we applied QE to line 2: We went to line 2, moved the tilde to the other side, switched quantifiers, and derived line 4.

1. (∃x) (Fx) ⊃ (∃x) (Gx)

2. (x) ~ Gx / (x)~ Fx

3. ~ (∃x) Gx   QE 2

4. ~ (∃x) Fx   MT 1, 3

5. (x) ~ Fx   QE 4

In this proof, the application of QE to line 2 produced line 3. Notice that Line 3 is the negation of the *consequent* of line 1. This allowed us to apply MT to 1 and 3 in order to derive line 4. We then used QE to trade an existential quantifier for a universal quantifier.

►   **EXERCISE 18.4**   ◄

Using truth-functional inference rules, truth-functional replacement rules, the four new quantifier rules, and the Quantifier Exchange rule, supply proofs for the following valid arguments.

(1)* 1. ~ (x) (Sx) / (∃x) (Sx ⊃ Px)

(2) 1. (x) [(Bx v Cx) ⊃ Wx]

    2. ~ (x) (Hx v ~ Bx) / (∃x) (Wx)

(3) 1. (x) (Hx) ⊃ (∃x)(Sx)

    2. (x) (~ Sx) / (∃x)(~ Hx)

(4) 1. (∃x) (~ Hx) v (∃x)(~ Sx)

    2. (x) (Sx) / ~(x) (Hx)

(5)* 1. ~ (∃x) (Ax) / (x)(Ax ⊃ Bx)

(6) 1. (∃x) (Gx) v (∃x) (Hx & Sx)

    2. ~ (∃x) (Hx) / (∃x) (Gx)

(7)* 1. (x) (Sx & Gx) v (x) (Qx & Hx)

    2. ~ (x) (Qx) / (x)(Gx)

(8) 1. (x) (Jx) ⊃ (∃x) (~ Sx)

    2. ~ (x) (Sx) ⊃ (∃x) (~ Hx) / (x) (Hx) ⊃ (∃x) (~ Jx)

(9)* 1. (∃x) (Px v Gx) ⊃ (x) (Hx)

    2. (∃x) (~ Hx) / (x) (~ Px)

~ **(10) 1.** ~ (∃x) (Mx & ~ Gx)

    2. ~ (∃x) (Mx & ~ Jx) / (x)[Mx ⊃ (Gx & Jx)]

(11) 1. (x) [(Px & Qx) ⊃ Rx]

    2. ~ (x) (Px ⊃ R x) / (∃x) (~ Qx)

(12) 1. (∃x) (~ Hx) ⊃ (x) (Ax ⊃ Bx)

    2. ~ (x) (Hx v Bx) / (∃x) (~ Ax)

(13) 1. ~(x) (Ax ⊃ Bx)

    2. (x) (Dx ⊃ Bx) / (∃x) ⊃ (Ax & ~ Dx)

▶   **EXERCISE 18.5**   ◀

Symbolize the following arguments and prove each valid using natural deduction.

1. *Every cat is a mammal. No fish is a mammal. So, no fish is a cat.

2. No cat is a reptile. Some pets are reptiles. So, some pets are not cats.

3. Some cats are orange. Every cat is a mammal. So, some mammals are orange.

4. Every dog is a mammal. No airplane is a mammal. If no airplane is a dog, then no airplane barks at cats. So, no airplane barks at cats.

5.*Every musician is an artist. Every artist is a dreamer. Some high school dropouts are musicians. So, some high school dropouts are dreamers.

6. Charlie's car has a personality. Anything that has a personality is a person. So, Charlie's car is a person.

7.*Cats and dogs reason, learn, and love. Any creature that reasons, learns, and loves possesses an immortal soul and goes to be with God when it dies. Therefore, cats and dogs possess immortal souls and go to be with God when they die.

8. Anyone who loves hamburgers isn't a vegetarian. Wimpy loves hamburgers. So, Wimpy isn't a vegetarian.

9. Each event in one's life possesses eternal significance. A person's birth is one event in his or her life. So, a person's birth is an event of eternal significance.

10. Every hamburger sold by Dag's has Dag's special sauce on it. No other burger joint puts Dag's special sauce on its burgers. The burger Joe is eating does not have Dag's special sauce on it. Therefore, the burger Joe is eating is not a Dag's burger.

11. No member of the Revolutionary Communist Party is a Republican. Some Marxists are members of the Revolutionary Communist Party. So, some Marxists are not Republicans.

12.*All who hang out at the Hasty Tasty admire Mao Tse-tung. All who admire Mao Tse-tung belong to the Progressive Labor Party. Stephanie does not belong to the Progressive Labor Party. So, Stephanie does not hang out at the Hasty Tasty.

13. Anyone who frequents the Blue Moon Tavern reads the *Helix*. If Deane sells the *Helix*, then Deane reads the *Helix*. Deane sells the *Helix* but doesn't frequent the Blue Moon. (He only goes in there once in a blue moon.) So, some who read the *Helix* don't frequent the Blue Moon.

14.*Dragons live forever. Nothing that lives forever is to be feared. Puff is a dragon. So, Puff is not to be feared.

15. All songs written by either Lennon or McCartney are rock 'n' roll songs. Therefore, there is no song written by Lennon or McCartney that is not a rock 'n' roll song.

16. If that piece of varnished sewer sludge is a work of art, then anything is a work of art. That piece of varnished sewer sludge is a work of art. So, this glazed pile of dead bugs is a work of art.

17. If rocking-horse people eat marshmallow pies, then newspaper taxis are waiting to take you away. If the girl with kaleidoscope eyes meets you,

then rocking-horse people eat marshmallow pies. The girl with kaleido-scope eyes will meet you. So, newspaper taxis are waiting to take you away. (Do you need to use quantifiers to prove this valid?)

18. Either all things are created by God or all things are material. If all things are material, then life has no transcendent meaning. If God does not exist, then it's not the case that all things are created by God. So, if God does not exist, then life has no transcendent meaning. (Do you need to use quantifiers to prove this valid?)

19. Any senator or House member is a member of Congress. No anarchist is a member of Congress. George is an anarchist. So, George isn't a senator.

20. All aardvarks are mammals. All mammals are warm-blooded. Art is an aardvark. If Art is warm-blooded, then Art is not cold. So, Art is not cold.

21. All extraterrestrials are purple. No purple thing is dangerous. So, no extraterrestrials are dangerous.

22. All extraterrestrials are green and hairless. Some extraterrestrials are carbon-based creatures. So, some carbon-based creatures are green and hairless.

23. Some cars are expensive. All expensive cars are reliable. If some cars are reliable, then Joe will win his bet. So Joe will win his bet.

NAMING OUR SYSTEM

We shall name the natural deduction system we have been using QD (for "quantificational deduction"). QD consists of two parts:

1. The language QL

2. The natural deduction rules of Part IV

A **proof in QD** is a sequence of sentences of QL, each of which is either a premise or an assumption or follows from one or more previous sentences according to a QD deduction rule, and in which (a) every line (other than a premise) has a justification and (b) any assumptions have been discharged. An argument is **valid in QD** if and only if it is possible to construct a proof in QD whose premises are the premises of the argument and whose conclusion is the conclusion of the argument. An argument is **invalid in QD** if and only if it is not valid in QD. A proof whose last line is P and that contains no premises is a **premise-free proof** of P. If a premise-free proof of P can be constructed in QD, P is a theorem of QD. In this case, P is also called logically true, or quantifica-tionally true because P can be proven true using only the methods of quantifi-cational logic, without investigating the physical world.

CHAPTER 18 GLOSSARY

Existential generalization of a singular sentence   The sentence that results if we begin with a singular sentence that contains one or more constants or John Doe names and then prefix to this an existential quantifier and uniformly replace one or more occurrences of a constant (or John Doe name) with the variable occurring in the existential quantifier.

Existential quantification   A sentence of QL that begins with an existential quantifier whose scope is the entire formula.

Instantiation (or substitution instance) of an existential quantification   The formula that results if the existential quantifier is removed from an existential quantification and each of the variables bound by that quantifier is uniformly replaced by a constant or a John Doe name.

Instantiation (or substitution instance) of a universal quantification   The formula that results if you remove the quantifier from a universal quantification and uniformly replace each occurrence of the variable it binds with a constant.

John Doe name   A temporary name assigned to an unnamed individual for the purpose of reasoning about that individual.

Singular sentence of QL   A sentence of QL containing no quantifiers and no variables.

Universal generalization of a singular sentence   The formula that results if we take a singular sentence containing one or more constants or John Doe names and prefix to it a universal quantifier term, and uniformly replace each occurrence of a constant or John Doe name in the sentence with the variable contained in the quantifier.

Universal quantification   A sentence of QL that begins with a universal quantifier whose scope covers the entire sentence.

Chapter

19

INTERPRETATIONS, INVALIDITY, AND SEMANTICS

Suppose **P** is a formula of QL and suppose **P** contains one or more predicates, one or more variables, and one or more individual constants. An **interpretation** of **P** specifies three things:

- The universe of discourse or domain over which the variables of **P** range

- Which individual objects of that universe of discourse are designated by each of the constants occurring in **P**

- Which property or relation is designated by each predicate constant occurring in **P**.

An interpretation of a quantified sentence is the analogue in quantified logic of a truth-table row or a truth-value assignment in truth-functional logic. An interpretation specifies what objects or things the formula is about and assigns meaning to the individual constants and to the predicate constants of the formula.

Here is an example of an interpretation of a QL sentence:

Sentence: (x) (Rx ⊃ Bx)

Interpretation: Let the domain be trees, let Rx abbreviate *x is a redwood* and let Bx abbreviate *x is beautiful*.

On this interpretation, the QL sentence expresses the claim that all redwood trees are beautiful. Notice that the sentence expresses a true proposition on this interpretation. However, on the following interpretation, the QL sentence is false:

Let the domain be animals, let Rx abbreviate *x is a reptile*, and let Bx abbreviate *x is blue*.

On this interpretation, the sentence expresses the claim that all reptiles are blue. Notice that a QL sentence prior to an interpretation is functioning here as if it is a sentence form rather than a sentence constant.

We may also provide an interpretation for a *set* of quantified sentences. For example, here is a set of three QL sentences and an interpretation:

1. (x)(Bx ⊃ Gx)

2. (x)(Gx ⊃ Px)

3. (x)(Bx ⊃ Px)

Domain: animals

Bx: x is a bird

Gx: x is graceful

Px: x is peaceful

On this interpretation, the three QL sentences express the following:

1. All birds are graceful.

2. All graceful things are peaceful.

3. All birds are peaceful.

Consider again the three QL sentences:

1. (x)(Bx ⊃ Gx)

2. (x)(Gx ⊃ Px)

3. (x)(Bx ⊃ Px)

If we specify that the first two QL sentences above represent the premises of an argument and the third sentence represents the argument's conclusion, the three sentences form a quantified argument, and on the current interpretation, the argument is:

1. All birds are graceful.

2. All graceful things are peaceful.

3. Therefore, all birds are peaceful.

Here is another QL argument:

1. (x) Fx

2. (x)Gx / (x) (Fx & Gx)

If we assume a universal domain, and if we let Bx abbreviate *x is beautiful* and Gx abbreviate *x is grand*, then on this interpretation this argument becomes:

1. Everything is beautiful.

2. Everything is grand.

3. Therefore, everything is beautiful and grand.

INTERPRETATIONS OF
MULTIPLY QUANTIFIED SENTENCES

Interpretations of multiply quantified sentences typically involve relations between individual things. For example:

Sentence: $(x) (\exists y) (Gxy)$

Interpretation: Let the domain be the rational numbers, and let Gxy abbreviate *x is greater than y.*

On this interpretation, the QL sentence expresses the claim that for every rational number x, there is a number y such that x greater than y. In other words, for every rational number, there is a number less than it. Notice that this sentence expresses a true proposition on this interpretation. However, consider another interpretation of the same formula:

Domain: human beings

Gxy: x is the grandmother of y

On this interpretation, $(x) (\exists y) (Gxy)$ expresses the claim that for every person x, there is a person y such that x is the grandmother of y. In other words, every person is a grandmother. The sentence comes out false on this interpretation.

We may also provide an interpretation of a set containing one or more multiply quantified sentences. For example:

Sentences: 1. $(x)(Ax \lor Bx)$
              2. $(\exists x)(Ax)$ & $(\exists y)(By)$
              3. $(\exists x)(\exists y)(Gxy)$

Interpretation:

Domain: the real numbers

Ax: x is positive

Bx: x is negative

Gxy: x is the square root of y

On this interpretation, the three sentences read:

1. Each real number is either positive or negative.

2. There exists at least one positive number and there exists at least one negative number.

3. Some number is the square root of another number.

All three sentences are true on this interpretation. However, consider another interpretation of the same set of sentences:

Domain: subatomic particles

Ax: x is a lepton

Bx: x is a gauge boson

Gxy: x is the antiparticle of y

On *this* interpretation, the three sentences read:

1. Each subatomic particle is either a lepton or a gauge boson.

2. There exists at least one lepton and there exists at least one gauge boson.

3. Some particle is the antiparticle of another particle.

All three sentences do *not* together come out true on this interpretation, because not all subatomic particles are either leptons or gauge bosons.

Consider the following two sentences:

1. (y) (∃x) Gxy

2. (∃x) (y) Gxy

If the first is designated a premise, and the second is designated a conclusion, then this is a quantified argument:

1. (y) (∃x) Gxy / (∃x) (y) Gxy

Let the domain be integers and let Gxy abbreviate *x is greater than y*. On this interpretation, we get:

1. For any integer there is a greater integer.

2. Therefore, there is an integer that is greater than any integer.

USING INTERPRETATIONS TO SHOW INVALIDITY

In everyday life, we sometimes use an analogy to show someone that their argument is invalid. For example, suppose Joe argues, "Because all squares have four sides, it follows that all four-sided figures are squares." In reply, someone might say, "Joe, that's like arguing that because all dogs are mammals it follows that all mammals are dogs." Now, this argument is similar or analogous to Joe's argument, yet this second argument is obviously invalid for it has a true premise and a false conclusion. When Joe sees the *analogy* between the form of his argument and the form of this obviously invalid one, he may agree that his argument too has an invalid form. From this he may conclude that his argument is invalid. This method of showing an argument formally invalid is sometimes called "refutation by logical analogy."

According to this method, to show someone that his or her argument has an invalid form, cite an argument that has exactly the same form but that has obviously true premises and a false conclusion. Since the new argument is obviously invalid, then an argument with exactly the same form is invalid as well, which suggests that the person's original argument is invalid.

This suggests one method of showing that a quantified English argument is invalid: Symbolize the argument in QL and then specify an interpretation that makes the premises obviously true and the conclusion obviously false. This will show that the argument is formally invalid. For example, consider the following English argument:

1. Everything has a cause.

2. Therefore, something causes everything.

In QL, assuming the universal domain, this is:

$$(x) (\exists y) \, Cyx$$

$$\text{Therefore } (\exists y) (x) \, Cyx$$

Now, let us specify a different interpretation for this QL argument. Let the domain be persons and let Cyx abbreviate y is the mother of x. On this interpretation, the QL argument symbolizes:

1. Every person has a mother.

2. Therefore, there is someone who is the mother of everyone.

Now, this argument is clearly analogous to the previous English argument. The two arguments have exactly the same logical form (since they are both interpretations of the same QL argument). But the second argument has a true premise and a false conclusion, and is therefore obviously invalid. Since this invalid argument has the same form as the original argument, the original argument has an invalid form.

For another example, consider argument A:

Argument A:

$$(\exists x) \, Fx$$

$$(\exists x) \, Gx \, / \, (\exists x) \, (Fx \, \& \, Gx)$$

Suppose Joe insists that this is a valid argument. We can show Joe that it is invalid by presenting the following interpretation:

Domain: mammals

Fx: x is an elephant

Gx: x is a lion

On this interpretation, the argument is:

Argument B:

There exists at least one elephant.

There exists at least one lion.

Therefore there exists a beast that is both a lion and an elephant. (Perhaps a "liophant.")

On this interpretation, the argument has true premises and a false conclusion. Yet, argument B has the same logical form as argument A. This interpretation shows that an argument with the same logical form as A nevertheless has true premises and a false conclusion, which shows that argument A has an invalid form. So, against argument A, we might argue: Argument A is of the same logical form as argument B, and yet B is clearly invalid, therefore argument A is invalid.

This interpretation also constitutes a *counterexample* to argument A, for the interpretation shows us the possibility that A has true premises and a false conclusion. This method is sometimes called refutation by counterexample, since it involves showing an argument invalid by producing a counterexample to the argument. Let us examine a few more examples. Consider this argument:

$$(x) (Fx \supset Gx)$$
$$(x) (Fx \supset Hx)$$
$$/(x) (Gx \supset Hx)$$

An interpretation that shows this to be invalid is

Domain: automobiles

Fx: x is a Model T

Gx: x is a Ford

Hx: x is black

On this interpretation, the argument reads

All Model T's are Fords.

All Model T's are black.

Therefore, all Fords are black.

Here is an analogous argument:

All sound arguments are valid.

All sound arguments have true premises.

So, all valid arguments have true premises.

For one more example, the following argument is invalid:

$$(x) [(Ax \lor Bx) \supset Fx]$$
$$(\exists x) (Ax \ \& \ Fx)$$
$$(\exists x) (Bx \ \& \ Fx) \ / \ (x) (Fx \supset (Ax \lor Bx))$$

In order to see why, consider this interpretation:

Domain: subatomic particles

Ax: x is a proton

Bx: x is a neutron

Fx: x is a baryon

This interpretation makes the two premises true, for protons and neutrons are baryons. However, the conclusion comes out false on this interpretation, for the conclusion now says that anything that is a baryon is a proton or a neutron. But this is false, for protons and neutrons are not the only types of baryons in existence. Because we have found an interpretation that makes the premises true and the conclusion false, the argument is invalid.

▶   **EXERCISE 19.1**   ◀

Part I. Specify an interpretation showing that each QL argument below is formally invalid.

1. *(x) (Ax $\supset$ Bx) / (x) (Bx $\supset$ Ax)

2. (x) (Ax $\supset$ Bx)

(x)(Bx) / (x) Ax

3. (x)(Ax $\supset$ Bx)

(x) ~ Ax / (x) ~ Bx

4. (x) (Ax $\supset$ Bx)

($\exists$x) Ax / (x) Bx

5. *(x) (Ax $\lor$ Bx)

(x) Ax / (x) Bx

6. (x) (Ax ⊃ Bx) / (x) (Bx ⊃ Ax)

7. (x) (Ax ⊃ Bx) / (x) (~ Ax ⊃ ~ Bx)

8. (x) (Ax ⊃ Bx)

(∃x) Bx / (x) Ax

9. (∃x) Ax

(∃x) Bx / (∃x)(Ax & Bx)

10. *(∃x) (Ax v Bx)

(∃x) Ax / (∃x) Bx

11. (x) (Ax ⊃ Bx)

(x) (Cx ⊃ ~Ax) / (x) (Cx ⊃ ~Bx)

12. (∃x) Fx v (x) Gx

~Gb / Fb

13. (x) (Ax v Bx)

(x) (Bx ⊃ Cx) / (∃x)(Cx ⊃ Ax)

14. (x) (Gx ⊃ ~Hx)

(x) (Ax ⊃ ~Gx) / (x) (Ax ⊃ ~Hx)

15. (x) (Ax ⊃ Bx) / (x) Bx

16. (x) (Ax v Bx) / (x) Ax v (x) Bx

Part II. Each of the following arguments is invalid. For each argument, supply a refutation by logical analogy.

1. *Some extraterrestrials are green. Some green beings are friendly. So, some extraterrestrials are friendly.

2. No widgets are expensive. No expensive things are old. So, no widgets are old.

3. All aardvarks are mammals. Some mammals are old. Therefore, some aardvarks are old.

4. All actors are rich. All famous people are rich. So, all actors are famous.

5. *All widgets are round. No widgets are green. So, no green things are round.

6. All widgets are round. No widgets are green. So, no round things are green.

7. All human beings have free will. All beings that have free will are rational beings. So, all rational beings are human beings.

8. All humans are sinners. Some sinners are happy. So, some happy beings are humans.

SEMANTICS

Using the idea of an interpretation, we can now specify some important definitions.

- A QL sentence is a logical truth if and only if it is true on every interpretation.

- A QL sentence is a logical falsehood if and only if it is false on every interpretation.

- A QL argument is valid if and only if there is no interpretation on which its premises would be true and its conclusion false.

- A QL argument is invalid if and only if there exists at least one interpretation on which the premises would be true and the conclusion would be false.

- Two QL sentences are equivalent if and only if there is no interpretation that would make one true but the other false.

In the above definitions, we must presuppose one important qualification: every interpretation involves a *nonempty* universe of discourse. The case of an empty universe—a universe containing no individuals or objects—must be excluded or else absurd logical results will follow. For instance, *if* we assumed an empty universe of discourse, we would have to say that the following argument, which is clearly valid, has a true premise and a false conclusion:

$$(x) \ Fx$$

$$(\exists x) \ Fx$$

This argument is invalid on an empty universe interpretation because the premise is equivalent to $\sim (\exists x) \sim Fx$, and this is true on an empty universe interpretation. Yet, the conclusion is false on the empty universe interpretation. Because this constitutes an interpretation that makes the premise true and the conclusion false, the argument is invalid, assuming our definition of validity allows the possibility of an empty universe. Yet, the argument seems intuitively valid. In order to avoid this paradox, we must stipulate that the definitions above speak only of interpretations involving membered universes—universes that contain at least one individual or object.

THE MONADIC PREDICATE TEST

If a quantificational argument contains only monadic predicates, it is possible to translate the argument into a purely truth-functional form containing no quantifiers. When this is accomplished, a truth-functional test for validity may be performed. Thus, for a limited class of quantified arguments, namely, monadic predicate arguments, truth-functional decision procedures exist. Let us proceed in stages.

Assume that the variable x ranges over a finite domain, one that consists of only the objects a, b, c, d, e. If we say that *every x is F,* where F is a monadic predicate, we are saying that each and every member of the domain is characterized by F, which is to say

Fa & Fb & Fc & Fd & Fe

Assuming a three-member domain that consists of just the individuals a, b, and c, we may translate (x) Fx into

Fa & Fb & Fc

Call this truth-functional formula a "truth-functional expansion" of the quantified formula.

Next, suppose instead that at least one member of the same domain is F. That is, assume, $(\exists x)$ Fx. This is to suppose that at least one of a, b, c is characterized by F. In symbols, this is

Fa v Fb v Fc

Consequently, assuming this three member domain, $(\exists x)$ Fx may be translated into the following truth-functional expansion:

Fa v Fb v Fc

Consider two more cases for a three-member domain:

Formula:	Expansion:
(x)(Fx ⊃ Gx)	(Fa ⊃ Ga) & (Fb ⊃ Gb) & (Fc ⊃ Gc)
$(\exists x)$(Sx & Hx)	(Sa & Ha) v (Sb & Hb) v (Sc & Hc)

A general principle has been proven for quantificational arguments, although we won't provide the proof here. That principle is this. Suppose a quantificational argument contains no polyadic predicates and n different monadic predicates. Given that the number of monadic predicate letters in the argument is n, if there exists no interpretation yielding true premises and a false conclusion for a domain of 2^n objects or individuals, the argument is valid.

This result makes possible a decision procedure, a mechanical test, for any quantificational argument that contains only monadic predicates. For any such argument, we simply go to the case of a 2^n universe or domain, where n stands for the number of monadic predicates in the argument, we construct the appropriate truth-functional expansions, and we mechanically test for validity using the methods of truth-functional logic.

However, it has also been proven that there exists no such mechanical procedure for quantificational arguments that contain two-place or higher predicates. That is, no decision procedures exist for quantificational arguments containing polyadic predicates. This is not to say that a decision procedure for such arguments hasn't been discovered. Rather, it has been proven that no such procedure exists.

Let us test a monadic predicate argument for validity using the method of truth-functional expansions. Consider the following argument:

1. $(\exists x) (Sx \,\&\, Px)$

2. $(\exists x) (\sim Sx \,\&\, Px)$

3. therefore $(x)\, Px$

The test for validity requires a truth-functional expansion in a domain of 2^n individuals, where n represents the number of different monadic predicates appearing in the argument. Because the argument contains two such predicates, the domain must contain four individuals. The argument's truth-functional expansion for a four-member universe is

Premise 1: $(Sa \,\&\, Pa)\; v\; (Sb \,\&\, Pb)\; v\; (Sc \,\&\, Pc)\; v\; (Sd \,\&\, Pd)$

Premise 2: $(\sim Sa \,\&\, Pa)\; v\; (\sim Sb \,\&\, Pb)\; v\; (\sim Sc \,\&\, Pc)\; v\; (\sim Sd \,\&\, Pd)$

Conclusion: $Pa \,\&\, Pb \,\&\, Pc \,\&\, Pd$

(We are omitting some of the larger brackets in order to simplify the expression of the formulas.)

A *partial truth-table test* (Chapter 5) on these expansions would prove that the argument is invalid. If Sa, Pa, and Pb are assigned **T**, and Sb and Pc are assigned **F**, the premises are true and the conclusion is false. This shows that it is possible the argument has true premises and a false conclusion, which shows that the argument is invalid.

Here are two more examples. Consider the following argument:

1. $(x) \sim (Hx \,\&\, \sim Rx) \,/\, (x) \sim (Rx \,\&\, \sim Hx)$

The premise expansion is

$\sim (Ha \,\&\, \sim Ra) \,\&\, \sim (Hb \,\&\, \sim Rb) \,\&\, \sim (Hc \,\&\, \sim Rc) \,\&\, \sim (Hd \,\&\, \sim Rd)$

The conclusion's expansion is

$\sim (Ra \,\&\, \sim Ha) \,\&\, \sim (Rb \,\&\, \sim Hb) \,\&\, \sim (Rc \,\&\, \sim Hc) \,\&\, \sim (Rd \,\&\, \sim Hd)$

If Ha, Hb, Hc, and Hd are assigned **F**, and Ra, Rb, Rc, and Rd are assigned **T**, the premise is true and the conclusion false, which shows that the argument is invalid.

Here is another example:

2. $(\exists x) (Hx \,\&\, Gx) \,/\, (x) (Hx \,\&\, Gx)$

The premise expansion is

$(Ha \,\&\, Ga)\; v\; (Hb \,\&\, Gb)\; v\; (Hc \,\&\, Gc)\; v\; (Hd \,\&\, Gd)$

The conclusion's expansion is

$(Ha \,\&\, Ga) \,\&\, (Hb \,\&\, Gb) \,\&\, (Hc \,\&\, Gc) \,\&\, (Hd \,\&\, Gd)$

If Ha is assigned **T**, Ga is assigned **T**, and Hb and Gb are assigned **F**, the premise is true and the conclusion is false.

▶   **EXERCISE 19.2**   ◀

Construct truth-functional expansions showing that the following QL arguments are invalid.

1.* (∃x) (Ax & Px) / (x) ~ (Ax & ~ Px)

2. (∃x) (Ax & Jx) / (∃x) (Ax & ~ Jx)

3. (x) Hx / (∃x) (Hx & Gx)

4. (x) (Ax ⊃ Bx) / (x) (Bx ⊃ Ax)

5. (x) (Ax ⊃ Bx) / (x) (~Ax ⊃ ~ Bx)

6. (x) (Ax ⊃ Bx)

    (∃x) Bx / (x) Ax

7. (x) (Ax ⊃ Bx)

    (∃x) Ax / (x) Bx

8. (x) (Ax v Bx) / (x) Ax v (x) Bx

9. (x) (Ax ⊃ Bx)

    (x) Bx / (x) Ax

CHAPTER 19 GLOSSARY

Counterexample to a QL argument An interpretation under which the premises of the argument are true and the conclusion false.

Interpretation of a QL sentence A specification of three things: (1) the universe of discourse or domain over which the variables of the sentence range, (2) which individual objects of that universe of discourse are designated by each of the constants occurring in the sentence, and (3) which property or relation is designated by each predicate constant occurring in the sentence.

Refutation by logical analogy To show that a particular argument is formally invalid: produce an argument that has exactly the same logical form as the original argument but that also has obviously true premises and a false conclusion. Because the second argument is obviously invalid, and because it displays the same logical form as that displayed by the first argument, the form in question must not be a valid form. This way of showing an argument formally invalid is also sometimes called refutation by counterexample since the logically analogous yet invalid argument is a counterexample to the original argument.

Chapter

20

CONDITIONAL AND INDIRECT QUANTIFIER PROOFS

et us begin with proofs that contain no overlapping quantifiers and that require only the truth-functional *inference* rules, the QL rules, IP, and CP. In this first example, we employ the Conditional Proof rule.

1. (∃x)(Gx) ⊃ (y) (Gy ⊃ Jy) / (x) (Gx ⊃ Jx)

2.	Gu	AP
3.	(∃x) Gx	EG 2
4.	(y) (Gy ⊃ Jy)	MP 1, 3
5.	Gu ⊃ Ju	UI 4
6.	Ju	MP 2, 5

7. Gu ⊃ Ju   CP 2–6

8. (x) (Gx ⊃ Jx)   UG 7

Notice that UG was applied (on line 8) only *after* the assumed premise was discharged. This follows the third clause of the UG rule. We could not have generalized upon the John Doe name u before the CP assumption had been discharged.

Next, consider this argument:

1. (x) (Ax ⊃ B x)   / (x)(Ax) ⊃ (∃y) By

2.	(x)(Ax)	AP
3.	Ac	UI 2
4.	Ac ⊃ Bc	UI 1
5.	Bc	MP 3, 4

6. | (∃y) By   EG 5

7. (x) (Ax) ⊃ (∃y) By   CP 2–6

Here we assumed the *antecedent* of the conclusion, (x) Ax, derived the *consequent* of the conclusion, (∃y) By, and asserted the conclusion after dis-indenting.

The next proof contains a nested CP:

1. (x) [Gx ⊃ (Jx v Mx)]

2. (x) [(Ax & Gx) ⊃ ~ Jx]   / (x) [G x ⊃ (Ax ⊃ Mx)]

3. Gu ⊃ (Ju v Mu)   UI 1

4. (Au & Gu) ⊃ ~ Ju   UI 2

5. | Gu   AP

6.   Au   AP

7.   Ju v Mu   MP 3, 5

8.   Au & Gu   Conj 5, 6

9.   ~ Ju   MP 4, 8

10.   Mu   DS 7, 9

11.   Au ⊃ Mu   CP 6–10

12. Gu ⊃ (Au ⊃ Mu)   CP 5–11

13. (x) [Gx ⊃ (Ax ⊃ Mx)]   UG 12

Notice again that UG was applied (on line 13) only *after* the assumed premise was discharged. This follows the second clause of the UG rule. We could not have generalized upon the John Doe name u before the assumption had been discharged.

The following proof employs the Indirect Proof rule:

1. (x) (Ax ⊃ Dx)

2. (∃x)Ax / (∃x) Dx

3. | ~ (∃x)Dx   AP

4. | (x) ~ Dx   QE 3

5. | Av   EI 2

6. | Av ⊃ Dv   UI 1

7. | Dv   MP 5, 6

8. | ~Dv   UI 4

9. | Dv & ~ Dv   Conj 7, 8

10. (∃x)Dx   IP 3–9

ADDING TRUTH-FUNCTIONAL
REPLACEMENT RULES TO THE MIX

The following indirect proof uses DeMorgan's rule at line 7.

1. ~ (∃x)Bx	/ (∃x)(~ Bx v Hx)	
2.	~ (∃x)(~ Bx v Hx)	AP
3.	(x) ~ (~ Bx v Hx)	QE 2
4.	(x) ~ Bx	QE 1
5.	~ Bd	UI 4
6.	~ (~ Bd v Hd)	UI 3
7.	~ ~ (~ ~ Bd & ~ Hd)	DM 6
8.	~ ~ Bd & ~ Hd	DNeg 7
9.	~ ~Bd	Simp 8
10.	Bd	DNeg 9
11.	Bd & ~Bd	Conj 5, 10
12. (∃x) (~ Bx v Hx)	IP 2–11	

Here is a CP proof also employing the ever-popular DeMorgan rule:

1. (x) (Ax ⊃ Gx)		
2. (x) (Bx ⊃ Ax)		
3. (x) ~ (Dx & ~ Bx)	/ (x) (Dx ⊃ Gx)	
4.	Du	AP
5.	~ (Du & ~ Bu)	UI 3
6.	~ ~ (~ Du v ~ ~ Bu)	DM 5
7.	~ Du v ~ ~ Bu	DNeg 6
8.	~ Du v Bu	DNeg 7
9.	~ ~ Du	DNeg 4
10.	Bu	DS 8, 9
11.	Bu ⊃ Au	UI 2
12.	Au ⊃ Gu	UI 1
13.	Bu ⊃ Gu	HS 11, 12
14.	Gu	MP 10, 13
15. Du ⊃ Gu	CP 4–14	
16. (x) (Dx ⊃ Gx)	UG 15	

Here is a fairly complex indirect proof:

1. (x) ~ (Cx & ~ Bx)

2. (x) ~ (Bx & ~Ax) / (x)~ (Cx & ~ Ax)

3.	~ (x)~ (Cx & ~ Ax)	AP
4.	(∃x) (Cx & ~ Ax)	QE 3
5.	Cv & ~ Av	EI 4
6.	~ (Cv & ~ Bv)	UI 1
7.	~ (Bv & ~Av)	UI 2
8.	~ ~ (~ Cv v ~ ~Bv)	DM 6
9.	~ Cv v Bv	DNeg 8 (twice)
10.	~ ~ (~Bv v ~ ~ Av)	DM 7
11.	~Bv v Av	DNeg 10 (twice)
12.	Cv	Simp 5
13.	~ ~ Cv	DNeg 12
14.	Bv	DS 13, 9
15.	~ ~ Bv	DNeg 14
16.	Av	DS 15, 11
17.	~Av	Simp 5
18.	Av & ~Av	Conj 16,17

19. (x)~ (Cx & ~ Ax)   IP 3-18

► **EXERCISE 20.1** ◄

Use either the Conditional Proof rule or the Indirect Proof rule as you prove each of the following.

(1)* 1. (x) (Ax ⊃ ~ Bx)

2. (x) [Bx ⊃ (Hx & Ax)] / (∃x) (~ Bx)

(2) 1. (x) (Hx ⊃ Sx)

2. (x) (Sx ⊃ Gx) / (x) [Hx ⊃ (Sx & Gx)]

(3) 1. (x) [Ax ⊃ (Bx & Cx)] / (x) (Ax ⊃ Cx)

(4) 1. (x)[Ax ⊃ (Bx v Cx)] / (∃x) (Ax) ⊃ (∃x) (Bx v Cx)

(5) * **1.** (x) (Hx ⊃ Qx)

    2. (x) (Hx ⊃ Rx) / (x) [Hx ⊃ (Qx & Rx)]

(6) **1.** (x) [Hx ⊃ (Qx & Sx)]

    2. (x) [Px ⊃ (Rx & Mx)] / (x)(Sx ⊃ Px) ⊃ (x) (Hx ⊃ Mx)

(7) * **1.** (x)[Px ⊃ (Hx & Qx)] / (x)(Sx ⊃ Px) ⊃ (x) (Sx ⊃ Qx)

(8) **1.** (x) ~ (Hx & ~ Bx)

    2. (x) ~ (Bx & ~ Gx) / (x) ~ (Hx & ~ Gx)

(9) **1.** (x) [(Fx v Gx) ⊃ Px]

    2. (∃x) (~ Fx v Sx) ⊃ (x)(Rx) / (x) (Px) v (x) (Rx)

(10) * **1.** (∃x) (Px) v (∃x)(Qx & Rx)

    2. (x) (Px ⊃ Rx) / (∃x) (Rx)

(11) **1.** (∃x) (Fx) ⊃ (x) (Gx ⊃ Sx)

    2. (∃x) (Hx) ⊃ (x) (~ Sx) / (x) [(Fx & Hx) ⊃ ~ Gx]

(12) **1.** (x) (Jx ⊃ Px)

    2. (x) (Hx ⊃ Mx) / (∃x) (Jx v Hx) ⊃ (∃x)(Px v Mx)

(13) * **1.** (x) [(Hx v Px) ⊃ Qx]

    2. (x) [(Qx v Mx) ⊃ ~Hx] / (x) (~ Hx)

(14) **1.** (∃x) (Qx) ⊃ (x)(Sx)

    2. Qa ⊃ ~ Sa / ~ Qa

(15) * **1.** (∃x) (Px v Jx) ⊃ ~ (∃x) (Px) / (x) (~ Px)

(16) **1.** (∃x) (Hx) ⊃ (∃x) (Sx & Fx)

    2. ~ (∃x) (Fx) / (x) (~ Hx)

(17) **1.** (∃x) (Sx) ⊃ (∃x)(Hx & Jx)

    2. (x) (Fx ⊃ Sx) / (∃x) (Fx) ⊃ (∃x) (Hx)

(18) **1.** (∃x) (Px) ⊃ (∃x) (Qx & Sx)

    2. (∃x) (Sx v Hx) ⊃ (x) (Gx) / (x) (Px ⊃ Gx)

(19) **1.** (x) (Ax ⊃ Bx)

    2. Ab v Ac / (∃x) Bx

PUTTING QD ON A DIET:
A REDUCED SET OF QUANTIFIER RULES

We have been working with truth-functional rules, five quantifier rules, plus CP and IP. If a quantified argument is valid, it can be proven valid with these rules. However, all the valid arguments that can be proven valid with the rules we have been employing can also be proven valid using a reduced set of rules, namely, the set that results if we drop UG. However, if we drop this rule and work with the simpler system, the proofs become more complex and in many cases more difficult to complete. In most cases, we are forced to employ IP. For example, compare the following two proofs. The first proof employs UG:

1.	(x) Gx	
2.	(x) Sx	/ (x) (Gx & Sx)
3.	Gu	UI 1
4.	Su	UI 2
5.	Gu & Su	Conj 3, 4
6.	(x) (Gx & Sx)	UG 5

The second proof employs IP *in lieu of* UG:

1.	(x) Gx	
2.	(x) Sx	/ (x) (Gx & Sx)
3.	~ (x) (Gx & Sx)	AP
4.	(∃x) ~ (Gx & Sx)	QE 3
5.	~ (Gv & Sv)	EI 4
6.	~ ~ (~ Gv v ~ Sv)	DM 5
7.	~ Gv v ~ Sv	DNeg 6
8.	Sv	UI 2
9.	~Sv v ~Gv	Comm 7
10.	~ ~ Sv	DNeg 8
11.	~Gv	DS 9, 10
12.	Gv	UI 1
13.	Gv & ~ Gv	Conj 11, 12
14.	(x) (Gx & Sx)	IP 3–13

Notice that it takes a bit more work to derive the same result when we use IP in lieu of UG. But both proofs "work," that is, both proofs succeeded in proving the argument valid.

Here is one more example. The next proof employs UG:

1. (x)(Fx ⊃ Gx)

2. (x)(Gx ⊃ Sx) / (x)(Fx ⊃ Sx)

3. Fu ⊃ Gu   UI 1

4. Gu ⊃ Su   UI 2

5. Fu ⊃ Su   HS 3, 4

6. (x) (Fx ⊃ Sx)   UG 5

Now, if we drop UG from our tool kit of rules, we can still prove the argument valid, but we must use IP:

1. (x)(Fx ⊃ Gx)

2. (x)(Gx ⊃ Sx) / (x)(Fx ⊃ Sx)

3.	~ (x)(Fx ⊃ Sx)	AP
4.	(∃x)~ (Fx ⊃ Sx)	QE 3
5.	~ (Fv ⊃ Sv)	EI 4
6.	~ (~ Fv v Sv)	Imp 5
7.	~ ~ Fv & ~ Sv	DM 6
8.	Fv & ~ Sv	DNeg 7
9.	Fv ⊃ Gv	UI 1
10.	Gv ⊃ Sv	UI 2
11.	Fv	Simp 8
12.	~ Sv	Simp 8
13.	Gv	MP 9, 11
14.	~ Gv	MT 10, 12
15.	Gv & ~ Gv	Conj 13, 14

16. (x)(Fx ⊃ Sx)   IP 3–15

PROVING LOGICAL TRUTHS

Recall that within truth-functional logic, a formula is proven tautological or logically true if it can be proven using a premise-free proof. Formulas proven true with premise-free proofs are called logical truths or are said to be logically true because we can prove them true using purely logical procedures and without relying on scientific experiments or observations of the physical world.

If a quantified formula can be proven using a premise-free proof, it is a logical truth of quantificational logic. It is also said to be **quantificationally true**. A premise-free proof must start, of course, with either a conditional proof assumption or an indirect proof assumption.

Here, the formula (x) (Fx) ⊃ (∃x) (Fx) is proven logically true:

1.	(x)Fx	AP
2.	Fa	UI 1
3.	(∃x)Fx	EG 2
4. (x)(Fx) ⊃ (∃x)(Fx)	CP 1–3	

The formulas (∃x) (Ax ⊃ Ax) and (x) (Ax ⊃ Ax) are each obviously logically true. Notice how similar their proofs are:

1.	~ (∃x)(Ax ⊃ Ax)	AP
2.	(x) ~ (Ax ⊃ Ax)	QE 1
3.	~ (Aa ⊃ Aa)	UI 2
4.	~ (~Aa v Aa)	Imp 3
5.	~ ~Aa & ~Aa	DM 4
6.	Aa & ~Aa	DNeg 5
7. (∃x)(Ax ⊃ Ax)	IP 1–5	

1.	~ (x)(Ax ⊃ Ax)	AP
2.	(∃x) ~ (Ax ⊃ Ax)	QE 1
3.	~ (Av ⊃ Av)	EI 2
4.	~ (~Av v Av)	Imp 3
5.	~ ~Av & ~Av	DM 4
6.	Av & ~Av	DNeg 5
7. (x)(Ax ⊃ Ax)	IP 1–6	

▶   **EXERCISE 20.2**   ◀

Prove the following logical truths.

1. *(x)[(Hx v Px) v ~ Px]

2. (x) (Px v Qx) v (∃x) (~ Px v ~ Qx)

3. ~ (x) (Ax & Bx) ⊃ (∃x) (~ Ax v ~ Bx)

4. (∃x) (Px) v (x) (~ Px)

5.*(x) (Px) ⊃ ~ (∃x) (~ Px)

6. ~ (∃x) (Px & ~Px)

7.*(∃x) (Fx & Gx) ⊃ [(∃x) (Fx) & (∃x) (Gx)]

8. [(x) (Fx) & (x) (Gx)] ⊃ (x) (Fx & Gx)

9. (x) (Sx) v (∃x)(~ Sx)

10.*(x) [(Sx & Hx) ⊃ (Sx v Hx)]

11. (x) (Hx) ⊃ (∃x) (Hx)

12. (x)(Sx ⊃ Px) ⊃ [(∃x)(Sx) ⊃ (∃x)(Px)]

Chapter

21

PROOFS WITH OVERLAPPING QUANTIFIERS

W hen we deal with arguments that contain adjacent or overlapping quantifiers, we remove and add quantifiers one at a time. Examine the following proof:

1. (∃x)(∃y) (Ax & Bxy) / (∃x)Ax

2. (∃y) (Av & Bvy)   EI 1

3. Av & Bvv'   EI 2

4. Av   Simp 3

5. (∃x)Ax   EG 4

In this proof, at line 2, we applied EI to line 1. This involved two actions: we stripped away from line 1 the first existential quantifier (∃x), and then we replaced the variable x with the John Doe name v. At line 3, we applied EI to line 2, removed the quantifier (∃y), and replaced the variable y with the John Doe name v'. The rest of the proof was simple.

Here is another example of a proof containing overlapping quantifiers:

1. (x)(y) Kxy / Kab

2. (y) Kay   UI 1

3. Kab   UI 2

At line 2, we applied UI to line 1 and removed only the universal quantifier (x) and the x variable, instantiating with the constant a. Notice that we left the y variable and its quantifier untouched. Then, at line 3, we applied UI to line 2 and removed the (y) quantifier and instantiated y with b.

Here are some further examples for your inspection:

1. (x) (∃y) (Kxy)

2. (∃y)(Kay) ⊃ Ga / Ga

3. (∃y) Kay   UI 1

4. Ga   MP 2, 3

1. (∃x) (∃y) (Fxy) / (∃y) (∃x) (Fxy)

2. (∃y) Fvy   EI 1

3. Fvv′   EI 2

4. (∃x) Fxv′   EG 3

5. (∃y) (∃x) (Fxy)   EG 4

1. (x)(y) (Hxy)/ (y) (x) (Hxy)

2. (y) Huy   UI 1

3. Huu′   UI 2

4. (x) Hxu′   UG 3

5. (y)(x) Hxy   UG 4

The last two proofs illustrate a point made in Chapter 17, namely, that when two quantifiers of the same type appear next to each other, their order doesn't affect the truth-value of the sentence. However, as we also saw in Chapter 17, if universal and existential quantifiers appear next to each other, the order cannot arbitrarily be altered, for the order affects the truth-value of the sentence. That is, (x) (∃y) (Fxy) does not imply (∃y) (x) (Fxy).

Here is another example:

1. (x) (y) (Fxy ⊃ ~ Fyx)

2. (∃x) (∃y) Fxy / (∃x) (∃y) ~ Fyx

3. (∃y) Fvy   EI 2

4. Fvv′   EI 3

5. (y) (Fvy ⊃ ~ Fyv)   UI 1

6. Fvv′ ⊃ ~Fv′v   UI 5

7. ~ Fv′v   MP 4, 6

8. (∃y) ~ Fyv   EG 7

9. (∃x) (∃y) ~ Fyx   EG 8

When you work with sentences that contain overlapping quantifiers—sentences in which one quantifier appears within the scope of another quantifier—you can apply the Quantifier Exchange rule in the usual way. The following proof illustrates the proper procedure:

1. ~ (x) (∃y) (Hxy) / (∃x) (y) (~ Hxy)

2. (∃x) ~ (∃y) (Hxy)   QE 1

3. (∃x) (y) ~ (Hxy)   QE 2

Notice that as the tilde passes over each quantifier, it converts the quantifier it passes over into the opposite quantifier.

Here is a proof that combines QE with IP:

1. (x) (∃y) (Axy) ⊃ (x)(∃y) Gxy

2. (∃x)(y) ~ Gxy   / (∃x)(y) ~ Axy

    3.   ~ (∃x)(y) ~ Axy   AP

    4.   (x) ~ (y) ~ Axy   QE 3

    5.   (x) (∃y) Axy   QE 4

    6.   (x) (∃y) Gxy   MP 1, 5

    7.   (y) ~ Gvy   EI 2

    8.   (∃y) Gvy   UI 6

    9.   Gvv′   EI 8

   10.   ~ Gvv′   UI 7

   11.   Gvv′ & ~ Gvv′   Conj 9, 10

12. (∃x)(y) ~ Axy   IP 3–11

Notice that we applied EI before we applied UI. The first application of EI, at line 7, replaced the variable x with the John Doe name v. Next, at line 8, we applied UI and again instantiated with v. At line 9, we again applied EI, but this time we used a new John Doe name—v′. The next application of UI also used v′. This allowed us to derive the contradiction at line 11.

The following proof uses the CP strategy:

1.   (x) (∃y) (Fxy) / Pa ⊃ (∃y) Fay

    2.   Pa   AP

    3.   (∃y) Fay   UI 1

4.   Pa ⊃ (∃y) Fay   CP 2-3

We began by this proof by assuming the antecedent of the conclusion. At line 3, UI was applied to line 1: The universal quantifier was stripped away and the x variable was replaced by the constant a. This produced the consequent of the conclusion. We discharged our assumption and asserted the conclusion by CP.

PROPERTIES OF RELATIONS

In the appendix to Chapter 17 we examined the properties of reflexivity, transitivity, and symmetry. Let's put our understanding of transitivity to work. Suppose we wish to prove valid the following argument.

1. Carol's car is heavier than Pete's car.

2. Pete's car is heavier than Katie's car.

3. So, Carol's car is heavier than Katie's car.

If we let c designate Carol's car, p designate Pete's car, and k designate Katie's car, this argument goes into QL as

1. Hcp

2. Hpk / Hck

The English argument is obviously valid. However, the QL argument can't be proven valid with our rules. The problem is that the argument is only valid if we assume that being "heavier than" is a transitive relation. We subconsciously add this assumption to the English version of the argument when we evaluate that, which is why the English version seems obviously valid even though the QL version cannot be proven valid. Let us fix up the QL version by adding to the argument a premise stating that *heavier than* is a transitive relation:

1. Hcp

2. Hpk

3. (x) (y) (z) [(Hxy & Hyz) ⊃ Hxz] / Hck

The third premise says that if one thing is heavier than a second thing, and if the second thing is heavier than a third thing, the first is heavier than the third. Now, the argument can be proven valid.

4.	(y) (z) [(Hcy & Hyz) ⊃ Hcz]	UI 3
5.	(z)[(Hcp & Hpz) ⊃ Hcz]	UI 4
6.	(Hcp & Hpk) ⊃ Hck	UI 5
7.	(Hcp & Hpk)	Conj 1, 2
8.	Hck	MP 6, 7

The following argument provides another illustration that involves multiple quantification:

> *Goober knows Betty. Betty does not know Goober. So, it's not the case that if one person knows another, then the second person knows the first person.*

In QL, this is represented as follows (assuming a domain limited to persons):

1. Kgb

2. ~Kbg / ~(x) (y) (Kxy ⊃ Kyx)

The proof follows.

1. Kgb

2. ~Kbg / ~ (x) (y) (Kxy ⊃ Kyx)

3. | ~ ~ (x) (y) (Kxy ⊃ Kyx)   AP

4. | (x) (y) (Kxy ⊃ Kyx)   DNeg 3

5. | (y) (Kgy ⊃ Kyg)   UI 4

6. | Kgb ⊃ Kbg   UI 5

7. | Kbg   MP 1, 6

8. | Kbg & ~ Kbg   Conj 2, 7

9. ~ (x) (y) (Kxy ⊃ Kyx) IP 3–8

►   **EXERCISE 21.1**   ◄

Part A. Prove the following arguments valid using natural deduction.

(1)* 1. (x) (y)(Hxy ⊃ ~ Hyx)

       2. (∃x) (∃y) (Hxy) / (∃x) (∃y) (~ Hyx)

(2) 1. (∃x) (y) (Sxy) / (y) (∃x) (Sxy)

(3) 1. (x)(Fx ⊃ Bx)

       2. (x) (∃y) (Sxy v ~ Bx) / (∃x) (∃y) (Sxy v ~ Fx)

(4) 1. (∃x) (y) (Hyx ⊃ ~ Hxy) / ~ (x) (Hxx)

(5) 1. (x) (∃y) (Sx & Py) / (∃y) (∃x) (Sx & Py)

(6) 1. (x) (∃y) (Jx v Ry) / (∃y) (∃x) (Jx v Ry)

(7) * **1.** (x) (∃y) (Hxy) ⊃ (x) (∃y) (Sxy)

      2. (∃x) (y) (~Sxy) / (∃x) (y) (~ Hxy)

(8) **1.** (∃x) (y) (Pxy ⊃ Sxy)

      2. (x) (∃y) (~ Sxy) / ~ (x) (y) (Pxy)

(9) **1.** (∃x) (y) Mxy / (x) (∃y) (Myx)

(10) **1.** (x) (y) [(Wx & Lxy) ⊃ Lya]

      2. (x) (y) (Lxa ⊃ Lxy) / (x) (y) [(Wx & Lxy) ⊃ Lyx]

(11) **1.** (x) (∃y) (Px ⊃ Wy) / (x) (Px) ⊃ (∃y) (Wy)

(12) * **1.** (∃x) (∃y) (Kxy) / (∃y) (∃x) (Kxy)

Part B. Symbolize the following arguments and prove each valid using natural deduction.

1.* Every material thing has a cause. Not everything has a cause. Therefore, at least one nonmaterial thing exists.

2. Every person loves that which he or she makes. Each person makes his or her own enemies. So, each person loves his or her enemies.

3. Everything has a cause. If the universe has a cause, then a Creator of the universe exists. So, a Creator of the universe exists.

4. Everything has a cause. If God exists, then something does not have a cause. (Note: God could not have a cause.) So, God does not exist.

5. Ann is taller than Bob. Bob is taller than Ed. If x is taller than y, and y is taller than z, then x is taller than z. So, Ann is taller than Ed.

6. Pat is older than Ned. If x is older than y, then y is not older than x. Therefore, Ned is not older than Pat.

22

PROOFS WITH
IDENTITY

We introduced the identity sign in Chapter 17. In order to work with arguments containing identity signs, we will need natural deduction rules applicable to the identity operator. The rules for identity will be based on the two principles outlined in the following box:

The Principle of Self-Identity

Each thing is identical with itself

In symbols:

$$(x) (x = x)$$

This principle is a necessary truth that needs no argument.

The Principle of the Indiscernibility of Identicals

If x is identical with y, then whatever is true of x is true of y and whatever is true of y is true of x.

For example, because Robert Zimmerman *is* Bob Dylan, if it is true that Bob Dylan sang with Joan Baez, it is true that Robert Zimmerman sang with Joan Baez. If it is true that Robert Zimmerman was born in Minnesota, it is true that Bob Dylan was born in Minnesota, and so on.

The first rule for identity, shown in the following box, is based on the self-identity principle.

Identity A (Id A)

At any step in a proof, you may assert (x) (x = x).

Typically, after you have used this rule, you will want to use UI and instantiate $(x)(x = x)$ using the appropriate constant or John Doe name. Let us consider an example. Suppose you want to prove that the following argument is valid:

1. Pat is a neighbor of Chris, and Bob is a neighbor of Chris.

2. Therefore, someone is a neighbor of Chris, and that person is Bob.

The proof, using obvious abbreviations, employs Identity A:

1. Npc & Nbc / $(\exists x) [(Nxc) \& (x = b)]$

2. Nbc   Simp 1

3. $(x) (x = x)$ Id A

4. b = b   UI 3

5. Nbc & (b = b)   Conj 2, 4

6. $(\exists x) [(Nxc) \& (x = b)]$ EG 5

The second rule for identity, shown in the following box, is based on the indiscernibility principle.

Identity B (Id B)

If **c** and **d** are two constants or John Doe names and a line of a proof asserts that the individual designated by **c** is identical with the individual designated by **d**, you may carry down and rewrite any available line of the proof replacing any or all occurrences of **c** with **d** or any or all occurrences of **d** with **c**. A line of a proof is available unless it is within the scope of a discharged assumption.

The Identity B rule can be expressed formally as follows:

1. c = d

2. P

3. Infer: **P** [**c** // **d**]

where **P** is an available line of the proof containing one or more occurrences of one of the constants or John Doe names and **P** [**c** // **d**] is exactly like **P** except that one or more occurrences of **c** have been replaced with **d** or one or more occurrences of **d** have been replaced with **c**.

Here is an application of Identity B:

1.	$(x) (Ax \supset Bx)$	
2.	At	
3.	$(x) [Bx \supset (x = g)] \: / \: Ag \: \& \: Bg$	
4.	$At \supset Bt$	UI 1
5.	Bt	MP 2, 4
6.	$Bt \supset (t = g)$	UI 3
7.	$t = g$	MP 5, 6
8.	Ag	Id B 2, 7 (g replaced t)
9.	Bg	Id B 5, 7 (g replaced t)
10.	Ag & Bg	Conj 8, 9

The Identity B Rule may be informally summed up as follows:

> *Given an identity statement asserting that the constants **c** and **d** designate the same individual, one constant may replace the other in any available line.*

When we use Identity B, we can, if we wish, replace only some occurrences of a constant. For example:

1.	$a = b \: / \: b = a$	
2.	$(x)(x = x)$	Id A
3.	$a = a$	UI 2
4.	$b = a$	Id B 1,3

In this proof, the identity statement on line (1) gives us the right—specified in the Identity B rule—to replace b with a or a with b on any line. Therefore, we derived line 4 by replacing only the left side occurrence of a in 3 with b.

Here is another argument employing Id B: Robert Zimmerman was a singer and Ed Sullivan was not a singer. Robert Zimmerman is Bob Dylan. So, Ed Sullivan is not Bob Dylan. Using obvious abbreviations, the proof of validity follows.

1. Sz & ~Ss

2. $z = d \: / \: \sim (s = d)$

3.	$\sim \sim (s = d)$	AP
4.	$(s = d)$	DNeg 3
5.	Sz	Simp 1
6.	$\sim$ Ss	Simp 1
7.	Sd	Id B 2, 5
8.	$\sim$ Sd	Id B 4, 6
9.	Sd & $\sim$ Sd	Conj 7,8
10. $\sim (s = d)$	IP 3-9	

PROPERTIES OF THE IDENTITY RELATION

We can now use our natural deduction rules to prove that the identity relation is **reflexive, transitive,** and **symmetrical.** First, identity is reflexive. That is, each and every thing is identical with itself. The claim that everything is identical with itself is expressed within QL as

$$(x) (x = x)$$

The proof of this follows:

1.	$\sim(x) (x = x)$	AP
2.	$(x) (x = x)$	Id A
3.	$(x)(x = x) \ \& \sim (x)(x = x)$	Conj 1, 2
4. $(x)(x = x)$	IP 1–2	

Next, the relation of identity is symmetrical. In QL, the claim that identity is a symmetrical relation is

$$(x)(y)[(x = y) \supset (y = x)]$$

This formula reads, For any x and for any y, if x is identical with y then y is identical with x.

    The proof of this follows:

1.	$\sim(x)(y)[(x = y) \supset (y = x)]$	AP
2.	$(\exists x) \sim(y) [(x = y) \supset (y = x)]$	QE 1
3.	$(\exists x) (\exists y) \sim[(x = y) \supset (y = x)]$	QE 2
4.	$(\exists y) \sim[(v = y) \supset (y = v)]$	EI 3
5.	$\sim[(v = v') \supset (v' = v)]$	EI 4
6.	$\sim[\sim (v = v') \lor (v' = v)]$	Imp 5

7.	$\sim \sim (v = v') \,\&\, \sim (v' = v)$	DM 6
8.	$\sim \sim (v = v')$	Simp 7
9.	$\sim (v' = v)$	Simp 7
10.	$v = v'$	DNeg 8
11.	$\sim (v = v)$	Id B, 9, 10
12.	$(x)\,(x = x)$	Id A
13.	$v = v$	UI 12
14.	$(v = v) \,\&\, \sim (v = v)$	Conj 11, 13
15.	$(x)(y)[(x = y) \supset (y = x)]$	IP 1–14

Finally, the identity relation is transitive. That is,

$$(x)\,(y)\,(z)\,\{[(x = y) \,\&\, (y = z)] \supset (x = z)\}$$

This formula reads, For any x, for any y, and for any z, if x is identical with y and y is identical with z, then x is identical with z.

The proof is as follows.

1.	$\sim(x)\,(y)\,(z)\,\{[(x = y) \,\&\, (y = z)] \supset (x = z)\}$	AP
2.	$(\exists x) \sim (y)(z)\{[(x = y) \,\&\, (y = z)] \supset (x = z)\}$	QE 1
3.	$(\exists x)\,(\exists y) \sim (z)\{[(x = y) \,\&\, (y = z)] \supset (x = z)\}$	QE 2
4.	$(\exists x)\,(\exists y)\,(\exists z) \sim \{[(x = y) \,\&\, (y = z)] \supset (x = z)\}$	QE 3
5.	$(\exists y)\,(\exists z) \sim \{[(v = y) \,\&\, (y = z)] \supset (v = z)\}$	EI 4
6.	$(\exists z) \sim \{[(v = v') \,\&\, (v' = z)] \supset (v = z)\}$	EI 5
7.	$\sim \{[(v = v') \,\&\, (v' = v'')] \supset (v = v'')\}$	EI 6
8.	$\sim \{\sim [(v = v') \,\&\, (v' = v'')] \lor (v = v'')\}$	Imp 7
9.	$\sim \sim [(v = v') \,\&\, (v' = v'')] \,\&\, \sim (v = v'')$	DM 8
10.	$[(v = v') \,\&\, (v' = v'')] \,\&\, \sim (v = v'')$	DNeg 9
11.	$(v = v') \,\&\, (v' = v'')$	Simp 10
12.	$\sim (v = v'')$	Simp 10
13.	$v = v'$	Simp 11
14.	$v' = v''$	Simp 11
15.	$v = v''$	Id B, 13, 14
16.	$(v = v'') \,\&\, \sim (v = v'')$	Conj 12, 15
17.	$(x)\,(y)\,(z)\,\{[(x = y) \,\&\, (y = z)] \supset (x = z)\}$	IP 1–16

Using the identity rules plus any of the other rules introduced so far, prove each of the following arguments valid.

(1)* 1. Aa ⊃ Ha

    2. ~ Ha

    3. a = b / ~ Ab

(2) 1. Aa ⊃ Ha

    2. ~ (Ab ⊃ Hb) / ~ (a = b)

(3)* 1. (x) (Ax ⊃ Px)

    2. (x) (Px ⊃ Hx)

    3. Aa & ~ Hb / ~ (a = b)

(4) 1. Ab ⊃ Bb

    2. Rd ⊃ Sd

    3. Ab & Rd

    4. b = d / Bd & Sb

(5) 1. Hc ⊃ Kc

    2. Md ⊃ Nd

    3. Hc & Md

    4. c = d / Kd & Nc

(6) 1. (x) (Fx ⊃ Gx)

    2. FA

    3. Fb ⊃ ~Gb / ~ (a = b)

(7) 1. Ma ⊃ Wa

    2. ~Wa

    3. a = b / ~ Mb

(8) 1. Wa

    2. (x) [Wx ⊃ (x = a)]

    3. (∃x)(Wx & Bx) / Ba

| ▶ | **EXERCISE 22.2** | ◀ |

Symbolize the following arguments and derive the conclusions by natural deduction. You will need to use the Identity rules.

1.* If Jim is sick, then the store will be closed. Sue's husband is sick. Jim is Sue's husband. Therefore, the store will be closed.

2. Jim is either an adult or he is a minor. If Jim is a minor, then he cannot enter the contest. Betty's nephew is not an adult. Jim is Betty's nephew. Therefore, Jim cannot enter the contest.

3. Bob Zimmerman is a Boeing engineer, and Bob Dylan is not an engineer. Bob Dylan is identical with Robert Zimmerman. So, Bob Zimmerman is not identical with Robert Zimmerman.

4. Lou is Sue's aunt. Sue's aunt is Karen's mother. If x is identical with y, and y is identical with z, then x is identical with z. So Lou is Karen's mother.

5. Ann is Joe's boss. Rita's mother is Joe's boss. If x is identical with y, and z is identical with y, then x is identical with z. So, Ann is Rita's mother.

CHAPTER 22 GLOSSARY

Identity To say that x is identical with y is to say that x and y are one and the same entity.

Principle of the Indiscernibility of Identicals If x is identical with y, then whatever is true of x is true of y and whatever is true of y is true of x.

Principle of Self-Identity Each thing is identical with itself.

Reflexive relation A relation that possesses the following feature: if one thing (A) has the relation to something (B) or B has the relation to A, A has that relation to itself.

Symmetrical relation A relation that possesses the following feature: if one thing has the relation to a second thing, the second thing has that relation to the first thing.

Transitive relation A relation that possesses the following feature: if one thing bears the relation to a second thing and the second thing bears the relation to a third thing, the first must bear the relation to the third.

P a r t

V

MODAL
LOGIC

23 INTRODUCTORY MODAL LOGIC

> *"We need scarcely add that the contemplation in natural science of a wider domain than the actual leads to a far better understanding of the actual."*—*Sir Arthur Eddington,* The Nature of the Physical World, *1928.*

TO SHAVE OR NOT TO SHAVE: THAT IS THE QUESTION

When, in approximately 525 B.C., the Greek philosopher and mathematician Pythagoras (c. 580 B.C.–500 B.C.) discovered the theorem now known as the Pythagorean theorem, he and other ancient mathematicians noticed a difference between this abstract mathematical truth about right triangles and an ordinary truth such as that expressed by, for example, "Zeno has a beard."[1] The difference seemed to be this: If you try logically to suppose that the Pythagorean theorem is *false,* you end up in a **logical contradiction** (a statement of the form **P** & ~ **P**). But a logical contradiction is a paradigm of the impossible. And so it seemed, upon much reflection, that Pythagoras's theorem couldn't even possibly be false, which is to say that it *necessarily* is true. On the other hand, it is easy to see that the statement about Zeno and his beard might have been false—for it would have been false if Zeno had chosen to shave it off, and we can easily picture him shaving and thus ensuring that the statement about his beard is false.

---

[1] The Pythagorean theorem states that the square of the hypotenuse of a right triangle is equal to the sum of the squares of the legs.

The statement about Zeno's beard, although true, thus seemed *contingent* on Zeno's free choice to shave or not to shave. But the Pythagorean theorem didn't seem contingent in any way. Therefore, it was argued, the Pythagorean theorem cannot even be *thought* false without contradiction, whereas the statement about Zeno can easily be thought false without contradiction (if we imagine Zeno shaving).[2] Consequently, the Pythagorean theorem was called a *necessary* truth and the statement about Zeno's beard was called a *contingent* truth.

On the basis of this distinction between what is necessary and what is contingent, many ancient mathematicians and philosophers distinguished two types or "modes" of truth: necessary truths and contingent truths. A necessary truth was understood to be a truth that could not possibly have been false, and a contingent truth was understood to be one that might have been false. (This corresponds closely to ordinary terminology, for in everyday life, something is called "necessary" if it could not possibly have been otherwise, and something is called "contingent" if it is dependent on circumstances that might possibly have been otherwise and so it too might have been otherwise.) Necessary truths are also called "logically necessary truths" or "logical necessities."

Modal logic is the study of the modes of truth and their relationship to reasoning and argumentation. One of the central concepts in modal logic is the concept of a modal argument. A **modal argument** is an argument built out of *modal* sentence operators and the components to which these attach. The two most commonly used modal operators, which we shall examine, are *necessarily* and *possibly*. Many of the most interesting arguments in the history of philosophy, arguments about free will, the existence of God, the nature of consciousness, and more, are modal arguments. Our goal in this chapter is to introduce a formal language for modal logic; the next chapter introduces elementary modal proof techniques—formal procedures that will allow us to evaluate modal arguments precisely and systematically.

FIVE MODAL PROPERTIES

Logicians distinguish five important *modal properties* that a statement or proposition might possess:

1. If a proposition is true in at least one possible circumstance or situation, the proposition is **possibly true.**

2. If a proposition is false in at least one possible circumstance or situation, the proposition is **possibly false.**

3. If a proposition is true in at least one possible circumstance and is also false in at least one possible circumstance, the proposition is **contingent.**

[2] More precisely, if we suppose the Pythagorean theorem false, this supposition *implies* a logical contradiction. But a logical contradiction—a statement of the form **P** & ~ **P**—is a paradigm of the impossible. Because that which logically implies an impossibility must itself be impossible, it follows that the Pythagorean theorem couldn't even possibly be false. On the other hand, the statement about Zeno and his beard would have been false if Zeno had chosen to shave yesterday, and so that statement could have been false.

4. If a proposition is true in all possible circumstances, the proposition is **necessarily true.**

5. If a proposition is false in all possible circumstances, the proposition is **necessarily false.**

The following three sections discuss some examples from each of these categories.

POSSIBLE TRUTHS, POSSIBLE FALSEHOODS, CONTINGENCIES

Consider the proposition expressed by "Mount Rainier has snow on its summit." Is there a situation in which this is or would be true? Yes. It is true in the current circumstance or situation in which we find ourselves. The proposition therefore qualifies as possibly true. Consider next the proposition expressed by "A comet collides with the Moon." Is there a circumstance in which this would be true? Certainly. An astrophysicist could describe a possible circumstance or scenario in which such a collision would happen, and he or she could also describe the effect such a collision would have on our planet. The sentence, "A comet collides with the Moon," also expresses a possibly true proposition.

Let us return to the proposition expressed by "Mount Rainier has snow on its summit." Is there a possible situation in which this is or would be *false?* Yes. It would be false if the Earth's climate heated up enough to melt Mount Rainier's entire snowpack. Because we can describe a possible circumstance in which the proposition would be false, it qualifies as possibly false.

A contingent proposition has the following feature: there are possible circumstances in which it would be true and there are possible circumstances in which it would be false. Now, because there are possible circumstances in which it would be true and others in which it would be false, the proposition expressed by "Mount Rainier has snow on its summit" is also contingent. The truth-value of the proposition depends on (is "contingent upon") which circumstances prevail.

Examples of Possibly True Propositions

1. In 2,000 B.C., a 4.5 magnitude earthquake shook the North Pole.

2. Someone named Cosmo Kramer will be elected president in 2008.

3. Starbucks is headquartered in Seattle.

4. The moon has craters.

5. $2 + 2 = 4$

6. Bill Gates is a billionaire.

(Continued)

Examples of Possibly False Propositions

1. In 2,000 B.C., a 4.5 magnitude earthquake shook the North Pole.

2. Someone named Cosmo Kramer will be elected president in 2008.

3. Starbucks is headquartered in Seattle.

4. The moon has craters.

5. $2 + 2 = 7$

6. Bill Gates is a billionaire.

Examples of propositions that are contingent

1. In 2,000 B.C., a 4.5 magnitude earthquake shook the North Pole.

2. Someone named Cosmo Kramer will be elected president in 2008.

3. Starbucks is headquartered in Seattle.

4. The moon has craters.

5. Bill Gates is a billionaire.

"Scientists try to characterize a range of possible causes of evolution, and then try to determine which of these possibilities actually obtained. The actual is understood by first embedding it in the possible."—Elliot Sober, The Philosophy of Biology

NECESSARY TRUTHS

A necessary truth, if one existed, would be a proposition that is true in every possible circumstance or situation, false in none. In other words, such a proposition would be true and could not be false. But do such propositions exist? Consider the following and judge for yourself. (As you think through each statement, give the words their commonly accepted meanings, and think about what the sentences *mean*.)

1. $2 + 2 = 4$

2. $\sqrt{2}$ is an irrational number.

3. All triangles have three sides.

4. All bachelors are unmarried.

5. If something is red, then it is colored.

6. Either Garth Brooks is over thirty or it is not the case that Garth Brooks is over thirty.

7. Nothing is both red all over and green all over at the same time.

8. If A is taller than B, and B is taller than C, then A is taller than C.

9. For any sentences **P** and **Q**, if **P** is true, and if **P** implies **Q**, then **Q** must be true.

Given the standard meaning that our linguistic community attaches to each of the above sentences, each expresses a unique truth. That is, each expresses a true statement or proposition. Are there any possible circumstances in which any of the statements expressed above would be false? To answer this, try the following "thought experiment." In the case of each sentence above, reflect carefully on the standard meaning the sentence possesses. Consider the proposition expressed by the sentence. (In each case, the sentence itself could be given different meanings in different societies or in different worlds, but that has nothing to do with the truth or falsity of the unique proposition expressed by the sentence when the sentence is given the standard meaning *we* actually give it.) Finally, when you grasp the proposition actually expressed by the sentence, try to write a description of a possible situation or circumstance in which that proposition would be *false.* For any of the nine statements above, your description will be self-contradictory. In each of the cases, you cannot describe a possible circumstance in which the proposition would be false without also falling into a logical contradiction, that is, a statement of the form **P** & ~ **P.** If you grant that a contradiction is an impossibility, the above nine statements could not possibly be false. If so, then those statements are necessary truths.

The first five of the above propositions demonstrate two particular types of truths that we will examine more closely. Propositions 1 through 3 are *mathematical truths.* As noted, mathematicians from the earliest days of the discipline realized that the truths they were discovering were in some sense necessary. Mathematical truths, they believed, could not possibly be false. This is part of the reason mathematical truths were viewed with awe and even reverence when first discovered. (Pythagoras is said to have offered a great sacrifice to the gods upon discovering the theorem that now bears his name. Necessity seemed to him and to other ancient mathematicians to be a mark of the divine.)

Sentences 4 and 5 are called **analytic truths** because they have the following feature. If we apply logical analysis to the concept expressed by the subject term of the sentence, we see that if the subject term applies to (is true of) an individual thing, the predicate term must also apply. By logically analyzing the subject, the predicate automatically follows. Thus, through a process of logical analysis, without investigating the physical world, we see that the propositions expressed by these sentences are true. Consequently, such sentences are said to be "analytically true."

NECESSARY FALSEHOODS

If the nine numbered statements above are necessarily true, their *negations* are all necessarily false, for the negation of a statement is true if the statement negated is false, and false if the statement negated is true. If the nine necessary truths cannot possibly be false, their negations cannot possibly be true. Therefore, by negating each of the nine sentences in the list above, you can produce a list of nine necessary falsehoods.

In more general terms, because a contradiction is a paradigm of the impossible, any contradiction will be a necessary falsehood. Thus, the following are additional examples of necessary falsehoods:

Dr. Niles Crane is president, and it is not the case that Dr. Niles Crane is president.

Vinegar is an acid, and it is not the case that vinegar is an acid.

Japan is an island, and it is not the case that Japan is an island.

Africa's total population is exactly one million, and it is not the case that Africa's total population is exactly one million.

PUTTING STATEMENTS INTO SYMBOLS

▶ THE NECESSITY OPERATOR

In English, the word *necessarily* functions as a monadic (one-place) sentence operator. For example, if we begin with a sentence such as

2 + 2 = 4

and prefix to it *necessarily,* we produce the compound sentence

Necessarily, 2 + 2 = 4.

According to this, it is necessarily true that 2 + 2 = 4, which is to say that 2 + 2 = 4 could not possibly be false. In other words, this sentence, "It is necessarily true that 2 + 2 = 4," is used to attribute necessity to that which is expressed by 2 + 2 = 4.

In the language we are about to introduce, if we want to say that a statement **P** is necessarily true, we will put a box, □ in front of the **P** and write □**P**. This formula, read as "box **P**," means "**P** is necessarily true" or "It is necessarily true that **P**." If **P** is necessarily true, in all possible circumstances or situations, **P** would be true, that is, **P** would have been true no matter how things might have been. So, □**P** may also be translated "In all circumstances, **P** is true."

Because ~**P** indicates that **P** is false, if we want to say that **P** is necessarily false, we place a box in front of ~**P** and write □~**P**. This formula, pronounced "box not **P**," means "Necessarily, it is false that **P**" or "It is necessarily false that **P**," or more simply, "**P** is necessarily false." Now, if **P** is necessarily false, in all possible circumstances or situations, **P** is false, that is, **P** would have been false no matter how things might have been. Therefore, □~**P** may also be translated "In all circumstances, **P** is false."

▶ THE POSSIBILITY OPERATOR

In English, the word *possibly* also functions as a monadic operator. For example, if we begin with a sentence such as

Someone is 120 years old.

and prefix to it *possibly,* we produce the compound sentence

Possibly, someone is 120 years old.

This asserts that the previous sentence, "Someone is 120 years old," expresses something that is possibly true.

In our system of modal logic, if we wish to say that a statement **P** is possibly true, we will place a diamond in front of **P** and write ◊**P**. This formula, read as "diamond **P**," is translated "It is possible that **P** is true," or "It is possibly true that **P**," or more simply as "**P** is possibly true." ◊**P** may also be read as "**P** is true in some circumstances," "**P** might be true," "**P** is a possible truth," or "**P** is logically possible." If **P** is possibly true, then in at least one possible circumstance, **P** is true, that is, there is at least one possible circumstance such that if that circumstance had been actual, **P** would have been true. So, ◊**P** may also be translated "In at least one circumstance **P** is true."

▶ THE CONTINGENCY OPERATOR

In English, the word *contingently* also functions as a monadic operator. In our system of modal logic, if we want to say that a statement **P** is contingent, we will put the symbol ∇ (called nabla) in front of the statement. Thus, ∇ **P**, read as "nabla **P**," means "**P** is contingent" or "It is contingent that **P**." If **P** is contingent, in some circumstances **P** would be true, and in some circumstances **P** would be false. In other words, **P**'s truth-value depends on (is *contingent* upon) the circumstances that obtain or do not obtain.

Possible Worlds

Since the early 1960s, many logicians have employed the concept of a **possible world** in order to conveniently discuss modal ideas. A necessarily true proposition is often defined as a proposition true in all possible worlds; a contingent proposition is defined as a proposition true in some possible worlds but false in others; and a necessarily false proposi-

(Continued)

tion is defined as a proposition false in all possible worlds. What do these philosophers have in mind when they speak of a possible world? Let us approach their idea in stages.

Our planet orbits a star that sits about twenty thousand light-years in from the edge of an enormous disk-shaped collection of stars called the Milky Way galaxy. Our galaxy, which contains more than 100 billion stars, is located near the edge of a swarm of about two dozen galaxies, each with billions of stars, each trillions of miles apart. This cluster of galaxies, called the Local Group, is part of a "supercluster" of galaxies—a gigantic cluster of clusters of galaxies. Beyond our supercluster lie neighboring superclusters containing billions of additional galaxies.

When we speak of "our world," let this cover everything in the universe, no matter how far away in space and no matter how far away in time. So the term "our world" means "the entire universe." Now, consider this statement: *Our world might have been different from the way it actually is.* Instead of containing as many galaxies as it actually contains, it might have contained only half as many, or a fourth as many, or twice as many, and so on. Instead of containing protons, neutrons, and electrons, it might have contained other kinds of particles. Furthermore, it seems plausible to suppose that the laws of nature (the physical laws studied in physics, chemistry, and so on) might have been different from what they are. Cosmologists plug alternative laws of nature into their computer models of the universe and produce a picture of what the universe *would have been like* if the laws of physics had been different. And the alternatives are seemingly endless. You might have grown up in a different city, you might have slept an hour longer this morning, and so on.

Just consider the sequence that begins thus: the universe might have contained exactly 1,000 galaxies, it might have contained exactly 1,001 galaxies, it might have contained exactly 1,002 galaxies, it might have contained exactly 1,003 galaxies. . . . It seems there are *infinitely* many ways things might have been.

These infinitely many ways things might have been constitute an infinite number of ways a world could be. Furthermore, had one of these other ways been actual, the actual world—the way things actually are—would itself have been just another way things might have been. Consequently, the actual world—the universe we actually inhabit—also represents a way a world could be. Let us call each of these ways that things might have been or might be a *possible world*. These infinitely many ways that things might have been or might be constitute an infinity of possible worlds—an infinity of ways a world could be.

However, there seems to be one major difference between the actual world we inhabit and all the other possibilities: All the other possible worlds are nonactual possible worlds; the world we inhabit is the only actual world. A few philosophers disagree with this last statement, but the absolute difference between the actual and the merely possible seems to be firmly rooted in commonsense.

It is important to emphasize that other possible worlds are not distant planets or galaxies or universes far away in space. Remember that anything, no matter how far away in physical space–time, counts as part of what we are defining as the actual world. So other possible worlds are not located "out there" in physical space. They are merely possibilities, not actualities.

LOGICAL AND PHYSICAL POSSIBILITY DISTINGUISHED

The idea of a possible world allows us to state and clarify an important distinction—the classic philosophical distinction between logical possibility and physical possibility. This distinction also plays an important role in modern physics and in the physical sciences in general. We can approach the distinction by way of the following question: Could a human being swim the Atlantic Ocean—all three thousand miles or so of it—in five minutes? It seems, upon initial reflection, that such a swim would be impossible. The laws of physics won't allow anyone to swim that far that fast. At the required speed, which would be around thirty-six thousand miles per hour, the friction alone would destroy a human body. And besides, no matter how many cans of spinach a person eats, the human body doesn't have the strength or the stored energy for such a task. So, *given the laws of physics and the facts of human biology,* that long-distance swim would be impossible.

If you answered no to the question about swimming the Atlantic in five minutes, and if you reasoned along the lines just given, you had in mind the kind of possibility termed *physical possibility*. An event is **physically possible** if its occurrence would not violate a physical law of nature, and an event is **physically impossible** if its occurrence would violate a physical law. Physical possibility may now be explained in terms of possible worlds. Consider just those possible worlds that have the same physical laws of nature as does the actual world. Call such worlds "physically possible" worlds. We can say: Something is a physical possibility if and only if there is a physically possible world in which it happens. To go back to our earlier example, it is *not* physically possible that someone might swim the Atlantic in five minutes, for there are no

(Continued)

physically possible worlds in which someone swims thirty-six thousand miles per hour.

In contrast, consider this: Is it physically possible for someone to win the Washington State Lottery ten times in a row? Well, such a thing would be extremely improbable, but it would not violate any physical laws of nature. That is to say, in some physically possible worlds a person could win the Washington State Lottery ten times in a row. Thus, such an event is physically possible, although incredibly *improbable*. Notice the distinction here between possibility and probability. Some events are physically possible even though they are highly improbable. In short, it is physically possible that someone win the Washington State Lottery ten times in a row, but it is physically impossible that someone swim the Atlantic in five minutes.

The next step in this sequence requires that we broaden our thinking. *If* the laws of nature had been sufficiently different, and *if* the facts of human biology had been sufficiently different, it would be possible, in some sense, that someone swim the Atlantic in five minutes. This is an entirely different kind of possibility, corresponding to a much broader sense of the word *possible*. Suppose the physical laws of nature had been so very different that people could easily swim the Atlantic in five minutes. It seems plausible to suppose that is a way things might have been. There is certainly nothing contradictory in the idea. Therefore, it seems, we're speaking of a possible world. Of course, it's not a *physically* possible world. But it is a possible world nonetheless, for it's a way things might have been (had they been very, very different). So, there is a possible world in which someone swims the Atlantic in five minutes. If there is even one possible world in which an event occurs, we will say the event is *logically* possible. Thus, although the Atlantic swim is not physically possible, it is nevertheless logically possible. Logical possibility thus encompasses a broader range of possibilities than is encompassed by physical possibility.

This distinction between logical and physical possibility is of such crucial significance in logic and in science that we must examine it more closely for a moment. Logicians traditionally use the term *logically possible* for the broadest, most inclusive category of possibilities, the category that includes within itself all possibilities and all kinds of possibilities. So, every possible world counts as a logically possible world. It follows that every physically possible world, that is, every world that contains the same physical laws of nature as the actual world, counts as a logically possible world as well. However, it seems plausible to suppose that some logically possible worlds are not physically possible worlds, for some possible worlds, it would seem, have laws of nature different from those we find in the actual world. Although all physically possible worlds are also logically possible worlds, not all logically possible worlds are physically possible worlds.

And so, to sum this up, is it possible that someone might swim the Atlantic Ocean in five minutes? It's not physically possible, for in no physically possible world does someone swim the Atlantic in five minutes. But such a thing is logically possible, simply because in some possible world, someone can swim the Atlantic in five minutes. We will speak of *possible worlds* rather than *logically possible worlds* to refer to this most inclusive kind of possibility.

Some readers might be puzzled by our talking of events happening *in* various possible worlds. Does such talk really make any sense? What do we mean when we speak of something happening in a possible world? Perhaps the following will help make sense of such talk. We readily understand sentences such as

In the novel *Robinson Crusoe,* a man is shipwrecked on a deserted island.

and

In the story of Hansel and Gretel, two children get lost in the woods.

And so it seems that we understand what it is for an event to happen "in" a novel or a story. But a novel or story is essentially a description of a possibility. As such, a novel or story describes a possible world. Therefore, let us think of an event happening "in" a possible world in much the way we think of an event happening "in" a novel, a story, or a movie. The two senses of *in*—the possible-worlds *in* and the literary *in*—seem closely analogous, if not identical.

THE LIMITS OF THE POSSIBLE

When we consider possible worlds, does anything we can think of qualify as possible? That is, will anything we suggest, no matter how absurd, count as at least logically possible, or are there limits to what is possible? Let us discuss whether or not some suggestions might count as logically *im*possible. We will base much of our discussion on the following principle:

That which is self-contradictory is a paradigm case of the impossible.

The principle that contradictions are impossible, often called the **Law of Noncontradiction,** was introduced in Chapter 9. Our discussion of

(Continued)

possible worlds will presuppose this law or principle, a principle that is sometimes called a "law of thought" or a "law of logic." Because it plays such a pivotal role in this area of logic, let us briefly clarify the principle and say a few words in its defense.

Recall that an explicit logical contradiction or self-contradiction is a conjunction in which one conjunct is the negation of the other conjunct. Here are some examples of self-contradictory sentences.

1. Rita is taller than Fred, and it is not the case that Rita is taller than Fred.

2. Jim weighs 150 pounds, and it is not the case that Jim weighs 150 pounds.

It is extremely important that you understand the following: In order for a sentence, in a certain context, to qualify as an explicit self-contradiction, the sentence component that has a negation operator attached to it must have one definite meaning, and the other component sentence conjoined to it must have exactly the same meaning. For example, in sentence 2 above, the sentence "Jim weighs 150 pounds" must have exactly the same meaning in both of its occurrences in that conjunction. That is, for the sentence "Jim weighs 150 pounds, and it is not the case that Jim weighs 150 pounds" to count as a self-contradiction, each occurrence in that sentence of "Jim weighs 150 pounds" must refer to the same Jim, the same scale, the same moment in time, and so on.

Why should we count that which is self-contradictory as impossible? Why shouldn't we suppose that some possible worlds include or constitute self-contradictory states of affairs? Three sets of considerations suggest that the Law of Noncontradiction is indispensable. First, in the case of a contradiction, one conjunct expresses one thing, and then the other conjunct takes back or denies precisely that which the first conjunct expressed. The net result is that nothing is really expressed. The one conjunct cancels out what was said by the other conjunct, leaving nothing. No possible situation is described by a self-contradiction because a self-contradiction doesn't describe anything.

Second, Chapter 9 demonstrated that if a natural deduction proof begins with a contradiction, using undeniably valid inference rules, *any and every* conclusion validly follows. In other words, if someone accepts one contradiction as true, and reasons validly from that, any and every conclusion logically follows and therefore *every* statement is true. Thus, absurdities follow if we accept the truth of even one contradiction.

Third, it seems that effective human communication itself rests upon the Law of Noncontradiction. Consider what you would do in the following situation. You have accidentally swallowed poison and will die in fifteen minutes unless you get to a hospital. You know there's a hospital within four blocks, but you don't know which direction to head. Fortunately, you see a doctor and you ask, "Where's the hospital?" Now, suppose she says: "It's two streets north and it's *not* two streets north." Which direction do you go? Has she really communicated anything of value? Has she communicated anything at all?

Consequently, if we are going to communicate, it seems we must presuppose the Law of Noncontradiction. Nothing that counts as self-contradictory will also count as logically possible; anything self-contradictory counts as logically impossible. If we agree that a self-contradiction is an impossibility, there are no possible worlds in which, for example, a person is sixty-six years old and yet is not sixty-six at the same time, or in which an only child plays with her brother.

BEAM ME UP, SCOTTY

Here is food for thought. In the original *Star Trek* series, Kirk, Spock, and others would report to the transporter room, where they would be "beamed down" to a strange planet (where all sorts of dangers usually lay in wait). The transporter on the starship *Enterprise* had a "safe beaming range" of 19,500 miles. Is the *Enterprise*'s transporter system (with its redesigned "field generator matrix") physically possible? Is it logically possible? If you want to pursue these and related questions, an interesting place to begin is Lawrence Krauss's excellent book *The Physics of Star Trek* (Basic Books, 1996).

TRANSLATING ENGLISH SENTENCES INTO MODAL SYMBOLS

Let us draw some implications and place the information into our new symbols. Any proposition that is actually true is true at least in the current circumstance we find ourselves in, and so it is true in at least one circumstance. But if it is true in at least one circumstance, it is possibly true. So, any proposition that is actually true is also possibly true. In symbols:

$$\mathbf{P} \supset \Diamond\, \mathbf{P}$$

This formula reads "If **P** is actually true, then it is possible **P** is true." In everyday life, we often cite actual truth to prove possible truth. For instance, consider the following dialogue:

Ann: Is it *possible* that someone could eat thirty cheese-filled hot dogs in thirty minutes? It's not possible!

Sue: Oh yes, it is possible. I saw a guy *actually* do it last year at the state fair—the guy that won the cheese-filled-hot-dog-eating contest.

Ann: Oh my. Then I guess it *is* possible!

Sue is reasoning that if something is actual, this shows that it is also possible.

Continuing, any proposition that is actually false is false in the current circumstance and so is false in at least one circumstance. This is just to say that any proposition that is actually false is also possibly false:

$$\sim\!P \supset \Diamond \sim\!P$$

However, the reverse does not hold in either case. That is, not all possible truths are actually true, and not all possible falsehoods are actually false. For example, the sentence "Cosmo Kramer is president" is possibly true but is certainly not actually true; and "Mount Everest has snow" is possibly false but is obviously not actually false.

Next, every necessarily true proposition is also possibly true. Remember that a possibly true proposition is one that is true in *at least one* possible situation, and a necessary truth is a proposition that is true in all possible situations. So, because a necessary truth is true in all possible situations, it is true in at least one situation, and so every necessary truth counts as possibly true as well as necessarily true. In symbols:

$$\Box P \supset \Diamond P$$

And it follows that every necessarily true proposition is actually true as well. The argument for this is obvious. A proposition is actually true if it is true in the present circumstance. Because a necessary truth is true in all circumstances, it is true in this circumstance. Therefore, every necessary truth is actually true as well. In symbols:

$$\Box P \supset P$$

However, not all possible truths are necessarily true, and not all actual truths are necessarily true either. For instance, although it's actually true, and possibly true, that Los Angeles County has a population of more than 1 million people, that proposition is not necessarily true, for it's not true in every possible circumstance. In certain circumstances the proposition is false (for instance, if Los Angeles County were to be evacuated because of an earthquake or some other catastrophe).

Turning to necessary falsehoods, we may observe that each necessarily false proposition is both possibly false and actually false as well. Here is why. A possibly false proposition is one that is false in at least one circumstance. A necessary falsehood is a proposition that is false in all circumstances. Therefore, any necessary falsehood is false in at least one circumstance and so counts as possibly false as well as necessarily false. In symbols:

$$\Box\!\sim\! P \supset \Diamond \sim\! P$$

However, not every possible falsehood is necessarily false, and not every actual falsehood is necessarily false. Can you state an actual falsehood that is not necessarily false? Such a proposition would also be possibly true, of course.

A NAME AND SYNTAX FOR OUR MODAL LANGUAGE

We may now specify the syntax for a formal language for modal logic. The language will be called ML (for "modal language").

THE VOCABULARY FOR ML

1. Sentence constants:

   A, B, . . . Z.

2. Sentence operators:

   Monadic operators: ~ ◊ ∇ □

   Dyadic operators: & v ⊃ ≡ → ↔

3. Parenthetical devices:

   Parentheses: ( )

   Brackets: [ ]

   Braces: { }

THE GRAMMAR FOR ML

M1. Any constant is a sentence of ML.

M2. If **P** is a sentence of ML, then ~ **P**, □ **P**, ◊ **P**, ∇**P** are sentences of ML.

M3. If **P** and **Q** are sentences of ML, then (**P** & **Q**), (**P** v **Q**), (**P** ⊃ **Q**), (**P** ≡ **Q**), (**P** → **Q**), (**P** ↔ **Q**) are sentences of ML.

Any expression that contains only items drawn from the vocabulary of ML and that can be constructed by a finite number of applications of the rules of grammar M1 through M3 is a sentence or well-formed formula of ML. Nothing else counts as a sentence of ML.

If you will examine ML's syntax, you will notice that it is exactly the same as the syntax for TL with the exception of one minor change: the modal operators have been added to the lists of monadic and dyadic operators. Thus, the language ML "contains" the language TL in the sense that any sentence that is a wff of TL will automatically count as a wff of ML, although it's not the case that any wff of ML will count as a wff of TL. Also, in order to simplify things, we will continue the following practice:

> *You may leave off the outer parentheses if no monadic operator applies to those parentheses.*

That is, a sentence such as (A ⊃ J) may be abbreviated A ⊃ J. However, a sentence such as □ (H v E) may not be abbreviated in such a manner. That is, □ (H v E) may *not* simply be abbreviated as □ H v E.

► **EXERCISE 23.1** ◄

Part I. Let the proposition that Shoreline Community College (SCC) has a volleyball team be represented by S. Translate the following into the symbolization of ML:

1.*In every possible circumstance SCC has a volleyball team.

2. There is no possible circumstance in which SCC has a volleyball team.

3. In every possible circumstance it is false that SCC has a volleyball team.

4. It is false that in every possible circumstance SCC has a volleyball team.

5.*Not every possible circumstance is one in which SCC has a volleyball team.

Part II. Symbolize the following in ML.

1.*It is not a necessary truth that protons are heavier than electrons. (Let P abbreviate "Protons are heavier than electrons.")

2. It is a contingent truth that a universe actually exists. (Let U abbreviate "A universe actually exists.")

3. It is impossible that 2 + 2 = 99. (Let A abbreviate "2 + 2 = 99.")

4.*It is necessarily false that some triangles have seven sides. (Let A abbreviate "Some triangles have seven sides.")

5. It is possible that Bill Gates retires, but it is also possible that he doesn't. (Let B abbreviate "Bill Gates retires.")

6. Necessarily, it is false that the square root of two is rational. (Use S for "The square root of two is rational.")

7.*It is necessarily true that the derivative of a constant is zero. (Use D for "The derivative of a constant is zero.")

8. It is not possible that it is not true that 2 + 2 = 4. (Let A abbreviate "2 + 2 = 4.")

9. It is not possible that it is true that 2 + 2 = 7. (Let A abbreviate "2 + 2 = 7.")

10. *Necessarily, if the universe was created, then God exists. (Let U abbreviate "The universe was created" and let G abbreviate "God exists.")

11. It is false that it is contingent that $2 + 2 = 4$. (Let A abbreviate "$2 + 2 = 4$.")

12. *Necessarily, either the universe will expand forever or it will not expand forever. (Let U abbreviate "The universe will expand forever.")

13. Either it is necessary that the universe will expand forever or it is necessary that the universe will not expand forever. (Let U abbreviate "The universe will expand forever.")

14. Either it is a necessary truth that the universe was created or it is not a necessary truth that the universe was created. (Let U abbreviate "The universe was created.")

15. Either it is necessary that the universe was created or it is necessary that the universe was not created. (Let U abbreviate "The universe was created.")

16. *If materialism is true, then it is necessarily true that God does not exist. (Let M abbreviate "Materialism is true;" let G abbreviate "God exists.")

17. If it is necessarily true that God exists, then it is necessarily true that materialism is false.

18. If it is necessarily true that the universe will expand forever, then it is not possible that God exists.

19. If God exists, then it is a necessary truth that God exists.

20. *If God does not exist, then it is a necessary truth that God does not exist.

21. If God does not exist, then it is not possible that God exists.

22. If it is possible that the universe was created, then it is possible that God exists.

23. If it is possible that nothing exists, then it is contingent that the universe exists. (Let N abbreviate "Nothing exists" and let U stand for "The universe exists.")

LINKING MODAL OPERATORS

Our three monadic modal operators can be applied to each other or *concatenated* (Latin for "linked together") as in the following examples:

- The formula $\Diamond \sim A$ can be translated in a number of ways. Strictly speaking, we can read this:

It's possible that it's false that A

There's at least one possible circumstance or possible world in which A is false.

Or, more idiomatically:

It's possible A is false.

A is possibly false.

• The formula ~ ◊ A can also be translated in several ways. Strictly speaking, we can read this as:

It's not the case that it's possible that A.

It's not the case that there's even one possible circumstance or world in which A is true.

A is false in every possible circumstance.

Or more idiomatically:

It's not possible A is true.

A is logically impossible.

• The formula ~ ◊ ~ A is translated

It's not the case that there's even one possible world in which A is false.

It's not possible A is false.

• The formula ~ □ A is translated

It's not the case that A is necessarily true.

A is not a necessary truth.

• The formula ~ □ ~ A is translated

It's not the case that in every circumstance A is false.

A is not a necessary falsehood.

Notice that in the above cases, one operator operates on another, generating a new compound sentence in the process.

"IT AIN'T NECESSARILY SO," OR, TRADING A DIAMOND FOR A BOX AND A BOX FOR A DIAMOND

The proof system in the next chapter will require us to translate sentences that contain possibility operators into sentences that contain necessity operators, and vice versa. The basic logic underlying these transformations is based on two truths:

The necessary is that which cannot possibly be otherwise.

The possible is that which is not necessarily otherwise.

Therefore:

"It is necessarily so" is equivalent to "It is not possibly not so."

"It is not necessarily so" is equivalent to "It's possibly not so."

"It is possibly so" is equivalent to "It is not necessarily not so."

"It is not possibly so" is equivalent to "It necessarily is not so."

Incidentally, Aristotle (384–322 B.C.), in his treatise *De Interpretatione,* was the first logician to explicitly state these logical equivalencies. Over two thousand years ago, Aristotle developed the beginnings of modal logic.

Let us now apply these four modal principles. To say, "In every possible circumstance, A is true" is just to say, "There's not even one possible circumstance in which A is false." Thus,

$$\Box \text{ A is equivalent to } \sim \Diamond \sim \text{A}$$

If one person says, "There's a possible circumstance in which A is true" and another says, "It's not the case that in every possible circumstance A is false," the two have made equivalent statements. Therefore,

$$\Diamond \text{ A is equivalent to } \sim \Box \sim \text{A}$$

To say, "There's no possible circumstance in which A is true" is just to say, "In every possible circumstance A is false." Therefore,

$$\sim \Diamond \text{ A is equivalent to } \Box \sim \text{A}$$

Finally, the expression "It's not the case that in every possible circumstance A is true" is equivalent to "There's at least one possible circumstance in which A is false." Consequently,

$$\sim \Box \text{ A is equivalent to } \Diamond \sim \text{A}$$

With the above in mind, suppose we want to translate a formula containing a box into an equivalent formula that contains a diamond, or vice versa. Here is a rule that can be applied to a formula in order to trade a diamond for a box or a box for a diamond. The rule, which we'll call the Diamond Exchange rule, has three parts, as shown in the following box:

The Diamond Exchange Rule (DE)

1. Add a tilde to each side of the box or diamond.

2. Change the box to the diamond or the diamond to the box.

3. Cancel out any double negatives that result.

For example, suppose we want to translate ~ □ A into an equivalent formula containing the diamond. Here are the three steps :

1. Add a tilde to each side of the box or diamond: ~ ~ □ ~ A

2. Change the box to the diamond or the diamond to the box: ~ ~ ◊ ~ A

3. Cancel out any double negatives that result: ◊ ~ A

In this way, we have transformed ~ □ A into the equivalent formula ◊ ~ A. Here are additional examples:

$$□ \sim A \text{ is equivalent to } \sim ◊ A$$

$$\sim □ \sim A \text{ is equivalent to } ◊ A$$

$$\sim ◊ \sim A \text{ is equivalent to } □ A$$

$$◊ \sim A \text{ is equivalent to } \sim □ A$$

$$\sim ◊ A \text{ is equivalent to } □ \sim A$$

MODAL OPERATORS NEED SCOPE, TOO

When symbolizing sentences containing modal operators, it is important to pay attention to the scopes of the modal operators. The box, diamond, and nabla are monadic operators. The scope of one of these three operators will always be exactly what the scope would be if the operator were changed to a tilde. In the following examples, the scopes of the tildes are marked by underlining:

<u>~ A</u> v (B & G)   <u>~ (G & B)</u>

In the following, the scopes of the boxes are marked by underlining:
<u>□ A</u> v (B & G)   <u>□ (G & B)</u>
The scopes of the diamonds are marked with underlining below:

<u>◊ A</u> v (B & G)   <u>◊ (G & B)</u>

The scopes of the nablas are underlined in the following formulas:

<u>∇ A</u> v (B & G)   <u>∇ (G & B)</u>

If you are uncertain about the scope of a box, nabla, or diamond, find your answer by figuring the scope as if the operator were a tilde.

Recall that the main connective of a sentence is the operator of greatest scope. In each of the following, the main connective is the box:
□ A   □ (A v B)   □ (A ⊃ B)

And the main connective is the diamond in each of these:

◊ (A v B)   ◊ A   ◊ ~ R

However, in each of the following, the main connective is a truth-functional operator:

□ A v B   ◊ A ⊃ B   ◊ A & B

| ▶ | **EXERCISE 23.2** | ◀ |

1. From the following list, pair together formulas that have the same meaning.

   a. □ A   **e.** □ ~ A

   b. ~◊ A   **f.** ◊ A

   c. ◊~A   **g.** ~ □ ~A

   d. ~ ◊ ~A   **h.** ~ □ A

2. True or false?

   a.* Every possibly true proposition is also possibly false.

   b. Every necessarily true proposition is possibly true.

   c. Every necessarily false proposition is possibly false.

   d. Every possibly true proposition is necessarily true.

   e. Every necessary truth is possibly false.

   f.* Every contingent proposition is possibly false.

   g. Every possibly true proposition is also contingent.

   h. Every true proposition is necessarily true.

   i. Every necessarily true proposition is true.

   j. Every necessarily false proposition is false.

   k.* Every necessarily true proposition is contingent.

3. Symbolize each of the following and indicate in each case whether the sentence is true or false.

   a.* If A is true, then either A is necessarily true or A is contingent.

   b. If A is necessarily true, then it's not possible that A is false.

   c. If A is contingent, then it's not possible that A is false.

   d.* If A is true, then it's not possible that A is false.

e. If A is contingent, then either A is possibly true or A is possibly false.

f. If A is possibly true, then A is either necessarily true or contingent.

g. *If A is necessarily false, then A is not contingent.

h. If A is false, then A is not necessarily true.

i. *If A is necessarily false, then A is not possibly true.

j. If A is true, then A is not necessarily false.

k. If A is contingent, then A is not necessarily true.

l. *If A is contingent, then A is possibly true.

4.   True or False?

Let A abbreviate "2 + 2 = 4":

a. *◊ ~ A

b. ~◊ ~ A

c. ~ ∇ A

d. ◊ A

e. *□ A

5.   Let B abbreviate "Dr. Frasier Crane is Seattle's most popular radio personality." True or false?

a. *~ ◊ ~ B

b. ◊ ~ B

c. ∇ ~ B

d. ◊ B

e. *□ ~ B

MODAL RELATIONS

▶ CONSISTENCY AND INCONSISTENCY

Why does the prosecuting attorney ask the defense witness so many questions? For one thing, if the witness is weaving together a false story, it is hoped that he or she will say something that is *inconsistent* with something he or she said earlier. The more details put on the table and the longer the testimony, the greater the chance of an inconsistency. But why is an inconsistency so important here? If the witness's story is inconsistent, it must contain one or more falsehoods.

The concepts of **consistency** and **inconsistency** were introduced in Chapter 1. Two propositions are inconsistent just in case they cannot both be true. Within modal logic, if we wish to say that two propositions, A and B, are inconsistent, we write the symbolic equivalent of "There is no possible situation in which the two would be true together":

$$\sim \Diamond \,(A \ \& \ B)$$

This reads "It is not possible that A and B are both true." Notice that in this formula, the diamond applies to the conjunction as a whole, and the tilde applies to the diamond.

Recall that two propositions are consistent if and only if it is possible that both are true. To represent consistency, we simply apply a diamond to a conjunction. For instance, to assert that A and B are consistent, we write

$$\Diamond \,(A \ \& \ B)$$

This reads, "It is possible that both A and B are true."

▶ **IMPLICATION**

We first encountered the relation of **implication** in Chapter 1. In many respects, implication lies at the heart of logical theory. Within the context of modal logic, we may define this relation in two different but equivalent ways.

- One proposition implies a second proposition if and only if there is no possible circumstance in which the first is true while, in that circumstance, the second is false.

- One proposition implies a second proposition if and only if the second is true in all those possible circumstances—if any—in which the first is true.

In each of the following examples, the first proposition implies the second:

1. Susan is fifteen.
2. Susan is a teen.

1. Jim is taller than Fred.
2. Fred is shorter than Jim.

1. World War II began in 1939.
2. World War II began before 1960.

The Implication Test

Suppose you are considering two propositions, and you aren't sure whether or not the first implies the second. Here is a test you can apply. Ask yourself this question: Are there any circumstances in which: (1) the first is true and (2) the second is false?

If the answer is no, the first implies the second.

If the answer is yes, the first does not imply the second.

If we wish to assert that proposition A implies proposition B, we will write A → B. The symbol placed between A and B, called the arrow, abbreviates *implies*. For example, the proposition expressed by "Ann is taller than Jane" implies the proposition expressed by "Jane is shorter than Ann." In symbols, letting A abbreviate *Ann is taller than Jane* and J abbreviate *Jane is shorter than Ann*, this translates as A → J, with → abbreviating *implies*.

The concept of implication defined in modal logic, represented by the arrow, is different from the (truth-functional) concept associated with the horseshoe operator. In order to distinguish the two, the truth-function for the horseshoe is named "material implication" and the modal concept associated with the arrow is termed "strict implication."

▶ IMPLICATION AND OUR PURSUIT OF TRUTH

An understanding of implication and an ability to "see" the relationship—to follow it from proposition to proposition—can help us greatly in our quest for truth. For example, if you begin with a true proposition and, by a process of reasoning, trace the logical implications of that proposition, you will always arrive at additional true propositions. For if some proposition is true and implies a second proposition, the second proposition must be true as well. If we begin with propositions already known to be true and reflect upon the implications of those truths, we acquire new truths.

The Paradoxes of Implication

Our definition of implication makes perfectly good sense when we apply it to contingent propositions. For instance, most find it entirely natural to say that the following statement:

Pat is sixteen.

implies that

Pat is a teenager.

However, things are different when we apply the definition of implication to noncontingent propositions. When we do, we get genuinely paradoxical results. You can see this for yourself by performing the following simple logical experiment. On one line of a sheet of paper, on the left side, write any sentence that expresses a necessary falsehood. On the right side of the line, write any proposition-expressing sentence of your choice. Let the proposition expressed by the sentence on the left be A, and let the proposition expressed by the sentence on the right be B. Now, ask yourself this question:

Are there any circumstances in which A would be true and B false?

If you think about it, you will see that your answer is no. But, then, according to the definition of implication, A implies B. And if you think about the matter further, you will see that your answer would be the same no matter which necessarily false proposition you represented on the left and no matter what proposition you represented on the right. The general principle in this case is paradoxical:

   A necessarily false proposition implies any and every proposition.

That is, any necessarily false proposition implies every proposition. Thus, the proposition that $2 + 2 = 88$ implies the proposition that Kennedy is president. The proposition that $2 + 2 = 88$ also implies the proposition that $1 + 1 = 2$, and so on.

   Here is a second logical experiment that will uncover a second paradox. On one line of a sheet of paper, on the left side, write any proposition-expressing sentence. On the right side of the line, write any necessary truth of your choice. Call the proposition expressed by the sentence on the left G and the proposition expressed by the sentence on the right H. Now, ask yourself: Are there any circumstances in which G is true and H is false? The answer is again no. But, then, according to the definition of implication, G implies H. Further reflection shows that the answer is the same no matter which necessary truth you choose and no matter what the other proposition is. The general principle in this case too is paradoxical:

   A necessary truth is implied by any proposition.

That is, any and every proposition implies a necessary truth. For example, the proposition that Jimmy Carter is president implies the proposition that all triangles have three sides. The proposition that $2 + 2 = 1,003$ implies the proposition that $1 + 1 = 2$, and so on.

(Continued)

Most people find these two principles paradoxical, yet the two paradoxes follow from our definition of implication. Consequently, they are called the paradoxes of implication.

In everyday life, we normally look for the implications of contingent propositions. This is the context in which our intuitions about implication are formed. With contingent propositions, when one proposition implies another, the two have something to do with each other; the two are relevant to each other. For example, that Chris is age fifty implies that Chris is an adult. Notice that both statements are about Chris and both are about age. But when we extend the definition of implication to cases of noncontingent propositions, we get the paradoxical consequences in which one proposition can imply another proposition even though the first is not about the second, nor is relevant to it in any obvious way. Perhaps this consequence shouldn't be terribly surprising. After all, noncontingent propositions are very different in nature from contingent ones. The logic of the one is likewise different from the other.

However, the paradoxes are bothersome. Why isn't the definition of implication altered so as to eliminate these paradoxical consequences? Logicians have found that the definition of implication cannot be altered without resulting in other equally paradoxical logical results. Why not restrict implication to contingent propositions only? The problem with this suggestion is that logic makes extensive use of the implication relations obtaining between noncontingent propositions.

Incidentally, logic is not the only field containing paradoxes. Modern physics contains a number of puzzling and even bizarre paradoxes, including the twin paradox and the clock paradox within relativity theory and the Schrodinger's-cat paradox within the theory of quantum mechanics.

SOFTENING THE PARADOXES

We saw in Chapter 9 that within truth-functional logic, if we apply a natural deduction proof to an argument that has a contradictory premise, then any conclusion whatsoever follows. But this is just another way of saying that a contradiction implies any and every proposition. In Chapter 9 we also saw that we can prove a proposition to be tautological or necessarily true using a premise-free proof. In such a case, no matter what premises we add to the argument, the argument remains valid, for we can prove it valid without even using the premises. But this is another way of saying that any and every proposition implies a necessary truth. Perhaps in the light of these considerations, the paradoxes of implication will seem a little less paradoxical.

► EQUIVALENCE

You were introduced to the concept of **equivalence** in Chapter 1. Two propositions are equivalent if they imply each other. If two propositions imply each other, it would not be possible for one to be true while the other is false. For instance,

No insects are reptiles.

is equivalent to

No reptiles are insects.

In terms of possible circumstances, if one proposition is equivalent to a second proposition, it follows that

- In any circumstance in which one of the two is true, the other is also true, and in any circumstance in which one of the two is false, the other is also false.

When we say that two propositions are equivalent, we are not saying that they are identical. That is, we are not saying that they are the same proposition. We are simply saying that they imply each other.

Here are examples of pairs of equivalent propositions:

A. Jim is older than Pat.

   Pat is younger than Jim.

B. Sue is married to Sam.

   Sam is married to Sue.

If we wish to assert that propositions **P** and **Q** are equivalent, we will write **P** $\leftrightarrow$ **Q**. The symbol between **P** and **Q**, called the double arrow, abbreviates is *equivalent to.* The proposition that Ann is older than Jane is equivalent to the proposition that Jane is younger than Ann. In symbols this translates A $\leftrightarrow$ J, with A abbreviating *Ann is older than Jane,* $\leftrightarrow$ abbreviating *is equivalent to* and J abbreviating *Jane is younger than Ann.*

SCOPES OF THE DYADIC MODAL OPERATORS

Each of these new modal operators, arrow and double arrow, is a dyadic operator, because each is used to join two components into a compound. As you work with these operators, you must be able to recognize the scopes of the operators. The scope of a dyadic modal operator is always going to be exactly what the scope would be if the operator were a dyadic truth-functional operator

such as an ampersand or wedge. In the following examples, the scopes of the
arrows and double arrows are marked with brackets:

<u>(A → ~ B)</u> v (E & B)   <u>(F v D) → (J & M)</u>   <u>(A ↔ B)</u> & G   <u>D ↔ H</u>

If you are ever uncertain as to the scope of an arrow or double arrow, figure
the scope as if the operator were a horseshoe or ampersand. Recall that the main
connective of a sentence is the operator of greatest scope. Thus, in the following, the main connective is the arrow:

A → (~ B & E)   [(~ F v D) ⊃ R] → (J & M)

In these, the main connective is the double arrow:

(A & B) ↔ G   ~H ↔ (T & O)

SYMBOLIZING WITH DYADIC OPERATORS

Using these new symbols, plus symbols introduced earlier, we can now abbreviate additional modal truths. In the following examples, let **P** and **Q** stand for
any two propositions.

1. Recall the first paradox of implication: A necessary falsehood implies any
other proposition. If **P** is a necessary falsehood, **P** implies **Q** no matter what
proposition **Q** happens to be:

$$□ \sim P → (P → Q)$$

2. The second paradox of implication is the following: A necessary truth is
implied by any proposition. If **P** is necessarily true, **Q** implies **P** no matter what
proposition **Q** happens to be:

$$□ P → (Q → P)$$

3. Remember that two propositions are equivalent if and only if the truth-values of the two match in every circumstance. Suppose **P** and **Q** are both necessarily false. What follows? Both are false in every circumstance. But in that case,
they have matching truth-values in every circumstance, and this implies that
they are equivalent. If **P** and **Q** are both necessarily false, **P** is equivalent to **Q**:

$$(□ \sim P \,\&\, □ \sim Q) → (P ↔ Q)$$

Of course, if **P** and **Q** are both necessarily true, they are both true in every
circumstance, and so their truth-values match in every circumstance. Consequently, if **P** and **Q** are necessarily true, they are equivalent as well:

$$(□ P \,\&\, □ Q) → (P ↔ Q)$$

▶ **EXERCISE 23.3** ◀

Which of the following seem true?

1.*A → (A v B)

2. □ A → □ (A v B)

3. □ (A & B) → (A v B)

4. ◊ A v ◊ ~ A

5.*(□ A & □ B) → (A ↔ B)

6. □ A v ~ A

7. □ ~ A → ◊~ A

8. □ (A & B) → □ B

9.*[(A → B) & ◊ ~ B] → ◊ ~ A

10. □ (A v B) → (□ A v □ B)

11. ~ □ A v □ A

12.*◊ (A & B) → ◊ A

▶ **EXERCISE 23.4** ◀

Part I. Determine which of the following are true and which are false.

1.*If proposition A is actually true and proposition B is actually true, then A and B must be consistent.

2. If A and B are both necessarily false, then A and B must be consistent.

3. If A is necessarily false and B is necessarily true, then A must imply B.

4.*If A is actually false and B is actually false, then A and B must be inconsistent.

5. A necessary truth is consistent with any contingent proposition.

6. If A is necessarily true and B is contingent, then B must imply A.

7. If A and B are both necessarily true, then A and B must be consistent.

8.*If A and B are both necessarily true, then A and B must be equivalent.

9. If A and B are both necessarily false, then A and B must be equivalent.

10. If propositions A and B are both contingent, then A and B must be equivalent.

11.*If propositions A and B are both contingent, then A and B must be consistent.

Part II. Using ML, symbolize the eleven statements above.

MODAL OPERATORS
ARE NOT TRUTH-FUNCTIONAL

Philosophers have encountered numerous difficulties in developing a semantical theory for modal logic. Although the semantics for truth-functional logic was developed during the first half of this century, a semantical theory for modal logic wasn't developed until the late 1950s and early 1960s. If the modal operators had been truth-functional in nature, the semantics of modality would have been as easy as drawing up a few new truth-tables. Unfortunately, the semantics for the modal operators cannot be stated in terms of truth-functions, for the fundamental modal concepts are not truth-functional in nature. Essentially, this is because the truth-value of a sentence whose main operator is a modal operator is not a function of just the truth-value (or values) of its component (or components). That is, when the main operator is a modal operator, there is no determinate or functional relationship between just the truth-value of the components and the truth-value of the compound as a whole.

For example, suppose we know that a given proposition A is false. Given just this information, what is the truth-value of □ A? In this case □ A must be false. However, suppose all we know is that A has the truth-value true. Given *just this information,* what is the truth-value of □ A? There's no way to tell. In some cases, a sentence A will be true when □ A is true, and in other cases A will be true when □ A is false. Given just that A is true, nothing follows regarding □ A. No purely truth-functional relationship exists between A's truth and the truth value of □ A. Therefore, although we can fill in the second row of the table, we cannot fill in the first row of a truth-table for □ A.

A	□ A
T	?
F	F

Thus, the truth-value of □ A is not a function of the truth-value of A.

Suppose some sentence A is true. Given just this information, what is the truth-value of ◊ A? Clearly, ◊ A is true. However, suppose we are given just the information that A is false. Given just this, what is the truth-value of ◊ A? There's no way to tell. If A is false, the truth-value of ◊ A is not thereby determined. There is no functional relationship between A's falsity and the truth-value of ◊ A. Thus, if we were to try to write a truth-table for the diamond operator, although we could fill in the first row, we could not fill in row 2:

A	◊ A
T	T
F	?

If we try to specify truth-tables for the other modal operators, in each case we find rows that cannot be filled in. Thus, the modal operators are not truth-functional in nature. No truth-functions exist that can be associated with the modal operators.

CHAPTER 23 APPENDIX: *There's Nothing New Under the Sun*

The horseshoe and the arrow represent different logical concepts, namely the concepts of material implication and strict implication respectively. Material implication is, of course, a truth-function defined on a (finite) four-row truth-table. Strict implication, on the other hand, is a modal concept specified in this text in terms of an infinity of possible circumstances. The horseshoe and the arrow thus represent two ways to formalize the if-then or conditional operator, and philosophers disagree regarding which way best captures the semantics of conditional sentences.

Shakespeare once observed that there is nothing new under the sun. It is interesting that the contemporary debate between proponents of the truth-functional analysis of the if-then or conditional sentence and advocates of the modal analysis actually goes back to ancient times. A brief examination of the origins of this debate will introduce you to some of the characters who played a role in the early development of logical theory and help you see that, as abstract as logic may be, it was and is the concern of real live human beings.

In the ancient Greek world, several schools of logical theory developed, each associated with a major school of philosophical thought. One school of philosophical thought consisted of philosophers who taught and wrote in the tradition started by Aristotle (384–322 B.C.), the founder of formal logic. The logical theory associated with this school of thought is known as Aristotelian logic. The Megarian school of thought was founded by Euclides, a student of Socrates (469–399 B.C.) and a contemporary of Plato (428–348 B.C.). The Megarians were also interested in logical theory. In addition, a group of philosophers now known as the Stoics also made significant contributions to logic during the ancient period.

In the third century B.C., the Megarian philosophers Diodorus Cronus (d. 307 B.C.) and his pupil Philo engaged in a spirited and famous debate over the logical nature of the conditional sentence, with Diodorus advocating the semantics we now associate with the arrow and Philo advocating the semantics we now associate with the horseshoe.

Because the actual writings of Philo and Diodorus were not preserved, our source for their debate is the writings of a third-century philosopher, Sextus Empiricus. In the Outlines of Phyrronism (II, 110), Sextus contrasts the views of Philo and Diodorus as follows:

> *Philo says that a sound [true] conditional is one that does not begin with a truth and end with a falsehood, e.g., when it is day and I am conversing the statement "If it is day, then I am conversing." But Diodorus says it is one that neither could nor can begin with a truth and end with a false-hood. According to him, the conditional statement just quoted seems to be false, since when it is day and I have become silent, it will begin with a truth and end with a falsehood.[3]*

This passage clearly suggests that Philo advocated the truth-table analysis of the conditional operator, whereas Diodorus advocated the modal analysis.

The following passage, from another of Sextus' works, also indicates that Philo advocated the analysis represented by the modern day truth-table for the horseshoe. In *Against the Mathematicians*, Sextus presents Philo's account as follows:

> *Since, then, there are four possible combinations of the parts of a conditional—true antecedent and true consequent, false antecedent and false consequent, false and true, or conversely true and false—they say that in the first three cases the conditional is true (i.e., if the antecedent is true and the consequent is true, it is true; if false and false, it again is true; likewise for false and true); but in one case only is it false, namely, when-ever the antecedent is true and the consequent is false.[4]*

It seems that quite a number of philosophers joined the debate over the semantics of the conditional. Indeed, so many philosophers entered the fray, and the discussion became so intense, that Callimachus is reported to have written the following:

> *Even the crows on the roofs caw about the nature of conditionals.[5]*

And this was more than 2,300 years ago!

---

[3] This passage appears in William Kneale and Martha Kneale, *The Development of Logic* (Oxford: The Clarendon Press, 1962) p. 128.
[4] This passage is quoted in Benson Mates, *Elementary Logic* (New York: Oxford University Press, 1972) p. 214.
[5] Quoted in Kneale and Kneale, *The Development of Logic*, p. 128.

Some of the ancient logicians were interesting characters. For instance, Zeno, who was from Cyprus, did not make himself popular when, upon arriving in Athens, he proposed to reform the Greek language before he had even learned to speak it! According to legend, Diodorus finally committed suicide because he couldn't solve a logical puzzle Stilpo had given him. (Now that's taking logic seriously!) Chrysippus (c. 280–205 B.C.) the founder of Stoic logic, is supposed to have written to Cleanthes (a fellow Stoic philosopher), "Just send me the theorems; I'll find the proofs for myself." One of the old sayings was, "If there is any logic in Heaven, it is that of Chrysippus." According to Diogenes, Chrysippus wrote 311 books on logic. Incidentally, Cleanthes has been described as a "poverty-stricken prize fighter who came to Athens and entered Zeno's school, became its head, transmitted Zeno's doctrines without change, and eventually starved himself to death at the age of 99."[6]

In addition to being the first to formulate a truth-functional analysis of the conditional operator, the Stoic logicians were the first, as far as we know, to formulate what amount to truth-functional analyses of the *and, or,* and *not* operators. For instance, Sextus Empiricus described the Stoic truth-functional account of conjunction in the following way:

> *A conjunction is [true] when it has everything in it true . . . , false when it has one thing false.*[7]

This is clearly a summary of the ampersand's truth-table.

The Stoics were also the first logicians, as far as we know, to formulate natural deduction inference rules. (Natural deduction was introduced in Chapter 7.) They used numbers instead of variables for sentences. Following are several Stoic inference rules:

1. If the first, then the second;
   The first;
   Therefore, the second.

2. If the first, then the second;
   Not the second;
   Therefore, not the first.

3. The first or the second;
   Not the first;
   Therefore, the second.

4. Not both the first and the second;
   The first;
   Therefore, not the second.[8]

---

[6] See Mates, *Elementary Logic*, pp. 212–213.
[7] Quoted in John Rist, ed. *The Stoics* (Los Angeles: University of California Press, 1978) p. 17.
[8] These rules appear in Benson Mates, *Elementary Logic*, pp. 214–216. See also Kneale and Kneale, *The Development of Logic*, p. 163.

▶	**QUESTIONS**	◀

The following questions are for those who studied truth-functional natural deduction.

1. Which modern truth-functional natural deduction rule corresponds to the first Stoic rule above? Which corresponds to the second? Which to the third?

2. State the Hypothetical Syllogism rule in the Stoic fashion. State the Constructive Dilemma rule in Stoic form.

3. Which valid argument form corresponds to the first Stoic rule above? Which corresponds to the second? Which to the third?

CHAPTER 23 GLOSSARY

Analytically true statement   A statement with the following feature: If we apply logical analysis to the concept expressed by the subject term of the sentence, we see that if the subject term applies to (is true of) an individual thing, the predicate term must also apply. By logically analyzing the subject, the predicate automatically follows. Thus, through a process of logical analysis, without investigating the physical world, we see that the proposition expressed by the sentence must be true.

Consistency   Two propositions are consistent if and only if it is *possible* that both are true.

Contingent proposition   A proposition that is true in at least one possible circumstance and that is also false in at least one possible circumstance.

Equivalence   Two propositions are equivalent if and only if they imply each other, which is to say that it would not be possible for the two to differ as to truth and falsity.

Implication   One proposition implies a second proposition if and only if no possible circumstance exists in which the first is true while, in that circumstance, the second is false.

Inconsistency   Two propositions are inconsistent just in case they cannot both be true, that is, no possible situation exists in which the two would be true together.

Logical contradiction   A statement of the form $P \ \& \sim P$.

Logically possible   Possibly true.

Modal argument   An argument built out of *modal* sentence operators and the components to which these attach.

Modal logic   The study of the modes of truth, especially necessity and possibility, and their relationship to reasoning and argumentation.

Necessarily false proposition   A proposition that is false in all possible circumstances.

Necessarily true proposition A proposition that is true in all possible circumstances.

Physically impossible Inconsistent with the actual laws of physics (and of the laws of nature in general) that "govern" the physical universe.

Physically possible Consistent with the actual laws of physics (and of the laws of nature in general) that "govern" the physical universe.

Possible world A way the universe might be or might have been.

Possibly false proposition A proposition that is false in at least one possible circumstance or situation.

Possibly true proposition A proposition that is true in at least one possible circumstance or situation.

C h a p t e r

24

MODAL LOGIC: METHODS OF PROOF

A **modal argument** is an argument whose validity or invalidity depends on the arrangement of its modal operators. Many of the most intriguing arguments in the history of philosophy are modal in nature. A system of modal logic can help us understand and evaluate modal arguments. Such a system consists of two parts: (1) a formal language for abbreviating sentences that contain modal operators; and (2) a set of principles for the evaluation of modal arguments. Let us begin with some fundamentals.

FIVE MODAL PRINCIPLES

1. If a sentence **P** is a truth-functional tautology, the final column of **P**'s truth-table contains all **T**'s. Because collectively the rows of a truth-table represent all possibilities, it follows that **P** is true in all possible circumstances or possible situations. Consequently, every tautology is a necessary truth. Thus:

> *Principle 1: If **P** is a tautology, then **P** is necessarily true.*

2. If **P** is necessarily true, **P** is true in every possible circumstance or possible situation. It would follow that **P** is true in the actual circumstance. So, if **P** is necessarily true, **P** is actually true. Every necessary truth is actually true. Thus:

> *Principle 2: If **P** is necessarily true, then **P** is actually true as well.*

3. Suppose that **P** is necessarily true and that **P** implies **Q**. Because **P** is necessarily true, **P** is true in every possible circumstance. Because **P** implies **Q**, there is no circumstance in which **P** is true and **Q** false. It follows that **Q** must be true in every possible circumstance as well. For if there were to be even one circumstance in which **Q** is false, that would be a circumstance in which **P** is true and **Q** is false, which would mean, contrary to our assumption, that **P** does not imply **Q**. Thus:

> *Principle 3: If **P** is necessarily true, and **P** implies **Q**, then **Q** is also necessarily true.*

That is, whatever is implied by a necessary truth is itself also necessarily true.

4. It would seem that if **P** is true in a circumstance—call it circumstance 1—then in every circumstance it's true that **P** is true in circumstance 1. Suppose **P** is true in all possible circumstances. It would seem that from the perspective of any circumstance, it is true that **P** is true in every circumstance. So, if **P** is true in every circumstance, then in every possible circumstance it must be true that **P** is true in every circumstance. That is, if it is necessarily true that **P**, it is necessarily true that **P** is necessarily true. That is:

> *Principle 4: If **P** is necessarily true, then **P** is necessarily necessarily true.*

Some philosophers dispute this principle, although we shall not enter into that debate in this text. Principle 4 amounts to the claim that the necessity of a proposition is itself also a matter of necessity. That is, it is not a contingent matter if **P** is a necessary truth. Rather, if **P** is necessarily true, it is necessary that **P** is necessarily true.

5. Suppose that a sentence **P** is possibly true. That is, suppose **P** is true in at least one possible circumstance. It would seem that from the perspective of any circumstance, it's true that there is a circumstance in which **P** is true. But then it seems plausible to suppose that in every possible circumstance, it's true that there is a possible circumstance in which **P** is true. So, if **P** is possibly true, then it's necessarily true that **P** is possibly true. Thus:

> *Principle 5: If **P** is possibly true, then **P** is necessarily possibly true.*

Some philosophers dispute this principle as well, although we shall not enter into that discussion in this text. According to principle 5, possibility is also a matter of necessity.

These five principles of modality may be expressed more formally as follows:

Principle 1: If **P** is tautological, then □ **P**

Principle 2: □ **P** ⊃ **P**

Principle 3: [□ **P** & (**P** → **Q**)] ⊃ □ **Q**

Principle 4: □ **P** ⊃ □ □ **P**

Principle 5: ◊ **P** ⊃ □ ◊ **P**

A number of different systems of modal logic have been developed, each with a different set of natural deduction rules and each based on a different set of modal principles. Each system constitutes a unique way of representing formally what we believe about the modal operators and modal reasoning. Many philosophers argue that all five of the above modal principles are necessarily true. Consequently, the deduction rules they specify are all logical consequences of principles 1 through 5. The system of modal logic that is based on all five of the modal principles is known as the **system S5**. Some philosophers dispute the truth of the fifth modal principle and accept only the first four principles. (We shall not enter into this debate here.) Accordingly, they specify natural deduction rules that are designed to reflect only principles 1 through 4. The system of modal logic that is based on the first four of the modal principles above is called the **system S4**. The systems S5 and S4 were first formulated and named by C. I. Lewis (1883–1964), a philosopher who helped develop the beginnings of modern modal logic.

Although philosophers do not all agree on which system best represents the true logic of possibility and necessity, the system S5 is probably the most widely-used formal system of modal logic. The system S5 is "stronger" than the system S4 in the sense that any argument that can be proven valid using S4 rules can also be proven valid using S5 rules, but not vice versa. So, many arguments that can be proven valid in S5 cannot be proven valid in S4. Another way to put this is to say that S5 "contains" S4. This chapter will introduce formal proof techniques only for the system S5.

An **S5 natural deduction system** is a natural deduction system that (a) includes all of the truth-functional rules, (b) includes rules for handling the modal operators, and (c) is designed to reflect all five of the principles of modality presented above. In this chapter, we will use an S5 natural deduction system to evaluate modal arguments.

SIX INFERENCE RULES

Let us begin with some fairly obvious rules of inference that are also pretty easy to work with. Each reflects one or more of the five principles of modality. Recall the second of the five principles of modality: If **P** is necessarily true, then **P** is true. That is, □ **P** ⊃ **P**. The corresponding inference rule in this case is summarized in the box below:

The Box Removal Rule (BR)

From a sentence □ **P**, you may infer the corresponding sentence **P**.

In each of the following inferences, the second formula was derived from the first in accord with the Box Removal rule:

1. □ A	**1.** □ B	**1.** □ (A & B)	**1.** □ ~ (A v B)
2. A	**2.** B	**2.** (A & B)	**2.** ~ (A v B)

It is important that you understand the following. You cannot apply BR to a line of a proof unless (a) the line begins with a box, (b) a tilde has *not* been applied to the box, and (c) the scope of the box covers the entire rest of the line. In other words, the box must sit unencumbered on the far left of the line, and it must be the main connective—it must govern the formula as a whole. So, BR *cannot* be applied to the following lines:

- ~ □ H (The line begins with a tilde rather than a box.)

- □ A ⊃ R (The box only applies to A, so the scope of the box is not the whole line.)

- □ G & Z (The box only applies to G, so the scope of the box is not the whole line.)

- G v □ ~ R (The box only applies to ~ R, so the scope of the box is not the whole line.)

The following proof employs the Box Removal rule:

1.	□ (A ⊃ B)	
2.	□ (B ⊃ C)	
3.	□ (C ⊃ D) / A ⊃ D	
4.	A ⊃ B	BR 1
5.	B ⊃ C	BR 2
6.	C ⊃ D	BR 3
7.	A ⊃ C	HS 4, 5
8.	A ⊃ D	HS 6, 7

The Possibilization rule, summarized in the following box, allows you to bring down a line of a proof and prefix a diamond (the possibility operator) to the line.

> ### The Possibilization Rule (Poss)
>
> From a sentence **P** you may infer the corresponding sentence ◊ **P**.

In each of the following inferences, the second formula was derived from the first in accord with the Possibilization rule:

1. A	1. A v B	1. J & K
2. ◊ A	2. ◊ (A v B)	2. ◊ (J & K)

The Possibilization rule reflects the following truth: If **P** is true, it logically follows that **P** is also possibly true. That which is actually true is also possibly true. Indeed, as we saw in the last chapter, we reason in accord with this principle in everyday life. For example:

Jan: Is it really *possible* someone could eat twenty-five bean-and-cheese burritos in thirty minutes?

Bob: Sure, it's possible.

Jan: How do you know?

Bob: Because a guy *actually* did it over at the state fair last year—the guy that won the burrito-eating contest.

Bob's reasoning fits the following form: **P** is actually true, therefore, **P** is possible. And this seems to be valid reasoning.

When you apply this rule, it is important that you place the diamond on the left side of the line and you must ensure that the scope of the diamond covers the entire line. The following inferences are *not* correct applications of the rule:

1. A v B	1. H & G
2. ◊ A v B (incorrect)	2. ◊ H & G (incorrect)

The following proof employs the Possibilization rule:

1. ◊ A ⊃ □ ~ B
2. A & H
3. W ⊃ B / ~ W
4. A   Simp 2
5. ◊ A   Poss 4
6. □ ~B   MP 1, 5
7. ~ B   BR 6
8. ~ W   MT 3, 7

The additional rules summarized in the following box are similar to rules we have already learned:

Modal Modus Ponens (MMP): From **P** → **Q** and the corresponding sentence **P** you may infer the corresponding sentence **Q**.

Modal Modus Tollens (MMT): From **P** → **Q** and the corresponding sentence ~**Q**, you may infer the corresponding sentence ~**P**.

Modal Hypothetical Syllogism (MHS): From **P** → **Q** and the corresponding sentence **Q** → **R** you may infer the corresponding sentence **P** → **R**.

The names of these rules derive from the fact that each is a *modal version* of a corresponding truth-functional inference rule. These three rules are "modal versions" of the analogous truth-functional rules in the following sense: They are the same as the corresponding truth-functional rules except for the fact that the modal rules display the arrow, a modal operator, in the place where the truth-functional rules display the truth-functional horseshoe operator.

Here are some sample proofs using these new rules:

1. A & B
2. ◊ (A & B) → □ E
3. □ E → ◊ S / ◊ S
4. ◊ (A & B)   Poss 1
5. □ E   MMP 2, 4
6. ◊ S   MMP 3, 5

In the above proof, at line 4, we applied the Poss rule and tacked on a diamond. At line 5, we employed the modal version of Modus Ponens and brought down □ E. We then applied this same rule to lines 3 and 5 to bring down ◊ S.

Consider the following proof:

1. □ ~ ~ (E v S)
2. ◊ B → ~ (E v S)
3. ~ ◊ B → □ G / G
4. ~ ~ (E v S)   BR 1
5. ~ ◊ B   MMT 2, 4
6. □ G   MMP 3, 5
7. G   BR 6

Notice that in the above proof, MMT, the modal version of Modus Tollens, works just as Modus Tollens works, except for the presence of the arrow in place of the horseshoe.

THE POSSIBILITY TO NECESSITY RULE

The next inference rule allows us, under special conditions, to reason from possibility to necessity. However, in order to understand the rule, you will need to understand the concept of a modally closed formula. A sentence **P** of ML is **modally closed** just in case every sentence letter within **P** appears within the scope of a modal operator. Thus, each of the following sentences is modally closed:

□ A   □ (A & B)   ◊ (H v □ G)   D → (R v H)

And each of the following is not modally closed:

□ A v B   ◊A v B   P ⊃ Q   B ⊃ □ A

An important consequence of the five principles of modality (stated near the beginning of this chapter) is that every modally closed sentence of ML expresses a noncontingent proposition. If a formula is modally closed, the proposition expressed is either true in every circumstance or it is false in every circumstance, that is, it is either necessarily true or it is necessarily false. Another way to put this: its truth-value is invariant. So, if a modally closed formula is assumed true, it follows from this assumption that the proposition expressed is necessarily true, and if a modally closed formula is assumed false, it follows from this assumption that the proposition expressed is necessarily false.

Now, the inference rule summarized in the following box is valid.

The Possibility to Necessity Rule (P2N)

From: **P → Q**

   ◊ **P**

Infer: □ **Q**

Provided that the formula instantiating **Q** is itself a modally closed formula.

According to this rule, from a line of the form **P → Q** and a corresponding line ◊ **P**, we may infer the corresponding line □ **Q** provided that the formula instantiating **Q** is a modally closed formula.

If this rule of inference seems invalid, think of it this way. Suppose **P** implies **Q**. Then there is no possibility that **P** is true and **Q** false. Suppose further that a circumstance exists in which **P** is true. Because **P** implies **Q**, **Q** must be true in that circumstance also. However, if **Q** is modally closed, then **Q** is either true in every circumstance or true in no circumstance. Because we know that **Q** is true in at least one circumstance, it follows that **Q** must be true in *all* circumstances. We are therefore justified in adding a box to **Q** and asserting that **Q** is necessarily true.

The following proof employs the P2N rule:

1.	A → ◊ H	
2.	J ⊃ A	
3.	J & G	
4.	□ ◊ H → D / D	
5.	J	Simp 3
6.	A	MP 2, 5
7.	◊ A	Poss 6
8.	□ ◊ H	P2N 1, 7
9.	D	MMP 8, 4

Notice that we were allowed to apply the P2N rule to lines 1 and 7 because (a) on line 1, the formula is of the form **P → Q**, and (b) the formula instantiating **Q** in line 1 is indeed modally closed (that formula is diamond H), and (c) line 7 is a possibilization of A.

►   **EXERCISE 24.1**   ◄

Supply proofs for the following arguments. You will need to use truth-functional inference and replacement rules plus the first six modal rules.

(1)* 1. ◊ (H & S) ⊃ G
    2. ~R
    3. R v (H & S) / G

(2) 1. R → M
    2. H & R / M v G

(3) 1. ~H
    2. S → H
    3. ◊ ~ S ⊃ R / R v ◊ M

(4)* **1.** H → G

    2. G → (M & N)

    3. [H → (M & N)] ⊃ S / S

(5) **1.** ~R

    2. R v (A ⊃ B)

    3. (A ⊃ B) ⊃ (P → Q)

    4. ~Q / ~ P

(6)* **1.** A → ◊ S

    2. A / □ ◊ S

(7) **1.** □ R

    2. R ⊃ G

    3. ~G v M / ◊ M

(8) **1.** □ A → ◊ B

    2. H ⊃ ~ ◊B

    3. ~ H ⊃ G

    4. ~ G / ~ □ A

(9)* **1.** F → (□ A & □ B)

    2. G ⊃ F

    3. H & G / □ (□ A & □ B)

(10) **1.** G → ◊ M

    2. H & (G & R)

    3. □ ◊ M ⊃ A / ◊ A

(11) **1.** M v (H & W)

    2. M → ~ G

    3. □ (G & S)

    4. ◊ W ⊃ □ B / A ⊃ B

(12)* **1.** □ (A ⊃ B)

    2. H v A

    3. ~ (F v H) / B

(13) **1.** □ (A ⊃ B)

    2. B ⊃ G

3. ~ G v S

4. ~ (S v R) / ◊ ~ A

(14) * **1.** □ (A v B)

    2. B ⊃ H

    3. ~ R & ~ H / A

(15) **1.** □ □ □ A / A

(16) **1.** ◊ A / ◊ ◊ A

(17) **1.** ◊ A

    2. A → ◊ (E v B)

    3. ◊ (E v B) → H / H

(18) **1.** □ □ A / ◊ □ A

(19) * **1.** □ (A & B) / ◊ A & ◊ B

(20) **1.** G → □ G

    2. ~ G → □ ~G

    3. ◊ G / G

THE NECESSITATION RULE

The next rule is easy to work with, but the explanation behind the rule is a bit complicated. According to the third of the five principles of modality, whatever follows from a necessary truth is itself a necessary truth. That is,

$$[\Box \, \mathbf{P} \, \& \, (\mathbf{P} \to \mathbf{Q})] \supset \Box \, \mathbf{Q}$$

With that as background, suppose that we begin a line of reasoning with one or more *modally closed* formulas and, using only valid rules of inference, derive a formula **P**. What conclusion may be drawn concerning **P**? First, *if* the modally closed formulas we began with are true, they are necessarily true, for a modally closed formula is true in all circumstances if it is true in even one circumstance. But the third modal principle (from the list of five modal principles near the beginning of the chapter) assures us that whatever follows from a necessary truth is itself also necessarily true. Therefore, if **P** follows from one or more modally closed formulas, and if those modally closed formulas are true, they are necessarily true and as a consequence **P** is also necessarily true as well. The conclusion we may draw from this is that **P** is necessarily true. In this case, we may therefore prefix a box to **P**. This logic is incorporated into our next rule, the Necessitation rule.

The Necessitation rule will require a move called the "reiteration" of a formula. A formula that appears as a *whole* line of a proof is reiterated if it is copied from an earlier line and entered as a later line in the proof. Now, suppose that at some point in a proof you indent, reiterate one or more modally closed lines, and derive from *just those reiterated lines* a sentence **P**. (Such a sequence of indented lines is called a necessitation subproof.) Because **P** has been derived from only modally closed formulas, **P** must be necessarily true if the modally closed formulas are true. It follows that we are justified in prefixing a box to **P** in such a case. The corresponding rule is as follows:

The Necessitation Rule (Nec)

At any point in a proof, you may indent and construct a "necessitation subproof" in which every line is either justified by the *reiteration rule* (below) or follows from previous lines of the subproof by a valid rule of inference. You may then end the indentation, draw a vertical line in front of the indented lines (to mark them off from the rest of the proof), write down any line derived within the subproof, and prefix a box to that line. (As justification write "Nec" and the line numbers of the subproof.)

The Reiteration Rule (Reit)

You may reiterate into a necessitation subproof any entire line provided that the entire line consists of just one modally closed formula and the formula does not lie within the scope of a discharged assumption or a terminated necessitation subproof. (As justification write "Reit" and the formula's original line number.)

This rule sounds complicated, but it is actually quite easy to apply. Here is a simple example to begin with:

1. $\Box(A \to B)$

2. $\Box(B \to R) \; / \; \Box(A \to R)$

3. $\qquad\qquad \mid \Box(A \to B) \qquad$ Reit 1

4. $\qquad\qquad \mid \Box(B \to R) \qquad$ Reit 2

5. $\qquad\qquad \mid A \to B \qquad\qquad$ BR 3,

6. $\qquad\qquad \mid B \to R \qquad\qquad$ BR 4

7. $\qquad\qquad \mid A \to R \qquad\qquad$ MHS 5, 6

8. $\Box(A \to R) \qquad\qquad$ Nec 3–7

Notice that after completing a necessitation subproof, we mark the indented lines off as in a conditional or indirect proof.

A necessitation subproof serves to demonstrate that a particular formula follows validly from only modally closed premises. When we have derived a formula within a necessitation subproof, we have shown that it follows from modally closed formulas. Because modally closed formulas are necessarily true if true at all, and because only necessary truths follow from necessary truths, any formula derived within such a subproof must be necessarily true if the premises it is derived from are true. We are therefore justified in prefixing a box to any formula so derived. This is clearly a valid rule of inference.

The following proof contains another necessitation subproof:

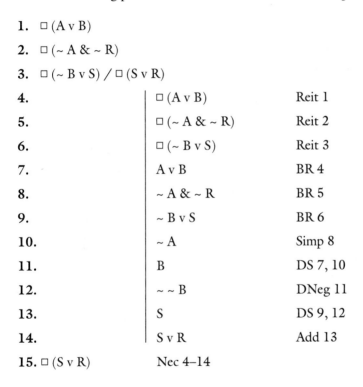

1.	□ (A v B)	
2.	□ (~ A & ~ R)	
3.	□ (~ B v S) / □ (S v R)	
4.	□ (A v B)	Reit 1
5.	□ (~ A & ~ R)	Reit 2
6.	□ (~ B v S)	Reit 3
7.	A v B	BR 4
8.	~ A & ~ R	BR 5
9.	~ B v S	BR 6
10.	~ A	Simp 8
11.	B	DS 7, 10
12.	~ ~ B	DNeg 11
13.	S	DS 9, 12
14.	S v R	Add 13
15.	□ (S v R)	Nec 4–14

▶   **EXERCISE 24.2**   ◀

Construct proofs for the following arguments. The Necessitation rule is recommended on these problems.

(1)   **1.** □ (A v B)

     2. □ ~ B

     3. □ A ⊃ □ E / E

(2) * **1.** □ (H ⊃ S)

    2. □ (S ⊃ R)

    3. □ (H ⊃ R) ⊃ G / ◊G

(3)   **1.** □ (~R ⊃ S)

    2. □ (~S v G)

    3. □ [(R v G) ⊃ H]

    4. ~ □ H v S / S

(4) * **1.** □ (A ⊃ B)

    2. □ (~ B v ~ A)

    3. G ⊃ ~ □ ~ A / ~G

(5)   **1.** □ (A & B)

    2. □ (B ⊃ S) / □ (H ⊃ S)

(6)   **1.** □ (H ⊃ S)

    2. □ (S ⊃ A)

    3. □ [◊ (H ⊃ A) ⊃ G] / □ (E ⊃ G)

(7) * **1.** A → B

    2. □ A / □ B

(8)   **1.** □ (A ⊃ B)

    2. □ (B ⊃ ~C) / □ ~ (A & C)

(9)   **1.** □ (A ⊃ B) / □ A ⊃ □ B (Hint: use CP)

(10) * **1.** □ A & □ B / □ (A & B)

(11)   **1.** □ [(A & B) & R] / □ A & (□ B & □ R)

(12) * **1.** ◊A / □ □ ◊ A

(13)   **1.** ◊ □ A / □ ◊ □ A

(14) * **1.** ◊ A / □ ◊ ◊ A

(15)   **1.** □ A / □ □ A

(16) * **1.** □ A / ◊ □ A

(17)   **1.** □ A & □ B / □ (A v B)

(18)   **1.** A → B

    2. □ A / □ B

(19)   **1.** A → B

    2. □ ~B / □ ~ A

(20) 1. □ A ∨ □ B

    2. ~ □ A / □ (A ∨ B)

(21)*1. □ (A & B) / □ A & □ B

FOUR MODAL REPLACEMENT RULES

Recall from Chapter 10 that if a sentence within an argument is removed and replaced by a sentence expressing an equivalent proposition, the validity of the argument will not be affected. That is, if the argument happened to be valid before the replacement, it will remain valid after the replacement. Our system will contain four replacement rules. The first three are fairly simple.

> ### Arrow Exchange (Arrow Ex)
>
> A sentence of the form **P → Q** may replace or be replaced with the corresponding sentence □ (**P ⊃ Q**).

    Essentially, the Arrow Exchange rule allows us to replace an arrow with a necessitated horseshoe, or vice versa. The rule is based on the fact that any sentence of the form □ (**P ⊃ Q**) is equivalent to the corresponding sentence (**P → Q**). If this sounds incorrect, consider the following. If a sentence of the form □ (**P ⊃ Q**) is true, then the corresponding sentence **P ⊃ Q** has the truth-value **T** in every possible circumstance. Because a conditional **P ⊃ Q** is false only when **P** is true and **Q** is false, this implies that no circumstance exists in which **P** is true and **Q** is false. Therefore, **P** implies **Q**. Therefore **P → Q** is true. Similar reasoning would show that if a sentence **P → Q** is true, then the corresponding sentence □ (**P ⊃ Q**) must be true. The two are therefore equivalent.

    The following proof employs arrow exchange:

1. A → B

2. B → S

3. ~ □ (A ⊃ S) ∨ □ R / R ∨ Z

4. A → S   MHS 1, 2

5. □ (A ⊃ S)   Arrow Ex 4

6. ~ ~ □ (A ⊃ S)   DNeg 5

7. □ R   DS 3, 6

8. R   BR 7

9. R ∨ Z   Add 8

> ### *Double Arrow Exchange (Double Arrow Ex)*
>
> A sentence of the form $P \leftrightarrow Q$ may replace or be replaced with the corresponding sentence $\square (P \equiv Q)$.

Double Arrow Exchange is a valid rule, for a sentence of the form $P \leftrightarrow Q$ implies the corresponding sentence $\square (P \equiv Q)$ and a sentence of the form $\square (P \equiv Q)$ implies the corresponding sentence $P \leftrightarrow Q$. Essentially, the Double Arrow Exchange rule allows us to replace a double arrow with a necessitated triple bar, or vice versa. The rule is based on the fact that any sentence of the form $P \leftrightarrow Q$ is equivalent to the corresponding sentence $\square (P \equiv Q)$. If this sounds incorrect, consider the following. If a sentence of the form $\square (P \equiv Q)$ is true, then the corresponding sentence $P \equiv Q$ has the truth-value T in every possible circumstance. Because a biconditional $P \equiv Q$ is false only when P and Q have differing truth-values, this implies that there is no possible circumstance in which P and Q have differing truth-values. Therefore, P and Q are equivalent. Therefore, $P \leftrightarrow Q$ is true. So if $\square (P \equiv Q)$ is true, then the corresponding sentence $P \leftrightarrow Q$ is true. Similar reasoning would show that if a sentence $P \leftrightarrow Q$ is true, then the corresponding sentence $\square (P \equiv Q)$ must be true. The two are therefore equivalent.

The following proof employs this rule:

1. $A \leftrightarrow B \; / \; A \supset B$

2. $\square (A \equiv B)$ Double Arrow Ex 1

3. $A \equiv B$ BR 2

4. $(A \supset B) \& (B \supset A)$ Equiv 3

5. $A \supset B$ Simp 4

> ### *The Modal Equivalence Rule (ME)*
>
> A sentence of the form $[(P \rightarrow Q) \& (Q \rightarrow P)]$ may replace or be replaced with the corresponding sentence $(P \leftrightarrow Q)$.

Modal Equivalence is obviously valid. A sentence of the form $P \leftrightarrow Q$ is true if and only if the corresponding sentence P implies the corresponding sentence Q and Q also implies P. The corresponding sentences $P \rightarrow Q$ and $Q \rightarrow P$ are true if and only if the corresponding sentence $P \leftrightarrow Q$ is true. A sentence of the form $P \leftrightarrow Q$ is thus equivalent to the corresponding sentence $[(P \rightarrow Q) \& (Q \rightarrow P)]$. Here is a proof using Modal Equivalence:

1. $A \leftrightarrow B \; / \; A \supset B$

2. $(A \rightarrow B) \& (B \rightarrow A)$ ME 1

3.	$(A \rightarrow B)$	Simp 2
4.	$\Box (A \supset B)$	Arrow Ex 3
5.	$A \supset B$	BR 4

Our fourth replacement rule, the Diamond Exchange rule, summarized in the following box, is essentially the same rule we introduced in Chapter 23.

The Diamond Exchange Rule (DE)

If a sentence **P** contains either a box or a diamond, **P** may be replaced by or may replace a sentence that is exactly like **P** except that the box has been switched for the diamond or vice versa in accord with the steps below:

a. Add a tilde to each side of the box or diamond.

b. Trade the box for a diamond or the diamond for a box.

c. Cancel out any double negatives that result.

Now, here are two proofs that employ equivalence rules. The first features DE:

1.	$\sim \Box \sim A$	
2.	$\sim \Box A$	
3.	$(\Diamond A \,\&\, \Diamond \sim A) \supset \Box H \,/\, H$	
4.	$\Diamond A$	DE 1
5.	$\Diamond \sim A$	DE 2
6.	$\Diamond A \,\&\, \Diamond \sim A$	Conj 4, 5
7.	$\Box H$	MP 3, 6
8.	H	BR 7

The next proof has two necessitation proofs nested within an indirect proof.

1.	$\Diamond (A \,\&\, B)$			
2.	$A \rightarrow G \,/\, \Diamond (G \,\&\, B)$			
3.		$\sim \Diamond (G \,\&\, B)$	AP	
4.		$\Box \sim (G \,\&\, B)$	DE 3	
5.			$\Box \sim (G \,\&\, B)$	Reit 4
6.			$\sim (G \,\&\, B)$	BR 5

7.		~G v ~ B	DM 6
8.		G ⊃ ~ B	Imp 7
9.	□ (G ⊃ ~ B)	Nec 5-8	
10.	G → ~B	Arrow Ex 9	
11.	A → ~B	MHS 2, 10	
12.		A → ~B	Reit 11
13.		□ (A ⊃ ~B)	Arrow Ex 12
14.		A ⊃ ~ B	BR 13
15.		~A v ~B	Imp 14
16.		~ (A & B)	DM 15
17.	□ ~ (A & B)	Nec 12-16	
18.	~ ◊ (A & B)	DE 17	
19.	◊ (A & B) & ~ ◊ (A & B)	Conj 1, 18	
20. ◊ (G & B)		IP 3-19	

In the above proof, the indirect proof sequence began at step 3 and concluded at step 20. Within that, the first necessitation proof began at step 5 and concluded at step 9; and the second necessitation proof began at step 12 and concluded at step 17.

VALIDITY IN S5

The system of modal logic consisting of (1) the language ML and (2) the S5 natural deduction rules of this chapter, shall be called the **system S5D** (for "S5 deduction"). A **proof in S5D** is a sequence of formulas of ML that is such that every formula is either a premise, an assumption, or follows from previous formulas by an S5D rule, and in which every line that is not a premise is justified and every assumption has been discharged. An argument is **valid in S5D** if and only if it is possible to construct a proof in S5D whose premises are the premises of the argument and whose conclusion is the conclusion of the argument. Any argument that is valid in S5D may also be said to be "valid in S5" or "S5 valid."

Finally, examine the following proofs. In each case, try to understand the justifications given for each line. This will help familiarize you with the proof procedures of the system S5D.

(1)   1. □ (A ⊃ E)

    2. □ (E ⊃ A)

    3. (A ↔ E) ⊃ □ R / R

4. A → E   Arrow Ex 1

5. E → A   Arrow Ex 2

6. (A → E) & (E → A)   Conj 4,5

7. A ↔ E   ME 6

8. □ R   MP 3, 7

9. R   BR, 8

(2) 1. ◊ S ⊃ □ Q

2. □ ∼ S ⊃ ◊ P

3. □ ∼ P / Q

4. ∼ ◊ P   DE 3

5. ∼ □ ∼ S   MT 2, 4

6. ◊ S   DE 5

7. □ Q   MP 1, 6

8. Q   BR, 7

(3) 1. ◊ (P & Q) ⊃ □ ∼ (R v S)

2. ◊ (R v S) / ∼ (P & Q)

3. ∼ □ ∼ (R v S)   DE 2

4. ∼ ◊ (P & Q)   MT 1, 3

5. □ ∼ (P & Q)   DE 4

6. ∼ (P & Q)   BR 5

(4) 1. □ P ⊃ □ Q

2. ◊ ∼ Q / ◊ ∼ P

3. ∼ □ Q   DE 2

4. ∼ □ P   MT 1, 3

5. ◊ ∼ P   DE 4

(5) 1. □ (P ⊃ Q)

2. ∼ ◊ Q / ∼ P

3. P → Q   Arrow Ex, 1

4. □ ∼ Q   DE 2

5. ∼ Q   BR, 4

6. ∼ P   MMT 3, 5

(6) **1.** □ (P ≡ Q)

    2. (P ↔ Q) → (Q ↔ R)

    3. □ (Q ≡ R) ⊃ □ □ □ S

    4. ~ □ A ⊃ ~ S / A

    5. P ↔ Q   Double Arrow Ex 1

    6. Q ↔ R   MMP 2, 5

    7. □ (Q ≡ R)   Double Arrow Ex 6

    8. □ □ S   MP 3, 7

    9. □ S   BR 8

  10. S   BR 9

  11. ~ ~ S   DNeg 10

  12. ~ ~ □ A   MT 4, 11

  13. □ A   DNeg 12

  14. A   BR 13

> ►   **EXERCISE 24.3**   ◄

Use natural deduction to establish the validity of the following arguments in S5. Replacement rules are recommended on these problems. You will need to use the Necessitation rule on some of these problems.

(1) * **1.** S ↔ Q

    2. H ⊃ Q

    3. F & H / S

(2) **1.** A ↔ G

    2. ~G

    3. ~A ⊃ □ E / E

(3) * **1.** □ ~A / ~ □ A

(4) **1.** ~ ◊ (A & ~B) / □ (A ⊃ B)

(5) **1.** A ⊃ □ B

    2. ~B / ◊~A

(6)* 1. A ↔ B

    2. A / B

(7) 1. A ↔ B

    2. B → G / A → G

(8) 1. A ↔ B

    2. □ ~B / □ ~A

(9) 1. A ↔ B

    2. □ ~A / □ ~B

(10) 1. A ↔ B

    2. ~ A / ~ B

(11)* 1. A ↔ B

    2. ~ B / ~ A

(12) 1. A → B

    2. B ⊃ D / A ⊃ D

(13) ~ ◊A / ◊ ~A

| ► | **EXERCISE 24.4** | ◄ |

Construct proofs for the following arguments. The Indirect Proof rule is recommended on these problems.

(1)* 1. H v S / ◊H v ◊S

(2) 1. □ (A ⊃ B)

    2. ◊ A / ◊B

(3) 1. □ A

    2. ~ □ B / ~ □ (A ⊃ B)

(4) 1. ◊ (A v B) / ◊A v ◊B

(5) 1. □ (A ⊃ B)

    2. ◊ ~ B / ◊ ~A

(6)* 1. ◊ A v ◊ B / ◊ (A v B)

▶   **EXERCISE 24.5**   ◀

Each of the following arguments is valid. Symbolize each argument and use S5 natural deduction to prove each valid. (Note: when symbolizing the if–then connective, use the arrow operator.)

1.* It's not possible that Spock and Kirk both beam down to the planet. If Spock doesn't beam down to the planet, then the mission will not be accomplished. If Kirk doesn't beam down to the planet, the mission will not be accomplished. Therefore, it's not possible that the mission be accomplished.

2. If Art belongs to the Peace and Freedom Party, then Art is a dangerous revolutionary and a threat to the entire political system. It's not possible that Art is a dangerous revolutionary. Therefore, it's not possible that Art belongs to the Peace and Freedom Party.

3. Bob either belongs to the SDS or to the PLP. If Bob is a staunch Republican, then it is impossible that Bob belongs to the PLP. Bob does not belong to the SDS. Bob is not a staunch Republican.

4. It's not possible that Glenn and Larry occupy the same place at the same time. If Glenn and Larry both deliver the closing speech, then Glenn and Larry will occupy the same place at the same time. If Glenn and Larry do not both deliver the closing speech, then Melissa and Val will be disappointed and so will Alan and Charlie, not to mention Ron, Mel, and Patty. Therefore, Patty will be disappointed.

5.* If Bob and Val are late, then Randy and Bill will wait. If Janet can't wait, then it's not possible that Randy will wait. Bob and Val will be late. So, Janet can wait.

6. If you have a job, then it is not possible you are unemployed. It is possible you are unemployed. So, you do not have a job.

7. It is not possible that Abe and Dave are both home. If it is possible Dave is home, then it is possible Ed is home. Therefore, necessarily, either Abe is not home or Dave is not home.

8. Necessarily, if Ann is home, then Dave is home. Necessarily, if Dave is home, then Bob is home. So, necessarily, if Bob is not home then Ann is not home.

9. If it is possible that Ann is home, then it is possible that Graham is home. Ann is home. If it is possible that Harry is home, then it is not possible that Graham is home. It is not possible that Harry is home.

10.* Necessarily, either Ann has a key to the building or Bob has a key. Necessarily, either Bob does not have a key or Ed does not have a key. It is not possible that it is not the case that Ed has a key. Therefore, necessarily, Ann has a key to the building.

11. Necessarily, water is wet and snow is cold. Therefore, it is necessary that water is wet and it is not possibly false that snow is cold.

12. If an absolutely perfect being exists, then it is necessarily true that an absolutely perfect being exists. If an absolutely perfect being does not exist, then it is necessarily false that an absolutely perfect being exists. It is not impossible that an absolutely perfect being exists. Therefore, an absolutely perfect being must exist. (Hint: Let A stand for "An absolutely perfect being exists.")

13. If it is true that tomorrow you will eat a taco, then it is necessary that you will eat a taco tomorrow. If it is true that tomorrow you will not eat a taco, then it is necessary that you will not eat a taco tomorrow. You will eat a taco tomorrow. If it is necessary that you eat a taco tomorrow, then you do not have free will. If you do not have free will, then you are not responsible for anything you do. Thus, you are not responsible for anything you do. (Let T abbreviate "You eat a taco tomorrow," let W stand for "You have free will," and let R abbreviate "You *are* responsible for anything you do.")

14. It is not possible that both God exists and evil exists. Evil exists. So, God does not exist. (Let G abbreviate "God exists" and let E abbreviate "Evil exists.")

15. *It is not possible that both God exists and evil exists. Necessarily, God exists. So, necessarily, evil does not exist. (Let G abbreviate "God exists" and let E abbreviate "Evil exists.")

PROVING THEOREMS OF S5

If a formula can be proven true using a premise-free proof, we say it is "logically true." In Chapters 9 and 11, we proved formulas logically true using premise-free conditional and indirect proofs. Within modal logic, we will use the same basic procedures to prove various *modal* logical truths. The following premise-free proof establishes that $A \supset \square \lozenge A$ is logically true:

1.	A	AP
2.	$\lozenge A$	Poss, 1
3.		$\lozenge A$	Reit 2
4.	$\square \lozenge A$	Nec 3
5.	$A \supset \square \lozenge A$	CP 1-4

Notice that this conditional proof contains a necessitation proof at steps 3 and 4.

We say formulas such as these are "logically true" because they can be proven true using just the techniques of logical theory without investigating the physical world, without conducting experiments, without using any procedures outside of logical theory. If it is possible to construct a premise-free S5D proof whose last line is the sentence **P**, **P** is logically true in S5D. A sentence that is logically true in S5D is also said to be a **theorem in S5D.**

ANOTHER INFERENCE RULE: THE TAUTOLOGY NECESSITATION RULE

According to the first of the five principles of modality at the start of this chapter, every tautology is necessarily true. Thus, if we can prove that a truth-functional sentence **P** is a tautology, we are justified in prefixing a box to **P** and thus asserting that **P** is necessarily true. Recall from Chapter 11 that we prove a statement tautological by proving the statement with a premise-free conditional or indirect proof sequence. If a truth-functional sentence can be proven tautological with a premise-free CP or IP, we may validly conclude that the sentence is a necessary truth and we may "box" it. The rule summarized in the following box incorporates this reasoning.

> **The Tautology Necessitation Rule (Taut Nec)**
>
> If a sentence **P** is proven tautological, we may infer from this the corresponding sentence □ **P**. (Write as justification "Taut Nec" and the line on which the tautology appears.)

Here are two simple examples of Tautology Necessitation:

(1) 1.   A   AP
    2.   A v ~A   Add 1
    3.   ~A v A   Comm 2
    4.   A ⊃ A   Imp 3
    5.   A ⊃ (A ⊃ A)   CP 1-4
    6.   □ [A ⊃ (A ⊃ A)]   Taut Nec 5

(2) 1.   A   AP
    2.   A v B   Add 1
    3.   A ⊃ (A v B)   CP 1–2
    4.   □ [A ⊃ (A v B)]   Taut Nec 3

Because every truth-functional tautology necessitated in this way can be proven with a premise-free S5 proof, every such formula is also a theorem of S5.

► **EXERCISE 24.6** ◄

Prove that each of the following is a theorem of S5.

1.*□ [~A ⊃ (A ⊃ B)]

2. □ [(A v B) ⊃ (~ B ⊃ A)]

3. □ [B ⊃ (A ⊃ B)]

4. □ {[(A ⊃ B) & A] ⊃ B}

5.*□ [(A ⊃ B) ⊃ (~ B ⊃ ~ A)]

6. □ {[(A ⊃ B) & ~ B] ⊃ ~ A}

7.*□ {[(A v B) & ~ A] ⊃ B}

8. □ {[(A ⊃ B) & (B ⊃ R)] ⊃ (A ⊃ R)}

9. □ {{[(A ⊃ B) & (R ⊃ S)] & (A v R)} ⊃ (B v S)}}

10.*□ {[(A v B) & (~ B v S)] ⊃ (A v S)}

11. □ {[(A & B) & (A ⊃ R)] ⊃ R}

12. □ [A ⊃ (A ⊃ A)]

13.*□ [(A ⊃ B) ⊃ ~ (A & ~ B)]

14. □ [(A v B) ⊃ (~ B ⊃ A)]

15.*A → A

16. □□ A → □ A

17.*(A → B) → (□ A ⊃ □ B)

18. □ A → □ (A v B)

19. A → (A v B)

20.*~ □ A → ◊ ~ A

21. □ [(A & B) ⊃ (A v B)]

22. ◊ ~ A → ~ □ A

23. □ ~ A → ~ ◊ A

24. ◊ A v ◊ ~ A

25.*~ ◊ A → □ ~ A

26. □ (A ≡ B) → (A ↔ B)

27. $(A \rightarrow B) \rightarrow \square (A \supset B)$

28. *$(A \rightarrow B) \rightarrow \square (A \rightarrow B)$

29. $(A \rightarrow B) \rightarrow \sim \Diamond (A \,\&\, \sim B)$

30. *$(A \rightarrow B) \rightarrow (\sim B \rightarrow \sim A)$

31. $(A \,\&\, \sim B) \rightarrow \sim (A \rightarrow B)$

32. $\sim \Diamond (A \,\&\, \sim A)$

33. $(A \leftrightarrow B) \rightarrow \square (A \equiv B)$

34. $\Diamond (A \leftrightarrow B) \rightarrow (A \equiv B)$

35. *$\square (A \supset B) \rightarrow (A \rightarrow B)$

36. $\sim (A \rightarrow B) \rightarrow \Diamond (A \,\&\, \sim B)$

37. $\Diamond (A \rightarrow B) \rightarrow (A \rightarrow B)$

CHAPTER 24 APPENDIX 1: Putting an S5 formula on a diet: S5 reduction

Each of the following four arguments presupposes the five modal principles presented at the beginning of this chapter:

1. If **P** is necessarily true, then in every possible circumstance it is true that **P** is necessarily true. That is, $\square \, \mathbf{P} \rightarrow \square \, \square \, \mathbf{P}$. Furthermore, if in every possible circumstance it's true that **P** is necessarily true, then **P** is necessarily true. That is, $\square \, \square \, \mathbf{P} \rightarrow \square \, \mathbf{P}$. One sentence is equivalent to another just in case the two imply each other. Because $\square \, \mathbf{P}$ and $\square \, \square \, \mathbf{P}$ imply each other, it follows that $\square \, \square \, \mathbf{P}$ is equivalent to $\square \, \mathbf{P}$.

2. If there is at least one possible circumstance in which **P** is true, then there is a circumstance in which it's true that there is a circumstance in which **P** is true. That is, $\Diamond \, \mathbf{P} \rightarrow \Diamond \, \Diamond \, \mathbf{P}$. Furthermore, if there is a possible circumstance in which it's true that there is a circumstance in which **P** is true, then there is a circumstance in which **P** is true. That is, $\Diamond \, \Diamond \, \mathbf{P} \rightarrow \Diamond \, \mathbf{P}$. Because $\Diamond \, \Diamond \, \mathbf{P}$ and $\Diamond \, \mathbf{P}$ imply each other, it follows that $\Diamond \, \Diamond \, \mathbf{P}$ is equivalent to $\Diamond \, \mathbf{P}$.

3. It is either true in all possible circumstances that **P** is a necessary truth, or it is true in no circumstance that **P** is a necessary truth. So, if there is even one possible circumstance in which it's true that **P** is necessarily true, then **P** is necessarily true. That is, $\Diamond \, \square \, \mathbf{P} \rightarrow \square \, \mathbf{P}$. Furthermore, if **P** is necessarily true, then there is at least one circumstance in which it's true that **P** is necessarily true. That is, $\square \, \mathbf{P} \rightarrow \Diamond \, \square \, \mathbf{P}$. Thus, because $\square \, \mathbf{P}$ and $\Diamond \, \square \, \mathbf{P}$ imply each other, it follows that $\Diamond \, \square \, \mathbf{P}$ is equivalent to $\square \, \mathbf{P}$.

4. If in every possible circumstance it's true that there is at least one circumstance in which **P** is true, then there is at least one circumstance in which **P** is true. That is, □ ◊ **P** → ◊ **P**. And if there is at least one possible circumstance in which **P** is true, then in every circumstance it's true that there is at least one circumstance in which **P** is true. That is, ◊ **P** → □ ◊ **P**. Therefore, because □ ◊ **P** and ◊ **P** imply each other, □ ◊ **P** is equivalent to ◊ **P**.

Two monadic operators are "iterated" if the first applies directly to the second. Thus, □ □ A contains iterated operators but □ (□ A & B) does not (because in the second formula, a parenthesis intervenes between the two modal operators). It follows from the four arguments set out above that

> *If we remove the left member of a pair of iterated monadic modal operators, the shortened formula represents a proposition equivalent to the one it formerly represented.*

Let us call a sentence shortened in this way a "reduced" sentence. If a sentence **P** reduces to a sentence **Q**, **P** and **Q** express equivalent propositions. We'll call a sentence of ML that contains no iterated operators a "fully reduced" modal sentence. Thus, the following formulas are all fully reduced:

◊ B   □ (◊ A & ◊ B)

And these formulas are not fully reduced:

□ □ A   □ ◊ A   ◊ ◊ □ B

Within S5, any sequence of iterated monadic modal operators in a formula, no matter how long, can be reduced to the last member on the right simply by reducing its operators two at a time until only one operator is left. The resulting reduced formula will express a proposition *equivalent* to the proposition expressed before the reduction. For example:

□ ◊ □ A	reduces to □ A
◊ ◊ ◊ □ ◊ A	reduces to ◊ A
◊ ◊ □ □ A	reduces to □ A
□ □ ◊ (□ □ A & ◊ ◊ B)	reduces to ◊ (□ A & ◊ B)

This may all be summed up in the replacement rule in the following box:

The Reduction Rule (Red)

Any sequence of iterated monadic modal operators in a formula may be reduced to the last member on the right, and the resulting reduced formula may replace the original formula anywhere within a proof. (Write as justification "Red" and the line of the reduced formula.)

The following two proofs employ the Reduction rule:

(1)   1. □ ◊ ◊ ◊ □ J ⊃ ◊ B

      2. ~ ◊ ~ J / ◊ B

      3. □ J   DE 2

      4. □ J ⊃ ◊ B   Red 1

      5. ◊ B   MP 3, 4

(2)   1. □ □ ◊J

      2. ◊ J → ~ □ B

      3. □ R → □ B

      4. ◊ ~ R → ◊ ~ S

      5. ◊ ~ G ⊃ □ S / G

      6. ◊ J   Red 1

      7. ~ □ B   MMP 2, 6

      8. ~ □ R   MMT 3, 7

      9. ◊ ~ R   DE 8

      10. ◊ ~ S   MMP 4, 9

      11. ~ □ S   DE 10

      12. ~ ◊ ~ G   MT 5, 11

      13. □ G   DE 12

      14. G   BR 13

► **EXERCISE. 24.7** ◄

Use the Reduction rule plus any of the other rules of our system to prove the following:

(1)* 1. ◊ □ A / □ A

(2) 1. □ □ ◊ A / ◊ A

(3) 1. ◊ ◊ A / ◊ A

(4) 1. □ □ □ Q ⊃ ◊ ◊ S

     2. ◊ S ⊃ □ P

   3. (□ Q ⊃ □ P) ⊃ ◊ R

   4. ~ ◊ R v □ A / A

(5)* **1.** □ ◊ □ P

   2. ◊ ~ P v □ Q

   3. Q ⊃ S / S

(6) **1.** ◊ □ ◊ P v □ Q

   2. ◊ ~ Q

   3. ◊ P ⊃ □ P / P

(7) **1.** □ ◊ ◊ □ P

   2. Q ⊃ ~ P

   3. ~ □ Q ⊃ □ S / S

(8) **1.** □ □ P → ◊ Q

   2. ◊ □ P

   3. ~ ◊ Q v □ R

   4. R → □ S / S

(9) **1.** ◊ ◊ ◊ ◊ □ P

   2. P → Q

   3. Q → R / R

(10) **1.** □ ◊ ◊ P

   2. □ S → □ ~ P

   3. ◊ ~ S → □ G / G

CHAPTER 24 APPENDIX 2 : *The modal fallacy*

We often express a relation of implication by saying, "If so and so, then it is *necessary* that such and such." Or we sometimes say, "If so and so, then it *must* be that such and such." For example, "If Rita is the mother of three children, then it is necessary that Rita is a mother."

   Now, this sentence about Rita is ambiguous. On one interpretation, the necessity applies only to the *consequent* of the conditional. According to this interpretation, if Rita has three children, then it is a *necessary truth* that she is a mother. On a second interpretation, the necessity applies to the conditional relationship as a whole, that is, to the *relation* between the antecedent and the consequent. On this interpretation, if Rita has three children, then it *necessarily follows* that she is a mother. The sentence on this second interpretation attributes necessity to the relation between the antecedent and the consequent, rather than to the consequent alone. The sentence on the first interpretation attributed necessity to the consequent.

The first interpretation should be symbolized: R ⊃ □ M, with R abbreviating Rita *has three children* and M abbreviating *Rita is a mother.* On the other hand, the second interpretation should be symbolized □ (R ⊃ M). Do you see the difference between these two interpretations?

On the first interpretation, the sentence is being used to assert that if Rita has three children it is then a necessary truth that Rita is a mother. On this interpretation, because Rita does have three children (let us suppose), she had children of necessity, that is, she could not possibly have *not* had children. This suggests that she did not have children of her own free will (assuming that if one could not have done otherwise then one is not acting of one's own free will). Thus:

1. R ⊃ □ M

2. R   / □ M

3. □ M   MP 1, 2

However, on the second interpretation, the assertion is something very different, namely: It is necessary that *if* she has three children, *then* she has at least one. This interpretation seems to sidestep the free will issue altogether.

The "modal fallacy" occurs when someone mistakenly transfers the necessity from the conditional as a whole to the consequent alone. Logicians in the Middle Ages were deeply interested in modal reasoning, and they catalogued numerous modal fallacies, including the fallacy just noted. In explaining the modal fallacy, they distinguished between the following two types of sentence:

Where **P** abbreviates the antecedent and **Q** the consequent:

- The "necessity of the consequence"—the sentence asserting the necessity of the connection between antecedent and consequent—symbolized □ (**P** ⊃ **Q**).

- The "necessity of the consequent"—symbolized **P** ⊃ □ **Q**.

Here is an argument that commits the modal fallacy. See if you can pinpoint the problem. (Or, perhaps you may wish instead to defend the argument.) This argument is adapted from an argument that was discussed in ancient philosophical circles.

1. Necessarily, if God foreknew that human beings would sin, then human beings would sin.

2. God foreknew that human beings would sin.

3. So, it is necessary that human beings sin.

4. So, when a person sins, he or she sins of necessity.

5. If someone does something of necessity, then he or she could not have done otherwise.

6. If someone does something, and could not have done otherwise, then he or she does not act of his or her own free will.

7. So, human beings do not sin of their own free will.

Do you see the place in this argument where the necessity is shifted from the conditional to the consequent? Is this a valid argument?

▶ ANOTHER MODAL FALLACY?

Another interesting argument, known as the "lazy" argument, was also discussed in ancient philosophical circles. The proponents of this argument claimed to have a proof that nobody has free will. Furthermore, they claimed, the argument proves this using purely abstract logical considerations alone.[1] Here is one way to put the argument:

1. Take any event in the future, for instance, the next lunch you will eat.

2. It is necessarily true that either you will eat a burrito for that lunch, or you will not eat a burrito for that lunch.

3. Therefore, it is necessarily true that you will eat a burrito for your next lunch, or it is necessarily true that you won't eat a burrito for your next lunch.

4. So, whichever one you do, you will do it of necessity, which means that you couldn't have done anything else.

5. If, in a situation, someone does something, and they couldn't have acted otherwise, then they do not act of their own free will.

6. Therefore, when you eat your next lunch, you will not eat of your own free will.

7. Furthermore, this reasoning applies to each minute and to each second of each day in your future. So, each thing you will ever do will be done of necessity. That is, each time you do something, it will be true that it was necessary that you do precisely that thing. Now, you cannot change or prevent what is necessarily true. So, nothing you ever do will be done of your own free will. Because this reasoning can be applied to each event in your future, your future is already predetermined and you do not have free will.

Before we analyze this argument, let us put it into symbols. Let B abbreviate *Pat's next lunch is a burrito,* and let F abbreviate *Pat has free will.*

---

[1] I am indebted to Richard Purtill's treatment of this argument in his *Logic, Argument, Refutation, and Proof* (New York: Harper & Row, 1979), Chapter 10.

1. □ (B v ~B)   (premise)
2. (□ B ⊃ ~F) & (□ ~B ⊃ ~ F)   (premise)
3. □ B v □ ~B   (inferred from premise 1)
4. ~ F v ~F   (inferred from 2,3)
5. ~F   Taut 4

Now, the above argument can be reconstructed for each possible event in Pat's future, and for each possible event in anyone else's future. The conclusion many have drawn from this is: Nobody has free will.

The objections to this argument often focus on the inference from 1 to 3. The proponent of the argument thinks that a sentence □ (**P** v **Q**) implies the corresponding sentence □ **P** v □ **Q**. However, if □ (**P** v **Q**) does imply □ **P** v □ **Q**, as the argument assumes, then an absurd consequence follows: it follows that *every* truth is a necessary truth. In other words, if this inference is indeed valid, this would abolish the distinction between contingent truth and necessary truth. Yet, the distinction between necessary and contingent truth is firmly rooted in common sense. Thus, most philosophers hold that the inference in question, the inference from a sentence of the form □ (**P** v **Q**) to the corresponding sentence □ **P** v □ **Q**, is invalid. Do you think the "Burrito" argument is valid?

CHAPTER 24 GLOSSARY

Modal argument   An argument whose validity or invalidity depends on the arrangement of its modal operators.

Modally closed   A sentence **P** of ML is modally closed just in case every sentence letter within **P** appears within the scope of a modal operator.

System S5D   The natural deduction system consisting of (1) the language ML and (2) the natural deduction rules of this chapter.

- **Proof in S5D**   A sequence of formulas of ML that is such that every formula is either a premise, an assumption, or follows from previous formulas by an S5D rule, and in which every line that is not a premise is justified and every assumption has been discharged.

- **Valid in S5D**   An argument is valid in S5D if and only if it is possible to construct a proof in S5D whose premises are the premises of the argument and whose conclusion is the conclusion of the argument. Any argument that is valid in S5D may also be said to be "valid in S5" or "S5 valid."

- **Theorem of S5D**   A formula that can be proven with a premise-free S5D proof.

VI

INDUCTION

Chapter

25 INDUCTIVE REASONING

et us begin by reviewing the key distinction between inductive and deductive argumentation. A **deductive argument** claims that

- If all the premises are true, the conclusion *must* be true; or,
- It is not possible that the premises are true and the conclusion is false.

An **inductive argument** claims that

- If all the premises are true, the conclusion is *probably* true, although it might be false; or,
- It is improbable that the premises are true and the conclusion is false.

A deductive argument has an air of definiteness—the premises, it is claimed, *guarantee* the conclusion. If the premises are true, it is claimed, we can be certain the conclusion is true. On the other hand, inductive arguments claim no such certainty or finality—they leave open the possibility that the conclusion might be false and only claim the conclusion is *probably* true.

So, within inductive reasoning, the conclusion is not conclusively established—it is not made certain—as in a sound deductive argument. However, although the conclusion of an inductive argument may not be conclusively shown, it can nevertheless be made probable, likely, or plausible, and on that ground may be reasonable to accept.

In this chapter, we examine three important types of inductive reasoning—*analogical argumentation, enumerative induction,* and *inference to the best explanation.* These three types of reasoning are common in everyday life. They are also widely employed in the fields of law, science, and business as well. We'll address the *hypothetico-deductive method*—the type of inductive reasoning by which scientific hypotheses are established or rejected—in the next chapter.

ANALOGICAL REASONING

Speaking very generally, one of the key features of inductive reasoning is this: A pattern is discovered and a prediction or a conclusion is drawn based on that pattern. In this way, we often reason from the known to the unknown. Inductive reasoning thus relies on our ability to spot a resemblance between things, that is, an ability to recognize a pattern. This general feature of induction comes out clearly in our first inductive argument form, analogical argumentation.

An **analogy** is a resemblance or similarity between two things. To draw an analogy between X and Y is to cite respects in which the two are similar. Here's an example of an analogy: Ollie's symptoms are similar to Mollie's symptoms. Both have no appetite, both have a temperature of 101 degrees, both feel nauseated, and both have red spots on their faces.

Analogical reasoning depends on comparison. Essentially, in an **analogical argument,** an analogy is asserted between two things or kinds of things, A and B; it is then pointed out that A has a particular feature and that B is *not* known *not* to have the feature, and it is then concluded that B probably also has the feature. The central idea in an analogical argument is that if two things (or kinds of things) have several features in common, and if it is observed that one of the two has an additional feature, it is likely the other has this feature too. In short, an analogical argument infers an additional similarity from a similarity.

In the example above, suppose that after Ollie and Mollie compare their symptoms, Ollie goes to the doctor and is diagnosed with chicken pox. We might then naturally conclude that Mollie probably also has chicken pox, because her symptoms are similar. This would be an example of analogical reasoning.

Let us put this into more formal terms. An analogical argument begins with an analogy between two or more things or groups of things. The analogy takes the following general form:

X has features ABCD.

Y has features ABCD.

The argument notes a feature of X that Y is *not* known *not* to have:

X also has feature E.

Y is not known *not* to have E.

The argument concludes

Y probably has feature E as well.

Analogical reasoning is at the heart of much of our everyday reasoning. Indeed, it is perhaps the most fundamental, common type of human reasoning. In each of the following cases, analogical reasoning is employed.

- A sixteen-year-old is buying his first car. Which kind of car should he buy? He reasons thus: Dad has always driven a Chevrolet, and they've always been reliable. If I buy the same type of car, my car will probably be as reliable as his. So, I should buy a Chevrolet.

- Because Jan is not happy with her dentist, she is wondering where she should go for her next crown. She reasons: Dr. Carey did a good job on my friend Oscar, and my friend Rollo raved about Dr. Carey. If I go to the same dentist, I'll probably get similar treatment. Consequently, I should go to Dr. Carey.

- A young child wonders, "What should I buy Dad for his birthday? Well, I got him a bright blue tie the past three years and he was really happy. I guess if I get him the same thing again, he'll probably really be happy, just like on the other birthdays. Therefore, I'll buy him a bright blue tie."

- Ten Clinton High School students with SAT scores above 1,000 did well at the University of Kansas. Katie also scored above 1,000 on her SAT. So, Katie probably will also do well when she attends the University of Kansas.

- A doctor reasons that her patient should take drug Z because other patients with the same symptoms were helped by drug Z.

- In the Summer of 1964, a bunch of kids went down to the record store to buy the new Beatles release, the album *A Hard Day's Night*. They reasoned: The last two Beatles albums were great. *A Hard Days Night* is a Beatles album, so, it probably will be great, too.

The general form of an analogical argument:

X has attributes ABCDE.

Y has attributes ABCD.

Y is not known not to have attribute E.

So, Y probably also has attribute E.

EVALUATING ANALOGICAL ARGUMENTS

Some analogical arguments are stronger than others. We judge the strength of an analogical argument by the following criteria:

1. Generally, the more characteristics in common between the things compared, the stronger the argument.

For example, compare the following two analogical arguments:

- Bob and Ann both took the same *three* math classes at Central High. Both earned straight A's. Bob just took the math portion of the University of Washington entrance exam and passed. Ann is taking the test tomorrow. It follows that Ann will probably pass as well.

- Bob and Ann both took the same *six* math classes at Central High. Both took *all* their classes from Mr. Mintek and both earned straight A's. Bob just took the math portion of the University of Washington entrance exam and passed. Ann is taking the test tomorrow. It follows that Ann will probably pass as well.

Isn't the second argument stronger than the first? In the second argument, the number of characteristics shared by Bob and Ann has been increased.

This raises an interesting question: Is it always the case that the larger the number of common characteristics, the stronger the argument? The answer is no, as the following case illustrates. Compare the next argument to the previous argument.

- Bob and Ann both took the same six math classes at Central High. Both took all their classes from Mr. Mintek and both earned straight A's. *In addition*, when Ann took calculus with Mr. Mintek during fall term, the cafeteria added two items to its menu. When Bob took calculus during spring term, the cafeteria added two items to its menu. Fall term, during Ann's algebra class, the president gave a nationally televised speech. Spring term, during Bob's algebra class, the president gave a nationally televised speech. Bob just took the math portion of the University of Washington entrance exam and passed. Ann is taking the test tomorrow. It follows that Ann will probably pass as well.

Notice that although this argument has an even larger number of similarities, it is no stronger than the previous argument. Why? The added similarities—the cafeteria's new food, the president's speech—aren't *relevant* to the conclusion. This brings us to the next criterion:

2. Generally, the more *relevant* the similar characteristics are to the characteristic mentioned in the conclusion, the stronger the argument.

The problem with the previous argument was that the additional similarities are not connected in some way to the conclusion. However, suppose we had added instead the following two similarities:

(1) Ann's calculus class used a text written by I. M. Amathnerd, and Bob's calculus class used the same text.

(2) Ann's geometry class had a high class average, and so did Bob's.

Now, these additional similarities (added to the premises) are relevant to the conclusion and therefore make the argument even stronger. But this raises the question: What makes a similarity a *relevant* one?

▶ RELEVANCE

Relevance generally involves either (a) a cause and effect connection, or (b) a statistical connection. In the previous example, a cause and effect connection exists between the text used in class and the successful test results. Thus, having the same text is a relevant similarity. Here is another example of *causal* relevance:

- Bob played bass in a heavy-metal band for five years in his twenties. The band frequently played at local high school dances, and now at the age of thirty Bob has significant hearing loss. Albert also played bass in a heavy-metal band for about five years in his twenties, frequently at high school dances, and Albert just turned thirty. Albert probably also has a significant hearing loss.

Notice that the similarities mentioned in the premises (playing bass and so on) are causally connected to the characteristic mentioned in the conclusion (hearing loss), for exposure to prolonged loud noise *causes* hearing loss. This again is causal relevance.

The following causally relevant similarity, if added to the "Bob and Albert" argument, would increase the argument's strength even further:

- Bob always stood in front of the amplifiers and never wore ear plugs.
- Albert also always stood in front of the amplifiers and never wore ear plugs.

Here is another example that illustrates causal relevance. Suppose I want a car with good gas mileage, and so I reason as follows: Joe's Toyota has good gas mileage, so, to get good mileage, I should buy a car like Joe's Toyota. This is an analogical argument. Now, let us suppose the Toyota I buy is the same color as Joe's car, has the same upholstery, the same stereo, and the same fluorescent lights underneath the frame. But suppose the similarities end there, for the model I buy has an engine twice the size of the engine in Joe's car. The problem is that the similarities between our two cars have no cause and effect relationship to gas mileage. My Toyota is like Joe's but in features irrelevant to mileage. My conclusion—that my new car will have good gas mileage—is therefore not warranted. If I am interested in getting good gas mileage, I need to focus on engine size, gross vehicle weight, and other features causally relevant to mileage.

Next, we will examine an example of *statistical* relevance:

- Ann earned a 3.8 cumulative average in high school and scored high on the SAT test. Bob also earned a 3.8 in high school and scored high on the SAT. Ann has done well in college. Therefore, it is likely that Bob will do well in college.

The characteristic mentioned in the conclusion is that of doing well in college. The characteristics mentioned in the premises (having good grades and a high

SAT score) are relevant to that mentioned in the conclusion. However, the characteristics cited in the premises—high grades and high SAT scores—do not *cause* good grades in college; rather they are statistically correlated with success in college. (That is, statistical evidence suggests that individuals with high grades and SAT scores tend to do well in college.) In other words, the characteristic mentioned in the premise does not *cause* the characteristic mentioned in the conclusion; they only share a statistical relationship.

Here is another example of statistical relevance:

- Fred is a college sociology instructor, he is active in the instructors' union, and he contributes to liberal causes. Jim is a college sociology instructor who is also active in the union and contributes to liberal causes. Fred is a Democrat. Therefore, Jim is probably also a Democrat.

The characteristics mentioned in the premises are relevant to the characteristic mentioned in the conclusion: being a Democrat. Again, the characteristics mentioned in the premises do not *cause* the characteristic mentioned in the conclusion; but a statistical relationship exists between them, based on statistical evidence that individuals with the stated characteristics tend to be Democrats. In other words, the characteristics cited in the premises are statistically relevant to the characteristic in the conclusion.

3.   The larger the number of primary analogates, the stronger the argument.

The **primary analogates** are the objects being compared that appear only in the premises; the **secondary analogate** is the object that also appears in the conclusion. Suppose Rita is trying to decide if she should buy a slice of pizza at Pete's Pizza Place. She reasons, "I know two people who said the pizza at Pete's Pizza Place is good. Because this slice is off of a Pete's pizza, it will probably also be good." Now, this argument is a fairly strong analogical argument. But suppose we alter the argument by changing the premise to "I know six people who said Pete's Pizza Place makes good pizza." The argument now contains six primary analogates, and is obviously stronger.

4.   The more diverse the primary analogates, the stronger the argument, provided that the diversity concerns features unrelated to the feature cited in the conclusion.

To illustrate this, let us compare the following two brief arguments:

- I know six people with Fords, and all six Fords have been reliable. The six Fords were all 1965 Ford Falcons. So, if I buy this Ford Galaxy, I'll have a reliable car.

- I know six people with Fords, and all six Fords have been reliable. One of the Fords was a 1965 Falcon, one was a 1969 Galaxy, one was a 1985 Escort, one was a 1990 Taurus, and one was a 1997 Ford truck. So if I buy this Ford Galaxy, I'll have a reliable car.

Wouldn't you agree that the second argument is stronger than the first? If the Fords represent a diverse range of years and models but all are reliable, it is more likely *all* Fords are reliable. That is, if the primary analogates vary widely in all sorts of features but all share the *relevant* feature or features, the secondary analogate will more likely share the relevant feature (or features), too. In the first argument, the worry arises: perhaps the only Fords that are reliable are the 1965 Ford Falcons. In the second argument, we find more reason to believe that all Fords are reliable, including the particular Ford under consideration.

5. The larger the number of relevant dissimilarities, the weaker the argument. (This is called the "degree of disanalogy.")

For example, suppose Sue argues, "Six of my friends have Chryslers, and all say their cars are reliable. So, the Chrysler I am about to buy will probably also be reliable." On the surface, this argument sounds fairly strong. However, suppose several performance-related differences exist between the Chryslers that her friends own and the Chrysler that she is about to buy. The Chryslers owned by her friends are all 1990s cars, which have performance-related improvements over the 1960s Chrysler she is looking at. Additionally, cars more than thirty years old often have more performance-related problems than relatively newer cars. The argument weakens when we take into account these relevant dissimilarities.

Of course, *irrelevant* dissimilarities do not weaken an argument. For example, in the previous argument, the dissimilarity concerned performance-related features. However, suppose that the only dissimilarity is this: Sue's friends all have brightly colored cars, and the car she is about to buy is plain white. This dissimilarity is not relevant—causally or statistically—to the feature mentioned in the conclusion (that is, reliability), and therefore does not weaken the argument.

6. The more specific the conclusion, the weaker the argument.

The more *specific* the conclusion, the more possible circumstances in which it is false, the "easier" it is for it to come out false, and consequently the weaker the argument. For example, to go back to the previous argument, suppose Sue's friends reported that their cars averaged one visit to a mechanic every two years. If Sue concludes not that her car will be generally reliable, but that her car will require a visit to the mechanic only once every two years, her argument's conclusion is more specific than if she had concluded merely that her car would be generally reliable. The argument is weaker with the more specific conclusion, for it is more likely her specific conclusion will turn out false. However, if she concluded merely that her car would generally be reliable, her argument has a less specific conclusion, it is more likely her conclusion will be true, and the argument is a stronger argument.

For another example, suppose that five of Lauren's friends have jeeps, and on average the five jeeps get 18 miles per gallon in city driving. Lauren buys a similar jeep. If she reasons that her jeep will probably average exactly 18 miles per gallon in the city, her analogical argument is not as strong as if she concludes that her jeep will get over 16 miles per gallon in the city. Lauren's analogical argument would be even stronger if she were to conclude just that her jeep will average at least 15 miles per gallon in the city. The more specific the conclusion, the weaker the argument.

ANALOGIES AS MODELS

A **model** is a simplified representation of a more complex object or state of affairs. Logician Daniel Bonevac calls analogical arguments "arguments by modeling."[1] He reasons that in an analogical argument we in effect construct a "model" of a situation or state of affairs, S, which is supposed to be similar to S. We analyze the model and conclude something about it, and we then conclude something similar about S. Because the model is simpler, it is usually clearer, more familiar, and easier to understand than the situation to which it is compared. For example, economists construct models of the economy. They study the models and make discoveries about the models. They conclude that the economy also has the features discovered in the models. Similarly, astronomers construct mathematical models of the universe, analyze those models, and make discoveries about, for instance, the *shape* of the universe.

▶   **EXERCISE 25.1**   ◀

Read the following argument and carry out the instructions below.

1.   Joe is fifty-five years old, 100 pounds overweight, and eats a fast-food diet that includes three Big Macs every day for lunch and usually something like pizza, fish and chips, or deep-fried burritos for dinner. Pete is fifty-six, 108 pounds overweight, and eats a fast-food diet that includes three Whoppers every day for lunch and usually something like a six-pack of corn dogs, or a large plate of deluxe nachos, or a large bag of Cheetos for dinner. Joe just had a heart attack. Therefore, Pete probably will have a heart attack.

   a. List several irrelevant dissimilarities that would have no bearing on the argument.

   b. List several additional relevant similarities that would strengthen the argument.

   c. Change the conclusion so that the argument is weaker.

   d. Change the conclusion so that the argument is stronger.

[1] Daniel Bonevac, *The Art and Science of Logic* (Mountain View, California: Mayfield, 1990) p. 415.

    e. Add a premise that makes the argument stronger.

    f. Add a premise that makes the argument weaker.

2. Consider this simple argument: "Professor Smith's Medieval History course last fall was extremely dry and extremely boring. Therefore, his class this fall will be like the previous one—extremely dry and extremely boring."

How do the following additional facts bear on this reasoning? Do they make the argument stronger? Weaker? Do they leave the strength of the argument unaffected? Consider each additional fact separately.

    a.*His class last fall met in room 1408 and this fall it will also meet in 1408.

    b. Last year's class met on Wednesdays and this year's class will also meet on Wednesdays.

    c.*Last year's class used the text written by Professor Smith himself. This year's class will, too.

    d. Last year's class was held just after lunch. This year's class is at 8:30 in the morning.

    e. Last year, Professor Smith was going through a divorce. This year, he is through with the divorce and is into the swinging singles scene.

    f.*His class last year was small. His class this year is large.

    g. Since last year, Professor Smith has experienced a religious conversion.

    h. This year the class will use a new textbook.

    i.*Last year, Smith did not know how to dance. This year Smith is taking swing dancing lessons.

    j. We change the conclusion to: His class this year will be very dry and very boring.

    k. We change the conclusion to: His class this year will be at least somewhat dry and boring.

    l.*We change the conclusion to: His class this year will not be terribly exciting.

    m. Professor Smith's class two years ago was very dry and boring.

    n. Professor Smith's class two years ago was not dry or boring at all.

    o.*Professor Smith's Ancient History course last year was also very dry and boring.

    p. Professor Smith's class four years ago was very dry and boring.

q. Professor Smith's class twenty years ago was very dry and boring.

r. * Professor Smith's class twenty years ago was not dry or boring at all.

3. Brigit has decided to buy a new television. She reasons, "Susan's RCA has an excellent picture, and Bob's RCA has an excellent picture. This RCA on sale will have an excellent picture, just like the others." How do the following additional facts bear on the argument? Do they make the argument stronger? Weaker? Do they leave the strength of the argument unaffected? Consider each additional fact separately.

   a. Brigit lives far out in the country, but Susan and Bob live in the city.

   b. Brigit will watch only what her antenna picks up, but Sue and Bob watch only cable.

   c. Brigit knows three others who own RCA televisions, and all three report excellent picture quality.

   d. Brigit knows three others, each with RCA televisions, each from a different part of the country, and all report excellent picture quality.

   e. Brigit revises her conclusion. She concludes that her new set will have at least a decent picture.

   f. The set Brigit plans to buy is like Bob's and Susan's only in two respects: It has the same color cabinet and the same screen size; it turns out not to be an RCA.

   g. Bob and Sue bought their television on a Sunday, Brigit plans to buy hers on a weekday.

   h. Brigit revises her conclusion. She concludes that her new set will have a good picture.

4. Spud's Fish and Chips, a Seattle institution since the 1950s, has decided it needs to hire a new fish-buyer. Smith, with a masters degree in fish-buying from the University of Washington, is offered the job. Spud's' CEO reasons, "The last time we hired a buyer from the UW fish-buying program, he increased business 10 percent because he selected such good fish. Smith should increase business 10 percent, too." How do the following additional facts bear on the argument? Do they make the argument stronger? Weaker? Do they leave the strength of the argument unaffected? Consider each additional fact separately.

   a. The previous fish-buyer had formerly worked for Skippers Fish and Chips (another Seattle institution) where he also increased business 10 percent. However, this will be the new buyer's first job.

   b. Smith previously worked for Ivar's Fish and Chips (yes, another Seattle institution) where he was fired for buying rotten fish.

   c. The old buyer was a philatelist; Smith's hobby is table tennis.

d. The earlier buyer was a workaholic, the new buyer is a member of Workaholics Anonymous and claims he is a recovering workaholic.

e. Graduates of the UW fish-buying program tend to be successful fish-buyers.

f. The CEO revises his conclusion to "Smith will likely increase sales by at least 5 percent."

g. The CEO revises his conclusion to "Smith at least will likely not cause business to drop."

5. Consider this argument: "The Lancaster New Grand Theater is showing a new Charlie Chaplin film. I've seen four Charlie Chaplin films and loved them all. I'll love this one just as much." Which of the following, if added to the argument, increase the argument's strength, which decrease it, and which leave it unaffected?

a.*All four Charlie Chaplin films that I've seen were comedies; this new film is not a comedy.

b. All four previous Charlie Chaplin films had the same director, but this new film has a new director.

c.*One of my favorite actors, Joe Blow, was in all four previous Charlie Chaplin films, but he is not in this new one.

d. I saw all four previous films on Wednesdays; this film will be shown on a Saturday.

e.*The previous four were watched in a small theater, this will be watched in a larger theater.

f. We change the conclusion to "I'll at least like this next film."

6. Here is another short argument: "I've taken three classes from Professor Jones. All were interesting. I expect his next course to be like the others—interesting." Which of the following, if added to the argument, increase the argument's strength, which decrease it, and which leave it unaffected?

a. The three previous classes were all in ancient philosophy, this next class is modern philosophy.

b. The previous three were ancient, medieval, and modern philosophy.

c. The previous three classes were all ancient philosophy, which is his specialty. The next class is existentialism, a field he says he dislikes.

d. We change the conclusion to "This class will at least not be boring."

e. All three of the previous classes were in the morning, this next class will be in the evening.

7. Here is another short argument: "We've always stayed at Bob Newhart's Inn the past ten years. Each time was wonderful. We expect this visit to be like the others—wonderful." Which of the following, if added to the argument, increase the argument's strength, which decrease it, and which leave it unaffected?

   a. *The previous visits were all in the fall, this visit will be in the summer.

   b. The inn has a new manager.

   c. *In previous visits, our room had a balcony, this time it is a ground floor room.

   d. On previous visits, we flew; this time we are driving.

   e. The previous visits were: three in the fall, three in the winter, and four in the spring.

   f. *We change the conclusion to "This visit will at least be tolerable."

   g. They've painted the rooms since we were last there.

   h. On previous visits, our rooms have been in the north section, the south section, the east section, and the west section.

   i. *On previous visits, our rooms have all been in the north section. This time, we are booked into the south section.

8. Return to problem 4 above and answer the following:

   a. Add a premise that makes the argument stronger.

   b. Add a premise that makes the argument weaker.

ENUMERATIVE INDUCTION

Imagine that a rock band has moved into the house next door to you. Every Wednesday night for the first month they practice their music from 6 p.m. until 11 P.M. The bass amp is so powerful that it rattles every dish in your house; the singing is so loud that your dog howls, and the lead guitarist is so loud it makes your ears hurt. (On top of that, all they play is old Ozzie Osborne numbers.) Now it is the first Wednesday of the second month and it is your turn to host the monthly meeting of your chess club. You conclude that the band next door will be playing again tonight, and that the club members will not be happy campers. Your reasoning is as follows:

1. The band practiced Wednesday night four weeks ago.

2. The band practiced Wednesday night three weeks ago.

3. The band practiced Wednesday night two weeks ago.

4. The band practiced Wednesday night one week ago.

Therefore, they will probably practice this Wednesday night.

This reasoning fits the form of an enumerative induction. In an **enumerative induction,** we use premises about observed individuals or cases as a basis for a conclusion about unobserved individuals or cases. In general terms, an enumerative induction has the following structure:

1. The premises list ("enumerate") cases. The cases can be individuals or classes of individuals.

2. On the basis of the cases enumerated in the premise or premises, a conclusion is drawn concerning another individual or a group or class of individuals.

Enumerative induction is inductive in nature because the argument claims only that the premises make the conclusion *probable.*

A classic case of this type of reasoning is the logic Galileo used in the derivation of his law of uniformly accelerated motion toward the Earth. Galileo measured the speeds of many different types of objects as they fell. After recording an acceleration of thirty-two feet per second squared in every case, he inferred that probably *all* dropped objects accelerate according to this formula.

An enumerative induction can take several different forms. In one, the cases cited in the premises are *individuals,* as in the following argument about cats:

Biggie is a cat and likes fried chicken.

Baby is a cat and likes fried chicken.

Worty is a cat and likes fried chicken.

Jinx is a cat and likes fried chicken.

Muff is a cat and likes fried chicken.

Therefore, probably, all cats like fried chicken

Notice that the premises list particular cases and the conclusion generalizes from these individual cases to a claim about *all* cats.

This type of enumerative induction is called **inductive generalization** because the conclusion is a generalization about *all* the members of a class or group based on the information enumerated in the premises. From premises that concern observed instances we reason to a conclusion that represents a generalization about all the members of a class. The general form of an inductive generalization is

The first A has feature G.

The second A has feature G.

The third A has feature G.

No A is known not to have feature G.

So, probably, all A's have feature G.

In a second type of enumerative induction, the cases in the premises are *classes* of individuals, as in the argument below:

All observed tabbies like fried chicken.

All observed calicos like fried chicken.

All observed Siamese cats like fried chicken.

All observed Manx cats like fried chicken.

So, probably, all cats like fried chicken.

These premises contain subgeneralizations—the particular instances have been put into groups (tabbies, calicos, and so on) instead of being listed individually.
      The general form of this type of enumerative induction is

All A's that are B are X.

All A's that are C are X.

All A's that are D are X.

So, probably, all A's are X.

      In the third type of enumerative induction, the particular instances might all be grouped together into one big category, as in the following example.

All observed crows are black.

Therefore, all crows—examined and yet to be examined—are black.

The general form of this type of inductive generalization is

All observed A's are G.

Therefore, probably, all A's are G

      So far, then, we have three general patterns an enumerative induction might take:

1.   The instances might be listed separately, as in the argument about the cats Biggie, Baby, and so on above.

2.   The instances might be grouped into subcategories, as in the argument about the types (tabbies, calicos, and so on) of cats.

3.   All the instances might be grouped together, as in the argument about the crows.

      In a fourth type of enumerative induction, the premises provide a listing of cases, but the conclusion is not explicitly about all of a group; rather, it is about the next case or individual we will observe. This type of argument is sometimes called induction by analogy because it resembles an analogical argument. For instance, someone reasons as follows:

The first skateboard rider I saw wore baggy pants.

The second skateboard rider I saw wore baggy pants.

The third skateboard rider I saw wore baggy pants.

The fourth skateboard rider I saw wore baggy pants.

Therefore, probably, the next skateboard rider I see will probably be wearing baggy pants.

The general form of this fourth pattern of enumerative induction is:

X has features A and B.

Y has features A and B.

Z has feature A.

So, probably, Z has feature B.

For another example, consider:

Swan 1 was white.

Swan 2 was white.

Swan 3 was white.

So, probably, the next swan observed will be white.

STATISTICAL INDUCTIVE GENERALIZATION

Consider the following argument:

> *We sampled eggs at Farmer Smith's chicken farm, and 80 percent of the eggs examined were grade A. Therefore, probably 80 percent of the eggs on Smith's farm are grade A.*

This is a statistical inductive generalization. The general form of this type of argument is

1. N percent of observed F's are G.

2. Therefore, probably N percent of F's—observed or not—are G.

Here is another example:

> *Half of all times this coin was tossed, it came up heads. Therefore, probably half of all tosses of this coin will be heads.*

Notice that the premises represent a *sample* of a larger group, a statistic concerning the sample is given, and from the sample the argument concludes something about the larger group.

The following factors determine the strength of enumerative and statistical inductions.

1. Sample size. In either of these two types of argument, the premises represent a *sampling* of a larger group. The premises claim that the entities in the sample have certain features, and the argument concludes that other entities in the larger population, or the population as a whole, probably also have those features. Generally, the larger the sample in relation to the overall group, the more likely the sample is representative of the group as a whole. Thus, the larger the sample in relation to the overall group, the stronger the argument. If the size of the sample is inadequate, the argument is weak.

2. Sample variation. Generally, the more varied the sample items, the more likely the sample is representative of the group as a whole. Thus, the more varied the sampled items, the stronger the argument; the more homogeneous the sampled items, the weaker the argument.

The following two examples illustrate these two criteria. First, suppose the results of a survey recently taken in the school cafeteria are announced in the campus newspaper:

> *Eighty percent of all students sampled said they liked the food at the student union building. Therefore, 80 percent of all students like the food at the student union building.*

However, suppose we discover that the survey was conducted only at breakfast. The students sampled were only those eating breakfast. The argument is now much weaker than if the survey had been conducted at lunch and dinner as well. Had the conclusion been based on lunch and dinner students as well, the sample (students surveyed) would have been more *varied* and probably more representative of the population as a whole, and the argument would have been much stronger.

Suppose next we discover that the survey was conducted only on a Monday. This makes the argument much weaker than if the survey had been conducted on all five weekdays. Why? Again, the sample would have been more varied and therefore probably more representative of the population as a whole.

Suppose a reporter from the school newspaper does some investigating and finds out that only five students were sampled, out of a student population of 3,000. The sample is obviously too small in relation to the whole population. A sample of 100 randomly selected students would probably have been far more representative and consequently would have produced a much stronger argument.

Here is another example. A public opinion research firm takes a poll to measure public opinion on the legalization of abortion. They survey 100 people at Saint Joseph's Catholic Church, find that 90 percent oppose abortion, and conclude from this that probably 90 percent of the American public oppose the legalization of abortion. This is obviously a very weak argument. Their argument would have been stronger if they had surveyed a more varied group of people. That is, if they had surveyed people from many walks of life, many age groups, many socio-economic classes, and so on, then their sample would have been more varied and their argument would have been stronger.

If the sample is unrepresentative, it is a **biased sample.** A researcher who generalizes about a population as a whole on the basis of an unrepresentative sample, commits the fallacy of hasty generalization (See Chapter 13).

▶ **EXERCISE 25.2** ◀

---

1. Read the following argument and list of additional items. Would the addition of each item strengthen the argument, weaken the argument, or leave the strength of the argument unchanged? Consider each addition separately from the rest.

   The Bards, a local rock band, needs a new sound system. The system must handle four singers, and it must be powerful enough to fill up a school gym. Friends in another band, the Dynamics, have a brand Z sound system that works well, so, the Bards decide to buy a brand Z system. They reason that the brand Z sound system worked fine for the Dynamics, so it will probably work fine for the Bards.

   a.*The Dynamics regularly play in school gyms.

   b. The Bards know another band that uses and likes the same brand sound system.

   c.*The Dynamics have used their sound system in many different settings and it has always worked well.

   d. The conclusion is changed to: The sound system will work perfectly.

   e.*The conclusion is broadened to: The sound system will be adequate, though not always perfect.

   f. The Bards play rock 'n' roll, but the Dynamics actually play only new wave polka music at senior centers.

2. On a visit to the Kingdom of Og on the Isle of Grog, all ten adults observed at dinner drank coffee with their meal. The conclusion drawn is that all adult Ogians drink coffee with dinner. Does each suggested alteration strengthen the argument, weaken it, or does it leave the argument unchanged? Consider each alteration independently of the others.

   a. All ten observed adults were adult men.

b. Five of ten adults were women, five were men.

c. All Ogians were observed on a Sunday.

d. We change the conclusion to: All Ogians drink coffee.

e. The ten adults were observed over period of seven different days.

f. All of the adult men observed wore unusually long beards, whereas no other Ogians wore beards.

g. We change the conclusion to: All adult Ogians drink coffee with their meals.

h. We change the conclusion to: All Ogians drink coffee with their dinners.

i. We change the conclusion to: Many Ogians drink coffee with their dinner.

3. You take your car to Latka's Total Car Care on three different occasions and each time your car is fixed. You conclude that this time your car will again be fixed. Does each alteration strengthen or weaken the original argument? Does it leave the argument's strength unaffected?

a. The earlier three visits were for transmission work; this one is for body work.

b. The shop has a new manager.

c. You have actually taken your car in eight times.

d. The earlier visits were for mechanical work, this time the problem is electrical.

e. The earlier visits were for your Ford Escort; this time you have a new car, a Jeep.

f. The earlier visits were for 1960s models; now you have a 1998 model.

g. You conclude instead that your car will probably be returned in absolutely perfect condition.

h. You conclude instead that you will probably be satisfied.

i. Since your last visit, Latka has sold the shop. Putty is the new owner and will personally work on your car.

4. You buy four sandwiches at Doug's Deli and all four are awful. A week later, you decide to go back, but you also figure that the next one you buy will probably also be awful. Does each of the following alterations strengthen or weaken the original argument? Does each leave argument strength unaffected?

    a. All four awful sandwiches were made by the same guy.

    b. The four awful sandwiches were made by four different employees.

    c. All four were bought on Sunday; this visit will be on a Tuesday.

    d. All four were ham; this one will be turkey.

    e. You conclude instead that the next sandwich won't be good.

    f. The four awful sandwiches were made by Doug himself.

    g. The four awful sandwiches were made by a new employee.

    h. The health department has since inspected and fined Doug's Deli.

5. One hundred college students from four colleges were surveyed, and 90 percent approved of the president's performance. Does each of the following alterations strengthen or weaken the original argument? Does each leave the argument's strength unaffected?

    a.*The colleges were all Catholic colleges.

    b. Actually, only ten students were surveyed.

    c.*One thousand students were surveyed.

    d. All surveyed were Capricorns.

    e.*All surveyed had grades of 3.8 or above.

    f. Ninety percent of those surveyed said they regularly read novels.

    g. Those surveyed were freshmen, sophomores, juniors, and seniors.

    h.*Those surveyed were seniors only.

    i. Those surveyed were from large cities only.

    j. Those surveyed were from the South only.

    k. Those surveyed were from all regions of country.

    l. All those surveyed like Mars bars.

    m. We change the conclusion to: Exactly 90 percent approve.

    n. We change the conclusion to: At least 70 percent approve.

    o. We change the conclusion to: The majority approve.

    p. Students were surveyed at three state universities, three private liberal arts colleges, three Ivy League colleges, and three religious colleges.

6. Consider the following argument: "All my cats like fried chicken. Therefore, all cats like fried chicken." Does each of the following alterations strengthen or weaken the original argument? Does each leave the argument's strength unaffected?

a. I actually only have one cat.

b. I have seven cats.

c. My cats are all Manx cats.

d. One cat is a calico, one is a rag-doll cat, one is Siamese, and one is a tabby.

e. My cats were all strays.

f. I lied. I have no cats.

g. I've always been allergic to cats.

h. I change the conclusion to: Most cats like fried chicken.

i. I change the conclusion to: Most cats like chicken.

j. I change the conclusion to: Most cats like meat.

k. I change the conclusion to: Most cats like fried food.

7. Your school's football team has played ten games so far this season and has won nine. You conclude that it will probably win the next game. Does each of the following alterations strengthen or weaken the original argument? Does each leave the argument's strength unaffected?

a. It rained hard in the other ten games. Sun is forecast for this game.

b. The next team you play has only played one game this year.

c. You won four games instead of eight.

d. Your star quarterback broke his leg snowboarding.

e. The next team has played twenty and won twenty.

f. The previous ten games were at home, the next is across the state.

g. The previous five games were all against the same team.

h. The next team you will play won every game by at least a twenty-point margin.

i. Your team has new uniforms.

8. Pete and Sally have flown from Kodiak to Seattle ten times. Each trip has taken three hours. They conclude that the next flight will probably also take three hours. Does each of the following alterations strengthen or weaken the original argument? Does each leave the argument's strength unaffected?

a.*Actually, Pete and Sally have made more than 100 trips, each lasting three hours.

b. Actually, they've only made a total of four trips.

c. *They conclude instead that the next flight will take at least three hours.

d. They conclude instead that next flight will take at least two and one-half hours.

e. They conclude that the next trip will take three to five hours.

f. *The previous flights were on Alaska Airlines; the next flight is on a new airline.

g. The previous flights were in winter, the next flight will be in summer.

h. They got meals on all previous flights. The next flight will have no meals.

i. *The previous flights were all on 727s, the next flight will be on a 747.

j. The previous flights had strong headwinds, this flight will have no headwind.

INFERENCE TO THE BEST EXPLANATION

An **inference to the best explanation** is an inductive argument that fits the following general format:

1. The argument cites one or more purported facts, which, it is claimed, are in need of explanation.

2. Possible explanations of the facts are considered.

3. It is argued that a particular explanation is the best or most reasonable explanation of the facts.

4. It is concluded that this explanation is *probably* the correct explanation.

This type of argument claims only that if its premises are true its conclusion is *probably* true. An *inference to the best explanation* type of argument is inductive because it does not aim to prove that its conclusion *must* be true, it only aims to show that its conclusion is likely or probable.

For example, suppose a bank is robbed. The bank robber was seven feet tall, had long red hair, a pointy red beard, and spoke with a strong Chicago accent. Suppose that an hour later, a man goes into a nearby car dealer's showroom and tries to pay cash for a brand new Cadillac. The man is seven feet tall, has long red hair, a pointy red beard, and speaks with a strong Chicago accent. There are many possible explanations, but which explanation is the most reasonable? The best explanation of the facts is that the tall man at the car dealership is the man who earlier robbed the bank.

In everyday life, we often give arguments in the form of an inference to the best explanation. For example: Mom comes into the playroom to find her lipstick smeared all over the wall. "How did that happen?" she says. She looks down to see little Joey with lipstick all over his face and fingers. A hypothesis quickly forms. What do you think the best explanation probably is?

Arguments given in the courtroom also often take the form of an inference to the best explanation. For instance, in the "trial of the century," Christopher Darden and Marcia Clarke argued that the best explanation of the sum of the evidence was that O. J. Simpson committed murder. On the other hand, Johnnie Cochran argued that the best explanation is that Detective Mark Fuhrman and others in the Los Angeles police department framed Simpson for the crime.

Scientific arguments often take the form of an inference to the best explanation. For example, at the beginning of the twentieth century, physicists who relied on Newtonian physics couldn't explain certain atomic phenomena. When Einstein put forward the Special Theory of Relativity in 1905, it became clear to physicists that Einstein's theory offered the best explanation of the phenomena in question. The reasoning in favor of the theory is complicated, but the fact that it provided the best explanation of the phenomena in question gave physicists an excellent reason to suppose that Einstein's Special Theory is the correct explanation. That is, the fact that the Special Theory provided the best explanation gave physicists a very good reason to suppose it *true*. The fact that a hypothesis provides the best explanation of a phenomena is a good reason for accepting the hypothesis.

One way to critique an inference to the best explanation is to come up with a better explanation and an argument for the superiority of this new explanation.

WHAT MAKES ONE EXPLANATION
BETTER THAN ANOTHER?

No set of universally agreed upon criteria exists that determines in a precise way which explanation is best. The evaluation of an explanation and the comparison of rival explanations is an inexact science, so to speak. However, certain criteria are widely accepted. When we evaluate potential explanations and compare rival explanations, we often apply the following criteria.

1. A good explanation is *internally consistent*. An internally consistent explanation contains no self-contradictory elements.

2. A good explanation is *externally consistent*. An externally consistent explanation does not contradict already established facts and already proven theories.

3. A good explanation explains the widest possible range of relevant data. If all else is equal, the potential explanation that explains more of the relevant data is preferable. The more completely the explanation explains the data, the better the explanation.

4. If two potential explanations explain the same range of data and are otherwise equal, except that one explanation is simpler than the other, the simpler

explanation is preferable. One explanation is simpler than another if it makes reference to fewer entities or contains fewer explanatory principles or explanatory elements.

Regarding this last criterion, sometimes called the *principle of economy,* consider the following brief example. Suppose a criminologist examines a crime scene and finds sixty shoe prints, all made by size twelve Ace brand tennis shoes. Several hypotheses suggest themselves: Perhaps *one* person wearing size twelve Ace shoes left the prints. Perhaps *two* persons, each wearing identical Ace shoes, left the prints. Perhaps *three* persons, each wearing identical Ace shoes, left the prints, and so on. If each hypothesis is equally consistent with the evidence, the simpler hypothesis is the preferred choice. In this case, the simpler hypothesis is the "one person" theory, since it posits one crook while the others posit multiple crooks.

> ► **EXERCISE 25.3** ◄

1. Construct an inference to the best explanation argument in support of each of the following claims. (Some of these may require some research.)

   a. The Earth is round.

   b. Caffeine is a stimulant.

   c. Lee Harvey Oswald assassinated President Kennedy alone.

   d. Oswald didn't do it alone.

   e. O. J. Simpson did not commit murder.

   f. Simpson did commit murder.

   g. Elvis is still alive.

   h. Elvis is no longer alive.

   i. UFOs are from another planet.

   j. UFOs are not from another planet.

2. Suppose someone is mad at you. Give an inference to the best explanation in support of your claim that the person is feeling angry.

3. Give an inference to the best explanation argument for the claim that your instructor is a living being rather than a computerized robot.

4. Pick a story from *The Adventures of Sherlock Holmes,* find an inference to the best explanation, ("It's elementary, my dear Watson") and explain Holmes' reasoning.

5. Pick an episode of a television detective story (or pick a detective movie) and summarize an inference to the best explanation argument employed by the detectives.

6. Choose an episode of *Star Trek*, find an example of an inference to the best explanation argument, and summarize the reasoning.

CHAPTER 25 GLOSSARY

Analogical argument   An argument in which we (1) assert an analogy between two things or kinds of things, A and B, (2) we point out that A has a particular feature and that B is *not* known *not* to have the feature, and (3) we conclude that B probably also has the feature.

Analogy   A resemblance or similarity between two or more things.

Biased sample   A sample of a larger population that is unrepresentative of that larger population.

Deductive argument   An argument that claims, explicitly or implicitly, that *if* the premises are true the conclusion must be true.

Enumerative induction   Argument in which premises about observed individuals or cases are used as a basis for a generalization about unobserved individuals or cases.

Inductive argument   An argument that claims, explicitly or implicitly, that *if* the premises are true the conclusion is probably true, although it might nevertheless be false.

Inductive generalization   An enumerative induction in which the conclusion *generalizes* about all of a class of things based on the information enumerated in the premises. From premises concerning observed instances we reason to a conclusion that represents a generalization about *all* the members of a class.

Inference to the best explanation   A type of argument that (a) cites one or more facts that need explanation, (b) canvasses possible explanations, (c) puts one explanation forward as the best explanation, and (d) concludes that that explanation is probably the correct (or true) explanation.

Model   A simplified representation of a more complex object or state of affairs.

Primary analogates   In an analogical argument, the objects being compared that appear only in the premises.

Secondary analogate   In an analogical argument, the object that appears in the premises and also appears in the conclusion.

Chapter

26 SCIENTIFIC REASONING

I
n everyday life, when we are puzzled by something that happens, we often construct hypotheses. A **hypothesis** is an explanation offered for a phenomenon. For example, suppose one sunny afternoon you are listening to your stereo, and it suddenly goes dead. You naturally begin formulating hypotheses. Maybe the power is out. Maybe a circuit breaker went out. Maybe the stereo is broken.

Next, you *test* the hypotheses to see which is correct. You reason that if the household power is out, no electrical devices will be on. You have derived a prediction from your first hypothesis. Now you check it out by looking around and perhaps turning on a light switch. Suppose the light comes on, contrary to the prediction. This shows your first hypothesis is false. Next, you derive a prediction from your second hypothesis: If the circuit is out, the refrigerator will be running but the light next to the stereo won't work. You check out this prediction. Suppose the refrigerator is running, but the light next to the stereo won't work. The prediction was correct. This is evidence in favor of your second hypothesis. Now, hot on the trail of this second hypothesis, you derive another prediction from it: If the circuit is out, when you go to the electrical panel you will find one of the circuit breakers in the off position. This is exactly what you find, and your second hypothesis now seems the likely explanation of the stereo's sudden failure.

When we reason in this way, we are engaging in a type of reasoning called *hypothetical induction*. We think up a hypothesis, deduce a prediction, and test the hypothesis by looking to see if the prediction is correct. If we find the predicted result, this counts as evidence in favor of our hypothesis, although it is not conclusive evidence and does not prove our hypothesis *must* be true. If we don't find the predicted result, the evidence is against our hypothesis.

The basic thought process we have been describing is used in science as well as in everyday life. Scientists also propose hypotheses and test their hypotheses by deriving predictions and looking for predicted results. Thus, in *Physics and Reality,* Albert Einstein said that science is "nothing more than a refinement of everyday thinking." Our goal in this chapter is to try to understand the basics of these "refinements."

SCIENTIFIC REASONING

The general process of hypothetical reasoning used in science is often called the **hypothetico-deductive method,** for it centers around *hypotheses* and predictions *deduced* from them. Using the hypothetico-deductive method, scientists seek to discover the general physical laws of the universe. Boiled down to essentials, this method involves the following five steps:

Step 1   Scientists encounter a puzzling phenomenon that needs an explanation.

Step 2   A hypothesis is proposed that would explain the phenomenon in question.

Step 3   Scientists ask, if the hypothesis is true, what facts about the world can we expect to *observe?* That is, what *observational facts* does the hypothesis imply or predict? In this way, they derive **observational predictions** from the hypothesis.

Step 4   Scientists next test the hypothesis. They make observations and see if they observe what the theory predicted they would observe.

Step 5   Finally, they accept, reject, or revise the hypothesis on the basis of the test.

These five steps briefly outline a complicated process. Now, let us fill in some details and discuss a few examples.

COMMENTS ON THE STEPS

► STEPS 1 AND 2

Puzzling over the facts and formulating an explanatory hypothesis is a creative activity that requires imagination, guesswork, inspiration, and a basic knowledge of how the world works. No hard and fast rules tell the scientist how to create hypotheses. For example, in 1900, the German physicist Max Planck (1858–1947) showed that when atoms are heated they absorb or emit energy in discrete units or particle-like bits that are all set either at a certain minimum quantity or at integral multiples of that minimum quantity. After he learned of Planck's discovery, Albert Einstein (1879–1955) began thinking deeply about the nature of light. Einstein's general knowledge of how the world works and his knowledge of current research soon helped him "think up" the hypothesis that light is particulate in nature.

However, Friedrich Kekule's hypothesis on the structure of the benzene molecule came to him during a dream. In 1858, Kekule (1829–96), a German chemist who laid much of the groundwork of organic chemistry, had shown that carbon atoms link together in long chains. One night in 1865, Kekule dreamed that the benzene molecule was a snake biting its own tail and whirling in a circle. That dream image gave him the concept of the six-carbon benzene ring, and a hypothesis was born. The point is that not all hypotheses originate in the same way; they take root in some creative area of the human mind where not all thought-processes are rule-governed.

Regardless of how it was thought up, when a scientific hypothesis is presented to the scientific community for testing, it is subjected to standardized scientific tests. Furthermore, these tests are performed by many different researchers from many locations, and many of these researchers will have no vested interest in the hypothesis being true. Indeed, some may actively try to prove the hypothesis false, which brings us to the next step.

▶ STEP 3

When the hypothesis is on the table, scientists derive a set of predictions from the hypothesis by asking the following question: *If* the hypothesis is true, what can we expect to observe about the world? Scientists employ deductive or inductive hypothetical reasoning in answering this question—and thus in deriving the prediction. Often they use complicated mathematics or complicated logical reasoning.

When we think of *prediction* we naturally think of a claim about the future. Perhaps an image of a fortune teller gazing into a crystal ball flashes into the mind. However, the predictions deduced from a scientific hypothesis do not need to be about the future; they can be about the past, the present, or the future. Essentially, all that science requires is that the predictions be about the results of observations that can be made, and the observation can concern facts related to the past, present, or future.

For example, when Charles Darwin (1809–82) proposed his theory of evolution, one of the implications of his theory was that species had changed slowly and *gradually* in the past. This implied that the fossil record should show a sequence of small, gradual changes with no large and sudden leaps between groups of animals. The prediction derived from the theory was therefore that a *current observation* of the complete fossil record should find small, gradual changes rather than large and sudden changes.

▶ STEPS 4 AND 5

When scientists test a hypothesis, the test may *confirm* the hypothesis or it may *disconfirm* the hypothesis. The process by which a scientific hypothesis is shown to be probably true is called the **process of confirmation.** The process by which a scientific hypothesis is shown to be probably false is called the **process of disconfirmation.** These two processes, which lie at the heart of the scientific method, differ in important ways, and in the next section, we shall look more deeply into each process. Now, we can more precisely define the hypothetico-deductive method as the method used in the sciences to confirm and disconfirm theories.

CONFIRMING AND DISCONFIRMING SCIENTIFIC HYPOTHESES

The processes of confirmation and of disconfirmation are both illustrated in an interesting episode from the history of science, the discovery of the cause of "childbed fever" by the Hungarian-born doctor Ignaz Semmelweiss (1818–65). While working at Vienna General Hospital during the 1840s, Semmelweiss found the mortality rate in the maternity division unusually high. Many women who gave birth in that division came down with puerperal or "childbed" fever, and a large percentage died from the disease. However, the death rate in the first maternity division was almost three times as high as the rate in the second maternity division. Semmelweiss began searching for the cause.[1]

When Semmelweiss began his investigation, one widely accepted hypothesis was that an epidemic was being transmitted through the air. However, Semmelweiss rejected this hypothesis by reasoning that if an airborne epidemic were the cause, it would affect both divisions equally, and it would also affect the city of Vienna. Yet no epidemic of childbed fever was detected outside the hospital's maternity division, and furthermore the two divisions had distinctly different disease rates.

Semmelweiss's reasoning may be cast in the form of the following argument:

1. If the "airborne epidemic" hypothesis is true, then the disease would affect both divisions equally, and it would also affect the larger city. (This is a prediction.)

2. The disease does not affect both divisions equally and does not affect the larger city. The prediction, in other words, is observed to be false.

3. So, the hypothesis is false.

Notice that this argument fits the Modus Tollens form (Chapter 6):

1. If **P** then **Q**

2. Not **Q**

3. Therefore not **P**

---

[1] My treatment of Semmelweiss's discovery is indebted to the account in Robert Churchill's *Becoming Logical: An Introduction to Logic* (New York: St. Martin's, 1986), Chapter 8.

The argument is therefore a valid deductive argument. If the premises are true, the conclusion *must* be true. The evidence, in this case, disconfirms the hypothesis, which means that it gives us reason to suppose the hypothesis is false.

Semmelweiss considered, tested, and rejected other hypotheses. For instance, someone had proposed that overcrowding was the cause of the illness. However, Semmelweiss investigated and found that the second division was actually more crowded than the first division (primarily because women were trying to avoid the notorious first division). Thus:

1. If overcrowding is the cause, then the first division should be more densely crowded than the second division.

2. The first division is *not* more densely crowded.

3. Therefore, the hypothesis is false.

Semmelweiss tested one interesting hypothesis. When a patient in the hospital's sickroom was near death, a priest, accompanied by an attendant ringing a bell, would arrive to administer the last rites of the Catholic Church. The layout of the hospital required that the priest walk through the first division but not through the second division. The hypothesis was this: Perhaps the fear caused by the appearance of the priest predisposed the women in the first division to become vulnerable to the fever. In order to test this hypothesis, Semmelweiss had the priest change his route. The change had no effect on the mortality rate. Thus, against the hypothesis, one could reason: If the hypothesis is correct, changing the priest's route should change the mortality rate. Changing the route did not change the rate. The hypothesis is false.

Semmelweiss noticed a curious fact: In the first division, women delivered their babies while lying on their backs, but in the Second division, they delivered while lying on their sides. Perhaps the position during birth was the cause of the illness. Semmelweiss had the two divisions switch their procedures. However, changing the procedures and having the first division adopt the lateral position did not affect the mortality rate. If the hypothesis were correct, changing the delivery procedures should have changed the mortality rate. Changing the procedures did not change the rate. The hypothesis was deemed false.

In the case of each of these tentative hypotheses, Semmelweiss rejected the hypothesis by reasoning: If this hypothesis were true, certain specified events should be observed. These events did not occur or were found not to have occurred. Therefore the hypothesis was false. In terms of truth-functional logic, such reasoning follows the Modus Tollens form:

1. If **P** then **Q**

2. Not **Q**

3. So, not **P**

This is, of course, a deductively valid form of reasoning. If the premises are true, the conclusion *must* be true.

Then, in 1847, Professor Kolletschka, an anatomy teacher, received a puncture wound from a scalpel while performing an autopsy at the hospital. The professor soon became extremely sick, and, as he lay dying, showed several of the symptoms of childbed fever. Reasoning by analogy, Semmelweiss formulated a new hypothesis: matter from the cadaver in the autopsy room was the cause of childbed fever. Upon investigating the matter, Semmelweiss discovered that doctors would routinely perform autopsies on dead bodies, wash their hands, and go immediately into the first maternity division to deliver babies and treat patients. The women in the second maternity division, on the other hand, were taken care of by midwives rather than by doctors, and the midwives did not perform autopsies.

Semmelweiss hypothesized that the soaps being used after the autopsies did not remove the infectious material from the hands of the doctors. To test his hypothesis, Semmelweiss had the doctors wash with a solution of chlorinated lime, a far more powerful cleaning agent, before they treated the women in the first division. The first division's mortality rate quickly plummeted. Indeed, it fell to a level *below* the level of the second division. Semmelweiss now had an argument in *favor* of a hypothesis: If this new hypothesis was true, washing with chlorinated lime should cut the mortality rate in the first division. The chlorinated lime wash did indeed cut the mortality rate. Therefore the hypothesis was deemed probably true.

Notice that this argument for the hypothesis is an inductive argument—the claim is that the premises make the conclusion *probable* rather than certain. The evidence, in this case, is said to *confirm* Semmelweiss's hypothesis, which means that it gives us reason to suppose the hypothesis is probably true. No one claims that the evidence proves the hypothesis with certainty, that is, that the hypothesis *must* be true. Rather, the claim is that the observations make the hypothesis likely, they make it probable, they make it the most reasonable explanation.

After researchers derive predictions from a hypothesis, each prediction that they observe or find to be true is called a *confirming instance* of the hypothesis (because the observation "confirms" the truth of the hypothesis). The more confirming instances, the more probable the hypothesis. Some hypotheses are confirmed over and over again, year after year. Such hypotheses have a high level of confirmation; that is, such hypotheses are most probably true.

Throughout his investigation, Semmelweiss followed the four steps of the hypothetico-deductive method noted above. He formulated a hypothesis. He derived predictions from the hypothesis. He tested the hypothesis to see if the predictions were true. He accepted, rejected, or modified the hypothesis based on the evidence he observed.

The story of Semmelweiss's discovery provides a general introduction to the processes of scientific confirmation and disconfirmation. However, some complications and refinements deserve a bit of attention after the general logical groundwork has been laid. The following two episodes from the history of science will allow us to look at confirmation and disconfirmation in more depth.

THE CONFIRMATION OF
A SCIENTIFIC HYPOTHESIS

▶ THE CASE OF NEWTONIAN PHYSICS

Isaac Newton (1642–1727) revolutionized physical science. A small amount of background will help us appreciate how truly revolutionary Newton's theory was in its day. Let us begin with Kepler's discovery of the mathematical laws of planetary motion. After he analyzed mountains of data collected earlier by Tycho Brahe, Johannes Kepler (1571–1630), an astronomer and skilled mathematician, came up with a truly revolutionary hypothesis, a hypothesis only a mathematician might originate. Kepler hypothesized that the planets move in ellipses—oval shaped orbits—with the sun at one focus of the ellipse. This was a revolutionary hypothesis, for it had previously been thought that the planets must move in perfect circles. Kepler also hypothesized that the planets orbit at varying speeds, with the speed changing so that a line drawn between the sun and a planet sweeps out an equal area of the ellipse in an equal period of time. Kepler put his hypothesis into clear and exact terms by describing the motions of the planets in terms of three precise, mathematical laws known now as Kepler's Laws.

One of the big questions of the day suddenly became: Why would the planets move in elliptical orbits? Why not perfect circles? Isaac Newton solved this problem. Using precise mathematics, Newton specified an equation that described a law of gravitation and additional equations for three laws of motion. He showed, again using precise mathematics, that *if* the planets obey this particular law of gravitation and the three laws of motion, *then* they must move in the elliptical orbits described by Kepler's laws. In short, Newton showed mathematically how his proposed law of gravity and laws of motion would produce elliptical planetary orbits that conform to all three of Kepler's laws.

Newton thus proposed an astonishing and revolutionary hypothesis: Gravity from the sun moves the planets in their orbits. The same general force that makes an apple fall to the ground on earth also makes a planet orbit the sun (and the moon orbit the Earth). Newton solved a problem that had eluded astronomers for thousands of years: What explains the motion of the planets? The question of the day now became: Is Newton's hypothesis true?

▶ HALLEY'S TEST OF NEWTON'S HYPOTHESIS

The Newtonian hypothesis may be defined as the claim that the heavens conform to the equations specified by Newton, the equations for the law of gravity and the three laws of motion. Edmond Halley (1656–1742), a British astronomer, proposed one of the first important tests of the hypothesis. His test concerned a small, seemingly insignificant object: a comet.[2]

---

[2] My account of Halley's test is indebted to the account in Ronald Giere's *Understanding Scientific Reasoning* (New York: Holt, Rinehart, and Winston, 1979).

In the late seventeenth century, comets were thought to be unpredictable, mysterious heavenly objects. Some thought comets were supernatural omens of the future, some thought they were evil demons. In 1682, Halley observed a comet and recorded quite a bit of data describing part of the comet's path. In 1695, a few years after Newton's theory had been published, Halley asked an extremely significant question: If the solar system is governed by Newton's laws, and if the 1682 data on a small part of the comet's path are correct, what should the rest of the comet's orbit be? Using Newton's laws plus his 1682 data, Halley deduced the complete orbit. Halley's result was based on the hypothesis that Newton's laws actually do govern the heavens.

However, in deriving his predictions, Halley had to make several *assumptions*. For instance, he knew Jupiter's gravitational pull would affect the comet, but he assumed that this was too small to have a significant effect and therefore left Jupiter's influence out of his calculations. (Jupiter's gravitational influence would also have been extremely difficult to factor into the equations.) In making his calculations, Halley also assumed that no outside or unknown forces are interfering with the comet. Halley concluded that the comet travels in an elongated ellipse and takes seventy-five years to orbit the sun. On the basis of this, Halley made two predictions: (a) his comet has been around many times before, and if historical records are searched, comet reports fitting his data should be found at seventy-five-year intervals; and (b) the comet should return in December 1758.

A search of the records did indeed find comet reports at seventy-five-year intervals back to 1305. Halley's first prediction was correct. And although Halley died in 1743, the comet did indeed return in December 1758, as predicted. Indeed, it returned on Christmas Day 1758. The second prediction had also been borne out by the facts. Newton's theory had been "confirmed" by two successful predictions.

The success of this second prediction caused a sensation in Europe and converted many to the Newtonian hypothesis. Remember that comets had previously been thought to be unpredictable, supernatural omens.

Now, why did the arrival of Halley's comet constitute such convincing evidence for Newton's theory? Ronald Giere, a philosopher who specializes in the philosophy of science, proposes that the following logic supports Newton's hypothesis.[3]

1.   If the hypothesis is correct (that the solar system obeys Newton's laws), and if the initial data is correct (the astronomical data from 1682), and if the assumptions are correct, (that Jupiter has an insignificant effect on the comet and that no outside or unknown forces are interfering with the comet), then the comet will return in December 1758.

2.   However, if Newton's laws are *not* correct, but the data and assumptions *are* correct, then the prediction is extremely unlikely to come true, for it is unlikely

[3]Ronald Giere, *Understanding Scientific Reasoning* (Holt Rinehart and Winston, 1979).

that anyone could make a correct prediction of the comet's return without using the correct hypothesis. In other words, the prediction is a "long shot."

3. The prediction came out true. Therefore, the hypothesis, initial data, and assumptions are probably correct.

4. Therefore, the solar system probably is governed by the Newtonian laws.

Notice that this argument is not deductively valid; the premises do not *imply* the conclusion. However, the argument doesn't purport to be deductively valid. The claim is not that the conclusion *must* be true if the premises are true; rather, the claim is that the conclusion is *probably* true if the premises are true. And this claim seems to be correct. Thus, although the argument is not valid, it is inductively strong. In this case, we say the Newtonian system is "confirmed." However, because the argument is neither deductively valid nor deductively sound, the hypothesis is not proven, it is not guaranteed true.

Let us now put this reasoning into a more formal schema:

Let H stand for the hypothesis that the solar system obeys Newton's laws.

Let ID stand for the *initial data* used in the calculations—primarily the 1682 astronomical data presupposed by Halley.

Let AA stand for the *auxiliary assumptions* Halley had to add to his reasoning, including (a) Jupiter has an insignificant effect and its influence can therefore be disregarded; (b) no outside or unknown forces are interfering with the comet.

Let P stand for the *prediction,* namely, that the comet will return in December 1758.

The "logic of confirmation," as Giere construes it, may now be summed up this way:

1. If H and ID and AA are all true, then P is true.

2. If H is false and ID and AA are both true, then probably P is false. (That is, it is unlikely the prediction will be true if H is false, for it is unlikely anyone could make the prediction without using the correct hypothesis.)

3. P is true.

4. So, H and ID and AA are probably true.

5. Therefore, H is probably true.

Again, this argument is not valid. The premises do not imply the conclusion. The hypothesis H is confirmed rather than proven or guaranteed.

In even more general terms, the confirmation process may be represented by this basic pattern:

1. An observational prediction (P) is derived from the conjunction of the hypothesis, data on initial conditions, and auxiliary assumptions.

2. Guided by the prediction, observations are made and the prediction is verified. P is found to be the case.

3. It is concluded that the hypothesis is probably true.

The observation of P is called a "confirming instance" of the hypothesis (and the initial data and assumptions). The confirming instance counts as evidence in support of the hypothesis and the test *confirms* the hypothesis. When a test confirms a hypothesis, this constitutes inductive support for the hypothesis—it makes the hypothesis probably true to some degree. Each time this process is repeated and an additional prediction of the hypothesis is verified, the hypothesis becomes more probable. That is, each confirming instance is additional evidence in favor of the hypothesis. A highly probable hypothesis is one that has survived many such tests. The more confirming instances, the more probable the hypothesis.

THE DISCONFIRMATION OF A SCIENTIFIC HYPOTHESIS

▶ THE CASE OF PHLOGISTON THEORY

What is combustion? This question was one of the great issues debated during the rise of modern science. When a metal burns, it turns to a powdery substance. When cloth burns, it turns to ashes. Upon observation, something seems to leave a burning thing, leaving behind only ashes or powder. What is this unknown something?

In the eighteenth century, the commonly accepted hypothesis of combustion was a theory proposed in the late seventeenth century by the German chemist Becher. According to Becher's theory, combustible materials contain a substance named "phlogiston." Becher posited that when something burns, phlogiston is given off. According to this theory, fire is actually a substance—phlogiston—leaving a burning thing.

Becher's phlogiston theory seemed to explain many things associated with combustion. For instance, in many cases, a material does lose weight when it burns, as if something has left the substance. Furthermore, burning materials do seem to give off something into the air when they burn down to an ash. Phlogiston theory even had an explanation for why smothering puts a fire out. According to the theory, smothering puts the fire out by holding the phlogiston in. Because the phlogiston is prevented from escaping, the fire dies. Why does the flame go out when a bell jar is placed over a burning candle? According to phlogiston theory, placing a bell jar over a candle extinguishes the flame because the air inside the sealed bell jar can hold no more phlogiston. Becher's theory also explained why heating a substance can cause it to burn: Heating causes flames because heating drives phlogiston off into the air. And why does cooling something put a fire out? Cooling makes the phlogiston less volatile, according to the theory.

Although today phlogiston theory may seem obviously wrong to us, it had a lot of support in the eighteenth century—it explained many of the facts about combustion. It was not at all obviously false.

However, phlogiston theory faced a serious problem. When a metal burns, the substance left behind, (then called a "calx," now known as a metallic oxide) weighs more than the original metal. How could the weight increase if something (phlogiston) is given off? One desperate suggestion put forward by advocates of phlogiston theory was: Perhaps phlogiston has negative weight.

Another problem with phlogiston theory concerned the stuff we use today on the backyard barbecue. When chemists heated a quantity of calx with charcoal, it was found that the calx turns back into the original metal. How could this be? Phlogiston theorists proposed the following hypothesis: Charcoal must be rich in phlogiston and when the phlogiston flows back into the calx, the original metal is reconstituted.

▶ LAVOISIER SHOWS PHLOGISTON THEORY FALSE

In 1775, the French chemist Lavoisier (1743–94) was experimenting with Mercury. After heating it in a closed glass container, and after making precise measurements, he discovered that the volume of air in the container actually decreased. He also discovered something else: when the mercury was heated, a red powder (the calx) formed. Surprisingly, the mercury plus the calx (oxide) that formed on the surface of the mercury weighed *more* than the original mercury. Now, phlogiston theory predicts three things: (a) the mercury and the calx should weigh less than original mercury (because phlogiston has been driven off); and (b) the air in the flask should increase in volume (because it contains the phlogiston that was driven off); and (c) the weight of the phlogiston in the air should correspond to the weight lost by the mercury.

Lavoisier pondered why a burned substance weighed more than the original. Burning seemed to be caused by something in the air uniting with the mercury rather than leaving the mercury. Burning did not seem to be caused merely by something from within the mercury itself. Lavoisier concluded that something has disappeared from the air and combined with the metal to produce the calx.

In another experiment, Lavoisier reheated the calx in an enclosed space. A gas (oxygen) was given off, and the original metal was regenerated. This new gas, he reasoned, must be what disappeared from the air and combined with the mercury, for the measured amount of the gas and the weight lost by the calx are the same. On the basis of his experiments, Lavoisier rejected phlogiston theory.

Following the analysis of Ronald Giere, the general pattern of reasoning in this case can be put into the following format:

1. If phlogiston theory is correct, and if the initial data is correct (the weights of the substances as recorded in the laboratory, and so on), and if the assumptions are correct (such as that no outside forces interfered, that phlogiston would not have negative weight), then the mercury should lose weight when it

burns. Thus, the prediction is that the mercury will lose weight when it is burned.

2.   The mercury did not lose weight when it was burned. The prediction was thus falsified.

3.   Therefore, it is not the case that (a) phlogiston theory, (b) the initial data, and (c) the assumptions are all true. In other words, one or more of (a), (b), (c) are false.

4.   The initial data (b) and the assumptions (c) are correct.

5.   So, phlogiston theory (a) must be false.

Let us put this reasoning into a more formal outline:

Let H abbreviate the hypothesis under consideration, phlogiston theory.

Let ID abbreviate the initial data at the start of the test: the original weight of the mercury, the measured volume of air, the measurements made as the mercury was heated in the lab, and so on.

Let AA abbreviate the auxiliary assumptions, that is, that nothing interferes from outside, that phlogiston does not have negative weight, and so on.

Let P abbreviate the prediction derived from phlogiston theory.

The logic of disconfirmation may now be summed up this way:

1.   If H and ID and AA are all true, then P.

2.   P is false. (That is, the prediction failed to come true.)

3.   So, it is not the case that H and ID and AA are all three true.

4.   ID and AA are true.

5.   So, H must be false.

This argument is deductive in nature. The claim is that the conclusion *must* be true if the premises are true. Furthermore, the claim is correct. The argument is therefore valid. The premises imply the conclusion. In this case, given that the premises are true, the hypothesis H is said to be "disconfirmed."

In more general terms, the disconfirmation process may be represented by this basic pattern:

1.   An observational prediction (P) is derived from the conjunction of the hypothesis, data on initial conditions, and auxiliary assumptions.

2.   Guided by the prediction, observations are made and the prediction is found to be false. That is, P is found not to be the case.

3.   It follows that the hypothesis, initial data, and assumptions cannot all be true.

4. The initial data and assumptions are confirmed as true.

5. It is concluded that the hypothesis must be false.

If P is predicted and it is found that P is not the case, P is said to be a "disconfirming instance" of the hypothesis and assumptions, and counts as evidence against H. The test in this case *disconfirms* the hypothesis.

> Lavoisier is known as the Father of Modern Chemistry. Shortly after the French Revolution (1789), because he had been a tax collector, Lavoisier was called before the French Revolutionary Tribunal. After a trumped up trial, the great chemist was condemned to death. When he went to the guillotine on May 8, 1794, in Paris, one of the revolutionary judges said, "The Republic has no need of men of science." But someone present whispered, "Only a moment to cut off that head, and a hundred years may not give us another like it."

If the implications or predictions of the hypothesis are found to be true, this is said to confirm the hypothesis (plus the initial data and auxiliary assumptions). If the implications are found to be false, this disconfirms the hypothesis (assuming that the initial data and auxiliary assumptions are true). A hypothesis is supported inductively by successful predictions; it is refuted deductively by failed predictions.

THE FACT OF THE CROSS

In some situations, one hypothesis implies one prediction, a competing hypothesis implies a prediction that contradicts the other hypothesis's prediction, and a single experiment is designed in such a way that one of the two hypotheses must be disconfirmed and the other confirmed. Isaac Newton called such an experimental result the "fact of the Cross." Here is an interesting example.

One of the questions that puzzled the early scientists was: Does light consist of waves rippling through an invisible medium, or is it a stream of tiny particles? According to the corpuscular theory of light published by Isaac Newton in 1704, light is a stream of extremely tiny particles ("corpuscles").

Alternatively, the wave theory proposed by Christiaan Huygens in 1690 asserted that light consists of spherical waves spreading out from a luminous object like ripples spreading out from a rock that has just been dropped into a pond.

Using data on initial conditions and auxiliary assumptions that both sides agreed on, physicists showed that the Newtonian hypothesis predicts that light travels faster in water than in air. Furthermore, they showed that the speed in water should exceed the speed in air by a precisely calculated amount. However, if the wave theory is true, using the initial data and auxiliary assumptions, the speed of light should be less in water than in air, also by a precisely calculated amount. In 1850, instruments were designed that could finally settle the issue. The velocity of light turned out to be lower in water than in air—by the exact amount calculated by the wave theorists.[4]

WHAT MAKES ONE HYPOTHESIS
BETTER THAN ANOTHER?

Scientists often must compare hypotheses and judge which is more worthy of investigation and which is more likely to be true. No set of universally agreed upon criteria exists that determines in a precise way which hypothesis constitutes the best explanation. The evaluation of a hypothesis and the comparison of rival hypotheses is not an exact science, so to speak. However, several widely accepted criteria do exist. When researchers evaluated potential hypotheses or explanations and compare rival hypotheses, they often apply the following criteria.

1. A scientific hypothesis should be *testable*. That is, a researcher should be able to derive observational predictions from the hypothesis. A hypothesis that yields predictions is said to have "predictive power."

2. A scientific hypothesis should be *internally consistent*. A hypothesis is internally consistent if it does not contain any self-contradictory elements.

3. A scientific hypothesis should be *externally consistent*. A hypothesis is externally consistent if it does not contradict already established facts and already proven theories.

4. A scientific hypothesis should fit the facts. That is, the predictions derived from the hypothesis should be borne out by observation.

5. A scientific hypothesis should be *internally coherent*. That is, its parts should be rationally interconnected, they should fit together in a reasonable way.

6. A scientific hypothesis should be *fruitful*. That is, it should suggest new ideas for future research.

7. A scientific hypothesis should explain the widest possible range of relevant data. The more completely the hypothesis explains the data in question, the better the hypothesis. An inadequate hypothesis fails to explain relevant data. Thus, if one potential hypothesis explains more of the relevant data than another, it is preferable, all else equal. A hypothesis that explains data has *explanatory power*. The more data it explains, the greater its explanatory power.

8. If two potential hypotheses explain the same range of data and are otherwise equal, except that one hypothesis is *simpler* than the other, the simpler hypothesis is preferable. One hypothesis is *simpler* than another if it hypothesizes or "posits" fewer entities or contains fewer explanatory principles or

---

[4] I owe this example also to Ronald Giere, Ibid., p. 112.

explanatory elements. In other words, an explanation should posit only what it must, and no more than it must. This is sometimes called the **principle of economy.** (It is also sometimes called the principle of parsimony.)

The principle of economy was first explicitly formulated by the philosopher William of Occam (1285–1347), in the fourteenth century. Occam put the principle this way: "What can be explained with fewer terms is explained in vain with more." This is usually interpreted to mean that when we are choosing between two competing explanations of a phenomenon, the simpler explanation—generally the explanation postulating fewer explanatory entities or explanatory principles—is preferable, all else equal. This principle is called "Occam's razor" because it requires that we "shave" our explanatory theories down to a minimum of entities and a minimum of complexity.

> Some philosophers have argued in support of Occam's razor as follows. If two potential explanations equally explain the same phenomena, and if one explanation is simpler than the other, if we adopt the more complicated explanation, we postulate additional entities with no gain in explanatory power. As a result, we take on an increased risk of error (because we might be wrong about the extra entities) with no compensating gain in explanatory power. Unnecessary elements in a hypothesis thus increase the possibility of falsehood with no balancing gain in explanatory power. If we follow the principle of economy (Occam's razor) and eliminate unnecessary explanatory elements, in the long run we will minimize the risk of error with no loss in explanatory power.

CASE STUDIES

▶ TORRICELLI'S "SEA OF AIR" HYPOTHESIS

It has been known since ancient times that if you want to drain liquid from one end of a barrel you must have an opening at the other end. Children are aware of a similar phenomenon: If you suck some root beer up a straw and quickly place your thumb over the end of the straw, the root beer won't drain out.

From ancient times until the seventeenth century, this phenomenon was explained by the "full universe" hypothesis originated by Aristotle. According to this theory, nature is completely full of itself. That is, nature is so jam-packed full that it will not allow a vacuum to form. The central slogan of this theory was "nature abhors a vacuum." Evidence that supported this theory was gained by trying to make liquid flow out of the bottom of a keg without a hole in the top. It was observed that unless the hole at the bottom is large enough to admit air, nothing will come out of the keg. It can't, because the flowing liquid would leave behind an enclosed empty space inside the keg and a vacuum would form. However, if a hole is cut in the top of the keg, air flows in as the liquid drains. It was thought that this phenomenon supports the full universe hypothesis.

Supporters of the full universe theory reasoned: Since nature abhors a vacuum, it won't allow the liquid to escape unless a hole is cut in the top of the keg. This explanation, it was argued, makes sense of the phenomenon.

Even so, in the seventeenth century certain observations seemed to contradict the full universe hypothesis. For example, in approximately 1641, an interesting experiment was performed in Rome. A glass container was attached to the top of a forty-foot pipe, this was strapped upright to a tower, and the whole tube was filled (from the top) with water. The bottom of the tube had a valve that was closed so that the water couldn't run out. The top was sealed shut. When the valve at the bottom was opened, the water rushed out, leaving a vacuum behind.

Engineers attempted to create huge suction pumps that would drain water out of a reservoir and up over a nearby hill. However, they discovered that a suction pump will not raise water above a certain height. After a certain point, the water stops rising and a vacuum seems to form.

It began to look as if something was wrong with Aristotle's "full universe" theory. Galileo's student, Evangelista Torricelli (1608–47), began searching for a better theory. He noticed that a suction pump, which uses a piston to draw water out of a well, could raise water only thirty-three feet above the surface of the well. In 1642, in a stroke of genius, Torricelli came up with a hypothesis that would explain this puzzling fact. He hypothesized that the earth is surrounded by a "sea of air." This sea of air actually has weight, just as a sea of water has weight. Torricelli further reasoned that if the sea of air has weight, it should exert pressure ("air pressure") on all objects within it—just as water exerts pressure ("water pressure") on someone submerged in it. This pressure on the surface of the well must be what pushes the water up the pipe when the suction pump's piston is raised. Furthermore, the height of the column of water in the suction pump must be related to the total pressure of the air.

This hypothesis gave Torricelli an intriguing idea. Torricelli reasoned that because mercury is fourteen times heavier than water, the sea of air ought to hold up only a twenty-nine-inch column of mercury (because it holds up a thirty-three-foot column of water). If a tube of mercury is sealed at one end, inverted, and the open end is placed in a bowl of liquid mercury, some of the mercury in the tube should flow out and into the bowl, but the weight of the sea of air ought to push down on the surface of the bowl of mercury with enough force to hold the rest of the mercury in the inverted tube, thus creating a vacuum inside the tube. An equilibrium should be reached where the weight of the mercury in the tube is equal to the weight of the sea of air pressing down on the bowl. This device, Torricelli realized, could measure fluctuations in atmospheric pressure. Torricelli had invented the mercury barometer.

With the above reasoning, Torricelli derived a prediction from his theory: the weight of the atmosphere should support a twenty-nine-inch column of mercury in the inverted tube, with the rest of the tube a vacuum. This prediction was easily verified.

After reading about Torricelli's sea-of-air hypothesis, the French philosopher Blaise Pascal (1623–62) realized another implication of the hypothesis:

just as water pressure diminishes as you ascend from the bottom of a lake toward its surface, air pressure should diminish if we ascend to the surface of the sea of air. Consequently, Pascal argued that if Torricelli's hypothesis is true, air pressure must decrease at higher altitudes, because at a higher altitude one is closer to the surface of the sea of air, and less air presses down. The following prediction was consequently derived from Torricelli's hypothesis: the height of a mercury column in an inverted tube should decrease as the device is carried up a mountain.

In 1648, Pascal had his brother-in-law, Perrier, perform the experiment. While the "Torricelli barometer" was carried up a mountain, an identical barometer on the ground (a *control*) was watched continuously. The device on the ground never changed, but the device that was carried up the mountain behaved as predicted. At the top of the mountain, the column in the primitive barometer had descended. The experiment was repeated five times while the observer at the bottom of the mountain watched the control tube. Each time, the same result was observed. These experiments provided dramatic confirmation of Torricelli's sea-of-air hypothesis.

▶ RUMFORD'S HYPOTHESIS: HEAT IS MOTION

If two material objects are brought together, and one has a higher temperature, the two will eventually reach the same temperature. It seems commonsense to suppose some sort of fluid substance flows from the warmer to the cooler object, just as water flows from a higher level to a lower level. In the Middle Ages, this phenomenon was explained by the "caloric theory." According to this hypothesis, heat is an invisible substance, called "caloric," that actually flows from the warmer to the cooler body. Caloric theory seemed to answer many questions about heat. It explained why two bodies reach a temperature equilibrium, and it explained why bodies expand when heated. (They supposedly expand as they fill with caloric.)

However, contradictory evidence abounded. For instance, according to caloric theory, red hot iron should weigh more than cold iron, because it is full of caloric. Yet after careful measurements, it was clear that hot metal weighs no more than cold metal. In the face of this contrary evidence, defenders of caloric theory suggested that caloric might be a weightless substance. This would explain why the hot iron does not weigh more than the cold iron.

Count Rumford (1753–1814) was supervising the boring of cannon for the British Artillery when he noticed that hot chips of metal fell off the borer. Rumford asked a simple question: where does the heat come from? The borer had no fire or obvious source of heat. Furthermore, the heat generated by the borer seemed to be inexhaustible. So, Rumford reasoned, heat cannot possibly be a material substance, because matter cannot be created out of nothing. Heat could only be motion, Rumford reasoned, given the way it is generated and communicated. Rumford's hypothesis is now known as the *kinetic theory of heat*.

More precisely, material objects are made of many small particles—atoms and molecules—moving in many directions, at many different speeds, and the modern understanding, which we owe to Rumford, is that heat is the average motion or mean kinetic energy of these particles. Heat increases as the average speed of these particles increases.

Sir Humphry Davy (1778–1829) performed an experiment that confirmed Rumford's hypothesis. The caloric theory implied that if two pieces of ice are kept below freezing in a vacuum and are rubbed vigorously together, they should not melt. On the other hand, Rumford's theory implied that the ice will melt. When the experiment was performed, the ice melted, just as Rumford's theory predicted.

▶ THE DISCOVERY OF NEPTUNE

During the eighteenth century, astronomers used Newton's laws of motion and his law of gravity to compute the exact orbits of the known planets and to predict exactly where various heavenly objects would be at future dates. In 1781, the English astronomer William Herschel discovered Uranus, the seventh planet. The French physicist Laplace soon worked out the mathematics of the gravitational interaction between Uranus, Jupiter, and Saturn, and presented tables for their orbits. (This was far more than a matter of telescopic observation and involved complicated mathematics.) However, Uranus's orbit differed slightly from what it should be if Newton's laws govern the solar system. By 1840, these differences seemed to contradict the predictions derived from Newton's laws.

Supporters of Newton's laws hatched several possible explanations. Perhaps the law of gravity breaks down at that distance, they suggested. Or perhaps Uranus has an undiscovered satellite that causes it to diverge from its Newtonian orbit.

In 1843, the English astronomer Adams and the French astronomer Leverrier independently arrived at the hypothesis that turned out to be the correct explanation: the observed orbit of Uranus could be due to an undetected planet further out that is pulling Uranus's orbit off its expected track. They used Newton's laws to compute the exact orbit the undetected planet would have, and calculated where it should be if it and Uranus are governed by Newton's laws. This allowed them to offer a prediction. They predicted that if astronomers would observe a specified zone of the night sky, at high power, a planet with the specified characteristics and orbit would be found. The new planet, later named Neptune, was discovered in 1846—exactly where Adams and Leverrier had predicted it would be. This successful prediction, made with the help of Newton's laws, provided another confirmation of Newton's theory.

Notice that although Adams and Leverrier postulated an auxiliary assumption, they also showed how to check it out independently of the theory in question. In other words, the auxiliary assumption was *independently testable*—it could be tested without making any special assumptions about Uranus and without using Newtonian theory.

▶   **EXERCISE 26.1**   ◀

---

1. Write a short (two- to three-page) paper on an episode from the history of science. Identify: (a) the problem under investigation; (b) the hypothesis or hypotheses formulated; (c) the implications drawn from the hypothesis or hypotheses; (d) the test procedure used, (e) the result.

2. Write a short paper on one of the hypotheses formulated by Sherlock Holmes in one of the stories by Arthur Conan Doyle. Include the problem under investigation; the hypothesis or hypotheses, the implications drawn, the test procedures, and the result.

3. Watch an episode of a contemporary detective show or an old show such as *Dragnet, Colombo, Magnum P. I.,* or *The Rockford Files* and explain the hypothetical reasoning employed to solve the problem.

4. Find an account of a specific line of scientific research in the newspaper, in a magazine such as *Scientific American*, or in a science book. Analyze the logical structure in terms of the steps discussed in this chapter.

5. Take a detective story and analyze the hypothetical reasoning in terms of the steps outlined in this chapter.

6. Formulate a hypothesis that can be confirmed or disconfirmed by observational evidence. Give two observation statements: one should be an observation statement that would confirm the hypothesis if the observation statement were true; the other should be an observational statement that would disconfirm the hypothesis if the observation statement were true.

7. List confirming evidence for the following hypotheses:

   a. The Earth is spherical.

   b. The Earth rotates on its axis.

   c. Electricity is a form of energy.

   d. Stars are extremely far away.

   e. Bleach kills bacteria.

CAUSE AND EFFECT AND MILL'S METHODS

When a group of people all come down with the same illness, scientists from the health department often are called in to track down the likely cause. For example, a few years ago, a number of people in Washington State got sick, and several died. In their search for the cause of the illness, scientists quickly learned that all the victims had eaten the same food at the same restaurant chain. Eventually, the deaths were traced to a particular strain of E coli bacteria in undercooked hamburgers.

When scientists track down causes, the procedures they use are often based on a set of principles first formulated by the British philosopher John Stuart Mill (1806–73) in his *System of Logic* published in 1843. These principles, now known as **Mill's Methods,** state procedures for identifying the probable causes of effects. However, before we survey Mill's Methods, we must first clarify the concept of a *cause*.

CAUSE AND EFFECT

One of the key objectives of "causal" investigation is to discover, for a specified effect, the conditions under which the effect will occur and the conditions under which the effect will not occur. Likewise, a key question of causal reasoning is, under which conditions will the effect occur and under which conditions will the effect be absent? Thus, philosophers have found it illuminating to analyze causes in terms of underlying or antecedent conditions, and specifically in terms of two types of antecedent or underlying conditions: necessary conditions and sufficient conditions.

In brief, a **necessary condition** for some circumstance or effect E has the following feature: in the absence of this condition, E cannot occur. In other words, this condition must be present for E to occur. Let us state this more formally: N is necessary condition for effect E if the following is true:

E will not be present unless N is present; or

In the absence of N, E will not be present; or

Without N, E won't occur; or

E cannot occur when N doesn't hold.

For example, the presence of oxygen is a necessary condition for the operation of a gasoline engine, reaching age eighteen is a necessary condition for voting, a bachelor's degree is usually a necessary condition for entering graduate school.

A **sufficient condition** for an effect E has a different feature: when the sufficient condition is present, E must occur. For instance, standing openly in the rain is a sufficient condition for getting wet, drinking six tequila sunrises is (for most people) a sufficient condition for getting drunk. More formally: S is a sufficient condition for effect E if the following is true:

If S is present, E will occur.

If S is present, E is certain to occur.

In presence of S, E is certain to occur.

E must occur when S holds.

The concept of cause is not precise or clearly defined. Some consider the cause of an effect E to be the sufficient condition of E. In this sense of the word *cause,* the cause of the crash of KAL flight 007 was a Soviet missile striking the jetliner. However, others hold that a cause is a necessary condition for an effect. The sum of all the necessary conditions for an effect E, on this interpretation, is the *complete cause,* and one necessary condition is a *partial cause* of E. Thus, oxygen is a partial cause of a fire, the fuel is another partial cause, and so on.

We often distinguish between *proximate* and *remote* causes. For example, suppose an insurance company receives the following description of the sequence of events that led up to an accident:

1. Jim slammed on his breaks to avoid hitting a family of aardvarks crossing the highway.

2. Fred smashed into Jim.

3. Joe smashed into Fred.

4. Alan smashed into Joe.

This sequence of events is a causal chain in which each event in the series was caused by the preceding event. Event 4 was caused by event 3, 3 was caused by 2, and so on. Event 3 is the proximate (nearest) cause of event 4, the others are remote causes of 4. In general, the proximate cause of an effect is the last event before the effect; and the causes before the proximate cause (in the chain of cause and effect leading to the proximate cause) are the remote causes.

MILL'S METHOD OF AGREEMENT

Suppose you and two of your friends eat lunch in the school cafeteria. Later in the afternoon, all three of you begin getting sick. By 3:00, you are in and out of the restroom every few minutes. What made you sick? That is, what is the *cause* of your illness? Suppose the following table lists what you each ate:

You:	Friend 1:	Friend 2:
2 super-burritos	3 double-cheeseburgers	2 fishburgers
1 large fry	2 orders of onion rings	1 large nachos
1 taco	2 tacos	1 taco

What would you conclude is the probable cause? If you concluded that the cause is in the tacos, you probably followed a logical process known as Mill's Method of Agreement. This is one of a group of methods first formulated by the philosopher John Stuart Mill.

In order to illustrate this method, let us enter the information about the meals into a table as follows.

Case:	Antecedent Condition:	Phenomenon:
1	B, F, T	S
2	C, O, T	S
3	F, N, T	S

On this table:

C: Cheeseburger

F: Fishburger

T: Taco

B: Burrito

O: Onion rings

S: The person is sick.

Notice that the taco (T) is the only antecedent factor always present when the effect (sickness) is present. From this, we naturally conclude that the cause of the sickness is probably in the tacos. The basic idea here is that the probable cause of the effect E (the illness) is to be found in the one antecedent condition common to each case where the effect E is present.

This method, which Mill called the **Method of Agreement,** requires that we first draw up a list of *possible* causes. We use our knowledge of cause and effect connections to identify the various possibilities. We then try to find one antecedent causal factor in common to all cases of the effect. This common factor, if found, is identified as the *probable* cause or as part of the cause. Note that the conclusion is *not* that the condition singled out *must* be the cause; the conclusion is only that this is *probably* the cause (or is part of the cause).

Mill put the method into these terms:

> *If two or more instances of a phenomena under investigation have only one circumstance in common, the circumstance in which alone all the instances agree, is the cause (or effect) of the given phenomenon.*

The Method of Agreement identifies a probable *necessary condition* of an effect. The reasoning underlying the method is that if an antecedent condition is absent when the effect is present, the condition cannot be a necessary condition for the effect and can thus be crossed off the list of suspected necessary conditions. After a variety of underlying antecedent conditions have been crossed off the list, the remaining condition that is present in all cases in which the effect is present is probably a necessary condition of the effect. The instances or cases examined must all be alike in terms of the presence of the effect under investigation, and they should differ as much as possible in all other respects.

Public health departments often use this method when they try to track down the cause of the outbreak of a disease. For instance, during the 1980s, 144 people on a Japan Airlines flight got sick. After much investigation, scientists tracked down the common factor: All 144 had eaten omelets that had been prepared by a single cook. The cook had a cut on his hand and transmitted an infectious agent to the passengers.

Mill gave the following example to illustrate this method.

> *For example, let the effect a be crystallization. We compare instances in which bodies are known to assume a crystalline structure, but which have no other point of agreement; and we find them to have one, and as far as we can observe, only one, antecedent in common: the deposition of a solid matter from a liquid state, either a state of fusion or of solution. We conclude, therefore, that the solidification of a substance from a liquid state is an invariable antecedent of its crystallizations.*[5]

In every instance where a material assumes a crystalline structure, one common antecedent condition is found. This common antecedent is therefore probably the cause (in the sense of being a necessary condition) of the crystalline structure.

In general terms, the information for the application of the Method of Agreement will fit into a table somewhat like the following:

Case:	Antecedent conditions:	Event for which a cause is sought:
1	xyz	E
2	xuvy	E
3	xag	E
4	xfsg	E

Condition x is the only antecedent condition in common to cases of E, so, x is probably the cause or part of the cause of E.

In sum, according to the Method of Agreement, if we investigate two or more cases in which a phenomenon A occurs, and only one antecedent phenomenon, B, is found to be present in each case, then we may conclude that A and B are causally related, that is, that B is probably the cause (or part of the cause) of A.

▶ MISTAKES SOMETIMES OCCUR

Fred finally got tired of the hangovers and decided to do something about it. After studying Mill's Method of Agreement in his philosophy class, Fred reasoned: Monday, I drank scotch and soda and got drunk. Tuesday, I drank bourbon and soda and got drunk. Wednesday, I drank rum and soda and got drunk. The common factor is the soda. If I lay off the soda and drink my scotch, rum, and bourbon straight out of the bottle, I won't get drunk anymore.

What is Fred's error? If he had used our next rule, the Method of Difference, Fred might not have made his erroneous inference.

[5]John Stuart Mill, *A System of Logic*, 8th ed. (New York: Harper and Brothers Publishers, 1874) p. 279.

MILL'S METHOD OF DIFFERENCE

Suppose Jan and Pat have lunch at the school cafeteria. Later on, in biology class, Jan gets sick, but Pat feels fine. The following table gives the relevant information:

Student:	Foods eaten:	Effect:
Jan	A B C D E F	Sick
Pat	A B C D E	not sick

Jan and Pat ate the same lunch except for one item. Food F is the obvious suspect. Why? It is the factor present when the effect (the sickness) is present and absent when the effect is absent, and the two cases are otherwise similar in all relevant respects. We naturally conclude that food F probably caused the sickness.

The above reasoning conforms closely to Mill's **Method of Difference**.[6] According to this method, if an effect E is present in one case, and a closely similar case doesn't exhibit E, we look for the difference between the cases. The probable cause of the effect E is the circumstance or condition present when E occurs and absent when E doesn't occur, provided that the cases compared are otherwise alike in all or nearly all relevant respects. The one way the two differ is therefore the suspected cause. This, in its essentials, is Mill's "Method of Difference."

Mill stated the Method of Difference as follows:

> *If an instance in which the phenomenon under investigation occurs, and an instance in which it does not occur, have every circumstance in common save one, that one occurring only in the former, the circumstance in which alone the two instances differ is the effect, or the cause, or an indispensable part of the cause of the phenomenon.*[6]

In general, the information for applying the Method of Difference will fit into a table at least somewhat like the following:

Case:	Antecedent conditions:	Event for which a cause is sought:
1	abcw	E
2	abc	E does not occur

Condition w is probably the cause of E.

To illustrate the method, Mill gave the following example:

> *It is scarcely necessary to give examples of a logical process to which we owe almost all the inductive conclusions we draw in daily life. When a man is shot through the heart, it is by this method we know that it is the gun shot that killed him: for he was in the fullness of life immediately before, all circumstances being the same, except the wound.*[7]

---

[6]John Stuart Mill, *A System of Logic*, p. 280.
[7]Ibid, p. 281.

The Method of Difference identifies a probable sufficient condition of an effect. The reasoning underlying the method is as follows: If a condition is present when the effect did not occur, that condition cannot be a sufficient condition of the effect, and can therefore be eliminated from the list of suspected sufficient conditions. After a variety of possibilities have been eliminated, the condition that is present when the effect is present and absent when the effect is absent is probably a sufficient condition for the effect. The instances or cases being compared should be as much alike as possible, except for the effect under investigation.

For another example, suppose researchers are testing the effects of a drug. They begin with, let us suppose, 100 rats. The rats are divided into two identical groups, each consisting of fifty rats. Both groups receive the same food and in the same conditions. Drug D is administered to the rats in one group (the *test* group) and no drug is administered to the rats in the other group (the *control* group). The two groups are extremely similar except for the one difference: only the test group received the drug. Suppose the rats in the test group act nervous, while the rats in the control group do not. The conclusion is that the drug probably causes nervousness. The drug is the only difference, and this is evidence in support of the claim that the drug is the cause of the nervousness. The difference is the factor present when the effect is present and absent when the effect is absent.

In science, the Method of Difference is often implemented in the form of a **controlled experiment.** In such an experiment, scientists search for the cause of a particular effect. Two groups of individuals are compared. The two groups must be extremely similar except for the following difference: the test group has or is given the factor suspected to be the cause of the effect. The control group is similar to the test group except that it lacks the factor under investigation, that is, the suspected cause. Upon investigating the differences, a table similar to the following is constructed:

Case:	Antecedent Factors:	Effect is:
1	ABCDJ	Present
2	ABCE	Absent
3	ABCD	Absent
4	ABCG	Absent

The cases are all similar. Factor J is the suspected cause, because it is the only factor that has the following feature: the effect occurs when it is present, and is absent when it is absent.

In the nineteenth century, Louis Pasteur hypothesized that vaccination with anthrax virus makes an animal immune to anthrax disease. Reluctantly, Pasteur performed the following experiment. He had one group of animals vaccinated with the anthrax vaccine while the vaccine was withheld from a second, nearly identical group. He then administered a fatal dose of anthrax germs to each group. The vaccinated animals didn't contract the disease, the others did. This was evidence that the vaccination does indeed confer immunity to the disease. Pasteur employed Mill's Method of Difference.

The tests to find the cause of lung cancer employed the Method of Difference. Groups of smokers were compared with various groups of nonsmokers who were otherwise extremely similar to the smokers. The only discernible difference between the groups being compared was the practice of smoking. The groups of smokers had significantly higher rates of lung cancer, which was then attributed to the only difference: smoking.

The Method of Difference is also common sense. For instance, a cook wonders, "What difference will it make if I leave out ingredient X when I bake these scones?" So he bakes two batches of scones. The batches are similar except for one difference: one batch contains ingredient X, the other does not. He notes the difference the absence of X makes. If the scones with X have property P, and the scones without X lack property P, he concludes that ingredient X probably causes the scones to have property P.

In sum, according to the Method of Difference, if there is some antecedent phenomenon, B, present when phenomenon A is present and absent when A is absent, and the two cases are alike in every other relevant respect, then we may conclude that A and B are causally related, that is, that B is probably the cause of A.

Let us now compare the first two methods. The Method of Agreement is aimed at the identification of a necessary condition for an effect E, and it bids us search for the common factor in all cases of E. The common factor, if found, is cited as the probable cause or as part of the cause. On the other hand, the Method of Difference is aimed at the identification of a sufficient condition for an effect E, and it directs us to choose as cause the one respect in which a case where the effect E occurs differs from an otherwise relevantly similar case where E is absent.

THE JOINT METHOD OF
AGREEMENT AND DIFFERENCE

The **Joint Method of Agreement and Difference** combines both of the above methods. Suppose we first use the Method of Agreement to identify what is probably a necessary condition of the effect under investigation. We next follow the Method of Difference and look at similar cases where the effect is not present. Specifically, we look to see if the suspected causal factor is present in any of these new cases. If the suspected causal condition is missing in all of these new cases (where the effect is absent), and if this is the only difference, this will strengthen our conclusion that the condition is a necessary condition for the effect and it will also provide evidence that the suspected condition is a sufficient condition for the effect.

For example, suppose sensitive documents relating to an investigation of the White House have been leaked to the press. The president is furious and wants to know the source of the leaks. White House investigators discover that Gray is the only reporter who had access to all the leaked documents. By the Method of Agreement, they conclude that Gray is the leaker. Next, they find out that in each case of a similar document that was not leaked, Gray did not have access to the document. When a document was leaked, there was Gray; when a similar document was not leaked, Gray was not there. By the Method of Difference, the same conclusion is reinforced: Gray is the leaker. Combined, these constitute the Method of Agreement and Difference.

MILL'S METHOD OF RESIDUES

Suppose a store owner is trying to figure out what is causing a $100,000 loss in sales. The owner figures that the loss is caused by factor A, factor B, factor C, or factor D. These are the only possible sources of the loss. After much investigation, let us suppose that the owner determines that 10 percent of the loss is due to factor A, 20 percent is due to factor B, and 20 percent is due to factor C. She concludes that the remaining 50 percent of the loss must be due to factor D. Factor D is probably the cause of a $50,000 loss in sales. The store owner has employed Mill's **Method of Residues.**

In general, if we know that (1) A, B, and C are causal conditions responsible for effects X, Y, and Z; and (2) if A is found to be the cause of X; and (3) if C is found to be the cause of Y, we can figure that B, the *residual* factor, is probably the cause of Z.

For another example, suppose Joe's car has a horrible squeak somewhere in the dash. Because he knows a lot about how the dash is put together, Joe figures that the squeak must be due to one or more of three possible causes: A, B, C. Each of these three possibilities would require some sort of repair, but A and B would each be easy to fix, while C would require removing the entire dashboard. Joe fixes A, but the squeak remains. He fixes B, but the squeak still remains. Now he concludes the squeak is probably due to condition C, and he begins removing the entire dashboard. Again, the Method of Residues has been employed.

MILL'S METHOD OF CONCOMITANT VARIATION

Suppose a direct relationship is discovered between eating food X and stomach cancer. People who eat food X tend to develop stomach cancer. People who do not eat food X tend not to develop stomach cancer. The rate of stomach cancer *varies* in a regular way with the eating of food X. The more of X that a person eats, the more likely he will get stomach cancer. We would conclude from this that a causal connection probably exists between eating food X and stomach cancer. Because some people who have stomach cancer do not eat food X, and some people who eat food X do not have stomach cancer, we would not conclude that eating food X is the only cause of stomach cancer, but that it is probably part of the cause or is connected with the cause.

Mill's **Method of Concomitant Variation** looks for changes in one phenomenon that accompany or correspond to (are *concomitant* with) changes in a second phenomenon. If the magnitude of the change in the one varies along with the magnitude of the change in the second, this is evidence that the two phenomena are probably causally related—either one of the two causes the other, or some third factor is the cause of both.

The Method of Concomitant ("accompanying") Variation is often used in everyday life as well as in the laboratory setting. For instance, suppose a community college district discovers a correlation between changes in enrollment and changes in the unemployment index: When unemployment goes up, enrollment goes up; when unemployment goes down, enrollment goes down. The district officials conclude that unemployment is one of the causes of college enrollment.

Or perhaps a panhandler experiments with his spiel. He varies the number of times per hour that he says, "God bless you." One day he says it 200 times per hour, the next day 100 times per hour, and so on. He discovers that the more times he says, "God Bless you," the more money he makes. When he says this 200 times an hour, he makes an average of $30 an hour. When he only says it fifty times an hour he only makes an average of $5 an hour. He concludes that saying, "God bless you" causes people to make a donation.

Mill expressed the Method of Concomitant Variation this way:

> *Whatever phenomenon varies in any manner whenever another phenomenon varies in some particular manner, is either a cause or an effect of that phenomenon, or is connected with it through some fact of causation.*

However, a word of caution: The mere fact that two phenomena are correlated does not prove that they are causally related. For more on this point, see the discussion of False Cause fallacies in Chapter 13.

▶   **EXERCISE 26.2**   ◀

Identify the method used in each of the following examples.

1.*To test the quality of a new transmission additive, a company buys two identical cars, one red, the other blue. The red car gets the new additive, the blue car gets no additive. At the end of the test, they take apart the transmissions and measure the wear and tear. The red car has significantly less wear. They conclude that the additive reduced wear and tear on the transmission.

2. A Martian lands on this "third rock from the sun," sees a car for the first time, and wants to figure out what makes the car speed up and slow down. He gets in, manages to start the engine, and starts pushing buttons and turning the steering wheel. Finally, because he is on a level street, he notices a perfect correlation between the position of the gas pedal and the speed of the car: the closer the pedal is to the floor, the faster the car goes, the further from the floor, the slower the car's speed. The Martian concludes that pushing on the gas pedal causes the car to speed up and slow down.

3. Six pupils are extremely poor readers. The teacher investigates the background of each student and finds that each comes from a different family, each has a different socio-economic background, and each had a different previous school. However, all six lacked phonics instruction. The teacher concludes that phonics instruction causes good reading.

4. Many XYZ computers keep getting returned for repairs. The company looks into the matter and finds that the computers that are returned were produced in different years and were sold in different areas. All were sold in areas with high humidity. The company concludes that highly humid air damages something in the computers.

5.* A department store manager notices a correlation between the local employment rate and the shoplifting rate in her store. As the employment rate increased, the theft rate decreased and vice versa. The manager concludes that unemployment is part of the cause of shoplifting.

6. A patient developed an allergic reaction. When she eliminated foods x, y, and z from her diet, the reaction ended. She added the foods back one at a time. When she began eating food y, milk products, the reaction reappeared. She concluded that the reaction was caused by milk products.

7.*A doctor is treating five cancer patients. The only common factor is that all five were employed by the XYZ Chemical Company in a division that produces a chemical defoliant. The doctor concludes that the defoliant caused the cancer.

8. The Smith family moves from city A to city B and discovers that their clothes are cleaner. Yet they are using the same soap and the same washing machine. They conclude the change must be due to the water.

9. Two bushes are covered with aphids. The gardener sprays one with chemical X and leaves the other unsprayed. The next day, the aphids are gone from the sprayed bush, but they remain on the unsprayed bush. The gardener concludes that chemical X killed the aphids.

10.*Tax revenues are down. The city council wants to know the cause. Their economist discovers that some of the decline is due to the economic downturn, some is due to the increase in the interest rate. The economist concludes that the remaining decline must be due to increased cheating on taxes.

11. A resort is testing a new bass plug. They give the new plug to six of twelve fishermen. The six with the new plug all used different gear, and all caught fish. The other six fishermen didn't catch any fish. The resort concludes that the bass plug works.

12. A psychiatrist has ten patients unable to sustain prolonged and meaningful relationships with men. All ten have different backgrounds, all have different religious beliefs, all have different jobs, and so on. However, all the patients had in common the lack of a male parent figure during childhood. The psychiatrist concludes that this is part of the cause of the problem.

13. A metallurgist tries adding five different chemicals to a metal while attempting to produce a new alloy. All the added chemicals that contain chemical X produce an alloy that is resistant to corrosion. The other chemicals fail to produce a corrosion-proof alloy. The metallurgist concludes that the chemical X makes the alloy resistant to corrosion.

14. During late 1940s men from different parts of the country developed the same eye disease. Investigators discovered that all had worked in nuclear energy projects during World War II and all had been exposed to neutron beams. The investigators concluded that the cancer was caused by the neutron beams.

15. *During the late nineteenth century, scientists searched for the cause of rickets (a bone disease). At first rickets was thought to be due to poverty because many of those who had rickets were poor. Consequently, malnutrition was investigated as a possible cause. But then it was discovered that wealthy people often had rickets as well. Finally, a common factor was found: a lack of sunlight. The conclusion was therefore that rickets is caused by a lack of sunlight.

16. After Randy adjusts the timing belt on his car, installs new tires, and puts in a new carburetor, he notices that his car gets better gas mileage. Obsessed with finding the cause, he does three things. First, he puts the old carburetor back in and drives the car for a week, carefully recording the gas mileage each day. Next, he puts the old tires back on and drives the car for another week, again carefully recording the gas mileage each day. Finally, he puts the timing belt back to its old setting and again drives the car for a week, carefully recording the gas mileage each day. He decides the increased gas mileage is due to the new carburetor.

17. In 1854, John Snow sought out the cause of cholera. He discovered that all areas of London had approximately the same frequency of cholera except for one area, which had a significantly lower level of the disease. Upon investigation, it turned out that the drinking water for this one area was supplied by a company that used drained water that had no sewage in it. However, the water for the other areas of London was contaminated by raw sewage. Snow consequently proposed that cholera is caused by fecal contamination of drinking water.

18. Dr. Christiaan Eijkman sought the cause of the disease beriberi. He observed that hens in the courtyard of an asylum where he worked showed symptoms similar to the humans who had the disease. Could it be, he wondered, that the same cause is at work in both humans and chickens? Further investigation revealed that prisons that fed prisoners polished rice showed high rates of beriberi. In these prisons, the chickens were also fed polished rice. But prisons that fed unpolished rice had few cases. Polished rice was present when the disease was present, and absent where the disease was absent. Eijkman could find no other relevant difference. He concluded that beriberi was caused by the diet of polished rice. (Technically, the disease is caused by a dietary deficiency of B complex vitamins. Whole grains provide this vitamin, but when rice was polished, the polishing process removed some of the outer hull, and the outer hull contained much of the vitamin content.)

19. People used to think that bacteria came from within meat. In the nineteenth century, Pasteur performed the following test of the theory. He boiled meat broth in a flask to kill the bacteria and then sealed the container and let it sit for an extended time. When he examined it later, the broth remained free of bacteria. However, meat broth that had been exposed to the air became covered with bacteria and soon started to decay. His conclusion: the bacteria came from the air rather than from within the meat broth.

20. *A group of patients with disease x all have one thing in common (other than x): a dietary deficiency of vitamin z. The patients are given vitamin z and their symptoms clear up. The doctor concludes that the condition is caused by a lack of vitamin z.

```
▶              EXERCISE 26.3              ◀
```

Explain how you would investigate the following situations using Mill's methods.

1. A disproportionately large number of athletes are enrolled in Professor Smith's class. Why?

2. Joe has occasional stuttering attacks. Why?

3. *The divorce rate rose during the 1980s, then fell during the 1990s. Why?

4. Business at Pink's Hot Dogs rose during August and declined during September. Why?

5. You and four coworkers get sick after lunch. Why?

6. You plant six rose bushes. Three grow, three die. Why?

7. *You bake a cherry pie three times, and twice it turns out lousy. Why?

8. Two singers in a chorus get sore throats. Why?

9. *Two people on a camping trip become infested with fleas. Why?

10. Sometimes the latte you order is good, sometimes it is not good. Why?

11. Every time you get busy, you get sick. Why?

CHAPTER 26 GLOSSARY

Auxiliary assumptions   Assumptions presupposed in the derivation of a prediction from a hypothesis.

Controlled experiment   An experiment in which a cause of an effect is sought and in which the following occurs: (a) two groups of individuals are compared; (b) the two groups are extremely similar except for the following difference: one group, called the *test* group, has or is given the factor suspected to be the cause of the effect, and the other group, called the *control* group, is similar to the test group except that it lacks the factor under investigation, that is, the suspected cause; and (c) the Method of Difference is employed to determine the probable cause.

Hypothesis   An explanation offered for a phenomenon.

Hypothetico-deductive method   The method used in the sciences to confirm and disconfirm theories.

Mill's Methods   A set of principles first formulated by the British philosopher John Stuart Mill (1806–73) in his *System of Logic* published in 1843. These principles are used when seeking to determine the probable cause of a circumstance or effect.

- **Method of Agreement**   This method requires that we first draw up a list of *possible* causes; we then try to find one causal factor common to all cases of the effect; this is identified as the *probable* cause or as part of the probable cause. (The claim is not that the condition singled out *must* be the cause; the conclusion is only that this is *probably* the cause.)

- **Method of Difference**   This method requires that we examine a case where an effect E occurs and a similar case where E does not occur, and we then choose as the probable cause the one respect in which the case where the effect E occurs differs from the case where E is absent.

- **Joint Method of Agreement and Difference**   In this method, both the Method of Difference and the Method of Agreement are combined.

- **Method of Residues** This method is based on the following general principle: if we know that (1) A, B, and C are causal conditions responsible for effects X, Y, and Z; and (2) A is found to be the cause of X; and (3) C is found to be the cause of Y, we can figure that B, the *residual* factor, is probably the cause of Z.

- **Method of Concomitant Variation**   This method is based on the following general principle: if changes in one phenomenon accompany or correspond to (are "concomitant" with) changes in a second phenomenon, and if the magnitude of the change in the one varies along with the magnitude of the change in the second, the two phenomena are probably causally related—either one of the two probably causes the other, or some third factor is probably the cause of both.

Necessary condition A necessary condition for some circumstance or effect E is a condition that has the following feature: in the absence of this condition, E cannot occur, that is, this condition must be present if E is to occur.

Observational prediction A prediction about the results of observations that can be made, where the observation can concern facts related to the past, present, or future.

Principle of Economy (or principle of parsimony) A principle that states that if two potential hypotheses explain the same range of data and are otherwise equal, except that one hypothesis is *simpler* than the other, the simpler hypothesis is preferable. One hypothesis is *simpler* than another if it makes reference to fewer entities or contains fewer explanatory principles or explanatory elements. Also known as Occam's razor.

Process of confirmation The process by which a scientific hypothesis is shown to be probably true.

Process of disconfirmation The process by which a scientific hypothesis is shown to be probably false.

Sufficient condition A sufficient condition for an effect E has this feature: when the sufficient condition is present, E must occur. The sufficient condition S for an effect E is the condition such that it is all that is required for E to occur.

APPENDIX 1:
TRUTH-TREES

The truth-table method from Chapter 5 contains decision procedures for truth-functional logic. However, as noted in Chapter 7, truth-tables have one major drawback. As we evaluate bigger and bigger formulas, the size of the truth-table increases exponentially. A formula with two atomic components may require only a four-row table, but a formula that contains four atomic components requires a sixteen-row table, six components calls for a sixty-four-row table, and so on. Thus, when we try to evaluate complex formulas and complicated arguments, the truth-tables become so large that the method becomes impractical.

Several alternative methods have been developed that accomplish some of the same purposes but that are easier to use when it comes to complicated arguments. The method of natural deduction, covered in previous chapters, is one alternative to the truth-tables. A second alternative is a method known as the *truth-tree method*.

The method of truth-trees is closely related to the method of **reductio ad absurdum** or "indirect" proof. As noted in Chapter 9, mathematicians in ancient times proved that the square root of two is an irrational number. The type of proof they used, known as a reductio ad absurdum proof, employs the following logical strategy. Suppose you want to prove a claim **P**. You begin your proof by assuming the opposite of the claim you wish to prove. Begin your proof with ~ **P**. (In English, the opposite of a sentence is what results when you attach a negation operator to the sentence. For example, the opposite of "It is snowing" is "It is not the case that it is snowing.") Next, demonstrate that this assumption implies a contradiction. Now, because only a contradictory statement can imply a contradiction, this shows that the assumption, ~ **P**, is itself contradictory or necessarily false. This is said to "reduce" the assumption to an "absurdity." It follows that the opposite of the assumption, namely **P**—the claim you originally sought to prove—must itself be true.

Reductio ad absurdum proofs are also called **indirect proofs** because in such a proof, the conclusion is derived in an indirect way, by first deriving a contradiction from the denial or opposite of the conclusion and inferring the conclusion from that. As you will shortly see, the truth-tree method employs a similar logical strategy.

TRUTH-FUNCTIONAL TRUTH-TREES

We shall begin by specifying a series of rules.[1] After the rules are specified, we shall begin constructing devices called "truth-trees." We will put these logical devices to work doing two things:

1. Testing a sentence to see whether it is tautological, contradictory, or contingent.

2. Testing an argument to see if it is valid or invalid.

▶ RULES FOR THE AMPERSAND

Suppose we begin by making an assumption. Let us assume that a sentence **P & Q** is true. Instead of placing a **T** under the main connective, as we did in Chapter 5, let us simply place the sentence on the left side of a line as follows:

$$\textbf{P \& Q} \mid$$

If a sentence is placed on the left side of a line, let this represent the assumption that the sentence as a whole is true. Next, what must be the case regarding the components **P** and **Q** if this assumption is correct? That is, what is the logical consequence of our initial assumption? If a conjunction is assigned **T**, both of its conjuncts must also be assigned **T**, because a conjunction is true if and only if both conjuncts are true. Therefore, **P** and **Q** must each be true as well. Accordingly, we should extend the line and place **P** and **Q** both on the true side below **P&Q**:

$$\checkmark, \textbf{P \& Q} \mid$$
$$\textbf{P} \mid$$
$$\textbf{Q} \mid$$

The check beside **P & Q** records the fact that we have taken account of the information represented by that sentence's placement and drawn an inference from that. The general rule in this case, the Ampersand Left rule, is summarized in the following box.

---

[1] The tree rules of this appendix are based on rules presented in Daniel Bonevac, *Deduction* (Mayfield, 1987).

Ampersand Left

If a sentence **P & Q** is situated on the true side of a line, you may carry down **P** and **Q** and place them individually on the true side of the line under the sentence **P & Q**.

More formally:

$$✔ P \& Q \mid$$

$$P$$

$$Q \mid$$

Each of the following involves an application of Ampersand Left:

✔ A & B	✔ (A v B) & (G v E)	✔ ~(A ⊃ B) & C
A	A v B	~ (A ⊃ B)
B	G v E	C

Notice that we may apply this rule only to a sentence positioned on the left side of a line and only when the main connective is an ampersand.

Assume next that a sentence **P & Q** is false. Instead of writing an **F** under the main connective, as we did in the previous section, let us place the sentence on the right side of the line:

$$\mid P \& Q$$

If a sentence is placed on the right side of a line, let this represent the assumption that the sentence as a whole is false. The left side of the line represents truth and the right side of the line represents falsity. Now, what follows from our assumption in this case? To begin with, if our assumption is correct, it does not follow that **P** and **Q** are both false. A conjunction is false just in case one or the other or both conjuncts are false. Thus, it follows only that one, or the other, or both of **P** and **Q** are false. Let us represent this inclusive disjunction of alternatives by splitting the line into two branches and placing the **P** and **Q** as follows:

You may be wondering whether we also need a third branch in this case upon which **P** and **Q** are both assigned **F**. The answer is that no such branch is needed, for this possibility is already covered by the two branches that constitute the split. Remember that the split represents the inclusive disjunction that either **P** is false, or **Q** is false, or both are false.

The general rule in this case may be called Ampersand Right.

> ### *Ampersand Right*
>
> If a sentence **P & Q** is situated on the false side of a line, you may split the line and place **P** on the false side of one branch and **Q** on the false side of the other branch.

More formally:

Each of the following involves an application of this new rule:

In every case, Ampersand Right is to be applied only to sentences positioned on the right side of the line and only to sentences whose main connective is the ampersand.

Notice that when we use Ampersand Left we "stack" the sentences and when we use Ampersand Right we "split" them.

▶ RULES FOR THE WEDGE

Suppose we assume that a sentence **P v Q** is false:

$$| \; P \lor Q$$

What must be the case if our assumption is correct? A disjunction is false if and only if both disjuncts are false. Therefore, if our assumption in this case is correct, then **P** and **Q** must both be false:

$$
\begin{array}{l}
| \; P \lor Q \checkmark \\
| \; P \\
| \; Q
\end{array}
$$

The general rule, represented by the diagram immediately above, may be called Wedge Right.

> ### *Wedge Right*
>
> If a sentence **P v Q** is situated on the false side of a line, you may carry down **P** and **Q** and place them individually on the false side of the line under the sentence **P v Q**.

Here are some applications:

A v B ✔	(E & H) v ~ (D ⊃ S) ✔
A	E & H
B	~(D ⊃ S)

Assume next that a sentence **P v Q** is true:

$$P v Q \mid$$

What follows? It certainly does not follow that both **P** and **Q** are true. A disjunction is true just in case one or the other or both disjuncts are true. Consequently, if our initial assumption is correct, it follows merely that either **P** is true, or **Q** is true, or both are true. This consequence is represented by splitting the line as follows:

The split in this case represents the inclusive disjunction that either **P** or **Q** or both are true. Call the rule represented by this diagram Wedge Left.

> ### *Wedge Left*
>
> If a sentence **P v Q** is situated on the true side of a line, you may split the line and place **P** on the true side of one branch and **Q** on the true side of the other branch.

For example:

✔ A v (B & C)

A   B & C

► RULES FOR THE HORSESHOE

Suppose we assume that a sentence **P ⊃ Q** is false. What follows? Well, if you look at the truth-table for **P ⊃ Q**, you will see that only one truth-value assignment results in an F in the final column, namely, that in which **P** is assigned true while **Q** is assigned false. This gives rise to the rule we will call Horseshoe Right:

> *Horseshoe Right*
>
> If a sentence **P ⊃ Q** is situated on the false side of a line, you may carry down **P** and place it on the true side and you may carry down **Q** and place it on the false side.

More formally:

$$
\begin{array}{c|c}
 & P \supset Q \; ✔ \\
\hline
P & Q
\end{array}
$$

For example:

$$
\begin{array}{c|c}
 & A \supset B \; ✔ \\
\hline
A & B
\end{array}
\qquad\qquad
\begin{array}{c|c}
 & E \supset (G \,\&\, S) \; ✔ \\
\hline
E & G \,\&\, S
\end{array}
$$

Suppose we assume next that a sentence **P ⊃ Q** is true:

$$P \supset Q \;\big|$$

What follows from this assumption? There are three possibilities:

1. If **P** itself is false, no matter what **Q** is—whether **Q** is true or false—**P ⊃ Q** is automatically true. This can be verified by examining the truth-table for the horseshoe. If **P ⊃ Q** is true, it might be that **P** is false.

2. If **Q** itself is true, no matter what **P** is, **P ⊃ Q** is true. This, too, we can verify by examining the horseshoe table. If **P ⊃ Q** is true, it might be that **Q** is true.

3. If **P** is false and **Q** is true, **P ⊃ Q** is true. If **P ⊃ Q** is true, it might be that **P** is false and **Q** is true.

If we split the tree in the following way, the split will represent all three possibilities:

The split records the fact that if **P** ⊃ **Q** is true, either **P** is false or **Q** is true, or both **P** is false and **Q** is true. Call the rule represented by this diagram Horseshoe Left.

> ### *Horseshoe Left*
>
> If a sentence **P** ⊃ **Q** is situated on the true side of a line, you may split the line and place **P** on the false side of the left branch and **Q** on the true side of the other branch.

Here are some applications of this rule:

▶ RULES FOR THE TRIPLE BAR

Suppose a sentence **P** ≡ **Q** is assumed true:

$$\mathbf{P} \equiv \mathbf{Q} \mid$$

Recall that a triple bar receives a T if and only if both sides have the same truth-value. Thus, if **P** ≡ **Q** is true, either **P** and **Q** are both true or **P** and **Q** are both false:

✔ **P** ≡ **Q**

P	**P**
Q	**Q**

Call this rule Triple Bar Left. Here is an application:

✔ A ≡ B

A	A
B	B

Consider next the case where a sentence **P** ≡ **Q** is assumed false:

$$\mid \mathbf{P} \equiv \mathbf{Q}$$

If $P \equiv Q$ is false, P and Q must not both have the same truth-value. Either P is true and Q is false, or P is false and Q is true. From this we derive the rule Triple Bar Right.

$$P \equiv Q ✔$$

$$P \qquad P$$
$$Q \quad Q$$

For example:

$$H \equiv S ✔$$

$$H \qquad H$$
$$S \quad S$$

▶ RULES FOR THE TILDE

The last rule to present is the simplest rule of all. If we assume a sentence ~P is true, we write

$$\sim P \mid$$

If ~ P is true, P must itself be false. The obvious consequence of our assumption is

$$✔ \sim P$$
$$P$$

Call this rule Tilde Left.

On the other hand, suppose we start by assuming that a sentence ~P is false:

$$\mid \sim P$$

It follows that P must itself be true:

$$\sim P ✔$$
$$P$$

Call this rule Tilde Right.

Before putting these rules to work, let us clarify some terminology. The line diagrams we've been constructing here are called truth-trees, and the ten rules developed above are truth-tree rules. A truth-tree has the shape of an upside-down tree. The top of the truth-tree is the tree's root. A branch of a truth-tree is a sequence of sentences that begins with a sentence at the top of the tree and continues to a tip of the tree. A tip of a truth-tree is formed by the last sentence at the bottom of a tree. Thus, a tree has as many branches as it has tips.

When a line running from the root to a tip turns into two branches, we say that the branch splits. The part of the tree above any splitting is the tree's trunk. Because the left side of any branch represents truth, we will call that side the *truth side*. The right side, representing falsity, will be called the *false side*. We will test sentences for the various logical properties by placing them on a tree's trunk. A sentence placed on the truth side will represent the assumption that the sentence is true, and a sentence placed on the false side will represent the assumption that the sentence is false.

When we have drawn a consequence from a sentence, we will dispatch the sentence by placing a check beside it. Dispatched sentences are "dead." Sentences not dispatched are "live." Let us say that a branch with the same sentence appearing "live" on both of its sides is closed. In other words, if a branch of a tree assigns both **T** and **F** to a sentence, the branch is a closed branch. We will mark closed branches with an X placed at their tip. If a branch is not closed, it is open. A completed branch consists only of atomic sentences and checked-off sentences. In other words, on a completed branch, everything that can be broken down and checked off has been broken down and checked off. If every branch of a tree is closed, we shall say the tree is a closed tree. If at least one branch is completed and open, the tree shall be called an open tree. Finally, if every branch is either closed or is completed and open, we shall say the tree is a completed tree.

TRUTH-TREE DECISION PROCEDURES

All of the material we have worked through so far has prepared us for the procedures developed in this section. Using these procedures, we will be able to test sentences and arguments for the standard logical properties introduced in Chapter 5.

▶ TAUTOLOGY TESTING

What is the logical status of (A & B) v ~(A & B)? In other words, is (A & B) v ~(A & B) tautological, contradictory, or contingent? Recall from Chapter 5 that a truth-functional sentence is tautological if and only if it cannot possibly be false, it is contradictory if and only if it cannot possibly be true, and it is contingent just in case it is possibly true and possible false. Let us begin by assuming (A & B) v ~(A & B) is false:

$$| \ (A \ \& \ B) \ v \ {\sim}(A \ \& \ B)$$

Next, we apply the tree rules to the main connectives of each sentence. Because the main connective is the wedge, the first rule to apply is Wedge Right:

$$(A \,\&\, B) \,v\, \sim(A \,\&\, B) \; ✔$$

$$A \,\&\, B$$

$$\sim(A \,\&\, B)$$

Next, let us apply Tilde Right to ~(A & B):

$$(A \,\&\, B) \,v\, \sim(A \,\&\, B) \; ✔$$

$$A \,\&\, B$$

$$\sim(A \,\&\, B) \; ✔$$

$$(A \,\&\, B)$$

In the next move, we apply Ampersand Left to the conjunction that sits on the true side:

$$(A \,\&\, B) \,v\, \sim(A \,\&\, B) \; ✔$$

$$A \,\&\, B$$

$$\sim(A \,\&\, B) \; ✔$$

$$✔ \,(A \,\&\, B)$$

$$A$$

$$B$$

Finally, when we apply Ampersand Right to the last undispatched sentence, the tree closes:

$$(A \,\&\, B) \,v\, \sim(A \,\&\, B) \; ✔$$

$$A \,\&\, B \; ✔$$
$$\sim(A \,\&\, B) \; ✔$$

$$✔ \,(A \,\&\, B)$$

$$A$$
$$B$$

$$A \qquad B$$

$$X \qquad X$$

We have now traced all the consequences of our initial assumption. The tree can extended no farther because all compound sentences have been broken down to their simplest components. The tree is therefore completed. Because each branch has ended in a *contradictory assignment of truth-values*, the completed tree has closed. This tells us that our assumption implies a contradiction. Because only that which is contradictory can imply a contradiction, this proves our assumption is contradictory. Note carefully: This proves our assumption is contradictory, but it does not prove the sentence at the top of the tree is contradictory. The assumption was that (A & B) v ~(A & B) is false. Because the assumption is contradictory, it follows that it is impossible that (A & B) v ~(A & B) is false. It follows that (A & B) v ~(A & B) is a tautology.

The general principle in the case of a tautology may be put as follows:

The Tautology Test

If a sentence is placed on the false side of a tree and the resulting completed tree is closed, the sentence is a tautology.

More formally:

$$\mathbf{P}$$

Closed: **P** is a tautology

If we had placed (A & B) v ~(A & B) on the true side of a tree, the completed tree would have had at least one open branch. An open branch tells us it is possible our assumption is correct. Thus, if we had assumed our sentence true—by placing it on the true side—our completed tree would have told us that the sentence is possibly true. In the following tree, we begin by assuming (A & B) v ~(A & B) is true. We apply Wedge Left and split the tree. The tree is completed with applications of Ampersand Left, Tilde Left, and Ampersand Right.

If a sentence is a tautology, it is a theorem of truth-functional logic. Such a sentence is also truth-functionally true or logically true because it can be proven true using just the methods of truth-functional logic and without performing scientific experiments or otherwise observing the physical world. If a sentence generates—according to the truth-functional rules of this chapter—a closed tree when placed on the false side of a tree, the sentence is a theorem of truth-functional logic.

▶ TESTING FOR CONTRADICTIONS

Let's test (E v B) & ~(E v B) to determine its logical status. We will start by assuming the sentence is true, which we may record by writing:

$$(E \text{ v } B) \ \& \sim (\ E \text{ v } B) \ \big|$$

The resulting tree, when complete, closes:

$$
\begin{array}{c}
\textbf{✔} (E \text{ v } B) \ \& \sim(E \text{ v } B) \ \big| \\[4pt]
E \text{ v } B \ \big| \\[4pt]
\textbf{✔} \sim(E \text{ v } B) \ \big| \\[6pt]
\big| \quad E \text{ v } B \\[4pt]
X
\end{array}
$$

Our tree has assigned **T** to E v B and **F** to E v B. Thus, the assumption that (E v B)&~(E v B) is true has shown itself contradictory. Therefore, (E v B)&~(E v B) cannot possibly be true, which is to say that (E v B) & ~(E v B) is contradictory. The general principle in the case of a contradiction is:

The Contradiction Test

If a sentence generates a closed tree when it is placed on the *true* side of a tree, then the sentence is contradictory.

More formally:

$$\textbf{P} \,\big|$$

Closed: **P** is a contradiction.

If we had placed this sentence on the false side of a tree, the tree would have had at least one open branch. Because an open branch represents the possibility that the assumption is correct, this tree would have told us only that (E v B) & ~(E v B) is possibly false:

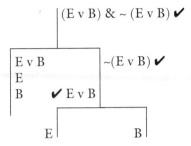

▶ WHEN TO CLOSE A BRANCH

When you build a tree, you may proceed in either of two ways: (1) you may stop and close off a branch as soon as the same sentence appears live on both sides, whether or not everything has been broken down into atomic sentences; or (2) you may extend each branch until everything has been broken down into atomic sentences. Thus, compare the following two trees:

```
✔ (E v F) & ~ (E v F)                    ✔ (E v F) & ~ (E v F)
        E v F                                    ✔ E v F
✔ ~(E v F)                               ✔ ~(E v F)
                                                              E v F ✔
              E v F                                           E
        X                                                     F

                                                E            F

                                                X            X
```

Whether we end a branch on this tree as soon as a contradiction arises, or whether we continue until every sentence has been broken down into atomic sentences, the result is the same: The branches close and the sentence is proven to be contradictory.

▶ TESTING FOR CONTINGENCY

Recall that a sentence is contingent just in case it is possible for it to be true and possible for it to be false. In order to prove a sentence contingent, we must prove two things: (1) it is possible for the sentence to be true; (2) it is possible for the sentence to be false. Consequently, it takes two trees to prove that a sentence is contingent. Let's test (A & B) v (A v B).

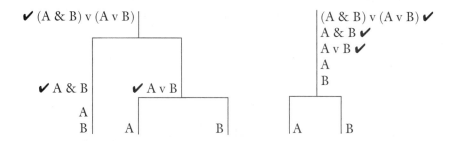

The tree on the left proves the sentence possibly true; the tree on the right proves the sentence possibly false. Putting these two conclusions together, we have proven (A & B) v (A v B) both possibly true and possibly false, which is to say that it is contingent. The general principle in the case of contingency is summarized in the following box.

The Contingency Test

If a sentence generates a completed open tree when placed on the true side and also generates a completed open tree when placed on the false side, then the sentence is contingent.

More formally:

$$ \mathbf{P} \mid \qquad \mid \mathbf{P} $$

Both open: **P** is contingent.

Now that we have completed a few trees, let us recall the rationale behind the tree method. The truth-tree method reflects the reductio ad absurdum strategy. We begin every tree with an assumption. This assumption is reflected in the initial positions that sentences receive at the top of the tree. When we place a sentence on a tree, we simply record an assumption concerning its truth-value. Now, as we apply the proper rules and extend the tree, we simply deduce the logical consequences of our assumption. A closed branch means that we have reached a contradiction. If every branch of the tree closes, this indicates that our assumption implies a contradiction. It follows from this that the opposite of our assumption must be true. In other words, if a tree closes, it says, in effect, "Your assumption is impossible." And if a tree turns out to be an open tree, it is saying, in effect, "Your assumption is possible."

For example, if the assumption is that a sentence **P** is false, and the tree closes, this tells us that **P** can't possibly be false. It follows that **P** is necessarily true. If the assumption is that a sentence **P** is true, and the tree closes, this tells us that it's impossible for that sentence to be true. It follows that the sentence is contradictory.

Sometimes it is tempting to treat the tree method as a game in which the goal is to get the same sentence live on both sides of the line. This can be a helpful approach. However, it is important that you understand the rationale behind the tree method and behind each of the tree rules. It is sometimes easier to build the trees if, instead of relying mechanically on the rules themselves, you think each move through logically.

▶   **EXERCISE A.1**   ◀

Use truth-trees to determine which of the following are tautological, which are contradictory, and which are contingent.

1. ~ P ⊃ (P ⊃ Q)   **9.** ~ (P v ~ P)

2. Q ⊃ (P ⊃ Q)   **10.** ~[P ⊃ (Q ⊃ P)]

3. ~ (P ≡ P)   **11.** P ≡ ~ P

4. P ⊃ ~ P   **12.** P ≡ (P v P)

5. ~ (P & ~ P)   **13.** (P & Q) ⊃ Q

6. P v (P & ~ P)   **14.** ~ (P & Q) v Q

7. P ⊃ (P v Q)   **15.** P & (Q & ~ P)

8. (P & ~ P) ⊃ Q

▶ PROPER SPLITTING

In the next section, you will be working with trees containing several sentences. In order to prepare for this, consider the following problem. Suppose you have a tree with three sentences placed as follows:

$$\begin{array}{c|c} P \lor Q & \\ A \lor B & Q \,\&\, P \end{array}$$

Let's first dispatch P v Q:

Now, how do we dispatch Q & P? When applying a rule to a sentence, we must write down the consequence on every branch underneath the spot on the tree occupied by the sentence being dispatched. Therefore, when we dispatch Q & P, we must split the tree on every tip underneath (Q & P):

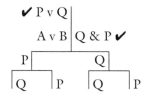

And when we dispatch A v B, we write down its consequences on each tip below where it sits:

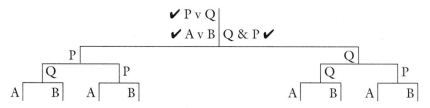

If you do not see why we must write down the consequences on every branch underneath the sentence we are applying a rule to, go back to the section on the rationale underlying the trees and think through once more the logic for the tree method. A consequence of an assumption must be placed at the tip of each branch underneath.

► PROVING AN ARGUMENT VALID OR INVALID

Suppose an English argument has been symbolized as follows:

1. (A v ~Q) v ~R

2. ~(A v S)

3. ~(~R v ~H)

4. ~Q

The sentences above the line are the premises and the sentence below the line is the conclusion. To determine this argument's validity, let us begin by assuming it is invalid. That is, we'll assume it is possible that the premises are true and the conclusion is false. In order to record this assumption, we place the premises on the true side and the conclusion on the false side:

$$(A v ~Q) v ~R$$
$$~(A v S)$$
$$~(~R v ~H) | ~Q$$

If this tree closes, this will indicate that our assumption is impossible. In that case, it would follow that it is impossible for the premises to be true and the conclusion false, which would prove the argument valid. However, if the tree is open, this will indicate that our assumption possibly is correct. In that case, it's possible for the premises to be true and the conclusion false, which would prove that the argument is invalid. The tree follows:

Because every branch of the completed tree closes, the tree is closed. The argument is therefore valid. The truth-tree test for deductive validity is summed up in the following box.

The Validity Test

Place the premises on the true side of a tree and the conclusion on the false side. If the completed tree closes, the argument is valid. If the completed tree is open, the argument is invalid.

More formally:

Premises	Conclusion

Closed: The argument is valid
Open: The argument is invalid

► FOUR COMMON PATTERNS OF REASONING

Let us now consider four patterns of reasoning we frequently follow in everyday thought.

THE DISJUNCTIVE SYLLOGISM PATTERN

Consider the following valid argument:
Argument A

1. Either we'll run or we'll swim.

2. We won't run.

3. So, we'll swim.

In TL, using obvious abbreviations, this becomes

1. R v S

2. ~R

3. S

This argument is an instance of the following valid argument form

$$P \lor Q$$

$$\sim P$$

$$Q$$

The following truth-tree proves the validity of argument A:

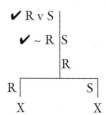

THE MODUS PONENS PATTERN

This argument is clearly valid:
Argument B

1. If it's raining, the roof is wet.

2. It's raining.

3. So, the roof is wet.

Using obvious abbreviations, this argument may be expressed in TL as

1. R ⊃ W

2. R

3. W

This instantiates the following valid argument form:

$$\begin{array}{c} P \supset Q \\ \underline{P} \\ Q \end{array}$$

The following tree proves the validity of argument B:

THE MODUS TOLLENS PATTERN

The following argument is valid, although it may not appear valid at first glance:
Argument C

1. If it's raining, then the roof is wet.

2. The roof is not wet.

3. So, it is not raining.

In TL, with obvious abbreviations, this is

1. R ⊃ W

2. ~W

3. ~R

If the two premises are both true, the conclusion must be true. This argument is an instance of the argument form known as Modus Tollens:

$$\begin{array}{c} P \supset Q \\ \underline{\sim Q} \\ \sim P \end{array}$$

The following tree proves argument C valid:

THE HYPOTHETICAL SYLLOGISM PATTERN

In everyday reasoning, we often link several things together in a chain of if-then links. For instance:

Argument D

1. If it rains, the roof gets wet.

2. If the roof gets wet, then the ceiling leaks.

3. So, if it rains, then the ceiling leaks.

In TL, this is

1. $R \supset W$

2. $W \supset L$

3. $R \supset L$

This argument instantiates the valid argument form Hypothetical Syllogism:

$$P \supset Q$$

$$Q \supset R$$

$$P \supset R$$

The validity of argument D is easily proven on a truth-tree:

► STRATEGY

A common question at approximately this point is, which sentences should I dispatch first? The order in which you proceed won't affect the outcome of the tree you are constructing. You may begin wherever you wish and dispatch sentences in whatever order you wish. However, your tree will be less complex, and therefore simpler to build, if you will follow two suggestions:

1. Stack before you split. Apply rules to sentences that do not require a split before you apply rules to those that do call for a split.

2. Close off a branch as soon as a contradiction appears. There's no need to continue developing a branch once it has closed.

For example, consider the following argument:

1. (P & Q) v (R v S)

2. ~(R ⊃ S)
_____
3. P v Q

In order to test this argument for validity, we may build the following tree:

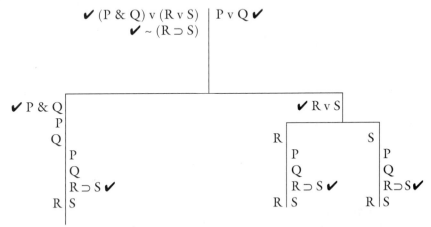

Because the completed tree is open, the argument is invalid.

If you trace the course of the reasoning from the top of the tree to the bottom, you will notice that we first broke down only those lines that did not require a split in the tree. We then broke down the line that required the split. If we had split the tree in the beginning, we would have had to break down all the other lines onto all three branches of our tree as follows:

Compare the above two trees. Both end up proving the same thing. No matter what order you follow in breaking sentences down on a tree, the result will ultimately be the same. That is, if a tree closes when you break sentences down in one order, it will close if you break the same sentences down in any other order; and if it is open when you follow one order, it will be open if you follow any other order. The only difference between breaking sentences down in one order rather than another will be in the complexity of the resulting tree. For example, if you compare the previous two trees, you will probably agree that although both reach the same result, the first is simpler and easier to construct and the second is more complex and harder to construct.

▶ **EXERCISE A.2** ◀

Test the following symbolized arguments for validity. In each case, the premises appear above the line and the conclusion appears below the line.

1. P ⊃ Q
─────
   Q ⊃ P

2. P ⊃ Q
   ~ Q
─────
   ~ P

3. ~ (P & Q)
─────────
   ~ P & ~ Q

4. Q
─────
   P v ~ P

5. ~ (P v Q)
─────────
   ~ P v ~ Q

6. P & ~ P
─────────
   Q

7. P v Q
   ~ P
─────
   Q

8. ~ (P & Q)
   P
─────────
   ~ Q

9. ~ (P & Q)
─────────
   ~ P & Q

10. P ⊃ Q
   ~ P
─────
   ~ Q

11. P ⊃ Q
   Q
─────
   P

12. ~ (P v Q)
─────────
   ~ P & ~ Q

13. P ⊃ Q
─────────
   ~ Q ⊃ ~ P

14. ~ P v ~ Q
─────────
   ~ (P v Q)

▶ **EXERCISE A.3** ◀

Symbolize each of the following and use truth-trees to test each for validity.

1. If Spock calls the ship, then Kirk will call the ship. If Kirk calls the ship, then Spock will call the ship. So, Kirk will call the ship if and only if Spock calls the ship.

2. If Ann and Bob both order burgers, then it's not the case that Elmer and Lulu will both order fish and chips. So, Ann and Bob won't both order burgers.

3. Either Edna brought the salad or Andy and Chris both brought salads. So, either Edna or Andy each brought a salad and either Edna or Chris each brought salad.

4. If Groucho sings, then if Chico sings then Harpo will play his harp. So, if Groucho and Chico sing, then Harpo will play his harp.

5. Mr. Ed speaks well. Therefore, either Mr. Ed speaks well or Muhammad Ali is the president of the United States.

6. Jane is an engineer. Melissa is a doctor. So, Jane is an engineer and Melissa is a doctor.

7. If the federal deficit increases, then Senator Smith's political future looks dim. So, either the federal deficit won't increase or Senator Smith's political future looks dim.

8. If Ilene swims, then Jolene will swim. So, if Jolene doesn't swim today, then Ilene won't swim today.

9. If Winnie wins a week's worth of the *Weekly,* then Monty will win a month's worth of the *Monthly.* If Daisy wins a day's worth of the *Daily,* then Yolanda will win a year's worth of the *Yearly.* Because Winnie will win her week's worth of the *Weekly* or Daisy will win her day's worth of the *Daily,* it follows that either Monty will win his month's worth of the *Monthly* or Yolanda will win her year's worth of the *Yearly.*

10. If the raindrops are one to two millimeters in size, the rainbow will be bright violet, green, and red but have little blue in it. If the raindrops are fifteen millimeters in size, the red of the primary bow will be weak. So, if the raindrops are one to two millimeters in size, then the red of the primary bow will not be weak.

11. If Elliott buys a pet turtle, then Lorraine will buy a pet hamster. Lorraine will buy a pet hamster. So, Elliott will buy a pet turtle.

12. If Joan puts roses in her garden, then Clyde will put daffodils in his garden. Clyde won't put daffodils in his garden. So, Joan won't put roses in her garden.

13. Either Katie will do an experiment with her chemistry set or she will look at a drop of pond water with her microscope. She will look at a drop of pond water with her microscope. So, she won't do an experiment with her chemistry set.

14. Either Nathan will look at some galaxies with his telescope or Katie will try to synthesize an exotic chemical in her chemistry lab. If Katie tries to synthesize an exotic chemical in her chemistry lab, Nathan will not look at some galaxies with his telescope. Nathan will therefore look at some galaxies with his telescope.

APPENDIX 2: TRUTH

What is truth? This ancient question raises two deep philosophical issues. First, what is it for something to be true? That is, what constitutes truth? Second, when someone says, "That's true," what sort of thing is it that is said to be true? The speaker seems to be attributing truth to *something*. What is this "something" that is being called true? And when someone says, "That's false," what is it that is supposedly false? Let us take up the second issue first.

THE BEARERS OF TRUTH AND FALSITY

Suppose someone says, "The moon has craters," and you reply, "That's true." When you say "That's true," what is it that is true? To what are you attributing truth? Or if you say, "That's false," what is it that is false? Truth and falsity seem to be characteristics of some entity. Of what entity? That which is true is called the *bearer of truth* or the *truth-bearer,* and that which is false is called the *bearer of falsity* or the *falsity bearer.* Our first issue is: What are the bearers of truth and falsity?

Consider the following three sentences, drawn from three different languages:

1. The Moon has craters.

2. La Luna tiene crateres.

3. Maraming maliliit na bulkau sa buwan.

The second and third sentences are accurate translations of the first sentence, "The Moon has craters." Each says that the Moon has craters. And this is true, by the way. Therefore, it seems that these three sentences all say the same thing, and, further, it seems that this one thing is itself a truth. In other words, these three sentences, drawn from three different languages, *express* one and the same thing, and the thing that the three express is true. But what is this *one* thing that is true here?

The thing that is the one truth in this case cannot, it seems, be the individual sentences themselves, for there are three sentences but one truth. Furthermore, we could erase all three sentences so that the space on the page is blank, but it would remain true that the Moon has craters. In other words, we could erase all three sentences, but the truth would remain. There seems to be no material object in this situation that could be identified as something that is true or as something that is the truth-bearer. The truth-bearer does not itself seem to be a material object.

Many philosophers believe we can make sense of this if we adopt the following hypothesis:

> *Whatever it is, that which is true in this case, that is, the truth-bearer, is not itself one of the three sentences. Furthermore, the truth-bearer is not itself a material object. The relation between this truth and the three sentences is this: the truth is something that can be* expressed *by each of the three sentences even though it is not itself one of the three sentences.*

If this is correct, then it seems that the thing that is true, the truth-bearer, is a very special type of entity, an entity that is unlike any ordinary entity, an entity that is an *abstract* entity of some sort. Let us briefly examine the idea of an abstract entity.

▶ MATERIAL VS. ABSTRACT ENTITIES

Many philosophers distinguish between *material* entities and *abstract* entities. A material entity is composed of matter. Matter is those things that science studies, namely, protons, electrons, photons, and such. Rocks, trees, clouds, dirt, and air are familiar examples of material entities. Each material entity occupies a volume of space and has a position in space and time. We detect such entities with our five physical senses and with physical instruments such as microscopes and telescopes. However, an abstract entity, according to those who distinguish abstract and material entities, is not composed of material particles, does not occupy a volume of space, and does not have a location in space and time. According to this view, an abstract entity is essentially a nonmaterial object that can be thought of or "grasped" by the mind, although it cannot be detected by the five physical senses or by physical instruments. Abstract entities are thus sometimes called "objects of thought." Philosophers who draw this distinction offer as a familiar example of an abstract entity the objects mathematicians study: numbers.

Do such things as abstract entities exist? *Absolute materialism* is the view that absolutely nothing exists except particles of matter and objects composed out of particles of matter. If you hold this view, you won't accept the existence of abstract entities.

▶ PROPOSITIONS

Philosophers generally use the term *proposition* to stand for something that is true or false. Thus, the bearers of truth, whatever they happen to be, are usually called propositions. (They are also sometimes called *statements*.) Using this terminology, the three sentences in the example above, all from different languages, nevertheless all express one and the same proposition, which is just to say that they express one and the same truth.

Propositions, if they are understood to be abstract entities, seem to be a very mysterious type of object. Although they are not physical or material and consequently occupy no location in physical space, our minds nevertheless apprehend them. After all, we think about various propositions, we believe them to be true or false, we hope this or that one is true, and we form various other judgments about them. This raises the question, how do our minds become aware of or "grasp" these abstract entities? How are propositions accessible to human thought?

Furthermore, how do such nonphysical objects fit into the modern scientific world-view, a view that seems to see everything as composed of purely physical parts? Anyone who believes in the existence of abstract entities faces this line of questioning. And these are difficult questions to answer. Indeed some philosophers argue that these difficulties make it hard to believe in the existence of abstract entities. Can we really accept a theory that posits such mysterious objects of thought? Let's explore this question for a moment.

▶ THE EXISTENCE OF THEORETICAL ENTITIES

Philosophers who hold that a proposition is an abstract entity consider propositions to be *theoretical entities,* as opposed to *observational entities.* An observational entity can be directly detected by the senses—it can be seen, heard, touched, and so on. A theoretical entity cannot be seen or directly detected, but, for theoretical reasons, we suppose it must exist.

Everyone has heard of the theoretical entities that physicists study, particles such as the electron, the proton, the neutron, and the quark. Consider, for example, the particles physicists call quarks. Physicists tell us that inside each proton and neutron there exist three quarks. However, we can't see or directly detect quarks. How do we know they even exist? Our best theories of the proton and neutron tell us each proton and neutron must contain three particles. In other words, quarks are referred to in our explanations of certain things. Although they can't be seen, reference to them helps *explain* various phenomena observed in the laboratory. And this is good reason to suppose that such objects truly exist.

In general, if reference to a theoretical entity helps make sense of some facet of the world, this constitutes a reason to suppose the theoretical object exists. Thus, one reason we suppose that electrons exist, to take another example from physics, is that although no one has ever seen one, we need to refer to electrons if we are to explain the operation of a battery, for example, or the flash of a lightning bolt, or the effect of a magnet. When we use the term *proposition,* we are assigning a name to the theoretical objects, whatever they are, that are true or false, so that we can *refer* to them.

Philosophers who believe propositions are abstract entities argue that we need to refer to propositions—conceived as abstract entities distinct from sentences—in order to explain or make sense of certain of our linguistic activities, specifically our attributions of truth and falsity. If reference to such abstract objects helps make sense of our world, this is good reason to suppose such objects actually exist. One reason for supposing the existence of propositions—understood as abstract objects—is that reference to them helps explain or make sense of our attributions of truth and falsity. Physicists and philosophers both posit theoretical entities in order to make sense of the world.

We can compare propositions, understood as abstract entities, to a similar type of abstract entity studied by mathematicians: numbers. We all think in terms of numbers, and we refer to them on a daily basis. But just what is a number?

Mathematicians distinguish numbers from numerals. Numerals are purely physical marks on paper, blackboards, computer terminals and so on. They are composed of molecules of graphite, ink, chalk or whatever. Because it is physical, a numeral has a size, a shape, a location, an age, a weight, and other various physical properties. Numbers, on the other hand, are not physical objects. The number three, for instance, which has been thought about since ancient times, is not located at some point in space, it is not composed of molecules of some substance, and it has no size, shape, weight, or age. Agreement is general among mathematicians that the number three is not located somewhere on Earth or in outer space.

Numerals *express* numbers. For example, the numerals *3, III,* and *3* each express the number three. The number itself is distinct from the numerals that express it, for in this example, there are three different numerals but only one number is expressed. Thus, mathematicians distinguish the number—an abstract entity—from the physical numerals that express it.

Similarly, philosophers who believe propositions are abstract entities distinguish between the proposition and the sentence that expresses the proposition. The distinction between numbers and numerals is therefore analogous to the distinction between propositions and the sentences that express propositions, and the relationship between numbers and the numerals that express them is analogous to the relationship between propositions and the sentences that express them.

Philosophers who accept the existence of propositional abstract objects address the following question to those who doubt the existence of such objects: If we can accept the existence of an infinity of abstract objects in mathematics called numbers—and most of us do—why should we balk at accepting an infinity of similar abstract objects in logic, objects we call propositions.[1]

WHAT IS TRUTH?

What is it for a proposition to be true? What characteristic does a true proposition have that makes it a *true* proposition? Let us begin by looking at a specific example. Suppose someone says, "Pat owns two cars." What do we mean when we say that the proposition expressed by this sentence is true? Or false? To begin with, if Pat actually does own two cars, the proposition expressed by "Pat owns two cars" is a true proposition. And if Pat does *not* actually own two cars, that proposition is a false proposition. More precisely:

> *The proposition expressed by "Pat owns two cars" is true if and only if Pat owns two cars. That proposition is false if and only if Pat does not own two cars.*

Reflecting upon the matter, it looks as if truth involves a relation between two separate entities. On the one hand, we have a proposition that in effect makes a claim about the way the world is, specifies a possible way the world might be, depicts a way the world might be. On the other hand, we have the way the world or reality really is, independent of or separate from the proposition. In the light of this, the following seems very plausible. When the proposition's claim or specification of the world *corresponds* to the way the world is, the proposition is true. When the proposition's claim or specification of the world does not correspond to the way the world is, the proposition is false.

For example, the sentence, "The Moon has craters" expresses a proposition that corresponds to the way things are. The sentence "The Sun is smaller in diameter than the Earth" expresses a proposition that does not correspond to or specify the way things are. After thinking about these examples, perhaps you will see the sense in which a proposition may "correspond" or fail to correspond to the way the world is.

If you are puzzled by the relationship between a proposition and the world, consider for a moment the relation between a map and the world. A map also specifies a way the world is, it makes a claim about the world. The map is "correct" just in case the way the world is corresponds to or "fits" the map's claim or specification; the map is incorrect otherwise. In this way, maps are somewhat like propositions. (Of course, one difference between propositions and maps is that maps represent the world by picturing it, whereas propositions do not depict reality by picturing it.)

And so, according to this account, truth is a property that a proposition has just in case it *corresponds* to the way the world is. Truth, in short, is the corre-

[1] For further reading on this issue, I recommend Bradley and Swartz, *Possible Worlds* (Indianapolis: Hackett, 1979) Ch. 2; and Michael Loux, *Metaphysics* (Indianapolis: University of Notre Dame Press, 1990) Ch. 5.

spondence of a proposition with reality. This account of truth goes back at least to Aristotle and is known as the *correspondence theory of truth*. It has been accepted by most philosophers throughout history as providing the best explanation of the matter, and it also constitutes a part of the common sense of humankind as well.[2]

▶ OBJECTIONS TO THE CORRESPONDENCE THEORY

Not all philosophers find the correspondence theory satisfactory. Some object by raising questions that they claim the theory cannot answer. For instance, what is the nature of the correspondence relation between a proposition and the world? It can't be a physical relationship, for propositions are not situated in space-time and so are not physically related to anything. And the correspondence relation isn't pictorial in nature, as when a map corresponds to some portion of the world by picturing that portion of the world, for propositions don't picture what they correspond to. In what sense, then, does a proposition "correspond"?

Here's another question for the correspondence theory. Consider the proposition expressed by "The moon has mountains." To what does this proposition correspond? To the universe as a whole? Or to the moon? Or only to the moon's mountains? One more question, which you can puzzle over on your own. Consider a negative proposition:

Santa Claus does not exist.

To what does this proposition correspond?

▶ TWO ALTERNATIVE THEORIES OF TRUTH

Although most philosophers throughout history have found the correspondence theory of truth "self-evident" or in some way intellectually indispensable, two alternative theories have been proposed and have won some support in certain parts of the philosophical world. According to the *coherence theory* of truth,

---

[2] If you will reflect upon the correspondence theory, you will notice that it seems to presuppose several things. Or at least, several assumptions seem to go along with the theory quite naturally. For instance, the theory speaks of the correspondence between two separate entities: a proposition and the world. It is natural to interpret this as follows. There exists a world of things out beyond our minds and distinct from propositions, such things as rocks, trees, mountains, stars, and so forth. That is, an "external world," a mind-independent reality, exists. A proposition may, or may not, correspond to this independent reality. So, reality has the character it has independently of the propositions our sentences express. For example, someone might say, "The Earth is flat." However, the Earth has its shape independent of what this person says. The search for truth, on the correspondence view, is the attempt to bring our thoughts into correspondence with a mind-independent, proposition-independent reality.

The view that material objects exist externally to us and independently of our thoughts about them is termed *metaphysical realism*. So, the correspondence theory of truth presupposes metaphysical realism. Some people claim that no such thing as reality exists. Such individuals favor an alternative theory of truth. An alternative to metaphysical realism is a view called *metaphysical idealism*. Roughly, according to this view, no objects exist apart from our thoughts. Objects exist, in other words, only as ideas in minds. One who accepts metaphysical idealism would not accept the correspondence theory of truth.

to say that a proposition is true is to say that it belongs to a coherent system of propositions. Briefly, a system of propositions is considered coherent if the propositions that belong to the system are consistent and stand in certain explanatory relations to one another. Thus, for coherence theorists, truth is a relation between propositions, and not, as in the correspondence theory, a relation between proposition and world.

The coherence theory of truth has its roots in the view—held by the philosophers in the idealist school of thought—that reality consists of one coherent system. The issue of why some philosophers hold a coherence theory of truth is a deep and difficult one, and any fair attempt at a summary would take us beyond the scope of this text.[3]

The coherence theory has always faced an objection that many philosophers consider decisive. It is possible to specify two opposing systems of propositions, such that each system seems coherent, yet such that one contradicts the other. Because the two systems are contradictory, they cannot both be true. If this is correct, a system of propositions, it seems, can be coherent yet fail to be true. The conclusion of this line of reasoning is that truth is not coherence.

A second alternative to the correspondence theory is the *pragmatic theory of truth*. According to this theory, advanced by philosophers in the school of thought known as pragmatism, to say that a proposition is true is to say that it is useful in some way. Truth, according to this view, is identified with usefulness, and a proposition is useful if our knowledge of it can serve a human interest.

The pragmatic theory of truth has its roots in the pragmatic theory of meaning. That theory ties meaning and truth to use. One objection that many philosophers have to the pragmatic definition of truth is that some propositions appear useful to some people, although they are clearly false. For instance, Hitler's racial theories were useful to him in his quest for power, yet those theories were most certainly false. Thus, many conclude that truth and usefulness are two different concepts. A further exploration of the pragmatic theory can be found in the writings of William James. [4]

---

[3] See Nicholas Rescher, *The Coherence Theory of Truth* (London: Oxford University Press, 1970).

[4] See William James, *Essays in Pragmatism* (New York: Macmillan, 1948), and *The Meaning of Truth* (Ann Arbor: University of Michigan Press, 1970).

ANSWERS TO SELECTED EXERCISES

The purpose of this section is to assist you in your work on the starred (*) problems in selected chapter exercises. Complete the corresponding chapter exercises and compare your answers to those listed below to see how well you are grasping the material. The order and the quantity of the answers varies from exercise to exercise.

CHAPTER 1

► EXERCISE 1.1 ◄

1. Not an argument.

5. Argument. Conclusion: The aardvark is warm-blooded.

7. Argument. Conclusion: That must be a DOS-based computer.

10. Not an argument.

► EXERCISE 1.2 ◄

1. Deductive

3. Inductive

5. Deductive

9. Inductive

▶ **EXERCISE 1.3** ◀

Part I.

1. Invalid

3. Valid

4. Valid

9. Invalid

11. Invalid

Part II.

1. Deductively valid

3. Inductively strong

5. Deductively valid

9. Inductively strong

▶ **EXERCISE 1.4** ◀

1. T

4. F

7. T

9. F

14. F

▶ **EXERCISE 1.5** ◀

1. Your pet is a reptile.

3. We won't take our bikes.

7. Anything that has a cause is not an act of free will.

12. All great pianists are great musicians.

CHAPTER 1 APPENDIX

1. Both individuals must be knights.

3. Both individuals must be knaves.

5. He answers "Yes."

10. Knight.

CHAPTER 2

▶ **EXERCISE 2.1** ◀

1. operator: and

sentences: "Juan . . . carpenter," "Rita . . . teacher."

5. operators: if-then, and, if-then

sentences: "First . . . problem," "it . . . logically," "Captain . . . attacked," "it . . . end"

10. operators: if-then, and, and

sentences: "Archie . . . tonight," "Archie . . . angry," "Michael . . . exasperated," "Archie . . . Meathead"

▶	**EXERCISE 2.2**	◀

1. False   **7.** False

4. True   **11.** True

CHAPTER 3

▶	**EXERCISE 3.1**	◀

1. (A & B) v D

4. J v (G & B)

7. ~(I & T)

11. (S & P) & ~ C

15. (~ A v ~ D) v (~ T & ~ J)

▶	**EXERCISE 3.2**	◀

1. F ⊃ (J v B)

4. [E & (L v H)] ⊃ M

7. ~ (N ⊃ T)

11. B ⊃ A   (B abbreviates: Joe buys a ticket; A abbreviates Joe will be admitted.)

14. T ⊃ (K & M)

17. [(H & F) ⊃ U] v (T ⊃ U)

U abbreviates: I will be filled up.

H abbreviates: I eat four hamburgers.

F abbreviates: I eat a fry.

T abbreviates: I eat three giant hot dogs.

CHAPTER 4

EXERCISE 4.1

1. Wff

5. Wff

9. Not a wff

13. Wff

16. Wff

EXERCISE 4.2

1. Second ampersand (&) from left

4. Wedge (v)

7. First wedge

EXERCISE 4.3

1. F

4. F

7. F

11. T

16. T

20. T

EXERCISE 4.4

1. T

4. T

7. T

11. F

16. T

EXERCISE 4.5

1. T

4. T

7. T

11. T

16. T

19. T

EXERCISE 4.6

1. T

4. T

7. T

11. T

16. F

18. F

CHAPTER 5

▶ **EXERCISE 5.1** ◀

1. Tautological

4. Tautological

7. Contradictory

11. Tautological

16. Tautological

20. Tautological

24. Tautological

28. Contingent

32. Contingent

▶ **EXERCISE 5.2** ◀

1. Invalid. The table is:

A B C	A v B	B v C	/ A v C
T T T	T	T	T
T T F	T	T	T
T F T	T	T	T
T F F	T	F	T
F T T	T	T	T
F T F	T	T	F ←
F F T	F	T	T
F F F	F	F	F

3. Invalid. The table is:

A B	A ⊃ B	/ B ⊃ A
T T	T	T
T F	F	T
F T	T	F ←
F F	T	T

7. Valid. The table is:

A B C	(A & B) ⊃ C	/ A ⊃ (B ⊃ C)
T T T	T	T
T T F	F	F
T F T	T	T
T F F	T	T
F T T	T	T
F T F	T	T
F F T	T	T
F F F	T	T

11. Valid. The table is:

A B C	(A & B) ⊃ C	~C	/ ~A v ~B
T T T	T	F	F
T T F	F	T	F
T F T	T	F	T
T F F	T	T	T
F T T	T	F	T
F T F	T	T	T
F F T	T	F	T
F F F	T	T	T

16. Invalid. The table is:

A B	~(A & B)	/ ~ A & ~ B	
T T	F	F	
T F	T	F	←
F T	T	F	←
F F	T	T	

► **EXERCISE 5.3** ◄
---

1. S ⊃ (~ J ⊃ R) / R ⊃ ~ J

　T T FT T T / T F FT

5. (A ⊃ R) R ⊃ (C v P) P ⊃ B / A ⊃ B

T T T   T T T T F   F T F   /   T F F

▶	**EXERCISE 5.4**	◀

1. A ⊃ (B v W)   B ⊃ S   / A ⊃ S

T T F T T   F T F   / T F F

4. ~(A & B)   ~ A   / B

T F F F   T F   / F

9. ~(A & B)   / ~ A

T T F F   / F T

12. A ⊃ B   W ⊃ S   B v S   / A v W

F T T   F T T   T T T   / F F F

▶	**EXERCISE 5.5**	◀

1.

B J	**B ⊃ J**	**~J**	**/ ~B**
T T	**T**	**F**	**F**
T F	**F**	**T**	**F**
F T	**T**	**F**	**T**
F F	**T**	**T**	**T**

Valid.

5.

G J D	**(G v J) ⊃ D**	**~D**	**~G & ~J**
T T T	**T**	**F**	**F**
T T F	**F**	**T**	**F**
T F T	**T**	**F**	**F**
T F F	**F**	**T**	**F**
F T T	**T**	**F**	**F**
F T F	**F**	**T**	**F**
F F T	**T**	**F**	**T**
F F F	**T**	**T**	**T**

Valid.

▶	**EXERCISE 5.6**	◀

1. Equivalent:

P Q	~(P v Q)	~P & ~Q
T T	F	F
T F	F	F
F T	F	F
F F	T	T

5. Not equivalent:

P Q	~(P & Q)	~P & ~Q
T T	F	F
T F	T	F
F T	T	F
F F	T	T

7. Not equivalent:

P Q	~(P & ~P)	(Q & ~ Q)
T T	T	F
T F	T	F
F T	T	F
F F	T	F

11. Equivalent:

P Q	P ⊃ Q	~P v Q
T T	T	T
T F	F	F
F T	T	T
F F	T	T

14. Equivalent

P	P & P	P
T	T	T
F	F	F

17. Equivalent:

P Q	P v Q	Q v P
T T	T	T
T F	T	T
F T	T	T
F F	F	F

20. Not equivalent:

P Q R	P ⊃ (Q ⊃ R)	(P ⊃ Q) ⊃ R
T T T	T	T
T T F	F	F
T F T	T	T
T F F	T	T
F T T	T	T
F T F	T	F
F F T	T	T
F F F	T	F

CHAPTER 6

▶   **EXERCISE 6.1**   ◀

b. 6, 12, 14 are instances of form b.

d. 1, 9 are instances of form d.

i. 2, 4 are instances of form i.

n. 18 is an instance of form n.

▶   **EXERCISE 6.2**   ◀

1. Invalid:

P Q	P ⊃ Q	Q	/ P	
T T	T	T	T	
T F	F	F	T	
F T	T	T	F	←
F F	T	F	F	

5. Valid:

P Q	P ⊃ Q	/ ~Q ⊃ ~P
T T	T	T
T F	F	F
F T	T	T
F F	T	T

7. Valid:

P Q	P v Q	/ Q v P
T T	T	T
T F	T	T
F T	T	T
F F	F	F

11. Valid:

P Q	P ⊃ Q	Q ⊃ P	/ P ≡ Q
T T	T	T	T
T F	F	T	F
F T	T	F	F
F F	T	T	T

16. Valid:

P Q R	P v (Q & R)	/ (P v Q) & (P v R)
T T T	T	T
T T F	T	T
T F T	T	T
T F F	T	T
F T T	T	T
F T F	F	F
F F T	F	F
F F F	F	F

21. Valid

24. Invalid:

P Q R	P v Q	Q v R	/ P v R	
T T T	T	T	T	
T T F	T	T	T	
T F T	T	T	T	
T F F	T	F	T	
F T T	T	T	T	
F T F	T	T	F	←
F F T	F	T	T	
F F F	F	F	F	

► **EXERCISE 6.3** ◄

2. Contradiction

CHAPTER 7

► **EXERCISE 7.1** ◄

1. HS 1, 2

MP 3, 5

DS 4, 6

5. MT 3, 4

DS 2, 5

MP 1, 6

7. DS 1, 3

MT 4, 5

10. MP 1, 2

MP 1, 3

MT 1, 4

13. DS 1, 3

HS 2, 4

15. HS 1, 2

MP 3, 5

    MT 4, 6

17. HS 2, 3

    MP 1, 4

    MP 1, 5

18. DS 2, 3

    DS 1, 5

    HS 4, 6

▶   **EXERCISE 7.2**   ◀

(1) **1.** (A & B) ⊃ S

    2. H ⊃ R

    3. (A & B) / S

    4. S   MP 1, 3

(5) **1.** F ⊃ (H & B)

    2. A ⊃ ~(H & B)

    3. A / ~ F

    4. ~(H & B)   MP 2, 3

    5. ~ F   MT 1, 4

(7) **1.** J v ~S

    2. ~J

    3. S v G / G

    4. ~S   DS 1, 2

    5. G   DS 3, 4

(10) **1.** R ⊃ H

    2. H ⊃ S

    3. S ⊃ G

    4. (R ⊃ G) ⊃ F / F

    5. R ⊃ S   HS 1, 2

    6. R ⊃ G   HS 3, 5

    7. F   MP 4, 6

(13) **1.** (H v B) ⊃ ~ (S ≡ F)

     2. R ⊃ (H v B)

     3. ~ (S ≡ F) ⊃ I / R ⊃ I

     4. R ⊃ ~ (S ≡ F)   HS 1, 2

     5. R ⊃ I   HS 3, 4

(15) **1.** A ⊃ (B ⊃ E)

     2. A

     3. ~ E / ~ B

     4. B ⊃ E   MP 1, 2

     5. ~ B   MT 3, 4

(17) **1.** I v A

     2. (I v A) ⊃ ~ S

     3. ~J ⊃ [J v (I ⊃ S)]

     4. ~ S ⊃ [S v (J ⊃ S)] / A

     5. ~ S   MP 1, 2

     6. S v (J ⊃ S)   MP 4, 5

     7. J ⊃ S   DS 5, 6

     8. ~J   MT 5, 7

     9. J v (I ⊃ S)   MP 3, 8

   10. I ⊃ S   DS 8, 9

   11. ~ I   MT 5, 10

   12. A   DS 1, 11

(18) **1.** A ⊃ B

     2. B ⊃ W

     3. ~ W

     4. J ⊃ I

     5. S ⊃ ~I

     6. ~A ⊃ (J v Z)

     7. S / Z

     8. ~ I   MP 5, 7

9. ~ J   MT 4, 8

10. A ⊃ W   HS 1, 2

11. ~A   MT 3, 10

12. J v Z   MP 6, 11

13. Z   DS 9, 12

► EXERCISE 7.3 ◄

(1) 1. A ⊃ B

2. A v C

3. ~ B / C

4. ~ A   MT 1, 3

5. C   DS 2, 4

(3) 1. (A v B) ⊃ (B v C)

2. A v (B ⊃ C)

3. (B ⊃ C) ⊃ (A v B)

4. ~ A / B v C

5. B ⊃ C   DS 2, 4

6. A v B   MP 3, 5

7. B v C   MP 1, 6

(5) 1. R ⊃ A

2. N ⊃ R

3. ~A / ~ N

4. ~ R   MT 1, 3

5. ~ N   MT 2, 4

(7) 1. ~ E

2. L ⊃ C

3. E v L / C

4. L   DS 1, 3

5. C   MP 2, 4

(13) 1. M ⊃ C

2. (M ⊃ ~ L) ⊃ (C ⊃ ~ D)

3. <u>C ⊃ ~ L / M ⊃ ~ D</u>
4. M ⊃ ~ L   HS 1, 3
5. C ⊃ ~ D   MP 2, 4
6. M ⊃ ~ D   HS 1, 5

CHAPTER 8

► EXERCISE 8.1 ◄

1. Simp 5
Add 6
MP 1, 7
MP 2, 8
Add 9
MP 3, 10
Add 6
MP 4, 12
Conj 11, 13

5. Simp 4
Add 7
MP 1, 8
DS 2, 9
MP 3, 10
Add 11
CD 5, 6, 12

6. Simp 1
Add 4
MP 2, 5
Simp 1
Conj 6, 7
MP 3, 8
Conj 9, 4

EXERCISE 8.2

(1) 1. A ⊃ B
  2. B ⊃ R
  3. <u>~R</u> / ~A & ~B
  4. ~B   MT 2, 3
  5. ~A   MT 1, 4
  6. ~A & ~B   Conj 4, 5

(5) 1. B ⊃ (S v R)
  2. S ⊃ J
  3. R ⊃ G
  4. <u>H & B</u> / J v G
  5. B   Simp 4
  6. S v R   MP 1, 5
  7. J v G   CD 2, 3, 6

(7) 1. H v ~ S
  2. [(H v ~S) v G] ⊃ ~M
  3. M v R / R v X
  4. (H v ~S) v G   Add 1
  5. ~M   MP 2, 4
  6. R   DS 3, 5
  7. R v X   Add 6

(10) 1. A ⊃ (J & S)
  2. B ⊃ F
  3. A
  4. <u>[(J & S) v F] ⊃ G</u> / G
  5. A v B   Add 3
  6. (J & S) v F   CD 1, 2, 5
  7. G   MP 4, 6

(13) 1. (A ⊃ B) & (A ⊃ C)
  2. A
  3. (B v C) ⊃ Z / Z

4. A ⊃ B	Simp 1
5. A ⊃ C	Simp 1
6. A ∨ A	Add 2
7. B ∨ C	CD 4, 5, 6
8. Z	MP 3, 7

(15) **1.** (A ⊃ B) & (S ⊃ I)

    2. J ⊃ R

    3. <u>(A ∨ J) & (S ∨ E)</u> / B ∨ R

4. A ⊃ B	Simp 1
5. A ∨ J	Simp 3
6. B ∨ R	CD 2, 4, 5

(17) **1.** S ⊃ [J & (I ∨ S)]

    2. [J ∨ (A & B)] ⊃ (~ N & ~ O)

    3. S & Z / ~N & J

4. S	Simp 3
5. J & (I ∨ S)	MP 1, 4
6. J	Simp 5
7. J ∨ (A & B)	Add 6
8. ~N & ~O	MP 2, 7
9. ~N	Simp 8
10. ~N & J	Conj 6, 9

(20) **1.** (A & S) ⊃ Z

    2. ~J ⊃ A

    3. ~J ⊃ S

    4. <u>~J & B</u> / Z ∨ I

5. ~J	Simp 4
6. A	MP 2, 5
7. S	MP 3, 5
8. A & S	Conj 6, 7
9. Z	MP 1, 8
10. Z ∨ I	Add 9

CHAPTER 9

▶	**EXERCISE 9.1**	◀

(1) **1.** J v (I & E)

    2. <u>J ⊃ E</u> / E

3.	~E	AP
4.	~J	MT 2, 3
5.	I & E	DS 1, 4
6.	E	Simp 5
7.	E & ~ E	Conj 3, 6
8. E	IP 3–7	

(5) **1.** ~B ⊃ C

    2. C ⊃ B / B

3.	~B	AP
4.	C	MP 1, 3
5.	B	MP 2, 4
6.	B & ~B	Conj 3, 5
7. B	IP 3–6	

(9) **1.** A

    2. A ⊃ B

    3. (A & B) ⊃ K

    4. K ⊃ G / G

5.	~G	AP
6.	~K	MT 4, 5
7.	B	MP 1, 2
8.	A & B	Conj 1, 7
9.	K	MP 3, 8
10.	K & ~K	Conj 6, 9
11. G	IP 5–10	

Part II.

This is 7.2. Problem 1:

1. (A & B) ⊃ S

2. H ⊃ R

3. A & B / S

4.	~S	AP
5.	~(A & B)	MT 1,4
6.	(A & B) & ~(A & B)	Conj 3,5
7. S	IP 4-6	

▶	**EXERCISE 9.2**	◀

(1) **1.** A ⊃ (B & C) / A ⊃ C

2.	A	AP
3.	B&C	MP 1, 2
4.	C	Simp 3
5. A ⊃ C	CP 2–4	

(3) **1.** J ⊃ (I ⊃ W)

2. (I ⊃ W) ⊃ (I ⊃ S) / J ⊃ (I ⊃ S)

3.	J	AP
4.	I ⊃ W	MP 1, 3
5.	I ⊃ S	MP 2, 4
6. J ⊃ (I ⊃ S)	CP 3–5	

(5) **1.** ~I v Z

2. Z ⊃ A / ~ ~I ⊃ (Z & A)

3.	~ ~I	AP
4.	Z	DS 1, 3
5.	A	MP 2, 4
6.	Z & A	Conj 4, 5
7. ~ ~I ⊃ (Z & A)	CP 3–6	

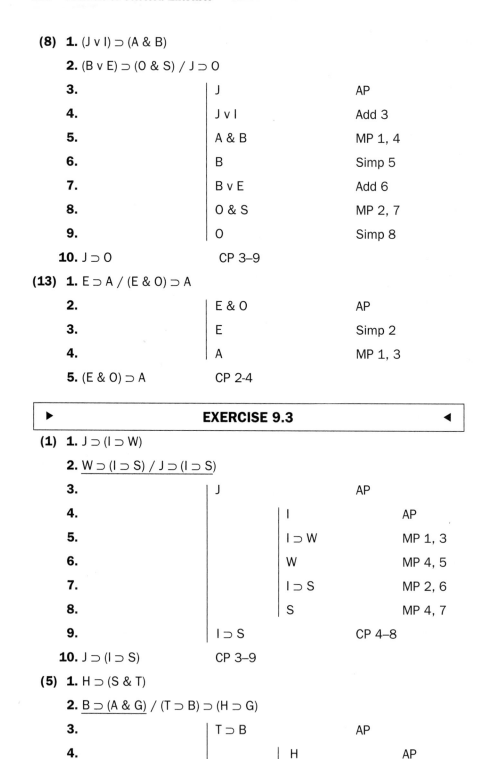

(8) **1.** (J v I) ⊃ (A & B)

    2. (B v E) ⊃ (O & S) / J ⊃ O

    3.   J   AP

    4.   J v I   Add 3

    5.   A & B   MP 1, 4

    6.   B   Simp 5

    7.   B v E   Add 6

    8.   O & S   MP 2, 7

    9.   O   Simp 8

   10. J ⊃ O   CP 3–9

(13) **1.** E ⊃ A / (E & O) ⊃ A

    2.   E & O   AP

    3.   E   Simp 2

    4.   A   MP 1, 3

    5. (E & O) ⊃ A   CP 2-4

EXERCISE 9.3

(1) **1.** J ⊃ (I ⊃ W)

    2. W ⊃ (I ⊃ S) / J ⊃ (I ⊃ S)

    3.   J   AP

    4.   I   AP

    5.   I ⊃ W   MP 1, 3

    6.   W   MP 4, 5

    7.   I ⊃ S   MP 2, 6

    8.   S   MP 4, 7

    9.   I ⊃ S   CP 4–8

   10. J ⊃ (I ⊃ S)   CP 3–9

(5) **1.** H ⊃ (S & T)

    2. B ⊃ (A & G) / (T ⊃ B) ⊃ (H ⊃ G)

    3.   T ⊃ B   AP

    4.   H   AP

5.		S & T	MP 1, 4
6.		T	Simp 5
7.		B	MP 3, 6
8.		A & G	MP 2, 7
9.		G	Simp 8
10.	H ⊃ G		CP 4–9
11. (T ⊃ B) ⊃ (H ⊃ G)	CP 3–10		

▶ **EXERCISE 9.4** ◀

(1) 1	[(A v B) & ~A]	AP	
2.	(A v B)	Simp 1	
3.	~A	Simp 1	
4.	B	DS 2, 3	
5. [(A v B) & ~A] ⊃ B	CP 1–4		
(5) 1.	[(A ⊃ B) & (A ⊃ I)]	AP	
2.		A	AP
3.		A ⊃ B	Simp 1
4.		B	MP 2, 3
5.		(A ⊃ I)	Simp 1
6.		I	MP 2, 5
7.		B & I	Conj 4, 6
8.	A ⊃ (B & I)	CP 2–7	
9. [(A ⊃ B) & (A ⊃ I)] ⊃ [A ⊃ (B & I)]		CP 1–8	

CHAPTER 10

▶ **EXERCISE 10.1** ◀

1. Dist 1

Simp 4

MP 2, 5

DNeg 6

DS 3, 7

Comm 8

5. Simp 1

MP 2, 6

MP 3, 7

Add 8

MP 4, 9

DNeg 10

DS 5, 11

► **EXERCISE 10.2** ◄

Part I.

(5) **1.** ~(A v B)

2. ~B ⊃ E

3. E ⊃ S / S

4. ~A & ~B   DM 1

5. ~B   Simp 4

6. E   MP 2, 5

7. S   MP 3, 6

(7) **1.** (H & S) v (H & P)

2. H ⊃ (G v M) / G v M

3. H & (S v P)   Dist 1

4. H   Simp 3

5. G v M   MP 2, 4

(10) **1.** (R ⊃ S) v ~G

2. ~ ~ G

3. (R ⊃ S) ⊃ ~ ~ P

4. H ⊃ ~P / ~ H

5. ~ G v (R ⊃ S)   Comm 1

6. R ⊃ S   DS 2, 5

7. ~ ~P   MP 3, 6

8. ~H   MT 4, 7

(13) 1. ~ (R & S)

2. ~R ⊃ Q

3. ~S ⊃ Z

4. ~Q

5. Z ⊃ (A ≡ B) / A ≡ B

6. ~R v ~S   DM 1

7. Q v Z   CD 2, 3, 6

8. Z   DS 4, 7

9. A ≡ B   MP 5, 8

(15) 1. ~ R ⊃ ~ S

2. ~~S

3. ~~R ⊃ H

4. (H ⊃ A) & (Z ⊃ O) / A v O

5. ~~R   MT 1, 2

6. H   MP 3, 5

7. H v Z   Add 6

8. H ⊃ A   Simp 4

9. Z ⊃ O   Simp 4

10. A v O   CD 7, 8, 9

(17) 1. S ⊃ ~ P

2. (P & S) v (P & F) / F

3. P & (S v F)   Dist 2

4. P   Simp 3

5. ~ ~ P   DNeg 4

6. ~S   MT 1, 5

7. S v F   Simp 3

8. F   DS 6, 7

(20) 1. ~ ~K

2. ~(A & B) & ~(E & F)

       3. (H & K) ⊃ [(A & B) v (E & F)] / ~ H

       4. ~ [(A & B) v (E & F)]   DM 2

       5. ~ (H & K)   MT 3, 4

       6. ~ H v ~K   DM 5

       7. ~ K v ~ H   Comm 6

       8. ~ H   DS 1, 7

(25) **1.** H ⊃ (E & P)

       2. A v (B & ~C)

       3. A ⊃ ~E

       4. ~C ⊃ ~P / ~H

       5. (A v B) & (A v ~C)   Dist 2

       6. A v ~C   Simp 5

       7. ~E v ~P   CD 3, 4, 6

       8. ~(E & P)   DM 7

       9. ~H   MT 1, 8

(27) **1.** ~ (A & B)

       2. ~ A ⊃ (P & Q)

       3. ~ B ⊃ (Q & S) / Q

       4. ~A v ~B   DM 1

       5. (P & Q) v (Q & S)   CD 2, 3, 4

       6. (Q & P) v (Q & S)   Comm 5

       7. Q & (P v S)   Dist 6

       8. Q   Simp 7

Part II.

(1) **1.** ~ (A & B)

       2. S ⊃ B

       3. S v ~J

       4. J / ~A

       5. ~A v ~B   DM 1

       6. ~ J v S   Comm 3

7. ~ ~ J	DNeg 4
8. S	DS 6,7
9. B	MP 2, 8
10. ~ B v ~ A	Comm 5
11. ~ ~B	DNeg 9
12. ~A	DS 10, 11

► **EXERCISE 10.3** ◄

1. Trans 1

MP 2, 5

Add 6

MP 3, 7

Exp 8

MP 4, 9

Trans 10

5. Imp 1

MP 2, 6

DNeg 7

MT 3, 8

MP 4, 9

Add 10

MP 5, 11

Taut 12

► **EXERCISE 10.4** ◄

Part I.

(1) **1.** H ⊃ ~S

2. (~H v ~S) ⊃ F

3. F ⊃ B / B

4. ~H v ~S   Imp 1

5. F   MP 2, 4

6. B   MP 3, 5

(5) **1.** ~(A & ~B) ⊃ C

    2. A ⊃ B

    3. <u>H v ~C</u> / H

    4. ~A v B   Imp 2

    5. ~(A & ~B)   DM 4

    6. C   MP 1, 5

    7. ~C v H   Comm 3

    8. ~ ~ C   DNeg 6

    9. H   DS 7, 8

(7) **1.** (J v J) ⊃ S

    2. J & F / S

    3. J ⊃ S   Taut 1

    4. J   Simp 2

    5. S   MP 3, 4

(10) **1.** (A & E) ⊃ S

    2. <u>(E ⊃ S) ⊃ F</u> / A ⊃ F

    3. A ⊃ (E ⊃ S)   Exp 1

    4. A ⊃ F   HS 2, 3

(13) **1.** J ≡ I

    2. ~ (J & I) / ~ J & ~ I

    3. (J & I) v (~J & ~I)   Equiv 1

    4. ~J & ~I   DS 2, 3

(15) **1.** (A ⊃ B) & (E ⊃ B)

    2. <u>~ (~A & ~E)</u> / B

    3. A v E   DM 2

    4. A ⊃ B   Simp 1

    5. E ⊃ B   Simp 1

    6. B v B   CD 3, 4, 5

    7. B   Taut 6

(20) **1.** A ⊃ B

    2. <u>A ⊃ ~B</u> / ~A

3. ~ ~ B ⊃ ~A   Trans 2

4. B ⊃ ~A   DNeg 3

5. A ⊃ ~A   HS 1, 4

6. ~A v ~A   Imp 5

7. ~A   Taut 6

(23) **1.** (J ⊃ I) ⊃ W

    2. W ⊃ ~ W / J

    3. ~ W v ~ W   Imp 2

    4. ~W   Taut 3

    5. ~(J ⊃ I)   MT 1, 4

    6. ~(~ J v I)   Imp 5

    7. ~ ~ (~ ~ J & ~ I)   DM 6

    8. J & ~I   DNeg 7

    9. J   Simp 8

(25) **1.** J ⊃ (S & H)

    2. W ⊃ (~S & ~H)

    3. J v W / H ⊃ S

    4. (S & H) v (~S & ~H)   CD 1, 2, 3

    5. S ≡ H   Equiv 4

    6. (S ⊃ H) & (H ⊃ S)   Equiv 5

    7. H ⊃ S   Simp 6

(28) **1.** (A ⊃ B) ⊃ G

    2. (I ⊃ B) ⊃ ~ G / ~ B

    3. ~ ~G ⊃ ~ (I ⊃ B)   Trans 2

    4. G ⊃ ~ (I ⊃ B)   DNeg 3

    5. (A ⊃ B) ⊃ ~ (I ⊃ B)   HS 1, 4

    6. ~(A ⊃ B) v ~ (I ⊃ B)   Imp 5

    7. ~(~ A v B) v ~(~I v B)   Imp 6

    8. (~ ~A & ~B) v (~ ~I & ~B)   DM 7

    9. (A & ~B) v (I & ~B)   DNeg 8

    10. (~ B & A) v (~ B & I)   Comm 9

11. ~B & (A v I)		Dist 10
12. ~B		Simp 11
(30)	**1.** (J ⊃ I) ⊃ (I ⊃ J)	
	2. (J ≡ I) ⊃ ~(A & ~B)	
	3. I & A / A & B	
	4. I	Simp 3
	5. I v ~J	Add 4
	6. ~J v I	Comm 5
	7. J ⊃ I	Imp 6
	8. I ⊃ J	MP 1, 7
	9. (J ⊃ I) & (I ⊃ J)	Conj 7, 8
	10. J ≡ I	Equiv 9
	11. ~(A & ~B)	MP 2, 10
	12. ~A v ~ ~B	DM 11
	13. A	Simp 3
	14. ~ ~A	DNeg 13
	15. ~ ~B	DS 12, 14
	16. B	DNeg 15
	17. A & B	Conj 13, 16

Part II.

(1)	**1.** ~A ⊃ ~B	
	2. (~B v A) ⊃ S / S	
	3. B ⊃ A	Trans 1
	4. ~B v A	Imp 3
	5. S	MP 2, 4

CHAPTER 11

EXERCISE 11.1

(4) **1.** (H v S) ⊃ (A & B)

  2. (B v F) ⊃ K

3. <u>H v F</u> / K

4.	~ K	AP
5.	~ (B v F)	MT 2, 4
6.	~B & ~F	DM 5
7.	~F	Simp 6
8.	F v H	Comm 3
9.	H	DS 7, 8
10.	H v S	Add 9
11.	A & B	MP 1, 10
12.	B	Simp 11
13.	~B	Simp 6
14.	B & ~ B	Conj 12, 13
15. K		IP 4–14

(7) **1.** (J & I) v E

2. <u>~E v I</u> / J ⊃ I

3.	~(J ⊃ I)	AP
4.	~(~J v I)	Imp 3
5.	~ ~(~ ~ J & ~I)	DM 4
6.	J & ~ I	DNeg 5
7.	~I	Simp 6
8.	I v ~E	Comm 2
9.	~E	DS 7, 8
10.	E v (J & I)	Comm 1
11.	J & I	DS 9, 10
12.	I	Simp 11
13.	I & ~ I	Conj 7, 12
14. J ⊃ I		IP 3–13

(12) **1.** J ⊃ (~I ⊃ S)

2. <u>(J ⊃ I) ⊃ S</u> / S

3.	~S	AP
4.	~(J ⊃ I)	MT 2, 3

5.		~(~J v I)	Imp 4
6.		~ ~ (~ ~J & ~I)	DM 5
7.		J & ~ I	DNeg 6
8.		J	Simp 7
9.		~I ⊃ S	MP 1, 8
10.		~ I	Simp 7
11.		S	MP 9, 10
12.		S & ~ S	Conj 3, 11
13. S		IP 3–12	

```
┌─────────────────────────────────────────────────┐
►              EXERCISE 11.3                      ◄
└─────────────────────────────────────────────────┘
```

(1)   **1.** A ⊃ (B ⊃ E)

   2. A ⊃ (H ⊃ E)

   3. ~ E / A ⊃ ~ (B v H)

4.		A	AP
5.		B ⊃ E	MP 1, 4
6.		H ⊃ E	MP 2, 4
7.		~H	MT 6, 3
8.		~B	MT 3, 5
9.		~B & ~H	Conj 7, 8
10.		~(B v H)	DM 9
11. A ⊃ ~ (B v H)		CP 4–10	

(3)   **1.** J ⊃ I

   2. (J & W) ⊃ E

   3. (I & E) ⊃ S / J ⊃ (W ⊃ S)

4.		J	AP
5.		I	MP 1, 4
6.		I ⊃ (E ⊃ S)	Exp 3
7.		E ⊃ S	MP 5, 6
8.		J ⊃ (W ⊃ E)	Exp 2
9.		W ⊃ E	MP 4, 8

10. | $\mid$ W ⊃ S | HS 7, 9

11. J ⊃ (W ⊃ S) | CP 4–10

(11) **1.** A ⊃ (B v C)

    2. E ⊃ S

    3. <u>B ⊃ C</u> /A ⊃ C

4.	A	AP
5.	B v C	MP 1, 4
6.	C v B	Comm 5
7.	~C ⊃ B	Imp 6
8.	~C ⊃ C	HS 3, 7
9.	C v C	Imp 8 and DNeg
10.	C	Taut 9
11. A ⊃ C	CP 4–10	

►   EXERCISE 11.5   ◄

(3)	**1.**	~[~ A v (B v A)]	AP
	2.	~ ~A & ~ (B v A)	DM 1
	3.	~ ~A	Simp 2
	4.	A	DNeg 3
	5.	~(B v A)	Simp 2
	6.	~B & ~A	DM 5
	7.	~A	Simp 6
	8.	A & ~A	Conj 4, 7
	9. [~A v (B v A)]	IP 1–7	
(8)	**1.**	~ [~(J & I) v ~(~J & ~I)]	AP
	2.	~ ~(J & I) &~ ~ (~J & ~I)	DM 1
	3.	(J & I) & (~J & ~I)	DNeg 2
	4.	J & I	Simp 3
	5.	~J & ~I	Simp 3
	6.	J	Simp 4
	7.	~J	Simp 5

8.	J & ~J	Conj 6, 7
9. ~(J & I) v ~(~J & ~I)	IP 1–8	
(11) 1.	~J	AP
2.	~J v ~I	Add 1
3.	~I v ~J	Comm 2
4.	~(I & J)	DM 3
5. ~J ⊃ ~(I & J)	CP 1– 4	
(14) 1.	A	AP
2.	A v ~I	Add 1
3.	~I v A	Comm 2
4.	I ⊃ A	Imp 3
5. A ⊃ (I ⊃ A)	CP 1– 4	
(23) 1.	A	AP
2.	A ⊃ B	AP
3.	B	MP 1, 2
4.	(A ⊃ B) ⊃ B	CP 2–3
5. A ⊃ [(A ⊃ B) ⊃ B]	CP 1– 4	

CHAPTER 12

▶   **EXERCISE 12.1**   ◀

2. Theoretical

11. a. Lexical

  c. Theoretical

  h. Lexical

  o. Lexical

▶   **EXERCISE 12.2**   ◀

2. b. Musical instruments, orchestral instruments, drums, snare drums, chrome snare drums, dented chrome snare drums

3. b. catcher, baseball player, ball player, athlete

4. c. An institution of higher learning, UCLA, University of Washington, Harvard, Princeton

5. a. tough, brave, determined, anti–Communist, military

►	**EXERCISE 12.3**	◄

1. Intensional

5. Extensional

7. Extensional

►	**EXERCISE 12.4**	◄

2. Synonymous definition

7. Enumerative definition

11. Analytical definition

►	**EXERCISE 12.5**	◄

1. d. "House" means "dwelling."

2. c. A jet is an aircraft such as a Boeing 747, an MD 80.

3. c. "Athletic" means "physically active and strong."

4. c. A magnet is any substance that passes the following test: Place the substance close to a small piece of iron. If the substance attracts the iron, then the substance is a magnet.

►	**EXERCISE 12.6**	◄

3. Genus: Deer; Difference: female

7. Genus: Sibling; Difference: Male

►	**EXERCISE 12.7**	◄

c. Rule 6. The definition is circular.

f. Rule 4. The definition is too figurative.

j. Rule 3. The definition is negative.

m. Rule 6. The definition is circular.

CHAPTER 13

► **EXERCISE 13.2** ◄
---

1. Tu Quoque

4. Petitio Principii

9. Ad Baculum

13. Red Herring

19. Circumstantial ad Hominem

23. Ad Ignorantiam

► **EXERCISE 13.4** ◄
---

1. Accident

5. Ad Verecundiam

10. False Dilemma

13. Accident

16. Hasty Generalization

20. False Cause

26. Weak Analogy

► **EXERCISE 13.6** ◄
---

2. Division

6. Amphiboly

11. Composition

17. Equivocation

21. Division

27. Composition

CHAPTER 14

► **EXERCISE 14.1** ◄
---

1. a. O   **b.** Particular negative

4. a. A   **b.** Universal affirmative

Answers to Selected Exercises 633

▶ **EXERCISE 14.2** ◀

1. a. F
   b. T
   c. F
5. a. Undetermined
   b. F
   c. Undetermined

▶ **EXERCISE 14.3** ◀

a. Valid
e. Valid
h. Invalid

▶ **EXERCISE 14.4** ◀

Part I.
1. Contraries

Part II.
1. Some extraterrestrials are not green.

▶ **EXERCISE 14.5** ◀

1. a. All drops of seawater are drops of water that taste salty.
   e. All heroin users are self-destructive individuals.
   j. All free persons are rational persons.
   k. All places he goes are places where there is a cloud of gloom over him.
2. a. No drops of seawater are drops of water that taste salty.
   e. No truly religious persons are charitable persons.
   i. No truly free persons are knowledgeable persons.
   k. No times on Earth are times it rains.
3. a. Some teenagers are energetic persons.
   e. Some persons are contented voters.
   h. Some persons are persons who saw smoke.
   j. Some places on Earth are places that are sacred.

4. a. Some politicians are dishonest persons.

 d. Some seniors are inactive persons.

 g. Some musicians are persons who cannot read music.

---

1. a. All creepy things are spiders. (Not equivalent)

 d. Some creepy things are not spiders. (Not equivalent)

2. a. No Corvairs are nonsafe vehicles. (Equivalent)

3. a. All nonhonest persons are nonpoliticians. (Equivalent)

CHAPTER 15

---

1. a. No mammals are reptiles.

 All dogs are mammals.

 So, no dogs are reptiles.

 Mood: EAE. Figure: 1

 e. Some hamburgers are fattening things.

 No Pink's hot dogs are hamburgers.

 So, no Pink's hot dogs are fattening things.

 Mood: IEE. Figure: 1

2. a. Major term: reptiles

 minor term: dogs

 middle term: mammals

 e. Major term: fattening things

 minor term: Pink's hot dogs

 middle term: hamburgers

3. a. valid

 e. invalid

4. No nuns are football players.

 Some teenagers are football players.

 So, some teenagers are not nuns.

EXERCISE 15.2

1.

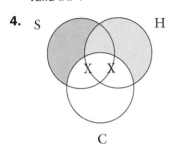

Valid IAI-4

4.

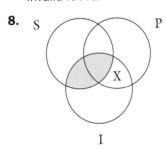

Invalid AAA-2

8.

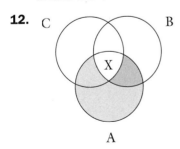

Invalid IEO-4

12.

C   B

X

A

Valid AAI3

17.

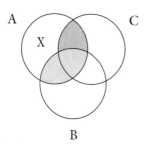

Invalid EEI-3

23.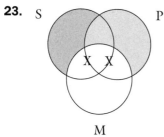

Invalid AAA-2

► **EXERCISE 15.3** ◄

1.

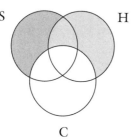

Valid

4.

Invalid

8.

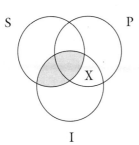

Invalid

12.

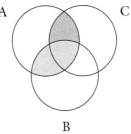

Invalid

17.

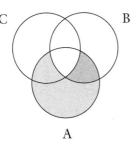

Invalid

23.

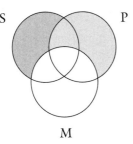

Invalid

1.   Standard form:

All persons fit to serve on a jury are sane persons.

All sane persons are persons who can do logic.

No sons of yours are persons who can do logic.

So, no sons of yours are persons fit to serve on a jury.

The two syllogisms in the chain:

All sane persons are persons who can do logic.

All persons fit to serve on a jury are sane persons.

So, all persons fit to serve on a jury are persons who can do logic.

All persons fit to serve on a jury are persons who can do logic.

No sons of yours are persons who can do logic.

So, no sons of yours are persons fit to serve on a jury.

Diagrams:

F: Persons fit to serve on a jury

S: Sane persons

L: Persons who can do logic

Y: Sons of yours

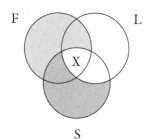

Valid

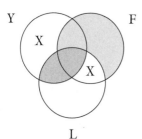

Valid

4. No athletes are lethargic individuals.

All chess players are athletes.

Some doctors are chess players.

Some doctors are not lethargic individuals. The two syllogisms:

No athletes are lethargic individuals.

All chess players are athletes.

So, no chess players are lethargic individuals.

No chess players are lethargic individuals.

Some doctors are chess players.

So, some doctors are not lethargic.

C: Chess players

L: Lethargic individuals

D: Doctors

A: Athletes

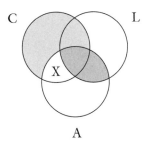

Valid

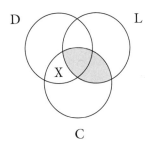

Valid

7. All A are B.

Some C are A.

All C are D.

So, some D are B. The two syllogisms:

All A are B.

Some C are A.

So, some C are B.

Some C are B.

All C are D.

So, some D are B.

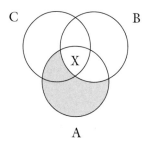

Valid

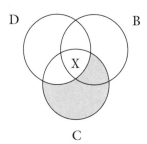

Valid

 EXERCISE 15.5 ◄

1. The government of Ruritania is a dictatorship. (missing premise)

4. Whales are mammals. (missing premise)

7. Tigers are cats. (missing premise)

11. All persons who run a mile every day are healthy persons. (missing premise)

> ▶   **EXERCISE 15.6**   ◀

1. That's like arguing: Some people are over 100 and some people are teenagers, so some people are both over 100 and in their teens at the same time.

4. That's like arguing: Some mammals are cats. No cats are dogs. So, some dogs are not mammals.

7. That's like arguing: All boxers are athletes. Some nuns are athletes. So, some nuns are boxers.

11. That's like arguing: No Republicans are Democrats. All Republicans are persons who belong to a political party. So, no persons who belong to a political party are Democrats.

CHAPTER 16

> ▶   **EXERCISE 16.1**   ◀

1. Bg   Abbreviations: Bx: x banks at the blood bank. g: Grandpa Munster

5. Sm ⊃ (Mc & Pl)   Abbreviations: Sx: x sings. Mx: x makes a face. Px: x pokes Curley in the eye. m: Moe. c: Curley. l: Larry

> ▶   **EXERCISE 16.2**   ◀

1. (x) Gx   ~(∃ x) ~ Gx

4. ~(x) Gx   (∃ x) ~ Gx

7. (x) (Bx ⊃ ~ Nx)   ~(∃ x) (Bx & Nx)

9. (x) (Vx ⊃ Lx)   ~ (∃ x)(Vx & ~ Lx)

> ▶   **EXERCISE 16.3**   ◀

1. De ⊃ (x) (Px ⊃ Lx) Abbreviations: e: Elaine. Dx: x dances. Lx: x laughs. Px: x is a person.

5. (x) Mx v (x) ~ Mx

9. (∃ x) (Mx & Px) & (∃x)(Mx & ~ Px)

14. (∃ x)[(Cx & Gx) & Ix]

Cx: x is a car. Gx: x is green

Ix: x is in the garage

17. (x) [Hx ⊃ (Rx v Lx)]

23. Bc ⊃ (x) (Sx ⊃ Lx)

Bx: x is broadcasting

Sx: x is a Seattleite

Lx: x is listening

c: Dr. Frasier Crane

CHAPTER 17

▶   **EXERCISE 17.1**   ◀

1. Tps & Osp

5. (∃ x) (Px & Kxx)   Abbreviations: Px: x is a person. Kxy: x knows y.

10. Fje   Abbreviation: Fxy: x is a friend of y.

13. (x)(Px ⊃ Dsx)   Abbreviations: Px: x is a person. Dxy: x dislikes y. s: Sam.

21. ~(x) (Px ⊃ Lxr)   Abbreviations: Px: x is a person. r: Raymond.

▶   **EXERCISE 17.2**   ◀

1. (x)(∃ y) (Kxy) ⊃ (∃ x) (y) (Kxy)

5. (∃ x) (y) Ryx   Rxy: x respects y

7. (∃ x) (∃ y) (Hxy) ⊃ Pg

Px: x is pleased

Hxy: x helps y

g: God

10. (x){[Hx & ~(∃y) (Hy & Lxy)] ⊃ Px} Abbreviations: Hx: x is human. Px: x deserves pity. Lxy: x loves y.

17. (x) Hx v (x) Ax Domain: persons. Abbreviations: Hx: x is happy. Ax: x is acting

▶   **EXERCISE 17.3**   ◀

1. b = n   b: Betty. n: the new winner.

5. ~(∃ x){Nx & (y) [(Ny & ~(y = x)) ⊃ Gxy]} Nx: x is a number. Gxy: x is greater than y.

7. (x) (Vx ⊃ Cx)

9. (∃ x) (y) {Dx & [Dy ⊃ (y = x)]}

16. (x) {[Px & ~(x = p)] ⊃ Lpx} & ~Lpp   Abbreviations: Px: x is a person. p: Pat. Lxy: x loves y.

► **EXERCISE 17.4** ◄

1. a. has a higher cholestoral level than, is richer than, owns more records than, can eat more burritos than, runs faster than.

2. b. nonsymmetric, nontransitive, irreflexive

l. reflexive, symmetric, transitive

CHAPTER 18

► **EXERCISE 18.1** ◄

1. UI 1

UI 2

HS 3, 4

UG 5

► **EXERCISE 18.2** ◄

(1) 1. (x)(Wx ⊃ Sx)

2. (x)(Sx ⊃ Px) / (x)(Wx ⊃ Px)

3. Wu ⊃ Su   UI 1

4. Su ⊃ Pu   UI 2

5. Wu ⊃ Pu   HS 3, 4

6. (x)(Wx ⊃ Px)   UG 5

(5) 1. (x)(Mx)

2. Hg / Hg & Mg

3. Mg   UI 1

4. Hg & Mg   Conj 2, 3

(7) 1. Sa ⊃ (x) Fx

2. Ha v Sa

    3. ~Ha

    4. (x) (Fx ⊃ Gx) / (x) Gx

    5. Sa   DS 2, 3

    6. (x)(Fx)   MP 1, 5

    7. Fu   UI 6

    8. Fu ⊃ Gu   UI 4

    9. Gu   MP 7, 8

   10. (x)(Gx)   UG 9

(10) **1.** (x) (Ax ⊃ Bx)

      2. (x) (Jx ⊃ Fx)

      3. (x) (Ax v Jx) / (x) (Bx v Fx)

      4. Au ⊃ Bu   UI 1

      5. Ju ⊃ Fu   UI 2

      6. Au v Ju   UI 3

      7. Bu v Fu   CD 4, 5, 6

      8. (x)(Bx v Fx)   UG 7

(12) **1.** (x)(Hx ⊃ Gx)

      2. (x)~Gx / (x)~Hx

      3. Hu ⊃ Gu   UI 1

      4. ~Gu   UI 2

      5. ~Hu   MT 3, 4

      6. (x)(~Hx)   UG 5

(16) **1.** (x) (Hx ⊃ Qx)

      2. Ha v Hb / Qa v Qb

      3. Ha ⊃ Qa   UI 1

      4. Hb ⊃ Qb   UI 1

      5. Qa v Qb   CD 2, 3, 4

(18) **1.** (∃x)(Fx) ⊃ (x)(Sx)

      2. (∃x)(Hx) ⊃ (x)(Gx)

      3. Fs & Hg / (x)(Sx & Gx)

      4. Fs   Simp 3

5. (∃x)(Fx)	EG 4
6. (x)(Sx)	MP 1, 5
7. Hg	Simp 3
8. (∃x)(Hx)	EG 7
9. (x)(Gx)	MP 8, 2
10. Su	UI 6
11. Gu	UI 9
12. Su & Gu	Conj 10, 11
13. (x)(Sx & Gx)	UG 12

▶ EXERCISE 18.3 ◀

(1)
1. (x) {Sx ⊃ [Px ⊃ (Bx & Qx)]}	
2. (∃x) Sx / (∃x) (Px ⊃ Qx)	
3. Sv	EI 2
4. Sv ⊃ [Pv ⊃ (Bv & Qv)]	UI 1
5. Pv ⊃ (Bv & Qv)	MP 3, 4
6. ~ Pv v (Bv & Qv)	Imp 5
7. (~Pv v Bv) & (~Pv v Qv)	Dist 6
8. ~Pv v Qv	Simp 7
9. Pv ⊃ Qv	Imp 8
10. (∃x) (Px ⊃ Qx)	EG 9

(4)
1. (x) [Hx ⊃ (Sx v Gx)]	
2. ~ Sb & ~ Gb / ~ Hb	
3. Hb ⊃ (Sb v Gb)	UI 1
4. ~(~ ~Sb v ~ ~Gb)	DM 2
5. ~(Sb v Gb)	DNeg 4
6. ~Hb	MT 3, 5

(6)
1. (x)[Hx ⊃ (Bx & Wx)]	
2. (∃y)(~By)	/ (∃z)~ Hz
3. ~Bv	EI 2
4. Hv ⊃ (Bv & Wv)	UI 1

5. ~Bv v ~Wv	Add 3
6. ~(~ ~Bv & ~ ~Wv)	DM 5
7. ~(Bv & Wv)	DNeg 6
8. ~Hv	MT 4, 7
9. (∃z) ~Hz	EG 8

► EXERCISE 18.4 ◄

(1) **1.** ~(x)Sx / (∃x)(Sx ⊃ Px)

2. (∃x)~Sx	QE 1
3. ~Sv	EI 2
4. ~Sv v Pv	Add 3
5. Sv ⊃ Pv	Imp 4
6. (∃x)(Sx ⊃ Px)	EG 5

(5) **1.** ~(∃x)(Ax) / (x)(Ax ⊃ Bx)

2. (x)~Ax	QE 1
3. ~Au	UI 2
4. ~Au v Bu	Add 3
5. Au ⊃ Bu	Imp 4
6. (x)(Ax ⊃ Bx)	UG 5

(7) **1.** (x)(Sx & Gx) v (x)(Qx & Hx)

2. ~(x)Qx / (x)Gx	
3. (∃x)~Qx	QE 2
4. ~Qv	EI 3
5. ~Qv v ~Hv	Add 4
6. ~(Qv & Hv)	DM 5
7. (∃x) ~ (Qx & Hx)	EG 6
8. ~(x)(Qx & Hx)	QE 7
9. (x)(Qx & Hx) v (x)(Sx & Gx)	Comm 1
10. (x)(Sx & Gx)	DS 8, 9
11. Su & Gu	UI 10
12. Gu	Simp 11

13. (x)Gx   UG 12

(9)   **1.** (∃x)(Px v Gx) ⊃ (x)(Hx)

     2. (∃x)~Hx / (x)~Px

     3. ~(x)Hx   QE 2

     4. ~(∃x)(Px v Gx)   MT 1, 3

     5. (x)~(Px v Gx)   QE 4

     6. ~(Pu v Gu)   UI 5

     7. ~Pu & ~Gu   DM 6

     8. ~Pu   Simp 7

     9. (x) ~Px   UG 8

►   **EXERCISE 18.5**   ◄

---

(1)   **1.** (x) (Cx ⊃ Mx)

     2. <u>(x) (Fx ⊃ ~Mx)</u> / (x) (Fx ⊃ ~Cx)

     3. Cu ⊃ Mu   UI 1

     4. Fu ⊃ ~Mu   UI 2

     5. ~ ~Mu ⊃ ~Fu   Trans 4

     6. Mu ⊃ ~Fu   DNeg 5

     7. Cu ⊃ ~Fu   HS 3, 6

     8. ~ ~Fu ⊃ ~Cu   Trans 7

     9. Fu ⊃ ~Cu   DNeg 8

    10. (x) (Fx ⊃ ~Cx)   UG 9

(5)   **1.** (x) (Mx ⊃ Ax)

     2. (x) (Ax ⊃ Dx)

     3. <u>(∃x) (Hx & Mx)</u> / (∃x) (Hx & Dx)

     4. Hv & Mv   EI 3

     5. Mv ⊃ Av   UI 1

     6. Av ⊃ Dv   UI 2

     7. Mv ⊃ Dv   HS 5, 6

     8. Mv   Simp 4

     9. Hv   Simp 4

10. Dv   MP 7, 8

11. Hv & Dv   Conj 9, 10

12. (∃x) (Hx & Dx)   EG 11

(7) **1.** (x)[(Cx v Dx) ⊃ Rx]   (note: Rx: x reasons, learns, loves)

2. (x)[Rx ⊃ (Px & Gx)] / (x) [(Cx v Dx) ⊃ (Px & Gx)]

3. (Cu v Du) ⊃ Ru   UI 1

4. Ru ⊃ (Pu & Gu)   UI 2

5. (Cu v Du) ⊃ (Pu & Gu)   HS 3, 4

6. (x)[(Cx v Dx) ⊃ (Px & Gx)]   UG 5

(12) **1.** (x)(Hx ⊃ Ax)

2. (x)(Ax ⊃ Px)

3. ~Ps / ~Hs

4. Hs ⊃ As   UI 1

5. As ⊃ Ps   UI 2

6. Hs ⊃ Ps   HS 4, 5

7. ~Hs   MT 3, 6

(14) **1.** (x) (Dx ⊃ Lx)

2. (x) (Lx ⊃ ~Fx)

3. Dp / ~Fp

4. Dp ⊃ Lp   UI 1

5. Lp ⊃ ~Fp   UI 2

6. Dp ⊃ ~Fp   HS 4, 5

7. ~Fp   MP 3, 6

CHAPTER 19

► **EXERCISE 19.1** ◄

Part I.

1. Domain: Numbers

Ax: x is even

Bx: x is rational

5. Domain: Persons

Ax: x is conscious at least once in x's life

Bx: x has never experienced consciousness

10. Domain: living beings

Ax: x is a human

Bx: x is a martian

Part II.
1. That's like arguing: Some birds are green. Some green things are fish. So some birds are fish.

5. That's like arguing: All humans are mammals. No humans are dogs. Thus, no dogs are mammals.

► **EXERCISE 19.2** ◄

1. (Aa & Pa) v (Ab & Pb)

~(Aa & ~ Pa) & ~(Ab & ~ Pb)

Ab: T Aa: T

Pb: F Pa: T

CHAPTER 20

► **EXERCISE 20.1** ◄

(1) 1. (x)(Ax ⊃ ~Bx)

2. (x)[(Bx ⊃ (Hx & Ax)] / (∃x)(~Bx)

3.	~(∃x)(~Bx)	AP
4.	(x)Bx	QE 3
5.	Bp	UI 4
6.	Bp ⊃ (Hp & Ap)	UI 2
7.	Hp & Ap	MP 5, 6
8.	Ap	Simp 7
9.	Ap ⊃ ~Bp	UI 1
10.	~Bp	MP 8, 9

11.	Bp & ~ Bp	Conj 5, 10
12. (∃x) ~Bx	IP 3–11	

(5) **1.** (x)(Hx ⊃ Qx)

2. (x)(Hx ⊃ Rx) / (x)[Hx ⊃ (Qx & Rx)]

3.	Hu	AP
4.	Hu ⊃ Qu	UI 1
5.	Qu	MP 3, 4
6.	Hu ⊃ Ru	UI 2
7.	Ru	MP 3, 6
8.	Qu & Ru	Conj 5, 7
9. Hu ⊃ (Qu & Ru)	CP 3—8	
10. (x)[Hx ⊃ (Qx & Rx)]	UG 9	

(7) **1.** (x) [Px ⊃ (Hx & Qx)] / (x) (Sx ⊃ Px) ⊃ (x) (Sx ⊃ Qx)

2.	(x)(Sx ⊃ Px)	AP
3.	Su ⊃ Pu	UI 2
4.	Pu ⊃ (Hu & Qu)	UI 1
5.	Su ⊃ (Hu & Qu)	HS 3, 4
6.	~Su v (Hu & Qu)	Imp 5
7.	(~Su v Hu) & (~Su v Qu)	Dist 6
8.	~Su v Qu	Simp 7
9.	Su ⊃ Qu	Imp 8
10.	(x)(Sx ⊃ Qx)	UG 9
11. (x)(Sx ⊃ Px) ⊃ (x)(Sx ⊃ Qx)		CP 2–10

(10) **1.** (∃x)(Px) v (∃x)(Qx & Rx)

2. (x)(Px ⊃ Rx) / (∃x)Rx

3.	~(∃x)Rx	AP
4.	(x)~Rx	QE 3
5.	~Ru	UI 4
6.	~Ru v ~Qu	Add 5
7.	~Qu v ~Ru	Comm 6
8.	~(Qu & Ru)	DM 7

9.	(x)~(Qx & Rx)	UG 8
10.	~(∃x)(Qx & Rx)	QE 9
11.	(∃x)(Qx & Rx) v (∃x)(Px)	Comm 1
12.	(∃x)Px	DS 10, 11
13.	Pv	EI 12
14.	Pv ⊃ Rv	UI 2
15.	Rv	MP 13, 14
16.	~Rv	UI 4
17.	Rv & ~Rv	Conj 15, 16
18. (∃x)Rx	IP 3–17	

(13) **1.** (x) [(Hx v Px) ⊃ Qx]

    2. (x) [(Qx v Mx) ⊃ ~Hx] / (x) (~ H x)

3.	~(x) ~Hx	AP
4.	(∃x)Hx	QE 3
5.	Hv	EI 4
6.	(Hv v Pv) ⊃ Qv	UI 1
7.	Hv v Pv	Add 5
8.	Qv	MP 6, 7
9.	Qv v Mv	Add 8
10.	(Qv v Mv) ⊃ ~ Hv	UI 2
11.	~Hv	MP 9, 10
12.	Hv & ~Hv	Conj 5, 11
13. (x)~Hx	IP 3–12	

(15) **1.** (∃x)(Px v Jx) ⊃ ~(∃x)(Px) / (x)~Px

2.	~(x)~ Px	AP
3.	(∃x) Px	QE 2
4.	Pv	EI 3
5.	Pv v Jv	Add 4
6.	(∃x)(Px v Jx)	EG 5
7.	~(∃x)Px	MP 1, 6
8.	(x)~Px	QE 7

9.	~Pv	UI 8
10.	Pv & ~Pv	Conj 4, 9
11. (x)~Px	IP 2–10	

► EXERCISE 20.2 ◄

(1) 1.	~(x)[(Hx v Px) v ~Px]	AP
2.	(∃x)~[(Hx v Px) v ~Px]	QE 1
3.	~[(Hv v Pv) v ~Pv]	EI 2
4.	~(Hv v Pv) & ~ ~Pv	DM 3
5.	~(Hv v Pv) & Pv	DNeg 4
6.	~(Hv v Pv)	Simp 5
7.	~Hv & ~Pv	DM 6
8.	~Pv	Simp 7
9.	Pv	Simp 5
10.	Pv & ~Pv	Conj 8, 9
11. (x)[(Hx v Px) v ~Px]	IP 1–10	

(5) 1.	(x) Px	AP
2.	~(∃x) ~Px	QE
3. (x)(Px) ⊃ ~(∃x)(~Px)	CP 1–2	

(7) 1.	(∃x)(Fx & Gx)	AP
2.	Fv & Gv	EI 1
3.	Fv	Simp 2
4.	Gv	Simp 2
5.	(∃x)Fx	EG 3
6.	(∃x)Gx	EG 4
7.	(∃x)(Fx) & (∃x)(Gx)	Conj 5, 6
8. (∃x)(Fx & Gx) ⊃ [(∃x)(Fx) & (∃x)(Gx)] CP 1–7		

(10) 1.	~(x)[(Sx & Hx) ⊃ (Sx v Hx)]	AP
2.	(∃x)~[(Sx & Hx) ⊃ (Sx v Hx)]	QE 1
3.	~[(Sv & Hv) ⊃ (Sv v Hv)]	EI 2
4.	~[~(Sv & Hv) v (Sv v Hv)]	Imp 3

5.	(Sv & Hv) & ~(Sv v Hv)	DM 4, DNeg
6.	Sv & Hv	Simp 5
7.	Sv	Simp 6
8.	~(Sv v Hv)	Simp 5
9.	~Sv & ~Hv	DM 8
10.	~Sv	Simp 9
11.	Sv & ~Sv	Conj 7, 10

12. (x)[(Sx & Hx) ⊃ (Sx v Hx)]   IP 1–11

CHAPTER 21

► EXERCISE 21.1 ◄

Part A.

(1) 1. (x)(y)(Hxy ⊃ ~Hyx)

2. (∃x)(∃y)(Hxy) / (∃x)(∃y)(~Hyx)

3. (∃y)(Hvy)   EI 2

4. Hvv′   EI 3

5. (y)(Hvy ⊃ ~Hyv)   UI 1

6. Hvv′ ⊃ ~Hv′v   UI 5

7. ~Hv′v   MP 4, 6

8. (∃y)(~Hyv)   EG 7

9. (∃x)(∃y)(~Hyx)   EG 8

(7) 1. (x) (∃y) (Hxy) ⊃ (x) (∃y) (Sxy)

2. (∃x) (y) (~Sxy) / (∃x) (y) (~ Hxy)

3. (∃x)~(∃y)(Sxy)   QE 2

4. ~(x)(∃y)(Sxy)   QE 3

5. ~(x)(∃y)(Hxy)   MT 1, 4

6. (∃x)~(∃y)(Hxy)   QE 5

7. (∃x)(y)(~Hxy)   QE 6

(12) 1. (∃x) (∃y) (Kxy) / (∃y) (∃x) (Kxy)

2. (∃y) (Kvy)   EI 1

3. Kv v′   EI 2

4. (∃x)(Kxv′)   EG 3

5. (∃y) (∃x) (Kxy)   EG 4

Part B.

(1) 1. (x) (∃y) (Mx ⊃ Cyx)

    2. ~ (x) (∃y) (Cyx) / (∃x) ~Mx

3.	~(∃x) ~Mx	AP
4.	(x) Mx	QE 3
5.	(∃x) ~ (∃y) (Cyx)	QE 2
6.	(∃x) (y)~Cyx	QE 5
7.	(y)~Cyv	EI 6
8.	Mv	UI 4
9.	(∃y)(Mv ⊃ Cyv)	UI 1
10.	Mv ⊃ Cv′v	EI 9
11.	Cv′v	MP 8, 10
12.	~ Cv′v	UI 7
13.	Cv′v & ~ Cv′v	Conj 11, 12
14. (∃x) ~Mx	IP 3-13	

CHAPTER 22

▶   **EXERCISE 22.1**   ◀

(1) 1. Aa ⊃ Ha

    2. ~Ha

    3. a = b / ~Ab

    4. ~Aa   MT 1, 2

    5. ~Ab   Id B 3, 4

(3) 1. (x) (Ax ⊃ Px)

    2. (x) (Px ⊃ Hx)

    3. Aa & ~ Hb / ~ (a = b)

    4. | ~ ~ (a = b)   AP

5.	(a = b)	DNeg 4
6.	Aa ⊃ Pa	UI 1
7.	Pa ⊃ Ha	UI 2
8.	Aa ⊃ Ha	HS 6, 7
9.	Aa	Simp 3
10.	Ha	MP 8, 9
11.	~Hb	Simp 3
12.	~Ha	Id B 11, 5
13.	Ha & ~ Ha	Conj 10, 12
14. ~(a = b)	IP 4–13	

▶   **EXERCISE 22.2**   ◀

(1)   **1.** Sj ⊃ S

     2. Ss

     3. j = s   / S

     4. Sj   Id B 2, 3

     5. S   MP 1, 4

   j: Jim. s: Sue's husband S: The store will be closed. Sx x is sick.

CHAPTER 23

▶   **EXERCISE 23.1**   ◀

Part I.

1. □ S   **5.** ~ □ S

Part II.

1. ~ □ P

4. □ ~A

7. □ D

10. □ (U ⊃ G)

12. □ (U v ~U)

16. M ⊃ □ ~ G

20. ~G ⊃ □ ~ G

▶   **EXERCISE 23.2**   ◀
---

2. a. F

   f. T

   k. F

3. a. $A \supset (\square A \vee \triangledown A)$ (true)

   d. $A \supset \sim \diamond \sim A$ (false)

   g. $\square \sim A \supset \sim \triangledown A$ (true)

   i. $\square \sim A \supset \sim \diamond A$ (true)

   l. $\triangledown A \supset \diamond A$ (true)

4. a. F   **e.** T

5. a. F   **e.** F

▶   **EXERCISE 23.3**   ◀
---

1. T   **9.** T

5. T   **12.** T

▶   **EXERCISE 23.4**   ◀

Part I.

1. T   **8.** T

4. F   **11.** F

Part II.

  1. $(A \& B) \supset \diamond (A \& B)$

  4. $(\sim A \& \sim B) \supset \sim \diamond (A \& B)$

  8. $(\square A \& \square B) \supset (A \leftrightarrow B)$

11. $(\triangledown A \& \triangledown B) \supset \diamond (A \& B)$

CHAPTER 24

▶   **EXERCISE 24.1**   ◀
---

(1) 1. $\diamond (H \& S) \supset G$

   2. $\sim R$

3. R v (H & S) / G

4. H & S   DS 2, 3

5. ◇ (H & S)   Poss 4

6. G   MP 1, 5

(4) 1. H → G

    2. G → (M & N)

    3. [H → (M & N)] ⊃ S / S

    4. H → (M & N)   MHS 1, 2

    5. S   MP 3, 4

(6) 1. A → ◇ S

    2. A / □ ◇ S

    3. ◇ A   Poss 2

    4. □◇ S   P2N 1, 3

(9) 1. F → (□ A & □ B)

    2. G ⊃ F

    3. H & G / □ (□ A & □ B)

    4. G   Simp 3

    5. F   MP 2, 4

    6. ◇F   Poss, 5

    7. □(□ A & □ B)   P2N 1, 6

(12) 1. □ (A ⊃ B)

    2. H v A

    3. ~ (F v H) / B

    4. ~F & ~H   DM 3

    5. ~H & ~F   Comm 4

    6. ~H   Simp 5

    7. A   DS 2, 6

    8. A ⊃ B   BR 1

    9. B   MP 7, 8

(14) 1. □ (A v B)

    2. B ⊃ H

    3. ~ R & ~ H / A

    4. A v B   BR, 1

    5. ~ H   Simp 3

    6. ~ B   MT 2, 5

    7. B v A   Com 4

    8. A   DS 6, 7

(19) **1.** □ (A & B) / ◇ A & ◇ B

    2. A & B   BR 1

    3. A   Simp 2

    4. B   Simp 2

    5. ◇ A   Poss 3

    6. ◇ B   Poss 4

    7. ◇ A & ◇ B   Conj 5,6

►   EXERCISE 24.2   ◄

(2) **1.** □ (H ⊃ S)

    2. □ (S ⊃ R)

    3. □ (H ⊃ R) ⊃ G / ◇G

    4. | □ (H ⊃ S)   Reit 1

    5. | □ (S ⊃ R)   Reit 2

    6. | H ⊃ S   BR 4

    7. | S ⊃ R   BR 5

    8. | H ⊃ R   HS 6, 7

    9. □ (H ⊃ R)   Nec 4–8

  10. G   MP 3, 9

  11. ◇G   Poss, 10

(4) **1.** □ (A ⊃ B)

    2. □ (~ B v ~ A)

    3. G ⊃ ~ □ ~ A / ~ G

    4. | □ (A ⊃ B)   Reit 1

    5. | □ (~ B v ~ A)   Reit 2

6.	A ⊃ B	BR 4
7.	~ B ∨ ~ A	BR 5
8.	B ⊃ ~ A	Imp 7
9.	A ⊃ ~ A	HS 6, 8
10.	~ A ∨ ~ A	Imp 9
11.	~ A	Taut 10
12. □ ~ A	Nec 4–11	
13. ~ ~ □ ~ A	DNeg 12	
14. ~ G	MT 3, 13	

(7) 1. A → B

   2. □A / □B

   3. □(A ⊃ B)   Arrow Ex 1

4.	□(A ⊃ B)	Reit 3
5.	□A	Reit 2
6.	A	BR 5
7.	(A ⊃ B)	BR 4
8.	B	MP 6,7
9. □B	Nec 4-8	

(10) 1. □A & □B / □(A & B)

2.	□A & □B	Reit 1
3.	□A	Simp 2
4.	□B	Simp 2
5.	A	BR 3
6.	B	BR 4
7.	A & B	Conj 5, 6
8. □(A & B)	Nec 2–7	

(12) 1. ◇ A / □ □ ◇ A

2.	◇ A	Reit 1
3. □ ◇ A	Nec 2	
4.	□ ◇ A	Reit 3
5. □ □ ◇ A	Nec 4	

(14) 1. ◇A / □ ◇ ◇A

2.	◇ A	Reit 1
3.	◇ ◇ A	Poss 2

4. □ ◇ ◇A   Nec 2–3

(16) 1. □A / ◇ □A

2. ◇ □A   Poss 1

(21) 1. <u>□ (A & B)</u> / □A & □B

2.	□ (A & B)	Reit 1
3.	A & B	BR 2
4.	A	Simp 3
5.	B	Simp 3

6. □ A   Nec 2–5

7. □ B   Nec 2–5

8. □A & □ B   Conj 6, 7

▶   **EXERCISE 24.3**   ◀

---

(1) 1. S ↔ Q

2. H ⊃ Q

3. F & H / S

4. □ (S ≡ Q)	Double Arrow Ex 1
5. S ≡ Q	BR 4
6. (S ⊃ Q) & (Q ⊃ S)	Equiv 5
7. Q ⊃ S	Simp 6
8. H	Simp 3
9. Q	MP 2, 8
10. S	MP 7, 9

(3) 1. □ ~A / ~ □A

2. ~ A	BR 1
3. ◇ ~ A	Poss 2
4. ~ □A	DE 3

(6) **1.** A ↔ B
 2. A / B
 3. □ (A ≡ B)   Double Arrow Ex 1
 4. (A ≡ B)   BR 3
 5. (A ⊃ B) & (B ⊃ A)   Equiv 4
 6. (A ⊃ B)   Simp 5
 7. B   MP 2, 6

(11) **1.** A ↔ B
 2. ~B / ~ A
 3. □ (A ≡ B)   Double Arrow Ex 1
 4. (A ≡ B)   BR 3
 5. (A ⊃ B) & (B ⊃ A)   Equiv 4
 6. A ⊃ B   Simp 5
 7. ~A   MT 2, 6

► **EXERCISE 24.4** ◄

(1) **1.** H v S / ◇H v ◇S
 2.   ~(◇H v ◇S)   AP
 3.   ~◇H & ~◇S   DM 2
 4.   ~◇H   Simp 3
 5.   ~◇S & ~◇H   Comm 3
 6.   ~◇S   Simp 5
 7.   □~H   DE 4
 8.   ~H   BR 7
 9.   □~S   DE 6
 10.   ~S   BR 9
 11.   S   DS 1, 8
 12.   S & ~S   Conj 10, 11
 13. ◇H v ◇S   IP 2–12
(6) **1.** ◇ A v ◇ B   / ◇ (A v B)

2.	~ ◊ (A v B)	AP
3.	□ ~ (A v B)	DE 2
4.	□ ~ (A v B) Reit 3	
5.	~ (A v B) BR 4	
6.	~ A & ~ B DM 5	
7.	~ A Simp 6	
8.	~ B Simp 6	
9.	□ ~ A	Nec 4–8
10.	□ ~ B	Nec 4–8
11.	~ ◊ A	DE 9
12.	~ ◊ B	DE 10
13.	~ ◊ A & ~ ◊ B	Conj 11, 12
14.	~ (◊ A v ◊ B)	DM 13
15.	(◊ A v ◊ B) & ~ (◊ A v ◊ B)	Conj 1, 14
16. ◊ (A v B)	IP 2–15	

► **EXERCISE 24.5** ◄

(1)
- **1.** ~ ◊ (S & K)
- **2.** ~S → ~M
- **3.** ~K → ~M / ~ ◊ M

4. □ ~ (S & K)	DE 1	
5.	□ ~ (S & K)	Reit 4
6.	~ (S & K)	BR 5
7.	~S v ~K	DM 6
8.	~S → ~M	Reit 2
9.	~K → ~M	Reit 3
10.	□ (~S ⊃ ~M)	Arrow Ex 8
11.	□ (~K ⊃ ~M)	Arrow Ex 9
12.	~S ⊃ ~M	BR 10
13.	~K ⊃ ~M	BR 11

14.   ~M v ~M   CD 7, 12, 13

15.   ~M   Taut 14

16. □ ~M   Nec 5–15

17. ~ ◇ M   DE 16

(5) **1.** B → (R & W)

   2. ~ J ⊃ ~ ◇ R

   3. B / J

   4. R & W   MMP 1,3

   5. R   Simp 4

   6. ◇ R   Poss 5

   7. ~ ~ ◇ R   DNeg 6

   8. ~ ~J   MT 2, 7

   9. J   DNeg 8

(10) **1.** □ (A v B)

   2. □ (~B v ~E)

   3. ~ ◇ ~E   / □ A

   4. □ E   DE 3

   5.   □ (A v B)   Reit 1

   6.   □ (~B v ~E)   Reit 2

   7.   □ E   Reit 4

   8.   (A v B)   BR 5

   9.   (~B v ~E)   BR 6

  10.   E   BR 7

  11.   ~ ~A v B   DNeg 8

  12.   ~ A ⊃ B   Imp 11

  13.   B ⊃ ~E   Imp 9

  14.   ~A ⊃ ~E   HS 12, 13

  15.   ~ ~ E   DNeg 10

  16.   ~ ~ A   MT 14, 15

  17.   A   DNeg 16

  18. □ A   Nec 5-17

(15) 1. ~ ◇ (G & E)

    2. □ G   / □ ~E

    3. □ ~ (G & E)   DE 1

    4.   □ G   Reit 2

    5.   □ ~ (G & E)   Reit 3

    6.   G   BR 4

    7.   ~ (G & E)   BR 5

    8.   ~ ~ G   DNeg 6

    9.   ~G v ~E   DM 7

    10.   ~E   DS 8,9

    11. □ ~E   Nec 4-10

►   EXERCISE 24.6   ◄

(1) 1.   ~A   AP

    2.   ~A v B   Add 1

    3.   A ⊃ B   Imp 2

    4. ~A ⊃ (A ⊃ B)   CP 1–3

    5. □ [~A ⊃ (A ⊃ B)]   Taut Nec 4

(5) 1.   A ⊃ B   AP

    2.   ~B ⊃ ~A   Trans 1

    3. (A ⊃ B) ⊃ (~B ⊃ ~A)   CP 1–2

    4. □ [(A ⊃ B) ⊃ (~B ⊃ ~A)]   Taut Nec 3

(7) 1.   [(A v B) & ~A]   AP

    2.   A v B   Simp 1

    3.   ~A   Simp 1

    4.   B   DS 2, 3

    5. [(A v B) & ~A] ⊃ B   CP 1–4

    6. □ {[(A v B) & ~A] ⊃ B}   Taut Nec 5

(10) 1.   (A v B) & (~B v S)   AP

    2.   A v B   Simp 1

    3.   ~B v S   Simp 1

    4.   ~A ⊃ B   Imp 2, DNeg

5. | B ⊃ S — Imp 3

6. | ~A ⊃ S — HS 4, 5

7. | A v S — Imp 6, DNeg

8. [(A v B) & (~B v S)] ⊃ (A v S) — CP 1–7

9. □ {[(A v B) & (~B v S)] ⊃ (A v S)} — Taut Nec 8

(13) 1. | A ⊃ B — AP

2. | ~A v B — Imp 1

3. | ~(~ ~A & ~B) — DM 2

4. . | ~(A & ~B) — DNeg 3

5. (A ⊃ B) ⊃ ~(A & ~B)   CP 1–4

6. □ {(A ⊃ B) ⊃ ~(A & ~B)} Taut Nec 5

(15) 1. | A — AP

2. A ⊃ A   CP 1

3. □ (A ⊃ A)   Taut Nec 2

4. A → A   Arrow Ex 3

(17) 1. | A → B — AP

2. | | □ A — AP

3. | | | □ A — Reit 2

4. | | | A — BR 3

5. | | | A → B — Reit 1

6. | | | B — MMP 4, 5

7. | | □ B — Nec 3–6

8. | □ A ⊃ □ B — CP 2–7

9. (A → B) ⊃ (□ A ⊃ □B)   CP 1–8

10. □ [(A → B) ⊃ (□ A ⊃ □B)] — Taut Nec 9

11. (A → B) → (□ A ⊃ □B)   Arrow Ex 10

(20) 1. | ~ □ A — AP

2. | ◊ ~A — DE 1

3. ~ □ A ⊃ ◊ ~A   CP 1–2

4. □ [~ □ A ⊃ ◊ ~A]   Taut Nec 3

5. ~ □ A → ◊ ~A   Arrow Ex 4

(25) 1. | ~ ◊ A   AP

2. | □ ~A   DE 1

3. ~ ◊ A ⊃ □ ~A   CP 1–2

4. □ (~ ◊ A ⊃ □ ~A)   Taut Nec 3

5. (~ ◊ A → □ ~A)   Arrow Ex 4

(28) 1. | A → B   AP

2. | A → B   Reit

3. | □(A → B)   Nec 2

4. (A → B) ⊃ □(A → B)   CP 1–3

5. □ [(A → B) ⊃ □(A → B)] Taut Nec 4

6. (A → B) → □(A → B)   Arrow Ex 5

(30) 1. | A → B   AP

2. | □ (A ⊃ B)   Arrow Ex 1

3. | □ (~B ⊃ ~A)   Trans 2

4. | ~B → ~A   Arrow Ex 3

5. (A → B) ⊃ (~B → ~A)   CP 1–4

6. □[(A → B) ⊃ (~B → ~A)] Taut Nec 5

7. (A → B) → (~B → ~A)   Arrow Ex 6

(35) 1. | □ (A ⊃ B)   AP

2. | A → B   Arrow Ex 1

3. □(A ⊃ B) ⊃ (A → B)   CP 1–2

4. □[□ (A ⊃ B) ⊃ (A → B)] Taut Nec 3

5. □(A ⊃ B) → (A → B)   Arrow Ex 4

▶   **EXERCISE 24.7**   ◀

(1) 1. ◊ □A / □A

2. □ A   Red 1

(5) 1. □ ◊ □ P

2. ◊ ~ P v □ Q

3. Q ⊃ S / S

4. □ P   Red 1

5. $\sim \Diamond \sim P$ DE 4

6. $\Box$ Q   DS 2, 5

7. Q   BR 6

8. S   MP 3, 7

CHAPTER 25

▶ **EXERCISE 25.1** ◀

2. **a.** Stronger

   c. Stronger

   f. Weaker

   i. Weaker

   l. Stronger

   o. Stronger

   r. Unaffected

5. **a.** Decreases

   c. Decreases

   e. Unaffected

7. **a.** Decreases

   c. Unaffected

   f. Increases

   i. Unaffected

▶ **EXERCISE 25.2** ◀

1. **a.** Strengthens

   c. Strengthens

   e. Strengthens

5. **a.** Weakens

   c. Strengthens

   e. Weakens

   h. Weakens

8. a. Strengthens

    c. Strengthens

    f. Weakens

    i. Weakens

CHAPTER 26

▶	EXERCISE 26.2	◀

1. Difference

5. Method of Concomitant Variation

7. Method of Agreement

10. Method of Residues

15. Method of Agreement

20. Method of Agreement and Difference

▶	EXERCISE. 26.3	◀

3. Method of Concomitant Variations: Search for a possible cause that also rose by a similar amount during the 1980s and fell by a similar amount during the 1990s.

7. Method of Difference. What was the difference between the two lousy pies and the one good pie? More specifically, what factor was present in the case of the good pie but absent in the cases of the lousy pies?

9. Method of Agreement. Search for a common situation. Where did they sleep? Where did they store their clothes? Where did they pitch their tents?

Index

Absolute materialism, 593–594
Abstract entities, 593–594
Abstraction, 112
Abusive ad hominem argument, 243, 244
Ad verecundiam fallacy, 256
Addition rule, 143–144, 158
Adverbs, 290–291
Affirmative sentences
  categorical sentences as, 349–350
  existential, 353–354
  explanation of, 280, 300
  particular, 280, 301
  universal, 280, 301, 350–352
Against the Mathematicians (Sextus), 474
Agreement
  Joint Method of Difference and, 560–561, 566
  Mill's Method of, 555–557, 566
Algorithms, 99
Ambiguity
  explanation of, 239, 263, 270
  removal of, 224
Ambiguous sentences
  explanation of, 48, 65
  interpretation of, 48–49
Ampersand
  explanation of, 46
  truth-trees and, 570–572, 578, 579
  use of, 49, 50
Amphiboly
  explanation of, 265, 270
  fallacy of, 265–266, 273
Analogical argument, 512, 534
Analogical reasoning
  explanation of, 512–513
  relevance and, 515–518
  strong arguments in, 513–514
Analogy
  categorical syllogisms and, 332–333
  explanation of, 512, 534
  fallacy of weak, 259, 273
  induction by, 525
  as model, 518
  refutation by logical, 327–328
  to show invalidity, 409–412
Analytic method, 92, 99
Analytic truths, 447
Analytical definitions, 234, 241
Analytically true statements, 447, 476
and with *or* and *not*, 52–54
Antecedent, 40, 111
any and *every*, 367–368
Appeal to force fallacy, 245–246, 271
Appeal to ignorance fallacy, 250, 272
Appeal to pity fallacy, 246, 271
Appeal to questionable authority fallacy, 256–257, 272
Appeal to the gallery fallacy, 247

Appeal to the people fallacy, 246–247, 271
Argument against the person fallacy, 243, 271
Argument forms
  explanation of, 105, 116
  invalid, 108, 110–112, 116
  valid, 108–110, 116
Arguments (*See also* specific types of arguments)
  analogical, 512–518 (*See also* Analogical reasoning)
  counterexamples to, 14–16
  explanation of, 1, 25
  inference to best explanation, 531–532
  modal, 444
  partial truth-table to show invalid, 92–94
  parts of, 1
  against the person, 243, 271
  quantificational, 338–362 (*See also* Quantificational argument)
  recognizing, 2–3
  testing validity of, 85–91
  valid vs. invalid, 13–19, 584–585
Aristotelian logic
  background of, 277, 278, 473
  categorical statements and, 297–298
  categorical syllogisms and, 310–315, 318–319
  equivalence rules for, 294–296
Aristotle, 234–235, 242, 277–278, 473, 549
Arrow Exchange rule, 491
Artificial languages, 31
Associative rule, 185–186
Assumption of existential import
  categorical sentences and, 297–299
  explanation of, 300
Atomic sentence of TL, 67, 76
Auxiliary assumptions, 566

Bacon, Francis, 44
Basic form, 3, 25
Becher, Johann Joachim, 544
Begging the question fallacy, 248–250, 271
Biased sample, 527, 534
Biconditional sentences
  explanation of, 42–44, 76
  translating, 59
Bonevac, Daniel, 518
Boole, George, 298
Boolean categorical, 300
Boolean interpretation
  categorical statements and, 299, 300
  categorical syllogisms and, 315–319
  explanation of, 298, 300
Borderline case, 225, 239
both, 55
Box Removal rule, 480–481

Brahe, Tycho, 541

Caloric theory, 551, 552
Carroll, Lewis, 224
Categorical sentences
  assumption of existential import and, 297–300
  existential affirmative, 353–354
  existential negative, 354
  explanation of, 278, 300–301, 349, 364
  forms of, 278–279, 301
  modern square of opposition and, 299–300
  producing contrapositive of, 295–296
  producing converse of, 294
  producing obverse of, 294–295
  quality and quantity of, 280–281
  traditional square of opposition and, 281–284
  translating sentences into standard form of, 287–292
  universal affirmative, 350–352
  universal negative, 352
Categorical syllogisms
  diagraming Aristotelian, 310–315
  diagraming from Boolean standpoint and, 315–319
  enthememes and, 325–326
  explanation of, 303, 332
  figure of, 303, 332
  logical analogy and, 327–328
  logical form of, 303–305, 332
  mood of, 304, 332
  rules for evaluating, 329–332
  sorites and, 320–322, 333
  testing sorites with Venn diagrams and, 322–324
  use of Venn diagrams to test validity of, 308–310
  Venn diagrams and, 306–308, 333
Cause and effect
  concept of, 554–555
  Mill's methods and, 553–554
Chrysippus, 475
Circumstantial ad hominem argument, 243, 245
Classes
  empty, 297, 301
  explanation of, 234, 239, 278, 301
Clinton, Bill, 376
Coherence theory of truth, 597–598
Commas, 49, 50
Commutative rule, 184–185
Complement of class C, 294–295, 301
Complete cause, 555
Component sentences
  within disjunction, 37
  explanation of, 33, 46

Compound sentences
  categorizing truth-functional, 73
  explanation of, 32, 33, 46
  truth-functional, 39, 46
Conclusions
  explanation of, 1, 25
  indicators of, 3, 4, 25
  natural deduction and, 125–128
  true, 16
Conditional operators, 40, 42
Conditional proofs
  explanation of, 182
  quantifier, 418–424
  with replacement rules, 214–215, 420–421
  rule for, 168–173, 214, 418
Conditional sentences (*See also* Biconditional sentences)
  categorical statements and, 291
  explanation of, 40, 76
  material, 42
  statements about necessary and sufficient conditions treated as, 61
  translating, 57–59, 291
  truth-function for horseshoe and, 40–42
Confirmation process (*See also* Scientific hypothesis)
  example of, 538–539
  explanation of, 537–538, 567
  Newtonian hypothesis and, 541–544
Conjunction
  explanation of, 33, 76
  truth-table for, 34
Conjunction operator
  explanation of, 33
  use of, 52
Conjunction rule, 141–143
Conjuncts
  explanation of, 33
  truth-values of, 33–35
Connotation
  conventional, 232, 240
  explanation of, 231, 240
  objective, 231–232, 240
  subjective, 231, 240
Connotative definitions, 233
Connotative meaning, 230–232
Consequent
  of conditional, 40
  fallacy of affirming, 111
Consistency
  explanation of, 20–21, 25, 476
  modal relations and, 464–465
Constants
  explanation of, 35, 46, 364
  individual, 340–341
  predicate, 340
  use of sentence, 47–48
Constructive dilemma rule, 144–146
Contingency operator, 449–451
Contingency test, 84, 581–583
Contingent detector, 83–84
Contingent propositions
  examples of, 446
  explanation of, 444, 476
Contingent sentences, 83, 99
Contingent statements, 23, 25
Contradiction detector, 82–83
Contradictions
  explanation of, 83, 100
  explicit, 181, 182
  law of noncontradictions and, 181–182
  logical, 443

as logical falsehoods, 84
reductio ad absurdum proofs and, 160
testing for, 83, 580–581
Contradictories
  explanation of, 301
  statements as, 282–283
Contradictory sentence form, 115, 116
Contrapositive, 295–296, 301
Contraries, 283, 301
Controlled experiments, 559, 566
Conventional connotation, 232, 240
Converse, 301
Copulas
  categorical sentences and, 289–290
  explanation of, 279, 301
Correspondence theory, 597
Counterexample
  to an argument, 14–16, 25, 94
  explanation of, 100
  to QL argument, 411, 417
  refutation by, 327

Davy, Sir Humphry, 552
Decision procedure, 99, 100
Declarative sentences
  explanation of, 5, 26
  propositions or statements expressed by, 6
Decomposition strategy for proofs, 148–149
Deductive arguments (*See also* Arguments)
  evaluation of, 11–13
  examples of, 7–8
  explanation of, 6–8, 26, 511, 534
  indicator words for, 8–9
  valid vs. invalid, 11–19, 26 (*See also* Invalid deductive arguments; Valid deductive arguments)
Deductive logic, 24, 26
Deductively sound arguments, 17–18, 26
Definiendum, 224, 240
Definiens, 224, 240
Definitions
  analytical, 234, 241
  enumerative, 233, 240
  explanation of, 223
  extensional or denotative, 233, 240
  forms of, 224
  function of, 224–226
  by genus and difference, 234–235, 240
  intensional or connotative, 233, 237–238, 240
  lexical, 226, 241
  operational, 234, 241
  ostensive or demonstrative, 233, 240
  persuasive, 227, 241
  precising, 227, 241
  stipulative, 226–227, 241
  by subclass, 233, 240
  synonymous, 233, 240
  techniques for constructing, 233–235
  theoretical, 227–228, 241
  types of meaning and, 230–232
DeMorgan, Augustus, 298
DeMorgan's rule
  CP proof using, 420

explanation of, 187–190
  indirect proof using, 420
Denotation, 231, 240
Denotative definitions, 233
Denotative meaning, 230–231
Diamond Exchange rule, 461–462, 493–494
Diodorus Cronus, 474, 475
Disconfirmation process (*See also* Scientific hypothesis)
  example of, 538–539
  explanation of, 538–540, 567
  phlogiston theory and, 544–547
Disjunction
  exclusive, 37–39, 46, 56
  explanation of, 37–39, 76
  inclusive, 37, 38, 46
Disjunction operators, 37
Disjunctive Syllogism form
  explanation of, 105–106, 119
  natural deduction and, 119–121
  truth-trees and, 586
  as valid argument form, 108–109
Disjuncts, 37–38
Distributed terms, 329, 332
Distribution rule, 190–191
Domain of variable, 372–373, 385
Domino arguments, 258–259, 273
Double Arrow Exchange rule, 492
Double negation rule, 186–187
Dyadic connective, 33, 46
Dyadic predicate expressions
  examples of, 365–366
  explanation of, 365, 385
  quantifiers combined with, 366–367

Eddington, Sir Arthur, 443
Einstein, Albert, 536–537
either, 55
Embedded sentences, 33, 46
Empty class, 297, 301
Enthememes
  categorical syllogisms and, 325–326
  explanation of, 19, 26, 325, 332
Entities
  abstract, 593–594
  material, 593–594
  theoretical, 594–596
Enumerative definitions, 233, 240
Enumerative induction
  explanation of, 522–523, 534
  types of, 523–525
Equivalence
  explanation of, 26, 100, 476
  logical, 21–22
  of propositions, 469
  testing pair of sentences for, 97–98
Equivalence rule, 200–201
Equivalence test, 98
Equivocation
  explanation of, 263, 270
  fallacy of, 263–265, 273
Euclides, 473
every and *any,* 367–368
Exclamatory sentences, 5, 26
Exclusive disjunction
  explanation of, 37–39, 46
  symbolizing, 56
Exclusive sentences, 291, 361
Existence, 360

Existential affirmative categorical sentences, 353–354
Existential fallacy, 331–332
Existential general sentences, 346–347, 364
Existential generalization, 389–390, 405
Existential import
assumption of, 297–300
of categorical sentences, 282
explanation of, 301
Existential instantiation
applications of, 391–393
explanation of, 390–391
Existential negative categorical sentences, 354
Existential quantification
explanation of, 347, 390, 405
instantiation or substantiation instance of, 390–391, 405
Existential quantifiers
adjacent quantifier order and, 374
next to universal quantifiers, 370, 371
Existential viewpoint, 282, 301
Explicit contradiction, 181, 182
Exportation rule, 198–199
Extension (*See* Denotation)
Extensional definitions, 233, 240
Extensional meaning of term, 231, 240

Fact of the Cross, 547–548
Fallacies (*See also* specific fallacies)
explanation of, 111, 116, 242, 270
formal, 111, 242, 270
informal, 242–273 (*See also* Informal fallacies)
of language, 263–268, 271, 273
of little evidence, 255–260, 271–273
of no evidence, 243–252, 270–272
Fallacy of accident, 255, 272
Fallacy of accidental correlation, 255, 273
Fallacy of affirming the consequent, 111
Fallacy of amphiboly, 265–266, 273
Fallacy of composition, 266–267, 273
Fallacy of denying the antecedent, 111
Fallacy of division, 268, 273
Fallacy of drawing affirmative conclusion from negative premises, 331
Fallacy of drawing negative conclusion from affirmative premises, 331
Fallacy of equivocation, 263–265, 273
Fallacy of exclusive premises, 331
Fallacy of false dilemma, 259–260, 273
Fallacy of hasty generalization, 257, 272–273
Fallacy of illicit major, 330
Fallacy of illicit minor, 330
Fallacy of irrelevant conclusion, 247–248, 271
Fallacy of special pleading, 260, 273
Fallacy of suppressed evidence, 273

Fallacy of undistributed middle, 330
Fallacy of weak analogy, 259, 273
False cause fallacy, 257–258, 272
Falsehoods, logical, 84, 100, 414
Figure, of categorical syllogisms, 303, 332
Final column, 78, 100
Formal fallacies (*See also* Fallacies)
examples of, 111
explanation of, 242, 270
Formal logic, 112
Formally invalid, 327, 332
Formally valid, 332
Full universe theory, 549, 550
Function, 39, 46

General sentences
categorical, 349–354, 356–357 (*See also* Categorical sentences)
existential, 346–347, 364
explanation of, 364
quantifier switch trick and, 354–356
symbolizing complicated, 358–359
universal, 342–346, 364
General terms, 338, 364
Genetic fallacy, 252, 272
Genus
definition by, 234–235
explanation of, 234, 240
Giere, Ronald, 545
Guilt by association fallacy, 244, 271

Halley, Edmond, 541–543
Halley's comet, 541–544
Health, 227
Heidegger, Martin, 238
Herschel, William, 552
Histories (Herodotus), 265–266, 273
Horseshoe
explanation of, 40, 46
truth-trees and, 574–575
truth-function for, 40–42
use of, 62, 63
Hypothesis, 535, 566 (*See also* Scientific hypothesis)
Hypothetical induction, 535
Hypothetical Syllogism form
errors in applying, 158
explanation of, 108, 124
modal, 483
natural deduction and, 124–125
truth-trees and, 588
as valid argument form, 109
Hypothetical viewpoint, 282, 301
Hypothetico-deductive method
explanation of, 511, 536, 566
steps in, 536–538

Ideals of intellectual interaction, 23–24
Identity
explanation of, 439
logic of, 376–377
rules for, 433–436
Identity signs
symbolizing with, 378–380
working with, 433
Imperative sentences, 5, 26
Implication
explanation of, 21, 26, 476
modal logic and, 465–466
paradoxes of, 466–468
Implication rule, 197–198

Inclusive disjunction, 37, 38, 39, 46
Inconsistency
explanation of, 20, 21, 26, 476
modal relations and, 464–465
Indeterminant sentence form, 116, 117
Indirect proof sequence, 162
Indirect proofs (*See also* Reductio ad absurdum proofs)
example of, 421
quantifier, 419–424
reductio ad absurdum proofs as, 160
with replacement rules, 211, 420–421
rule of, 162–166, 210, 419
using DeMorgan's rule, 420
Individual constants, 340–341
Individual variables, 340
Induction
by analogy, 525
enumerative, 522–525, 534
hypothetical, 535
Inductive arguments (*See also* Arguments)
evaluation of, 9–10
explanation of, 7, 8, 26, 511, 534
indicator words for, 8–9
strong, 9–10, 26
weak, 10, 27
Inductive generalizations
explanation of, 523, 534
statistical, 525–527
Inductive logic, 24, 27
Inductive reasoning
analogical reasoning as, 512–518
enumerative induction as, 522–527
inference to the best explanation as, 531–533
overview of, 511
Inductively sound arguments, 10, 27
Inference
to best explanation, 531–532, 534
explanation of, 139
valid, 118–119, 139
Inference rules (*See also* specific rules)
application of, 183
box removal, 480–481
explanation of, 119, 139
formulation of, 140
hypothetical syllogism, 483
Modus Ponens, 483
Modus Tollens, 483
possibilization, 481–482
Stoic, 475
universal, 386–389
valid, 119, 139
Informal fallacies (*See also* Fallacies; specific fallacies)
explanation of, 242–243, 270
of language, 263–268, 271, 273
of little evidence, 255–260, 271–273
of no evidence, 243–252, 270–272
Informal logic, 112
Instantiation
of existential quantification, 390–391, 405
of universal quantification, 386–388, 405

Intensional definitions, 233, 237–238, 240
Intensional meaning, 231, 241
Intention (*See* Connotation)
Intentional definitions, 237–238
Interpretation of QL sentence, 406–408, 417
Interrogatory sentences, 5, 27
Invalid argument forms, 108, 110–112, 116
Invalid deductive arguments
  examples of, 12–13
  explanation of, 11–12, 26
  valid vs., 13–19
Invalidity, 409–412

James, William, 598
Jason, Gary, 242
John Doe name
  existential instantiation and, 390–393
  explanation of, 387, 389, 405
  universal generalization and, 393–395
Joint Method of Agreement and Difference, 560–561, 566
Justification, 139

Kekule, Friedrich, 537
Kepler, Johannes, 541
Kepler's Laws, 541
Kinetic theory of heat, 551–552
Krauss, Lawrence, 455

Lavoisier, Antoine, 545–547
Law of noncontradiction
  explanation of, 181–182
  possible worlds and, 453–455
Leibniz, Gottfried Wilhelm, 91
Lewis, C. I., 480
Lexical definitions, 226, 241
Lexical meaning, 224, 241
Logic
  Aristotelian, 277, 278, 294–295, 297–298
  Bacon's, 44
  deductive, 24
  formal, 112
  function of, 1, 18
  goals of, 2
  of identity, 376–377
  inductive, 24
  informal, 112
Logical analogy
  categorical syllogisms and, 327–328
  refutation by, 332–333, 417
Logical consistency, 20
Logical contradiction, 443, 476
Logical equivalence, 294, 302
Logical falsehood
  contradictions as, 84
  explanation of, 100
  QL sentence as, 414
Logical form, 303–305, 332
Logical inconsistency, 20
Logical status, 84
Logical theory, 277–278, 473–475
Logical truths
  explanation of, 100, 219
  methods of proving, 424–425
  QL sentence as, 414
  tautologies as, 84, 178, 217–218
Logically possible, 451–453, 476

Main operator, 69, 76
Major premise
  in categorical syllogisms, 303, 330

explanation of, 332
Major term, 303, 332
Material biconditional operators, 43
Material conditional operators, 42
Material conditional sentence, 42
Material entities, 593–594
Materialism
  absolute, 593–594
  explanation of, 342–343, 364
Maxwell, James Clerk, 226
Meaning
  connotative, 230–232
  denotative, 230–231
Megarian school of thought, 473, 474
Metalanguage, 68–69, 76
Method of Agreement, 555–557, 566
Method of Concomitant Variation, 561–562, 566–567
Method of Difference, 558–560, 566
Method of Residues, 561, 566–567
Middle term
  in categorical syllogisms, 303, 330
  explanation of, 332
Mill, John Stuart, 59, 554
Mill's Methods
  of agreement, 555–557, 566
  of concomitant variation, 561–562, 566–567
  of difference, 558–560, 566
  explanation of, 554, 566
  of residues, 561, 566–567
Minor premise
  in categorical syllogisms, 303, 330
  explanation of, 332
Minor term
  in categorical syllogisms, 303
  explanation of, 332
ML (modal language), 457–458
Modal arguments
  evaluation of, 478
  explanation of, 444, 476, 478, 508
Modal Equivalence rule, 492–493
Modal fallacy, 505–508
Modal logic
  explanation of, 444, 476
  inference rules and, 480–484
  limits of the possible and, 453–455
  logical and physical possibility distinguished and, 451–453
  modal fallacy and, 505–508
  modal principles and, 478–480
  modal properties and, 444–445
  modal replacement rules and, 491–494
  necessary falsehoods and, 448
  necessary truths and, 446–447
  necessitation rule and, 487–489
  possibility to necessity rule and, 484–485
  possible truths, possible falsehoods, and contingencies and, 445–446
  proving theorems of S5 and, 499–500
  putting statements into symbols and, 448–451
  S5 reduction and, 502–504
  scopes of, 462–463

tautology necessitation rule and, 500
transformations and, 460–462
translating sentences into modal symbols and, 455–457
validity in S5 and, 494–496
Modal Modus Ponens rule, 483
Modal Modus Tollens rule, 483–484
Modal operators
  process of linking, 459–460
  scope of dyadic, 469–470
  symbolizing with, 470
  truth-functional logic and, 472–473
Modal relations
  consistency and inconsistency in, 464–465
  equivalence in, 469
  implication in, 465–466
  pursuit of truth and, 466–468
  softening paradoxes in, 468
Modally closed
  explanation of, 484, 508
  necessitation rule and, 487–488
Model, 518, 534
Modern square of opposition, 299–300
Modus Hypothetical Syllogism role, 483
Modus Ponens form
  application of, 183
  errors in applying, 158, 159
  explanation of, 106–107, 121
  modal, 483
  natural deduction and, 121–122
  truth-trees and, 586–587
  as valid argument form, 108–109
Modus Tollens form
  errors in applying, 158
  explanation of, 107–108, 123
  modal, 483
  natural deduction and, 122–124
  truth-trees and, 587
  as valid argument form, 109
Molecular sentence of TL, 67, 76
Monadic connective, 35, 46
Monadic predicate phrases
  explanation of, 365, 385
  validity test for, 414–417
Mood of categorical syllogism, 304, 332
Multiply quantified sentences, 408–409

Natural deduction
  Disjunctive Syllogism rule and, 119–121
  explanation of, 118–119, 128, 139
  Hypothetical Syllogism rule and, 124–125
  Modus Ponens rule and, 121–122
  Modus Tollens rule and, 122–124
  proofs and, 128–130, 468
  proving conclusion validly follows and, 125–128
  S5, 480
  Stoics and, 475
  system TD and, 128
Natural languages, 31–32
Necessarily false propositions, 445, 476
Necessarily false statements, 23, 27
Necessarily true propositions, 445, 477

Necessarily true statements, 21–23, 27
Necessary conditions
  explanation of, 60, 65, 554, 567
  symbolizing, 59–63
Necessary falsehoods, 448
Necessary operator, 448–449
Necessary truths, 446–447
Necessitation rule, 487–489
Negation, 35–36, 73, 76
Negation operators
  conjunctions and, 52
  in English, 51–52
  explanation of, 35
Negative sentences
  categorical sentences as, 350
  existential, 354
  explanation of, 302
  particular, 280, 301
  universal, 280, 301, 307–308, 352
Neptune, 552
Nested proofs, 175–178, 182
Newton, Sir Isaac, 541, 547
Newtonian hypothesis, 541–544
Non causa pro causa fallacy, 258
Noncontradiction law, 181–182
not with *and* and *or*, 52–54

Object language, 69
Objective connotation, 231–232, 240
Observational entities, 594
Observational prediction, 567
Obverse, 294–295, 302
Occam, William of, 549
On Sophistical Refutations (Aristotle), 242
One-place predicate phrase (*See* Monadic predicate phrase)
only, 360–362
only if, 62
Open sentences, 340, 364
Operational definitions, 234, 241
Opposition
  explanation of, 281, 302
  modern square of, 299–300
or
  inclusive and exclusive, 37, 38
  with *and* and *not*, 52–54
Order of increasing extension, 232, 241
Order of increasing intension, 232, 241
Ostensive definition, 233, 240
Outlines of Phyrronism (Sextus), 474
Overlapping quantifiers
  explanation of, 385
  proofs with, 427–431
  sentences with, 369–372

Parentheses, 49
Partial cause, 555
Partial truth-table, 92–94, 416
Particular sentences
  affirmative, 280, 301
  explanation of, 302
  negative, 280, 301
Pascal, Blaise, 550–551
Pasteur, Louis, 559
Persuasive definitions, 227, 241
Philo, 474
Philogiston theory, 544–547
Physically impossible, 451, 477
Physically possible, 451–453, 477
The Physics of Star Trek (Krauss), 455

Planck, Max, 536
Plato, 238, 473
Poisoning the well, 252, 272
Possibility operator, 449
Possibility to necessity rule (P2N), 484–485
Possibilization Rule, 481–482
Possible world
  explanation of, 449–451, 477
  law of noncontradiction and, 453–455
  logical possibility and physical possibility and, 451–453
Possibly false propositions
  examples of, 446
  explanation of, 444, 477
Possibly true propositions
  examples of, 445
  explanation of, 444, 477
Post hoc ergo propter hoc fallacy, 257–258
Pragmatic theory of truth, 598
Precising definitions, 227, 241
Predicate constants, 340
Predicate phrases
  dyadic, 365–367, 385
  explanation of, 365
  monadic, 365, 385
  singular sentences and, 338–341
Predicates, 288
Premises
  explanation of, 1, 27
  indicators of, 3–4, 27
  major, 303, 332
  minor, 303, 332
  suppressed, 325, 333
  true, 10, 16
Primary analogates, 516, 534
Principle of charity, 325, 332
Principle of economy, 532–533, 549, 567
Principle of parsimony, 549
Principle of self-identity, 433, 439
Principle of the indiscernibility of identicals, 435, 439
Principle of universal causation, 369–370
Proofs
  additional strategies for, 151
  common deduction errors for, 157–159
  conditional, 168–173, 182, 214–215
  decomposition strategy for, 148–149
  indirect, 160, 162–166, 210–211
  nested, 175–178, 182
  with overlapping quantifiers, 427–431
  reductio ad absurdum, 160–161, 182
  in S5D, 494–496, 508
  steps in learning to construct, 146–147
  in TD, 128–130, 139
  trial and error strategies for, 147–148
  working backward from conclusion to construct, 149–151
Propositions (*See also* specific propositions)
  explanation of, 6
  modal properties of, 444–445
  truth and, 594, 596
Proximate cause, 555
Pythagoras, 443

Pythagorean theorem
  explanation of, 393, 443
  as false, 444

QL
  abbreviatory convention for, 349, 382
  counterexample to argument in, 411, 417
  definitions related to, 414
  explanation of, 348, 364
  formation rules of, 348–349
  interpretation of sentence in, 406–408, 417
  singular sentence of, 389, 405
  vocabulary of, 348
Quantificational arguments
  categorical sentences and, 349–354
  denying existence and, 360
  explanation of, 338, 364
  general sentences and, 342–347
  only and, 360–362
  QL language and, 348–349
  quantifier switch trick and, 354–356
  singular sentences and, 338–341
  switching quantifiers on categoricals and, 356–357
  symbolizing complicated general sentences and, 358–359
Quantifier exchange rule, 429
Quantifiers
  categorical sentences and, 288–289
  dyadic predicate combined with, 366–368
  explanation of, 279, 302
  overlapping, 369–372, 385, 427–431 (*See also* Overlapping quantifiers)
  reduced set of rules for, 423–424

Reasoning (*See* Analogical reasoning; Inductive reasoning; Scientific reasoning)
Red herring fallacy, 250–251, 272
Reductio ad absurdum proofs, 160–161, 182, 569–570 (*See also* Indirect proofs)
Reduction rule, 503–504
Reflexive relation
  explanation of, 384, 385, 439
  identity as, 436–437
Reflexive sentences, 368
Refutation
  by counterexample, 327
  by logical analogy, 327–328, 332–333, 417
Reiteration rule, 488–489
Relations
  properties of, 430–431
  reflexive, 384, 385
  symmetrical, 383–385
  transitive, 382–383, 385
Relevance, 515–518
Remote cause, 555
Replacement rules
  associative, 185–186
  commutative, 184–185
  conditional proofs with, 214–215
  DeMorgan's, 187–190
  distribution, 190–191
  double negation, 186–187
  equivalence, 200–201
  explanation of, 183, 209

Replacement rules (*continued*)
exportation, 198–199
function of, 201
implication, 197–198
indirect proofs with, 210–211
modal, 491–494
tautology, 199–200, 216–218
transposition, 197
Rumford, Count, 226, 551–552

S5
natural deduction and, 480
proofs in, 494–496, 508
reduction and, 502–504
theorem of, 499–500, 508
validity in, 494–496, 508
Scientific hypothesis
case studies in, 549–552
confirmation and disconfirmation of, 538–540
confirmation of, 541–544
criteria to analyze, 548–549
disconfirmation of, 544–547
fact of cross and, 547–548
Scientific reasoning
cause and effect and, 553–555
Joint Method of Agreement and
Difference and, 560–561, 566
Mill's Method of Concomitant
Variation and, 561–562,
566–567
Mill's Method of Residues and,
561, 566–567
Mill's methods and, 554–560
overview of, 535–536
steps in, 536–538
Scope
of dyadic modal operators,
469–470
of modal operators, 462–463
of operators, 69, 76
Sea of air hypothesis, 549–551
Secondary analogates, 516, 534
Semantics
explanation of, 66, 76
interpretations and, 414
Semmelweiss, Ignaz, 340, 538,
539
Sentence connectives (*See* Sentence operators)
Sentence forms
contradictory, 115, 116
explanation of, 101–104, 116
indeterminant, 116, 117
substitution instance of, 102,
117
tautological, 114–115, 117,
178–180
Sentence operators
explanation of, 32, 33, 46
sentences with more than one,
48–51
Sentences
ambiguous, 48–49
biconditional, 42–44, 76
categorical, 277–300 (*See also*
Categorical sentences)
compound, 32, 33, 46
conditional, 40–42, 57–59, 61,
76, 291 (*See also* Conditional
sentences)
exclusive, 291, 361
function of, 5
general, 338, 342–347, 364 (*See
also* General sentences)
with more than one operator,
48–51
open, 340, 364
reflexive, 368

simple, 32, 33, 46
singular, 290, 338–341, 364
statements or propositions vs., 6
that deny existence, 360
truth-values of, 32
types of, 5–6
Sextus Empiricus, 474
Shakespeare, William, 473
Simple sentences, 32, 33, 46
Simplification rule
errors in applying, 157–159
explanation of, 140–141
Singular sentences
categorical statements and, 290
explanation of, 302, 338, 364
forms of, 338–339
of QL, 389, 405
quantificational argument and,
339–341
universal generalization of,
393–396, 405
Singular terms, 338, 364
Slippery slope fallacy, 258–259,
273
Snob appeal fallacy, 247, 271
Sober, Elliot, 446
Socrates, 473
Some, 279, 302
Sorites
example of, 320–322
explanation of, 320, 333
Venn diagrams to test, 322–324
Species, 234, 241
Square of opposition, 281, 302
Standard form
of categorical sentences, 288
of categorical syllogisms, 303,
333
explanation of, 302
Star Trek series, 455
Statements
contingent, 23, 25
declarative, 6
exceptive, 292
explanation of, 6
necessarily false, 23
necessarily true, 21–22
Statistical inductive generaliza-
tions, 525–527
Stipulative definitions, 226–227,
241
Stoics, 475
Straw man fallacy, 256, 272
Strong inductive arguments,
9–10, 24
Subalternation, 283–284
Subclass, 233, 240
Subcontraries, 283, 302
Subformula, 183, 209
Subimplication, 283–284
Subject, categorical sentences
and, 288
Subjective connotation, 231, 240
Substitution instance
of argument form, 105
explanation of, 102, 117
of universal quantification, 386
Sufficient conditions
cause and, 555
explanation of, 60, 65, 554, 567
symbolizing, 59–63
Suppressed premises, 325, 333
Syllogisms (*See* Categorical syllogisms)
Symbolizing
conditionals and biconditionals
and, 57–59
with dyadic operators, 470
exclusive disjunctions, 56

general hints related to, 55
with identity sign, 378–380
necessary and sufficient conditions, 59–63
and to *or* and, 52–54
overview of, 47–48
sentences about people,
374–375
sentences containing more than
one operator, 48–51
tildes and, 51–52, 54
Symmetrical relation
explanation of, 383–385, 439
identity as, 436
Synonymous definitions, 233,
240
Syntax, 66, 76
System of Logic (Mill), 59, 554
System S4, 480
System S5D, 480, 494 (*See also*
S5)

Tautological sentence form
explanation of, 114–115, 117
proof for, 178–180, 216–218
Tautologies
explanation of, 81, 100, 219
as logical truths, 84, 178,
217–218
proving, 216–218
using paper and pencil to make,
81–82
Tautology Necessitation rule,
500
Tautology rule, 199–200
Tautology test, 82, 577–580
TD (truth-functional deduction)
explanation of, 128, 139, 219
proof in, 128–130, 139
theorem of, 180, 182, 218,
219
valid in, 128, 139
Term complement, 295, 302
Terms
distributed, 329, 332
explanation of, 241
extensional meaning of, 231,
240
general, 338, 364
major, 303, 332
middle, 303, 330, 332
minor, 303, 332
singular, 338, 364
Theorem of S5D, 499–500, 508
Theorems
explanation of, 182
of TD, 180, 182, 218, 219
Theoretical definitions, 227–228,
241
Theoretical entities, 594–596
Tilde
explanation of, 36, 46
mathematical negative sign and,
54
truth-trees and, 576–578
use of, 51–52
TL (truth-functional language)
calculating truth-value of whole
from values of parts and, 69–73
explanation of, 66, 76, 337
formula as contingent and, 84
formula as contradiction and, 83
formula as tautology and, 82
grammar for, 66–68
vocabulary for, 66
to be, 376–377
Torricelli, Evangelista, 550, 551
Traditional square of opposition,
281–284

Transitive relation
 example of, 430
 explanation of, 382–383, 385, 439
 identity as, 436, 437
Transposition rule, 197
Trial and error strategies for proofs, 147–148
Triple bar
 explanation of, 43, 46
 truth-table for, 43–44
 truth-trees and, 575–576
Truth
 alternative theories of, 597–598
 analytic, 447
 bearers of, 592–596
 correspondence theory of, 597
 meaning of, 596
 necessary, 446–447
 relation between validity and, 15–16
Truth-tables
 alternatives to, 118, 569 (*See also* Natural deduction; Truth-trees)
 benefits and drawbacks of, 118, 569
 constructed to formula, 78–79
 contingency detector and, 83–84
 decision procedures and, 99
 eight-row tables and, 79–81
 example of, 34–35
 explanation of, 34, 46
 function of, 77
 for horseshoe operators, 42
 for inclusive disjunction, 38
 making a tautology detector and, 81–82
 making contradiction detector and, 82–83
 for negation, 36
 number of rows on, 81
 partial, 92–94, 416
 testing argument for validity and, 85–91
 testing pair of sentences for equivalence and, 97–98
 for triple bar, 43–44
Truth-trees
 ampersand and, 570–572, 578, 579
 closing a branch of, 581
 contingency testing and, 581–583
 contradiction testing and, 580–581
 explanation of, 569
 horseshoe and, 574–576
 patterns of reasoning and, 585–588
 proper splitting and, 583–584
 proving argument valid or invalid and, 584–585

strategy and, 590
 tautology testing and, 577–580
 tilde and, 576–578
 wedge and, 572–573, 577–579
Truth-function
 explanation of, 39, 46
 for horseshoe, 40–42
Truth-functional compound sentences, 39, 46
Truth-functional deduction (TD) (*See* TD, truth-functional deduction)
Truth-functional language (*See* TL)
Truth-functional logic
 as analytic, 92
 biconditional sentences and, 42–44, 76
 conditional sentences and, 40–42
 disjunction and, 37–39
 as formal logic, 112–113
 modal operators and, 472–473
 negation and, 35–36
 simple and compound sentences, sentence operators, and conjunctions and, 32–35
 truth-functions and truth-functional compound sentences and, 39
Truth-functional sentence operators, 39, 46
Truth-functional validity, 90–91
Truth-value
 of compounds, 33–34
 for negation, 35–36
 of sentences, 32
 of whole vs. parts, 69–73
Truth-value assignments, 78–80, 100
Tu quoque fallacy, 244, 271
Two-place predicate expression (*See* Dyadic predicate expression)

Universal causation principle, 369–370
Universal domain, 372, 385
Universal generalization
 examples of, 394–396
 of singular sentence, 393–396, 405
Universal installation rule, 386–389
Universal quantification
 explanation of, 405
 instantiation of, 386–388, 405
Universal quantifiers
 adjacent quantifier order and, 374
 categoricals and, 356–357
 explanation of, 347
 next to existential, 370, 371

translated into sentences containing existential quantifiers, 354–356
Universal sentences
 affirmative, 280, 301, 350–352
 affirmative categorical, 350–352
 explanation of, 302
 general, 342–346, 349, 364
 negative, 280, 301, 307–308
 negative categorical, 352
Universe of discourse
 explanation of, 372–374, 385
 nonempty, 414
Uranus, 552

Vagueness, 225, 241
Valid argument forms, 108–110, 116
Valid deductive arguments
 examples of, 12, 539
 explanation of, 26
 features of, 11
 invalid vs., 13–19
Valid inference, 118–119, 139
Valid inference rules
 explanation of, 119, 139
 formulation of, 140
Valid TD, 128, 139
Validity
 relation between truth and, 15–16
 in S5D, 494–496, 508
 testing argument for, 85–91, 110
 testing categorical syllogisms for, 308–310
 truth-trees and testing for, 584–585
 truth-functional, 90–91
Variables
 domain of, 372–373, 385
 explanation of, 35, 46, 364
 individual, 340
Venn, John, 298, 306
Venn diagrams
 explanation of, 306–308, 333
 testing sorites with, 322–324
 testing validity of categorical syllogisms with, 308–310

Weak inductive arguments, 10, 27
Wedge
 explanation of, 38, 46
 truth-trees and, 572–573, 577–579
Well-formed formula of TL, 67, 68, 76
William of Occam, 549
World Health Organization (WHO), 227

Zeno, 475